Transportation Geotechnics for Green, Digital, and Modern Infrastructures

This book looks into the future of transport infrastructures through green transition, digital transformation, and modern transport infrastructure. It is divided into three parts. The first part provides an overview of green transition in transport infrastructures (eco-design and life cycle assessment, recycled and alternative materials, and bio-based materials and technologies). This is preceded by an overview of related initiatives in the EU and the US. The second part describes intelligent and digital transformation aspects in geotechnical engineering, including geotechnical databases, sensing and monitorization, and machine learning tools and optimization. The third part presents transport infrastructure case studies that demonstrate the contribution of geotechnical engineering to creating smarter, more efficient, more resilient, and more sustainable infrastructures. With 14 chapters, this book serves as a reference tool for academics and engineers in the fields involved or interested in the design, construction, and maintenance of transportation geotechnical infrastructures.

Transportation Geotechnics for Green, Digital, and Modern Infrastructures

Edited by António Gomes Correia, Alexandre Pinto, José Neves, and Eduardo Fortunato

CRC Press
Taylor & Francis Group
Boca Raton London New York

CRC Press is an imprint of the
Taylor & Francis Group, an **informa** business

Cover: Earthworks during the construction of a railway line, Alentejo, Portugal.
Courtesy of National Laboratory for Civil Engineering, Lisboa, Portugal

First edition published 2026
by CRC Press
4 Park Square, Milton Park, Abingdon, Oxon, OX14 4RN

and by CRC Press
2385 NW Executive Center Drive, Suite 320, Boca Raton FL 33431

CRC Press is an imprint of the Taylor & Francis Group, an informa business

British Library Cataloguing-in-Publication Data
A catalogue record for this book is available from the British Library

ISBN: 978-1-032-86694-9 (hbk)
ISBN: 978-1-032-86973-5 (pbk)
ISBN: 978-1-003-53022-0 (ebk)

DOI: 10.1201/9781003530220

Typeset in Times New Roman
by codeMantra

Contents

Preface

Transportation geotechnics encompasses pavements, railways, harbors, foundations, and earthworks. This book summarizes 12 years of technical activities by the Portuguese Technical Committee of Transportation Geotechnics of the Portuguese Geotechnical Society, a mirror of TC202 "Transportation Geotechnics" of the International Society for Soil Mechanics and Geotechnical Engineering (ISSMGE). It also provides an overview of EU and US approaches to the transition toward sustainable transportation infrastructure, a global priority driven by the need to mitigate climate change, reduce carbon emissions, and minimize environmental impact. To achieve these goals, the challenge lies in balancing performance, durability, sustainability, and resilience. To address this, different players are adopting policy-driven environmental assessments and innovative material solutions to reduce the ecological footprint of transport systems. This book begins by exploring two of these key perspectives: The European Union's policy-based approach under the Green Deal (Chapter 1) and the United States' emphasis on low-carbon materials in construction (Chapter 2). Further developments in these areas are discussed in Part 1 (Chapters 3, 4, and 5).

Another critical transition is the digital transformation of geotechnical transportation infrastructure (Part 2), which enhances monitoring, predictive modeling, and decision-making through the Internet of Things (IoT), remote sensing, cloud computing, and automation. It also highlights the role of standardized data formats (AGS) in improving accuracy and integration. These topics are explored in Chapters 6, 7, 8, 9, and 10.

Finally, Part 3 (Chapters 11, 12, 13, and 14) addresses challenges and innovative solutions in maritime and road infrastructure projects.

A detailed but concise summary of all chapters is provided here.

Chapter 1 (Julien and Marin-Lamellet, 2026) explores the impact of the EU Green Deal on transport infrastructure, outlining its role in shaping research, innovation, and environmental assessment methods. A key aspect is the application of life cycle assessment (LCA), which provides a systematic approach to evaluating environmental impacts throughout an infrastructure project's life cycle. The chapter highlights challenges in harmonizing LCA methodologies, discusses certification tools like the HQE Sustainable Territories methodology, and presents case studies demonstrating LCA applications in transport infrastructure. The importance of collaboration among engineers, policymakers, and environmental experts is emphasized, as sustainable infrastructure requires input from multiple disciplines.

Chapter 2 (Huang et al., 2026) shifts the focus to the United States, where the push for sustainability has led to the development and application of low-carbon construction materials. Traditional materials, such as asphalt and concrete, contribute significantly to carbon emissions and energy consumption, prompting researchers to explore alternatives like recycled asphalt pavement, quarry fines, geopolymers, and recycled concrete aggregates. The chapter reviews the mechanical properties and durability of these materials, along with their potential to reduce environmental impact. Additionally, it examines federal and state policies that support the adoption of sustainable materials, such as the Low-Carbon Transportation Materials Program and the Carbon Reduction Program under the Federal Highway Administration. Despite progress, challenges remain, including cost considerations, material availability, technology readiness, and regulatory inconsistencies across states.

Part 1 follows the same framework as Chapters 1 and 2, exploring sustainable geotechnical engineering solutions, focusing on innovative soil stabilization techniques, recycled materials, and eco-design approaches to reduce environmental impact in transport infrastructure projects. Chapter 3 (Cristelo et al., 2026) addresses the bio-mediated stabilization (MICP, fungi-mediated techniques) and alkaline cements, offering sustainable alternatives to traditional methods in road and railway foundations. These methods influence soil-water characteristic curves (SWCC), increasing shear strength and contributing to climate resilience. However, challenges such as scalability, precursor availability, and environmental impact require further research. Chapter 4 (Freire et al., 2026) focuses on the use of recycled and alternative aggregates in transport infrastructure, emphasizing construction and demolition waste (CDW) and industrial byproducts like steel slag. Proper processing and quality assessment ensure these materials' suitability in geotechnical applications. Steel slag aggregates, particularly inert steel aggregate for construction (ISAC), have shown comparable or superior performance to natural materials in road pavement layers. Their adoption supports resource conservation, waste reduction, and a circular economy, with legislative frameworks playing a crucial role in their implementation. Chapter 5 (Venda Oliveira et al., 2026) integrates eco-design and LCA to help minimize environmental impact in geotechnical projects. A UK case study demonstrates LCA's role in sustainable material selection and infrastructure planning, promoting low-carbon alternatives. Together, all these approaches advocate for greener alternatives in geotechnical engineering, emphasizing the importance of research, regulation, and field application to ensure long-term resilience and sustainability in construction.

Part 2 explores the digital transformation of geotechnical engineering, emphasizing advancements in real-time monitoring, predictive modeling, and AI-driven decision-making. Chapter 6 (Miranda et al., 2026) shows how the integration of IoT-enabled sensors, remote sensing (InSAR, LiDAR, and photogrammetry), and cloud computing enhances infrastructure resilience and risk management. Cloud computing and automation further optimize data processing, enabling more efficient infrastructure monitoring. Chapter 7 (Pinheiro, et al., 2026) embraces emerging technologies, such as artificial intelligence (AI), machine learning (ML), and the observational method, that refine predictive models, optimize geotechnical processes, and enable dynamic adjustments in construction. Case studies highlight real-world applications, including retaining wall monitoring in Lisbon, tunnel construction in São Paulo, and aerial slope assessments in Northern Portugal, demonstrating how automation enhances safety and efficiency. Chapter 8 (Cruz et al., 2026) refers to how the importance of adoption of standardized geotechnical data formats (e.g., AGS) improves data accuracy, integration, and accessibility, promoting efficiency in design and construction. Chapter 9 (Tinoco et al., 2026) presents ML applications and how they are transforming transportation geotechnics, by application to case studies on slope stability classification using evolutionary neural networks (ENNs) and road surface condition prediction via random forest models. These approaches optimize maintenance planning and infrastructure management by refining predictive accuracy and reducing operational costs. Chapter 10 (Parente et al., 2026) uses digital twin optimization integrating AI, IoT, and real-time monitoring to enhance geotechnical decision-making, as demonstrated in an earthworks case study where real-time resource allocation minimized project delays using GPS sensors and geo-fencing. This dynamic re-optimization approach ensures adaptability to unexpected construction challenges, reinforcing the role of intelligent systems in modern geotechnical engineering.

Part 3 explores geotechnical engineering challenges and innovative solutions in maritime and road infrastructure projects. Chapter 11 (Watabe, 2026) covers predictive approaches for settlement estimation and ground improvement technologies essential for constructing manmade islands on soft seabeds. Chapter 12 (Ramos, 2026) examines structural challenges in seismically active ports, highlighting seabed dredging, quay wall reinforcement, and underpinning for a high-capacity vessel synchro lift using bored piles, jet grouting, and micropiles. Chapter 13 (Lopes and Silva, 2026) treats Port of Leixões, in Portugal, as a case study in harbor geotechnics, illustrating adaptive engineering over 120 years to mitigate structural issues related to its fossil valley foundation. Chapter 14 (Carreto et al., 2026) analyzes the observational approach in slope stability management, particularly for the A1 highway landslide near Condeixa, in Portugal, where monitoring groundwater levels and implementing drainage systems proved a cost-effective long-term solution. Across these case studies, innovation, monitoring, and multidisciplinary strategies play a crucial role in addressing complex geotechnical challenges in marine and road transportation infrastructures.

The Editors
António Gomes Correia
Alexandre Pinto
José Neves
Eduardo Fortunato

References

Carreto, J., Gonzalez, I., Martins, J.F., Cruz, M.Q., Fernandes, E.M., Caldeira, L. 2026. The Use of the Observational Approach in the Risk Management of an Excavation Slope on the A1 Motorway near Condeixa, During the Operation Phase, in Gomes Correia, A., Pinto, A., Neves, J., Fortunato, E., (Eds.), *Transportation Geotechnics for Green, Digital, and Modern Infrastructures,* CRC Press.

Cristelo, N., Salifu, E., Banerjee, A. 2026. Bio-and Waste-Based Solutions as the Next Generation of Soil Stabilisation Techniques, in Gomes Correia, A., Pinto, A., Neves, J., Fortunato, E., (Eds.), *Transportation Geotechnics for Green, Digital, and Modern Infrastructures,* CRC Press.

Cruz, B., Cruz, J., Branco, J., Pinto, L., Pinho, N., Freitas, L., Chaminé, H. 2026. Geotechnical Data from Ground Investigations: Insights Towards a Digital Geotechnics, in Gomes Correia, A., Pinto, A., Neves, J., Fortunato, E., (Eds.), *Transportation Geotechnics for Green, Digital, and Modern Infrastructures,* CRC Press.

Freire, A.C., Martins, I., Ferreira, C., Correia, E., Neves, J., Roque, A., Fortunato, E., Paixão, A., Rios, S. 2026. Use of Construction and Demolition Waste and Steel Slag, in Gomes Correia, A., Pinto, A., Neves, J., Fortunato, E., (Eds.), *Transportation Geotechnics for Green, Digital, and Modern Infrastructures,* CRC Press.

Huang, J., Puppala, A., Biswas, N. 2026. Carbon Friendly Construction Materials in Transportation Infrastructure Applications in the United States: Current Practices and Guidelines, in Gomes Correia, A., Pinto, A., Neves, J., Fortunato, E., (Eds.), *Transportation Geotechnics for Green, Digital, and Modern Infrastructures,* CRC Press.

Julien, A., Marin-Lamellet, C. 2026. The EU Green Deal, Actor Communities, and Environmental Commitment, in Gomes Correia, A., Pinto, A., Neves, J., Fortunato, E., (Eds.), *Transportation Geotechnics for Green, Digital, and Modern Infrastructures,* CRC Press.

Lopes, H.G., Silva, J. 2026. The Port of Leixões (Portugal): Over 120 Years of Knowledge, in Gomes Correia, A., Pinto, A., Neves, J., Fortunato, E., (Eds.), *Transportation Geotechnics for Green, Digital, and Modern Infrastructures,* CRC Press.

Miranda, T., Moutinho J., Dragos, I. 2026. Geotechnical Monitoring in the Digital Age: Integrating IoT, Remote Sensing, and Data Analysis, in Gomes Correia, A., Pinto, A., Neves, J., Fortunato, E., (Eds.), *Transportation Geotechnics for Green, Digital, and Modern Infrastructures,* CRC Press.

Parente, M., Amândio, M., Correia, A.G. 2026. Dynamic Resource Optimization in Earthworks, in Gomes Correia, A., Pinto, A., Neves, J., Fortunato, E., (Eds.), *Transportation Geotechnics for Green, Digital, and Modern Infrastructures,* CRC Press.

Pinheiro, M., Silva, A., Pereira, E. 2026. Exploring Cutting-Edge Approaches: Case Studies Unveiling New Trends and Tools in Real Geotechnical Data Analysis, in Gomes Correia, A., Pinto, A., Neves, J., Fortunato, E., (Eds.), *Transportation Geotechnics for Green, Digital, and Modern Infrastructures,* CRC Press.

Ramos, P.M. 2026. Underpinning and Strengthening of an Existing Port Infrastructure to Allow the Installation of a High-Capacity Vessel Synchrolift, in Gomes Correia, A., Pinto, A., Neves, J., Fortunato, E., (Eds.), *Transportation Geotechnics for Green, Digital, and Modern Infrastructures,* CRC Press.

Tinoco, J., Cortez, P., Toll, D., Silva, R. 2026. Creating Knowledge by Transforming Geotechnical Data Based on Machine Learning Algorithms, in Gomes Correia, A., Pinto, A., Neves, J., Fortunato, E., (Eds.), *Transportation Geotechnics for Green, Digital, and Modern Infrastructures,* CRC Press.

Venda Oliveira, P., Correia, A., Gervásio, H., Heitor, A. 2026. Eco Design and Life Cycle Assessment, in Gomes Correia, A., Pinto, A., Neves, J., Fortunato, E., (Eds.), *Transportation Geotechnics for Green, Digital, and Modern Infrastructures,* CRC Press.

Watabe, Y. 2026. Leanings from Case Studies of Large-Scale Manmade Island, in Gomes Correia, A., Pinto, A., Neves, J., Fortunato, E., (Eds.), *Transportation Geotechnics for Green, Digital, and Modern Infrastructures,* CRC Press.

About the editors

António Gomes Correia is an Emeritus Professor at the University of Minho and continues his research at the Institute for Sustainability and Innovation in Structural Engineering (ISISE), University of Minho. His activities involve research, teaching, and consulting in the general field of geotechnical and transportation engineering for 45 years. He has been a keynote speaker and member of the organizing/technical committees for many well-established international conferences. He is a co-founder of the International Society for Intelligent Construction (ISIC) and has founded the conference series on Transportation Geotechnics. He is also one of the founding editors for the international journal *Transportation Geotechnics* (from 2014) as well as for *Transportation Engineering* (from 2020). He was also editor of the journal *Geotecnia* (from 2016 to 2024). He is an author/co-author of over 200 publications indexed in SCOPUS.

Alexandre Pinto is the founder and technical manager of JETsj Geotecnia Lda and an invited Professor at the Department of Civil Engineering, Architecture and Environment, Instituto Superior Técnico, University of Lisbon, Portugal. His activities involve consulting and teaching in the general field of geotechnical engineering for 30 years. He is an author/co-author of 120 technical papers. He has been a keynote speaker and member of the organizing/technical committees for several conferences. He was the Vice President of the Portuguese Geotechnical Society from 2016 to 2020 and President from 2020 to 2024.

José Neves graduated in Civil Engineering (PhD, MSc, BSc). He is an Associate Professor in the Department of Civil Engineering, Architecture and Environment, Instituto Superior Técnico, University of Lisbon, Portugal. He is a senior research member of the Civil Engineering Research and Innovation for Sustainability (CERIS), Portugal. He belonged to the Board of Directors of the Portuguese Geotechnical Society (2020–2024). He is the President of the Portuguese Chapter of the International Geosynthetics Society (2024–2028). He is a member of the Academy of Pavement Science and Engineering (APSE), TC202 – Transportation Geotechnics (ISSMGE), and CEN/TC227/WG4 – Hydraulic bound and unbound mixtures (including byproducts and waste materials) of the European Committee for Standardization. He serves as an associate editor of the journal *Discover Applied Sciences*. The main domains of research and consultancy are pavement engineering and transportation geotechnics. He is the author/co-author of over 80 publications indexed in SCOPUS.

Eduardo Fortunato earned his PhD in Civil Engineering from the University of Porto and his Master of Science in Business Administration from the University Institute of Lisbon, Portugal. He is a Principal Researcher and Member of the Board of Directors at the

National Laboratory for Civil Engineering (LNEC) and an invited Full Professor at the University of Porto, Portugal. He is an Expert in Geotechnics and an Expert in Transport and Transportation Infrastructures from the Portuguese Engineers Association. He was the Vice President of the Portuguese Geotechnical Society (2020–2024), and he is a member of the TC202 – Transportation Geotechnics (ISSMGE). The main domains of his research and consultancy are transportation geotechnics and asset management. He is the Associated Editor of *Frontiers in Built Environment* (Transportation and Transit Systems section) and *Sustainability* (Sustainable Engineering and Science section). He is the author/co-author of over 80 publications indexed in SCOPUS.

Contributors

Margarida Amândio
BUILT CoLAB
Porto, Portugal

Aritra Banerjee
Department of Civil and Environmental Engineering
South Dakota State University
Brookings, South Dakota, USA

Nripojyoti Biswas
Department of Civil and Environmental Engineering
University of Massachusetts – Lowell
Lowell, Massachusetts, USA

João Branco
Department of Geotechnics
Mota-Engil – Engineering and Construction
Canelas, V.N. Gaia, Portugal

Laura Caldeira
Laboratório Nacional de Engenharia Civil
Lisbon, Portugal

Joana Carreto
Laboratório Nacional de Engenharia Civil
Lisbon, Portugal

Helder I. Chaminé
Laboratory of Cartography and Applied Geology (LABCARGA), Department of Geotechnical Engineering
School of Engineering (ISEP), Polytechnic of Porto
Porto, Portugal
Centre GeoBioTec|UA
Aveiro, Portugal

António Gomes Correia
Institute for Sustainability and Innovation in Structural Engineering (ISISE), ARISE, Department of Civil Engineering
University of Minho
Guimarães, Portugal

António A. Santos Correia
Institute for Sustainability and Innovation in Structural Engineering (ISISE), Department of Civil Engineering
University of Coimbra
Coimbra, Portugal

Eugénia Correia
Grupo Brisa, BGI - Brisa Gestão de Infraestruturas
Carcavelos, Portugal

Paulo Cortez
ALGORITMI Research Center, School of Engineering
University of Minho
Guimarães, Portugal

Nuno Cristelo
Department of Engineering
University of Trás-os-Montes e Alto Douro
Vila Real, Portugal

Beatriz Cruz
Department of Geotechnics
Mota-Engil – Engineering and Construction

Canelas, V.N. Gaia, Portugal
Laboratory of Cartography and Applied Geology (LABCARGA), ISEP
Porto, Portugal

Jorge Cruz
Department of Geotechnics
Mota-Engil – Engineering and Construction
Canelas, V.N. Gaia, Portugal

Miguel Quintela Cruz
Grupo Brisa, BGI - Brisa Gestão de Infraestruturas
Carcavelos, Portugal

Eduardo Mira Fernandes
Geovia, S.A.
Lisbon, Portugal

Cláudia Ferreira
Ouzo Engenharia
Porto, Portugal

Eduardo Fortunato
Laboratório Nacional de Engenharia Civil
Lisbon, Portugal

Ana Cristina Freire
Laboratório Nacional de Engenharia Civil
Lisbon, Portugal

Liliana Freitas
Laboratory of Cartography and Applied Geology (LABCARGA)
School of Engineering (ISEP), Polytechnic of Porto
Porto, Portugal
Department of Geotechnics
Mota-Engil – Engineering and Construction
Canelas, V.N. Gaia, Portugal

Helena Gervásio
Chemical Engineering and Renewable Resources for Sustainability (CERES), Department of Civil Engineering
University of Coimbra
Coimbra, Portugal

Isabel Gonzalez
Grupo Brisa, BGI - Brisa Gestão de Infraestruturas
Carcavelos, Portugal

Ana Heitor
School of Civil Engineering
University of Leeds
Leeds, UK

Jianxin Huang
Zachry Department of Civil and Environmental Engineering
Texas A&M University
College Station, Texas, USA

Agnès Jullien
Département GERS
Université Gustave Eiffel
Nantes, France

Hugo Guedes Lopes
APDL - Administração dos Portos do Douro, Leixões e Viana do Castelo, S.A.
Matosinhos, Portugal

Claude Marin-Lamellet
Département GERS
Université Gustave Eiffel
Nantes, France

Isabel Martins
Laboratório Nacional de Engenharia Civil
Lisbon, Portugal

Joana Ferreira Martins
Grupo Brisa, BGI - Brisa Gestão de Infraestruturas
Carcavelos, Portugal

Tiago Miranda
Institute for Sustainability and Innovation in Structural Engineering (ISISE), ARISE, Department of Civil Engineering
University of Minho
Guimarães, Portugal

Ionut Dragos Moldovan
Faculty of Engineering
Lusófona University
Lisbon, Portugal

João Moutinho
BUILT CoLAB
Porto, Portugal

José Neves
CERIS, Department of Civil Engineering, Architecture and Environment
Instituto Superior Técnico, Universidade de Lisboa
Lisbon, Portugal

Paulo J. Venda Oliveira
Institute for Sustainability and Innovation in Structural Engineering (ISISE), Department of Civil Engineering
University of Coimbra
Coimbra, Portugal

André Paixão
Laboratório Nacional de Engenharia Civil
Lisbon, Portugal

Manuel Parente
Institute for Sustainability and Innovation in Structural Engineering (ISISE), ARISE, Department of Civil Engineering
University of Minho
Guimarães, Portugal

Eduardo Pereira
Institute for Sustainability and Innovation in Structural Engineering (ISISE), ARISE, Department of Civil Engineering
University of Minho
Guimarães, Portugal

Marisa Pinheiro
Level-Geotechnical Company
Braga, Portugal

Nicole Pinho
Laboratory of Cartography and Applied Geology (LABCARGA)
School of Engineering (ISEP), Polytechnic of Porto
Porto, Portugal

Alexandre Pinto
Department of Civil Engineering, Architecture and Environment
Instituto Superior Técnico, Universidade de Lisboa
JETsj Geotecnia, Lda.
Lisbon, Portugal

Leandro Costa Pinto
Laboratory of Cartography and Applied Geology (LABCARGA)
School of Engineering (ISEP), Polytechnic of Porto
Porto, Portugal

Anand Puppala
Zachry Department of Civil and Environmental Engineering
Texas A&M University
College Station, Texas, USA

Paulo Matias Ramos
Teixeira Duarte, Engenharia e Construções, SA
Lisbon, Portugal

Sara Rios
Construct, Faculdade de Engenharia
Universidade do Porto
Porto, Portugal

António José Roque
Laboratório Nacional de Engenharia Civil
Lisbon, Portugal

Emmanuel Salifu
School of Sustainable Engineering and the Built Environment
Arizona State University
Tempe, Arizona, USA

André Silva
Measurand
São Paulo, Brazil

Joel Silva
APDL - Administração dos Portos do Douro, Leixões e Viana do Castelo, S.A.
Matosinhos, Portugal

Raquel Silva
Institute for Sustainability and Innovation in Structural Engineering (ISISE), ARISE, Department of Civil Engineering
University of Minho
Guimarães, Portugal

Joaquim Tinoco
Institute for Sustainability and Innovation in Structural Engineering (ISISE), ARISE, Department of Civil Engineering
University of Minho
Guimarães, Portugal

David Toll
School of Engineering and Computing Sciences
University of Durham
Durham, UK

Yoichi Watabe
Division of Civil Engineering
Faculty of Engineering
Hokkaido University
Sapporo, Japan

Chapter 1

The EU Green Deal, actor communities, and environmental commitment

Agnès Jullien and Claude Marin-Lamellet

1.1 Introduction

In September 2015, the 193 member states of the United Nations adopted the 2030 Agenda for Sustainable Development (www.agenda-2030.fr 2024). As part of the Voluntary National Review (VNR) process (hlpf.un.org 2024), countries conduct an annual review to assess their progress. The 2030 Agenda for Sustainable Development encourages member states to "conduct regular and inclusive reviews of progress at national and sub-national levels, which are country-led and country-driven". These VNRs monitor and evaluate the national implementation of the 2030 Agenda and its Sustainable Development Goals (SDGs). Reports are submitted to the High-level Political Forum (HLPF) on Sustainable Development, held annually in July under the auspices of the UN (United Nations) Economic and Social Council. VNRs provide information and data on how countries are measuring progress towards the 17 SDGs and their targets (Alibrahim et al. 2023).

In the same year, to address climate change and its adverse impacts, world leaders at the UN Climate Change Conference (COP21) in Paris reached a breakthrough agreement on the 12 December 2015 (www.un.org 2024): the historic Paris Agreement. The Agreement is a legally binding international treaty. It entered into force on 4 November 2016. Today, 195 Parties (194 states plus the European Union) have joined the Paris Agreement.

In November 2019, the European Parliament (European Commission 2019) declared a climate emergency, urging the European Commission to adapt all its proposals with the goal of limiting global warming to 1.5°C and ensuring significant reductions in greenhouse gas emissions. In response, the European Commission prepared the European Green Deal, a roadmap for Europe to becoming a climate-neutral continent by 2050. This European pact has been partly implemented through the European Union's funding programme for research and innovation (R&I), "Horizon Europe", which began in 2021. It supports R&I in sectors essential for ecological transition. Since 2014, the former Horizon 2020 programme had already launched calls aimed at advancing communities' progress towards these global goals, with specific Green Deal calls and dedicated budgets towards the end of its term.

As a result of these political commitments, today more than ever, the European landscape has become increasingly complex. Over the last decade, there has been a continuous shift from mono-disciplinarily approaches towards cross fertilisation among both scientific and political communities. European research projects addressing sustainability and supporting R&I programmes have become multidisciplinary while covering several domains. The need to address

DOI: 10.1201/9781003530220-1

transitions regarding environmental problems impacting society has brought a multiplicity of actors together, strengthening their interconnections.

This chapter provides a non-exhaustive overview of the recent trajectory and areas of interest of some communities that are today engaged or concerned with environmental commitment, specifically in the context of the Green Deal, with a focus on transportation geotechnical infrastructures.

1.2 Where we are: focus of the communities

1.2.1 Transport infrastructures communities' thematic focus

The political and scientific challenges of climate change and environmental impacts are interconnected due to an integrating trajectory based on co-construction of various communities since the 1990s, which was reinforced during the last decade. The communities considered in this chapter include scientific and policy communities focused on geotechnics for transport infrastructures. Additionally, industries contribute through the Horizon Europe programme. Within these ecosystems, European scientific communities collaborate to develop new services and solutions for society, offering knowledge and prototypes through projects of various Technology Readiness Levels. Politically, these communities are connected to the territorial ecosystems through local plans and programmes in individual countries. The competitiveness of the European industry has become a central concern for the European Commission, particularly dedicated to R&I programmes. The thematic focus related to transport infrastructures and the corresponding issues of the communities within the Green Deal framework is illustrated in Figure 1.1.

Academic transport communities are typically organized around specific scientific areas of interest and often gather at regular conferences, such as the Transport Research Arena and the European Transport Conference, or at networking workshops. Experts in these fields co-author reports, create standards, and produce scientific literature. Additionally, experts are clustered within international communities and European R&I networks, which write position papers

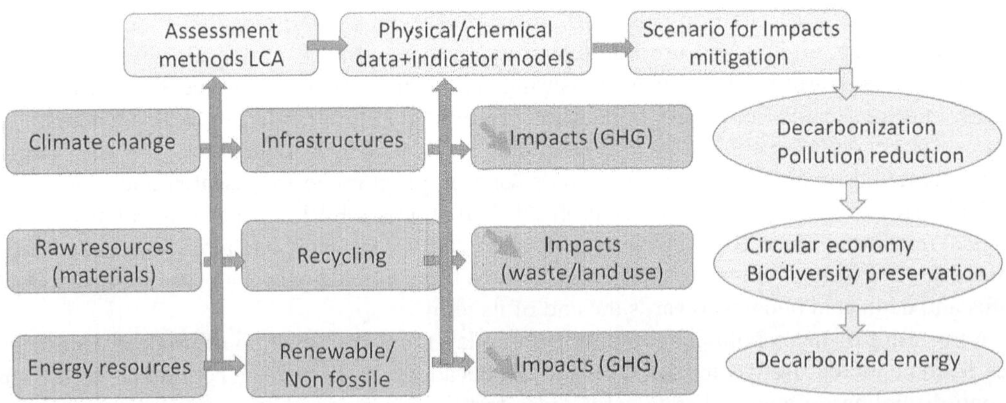

Figure 1.1 Environmental thematic areas and actions focus to proceed with infrastructures and GD issues.

and collaborate with the European Commission. Some of these transport networks are linked to the thematic areas and focus, as shown in Figure 1.1 (e.g., ERTRAC, ERRAC, ECTP, FEHRL, ECTRI). When addressing critical environmental challenges, experts from intergovernmental and international organizations play a key role. The next section explores their contributions to diagnosing the impact assessment process and potential areas for improvements.

1.2.2 Climate change experts and their actions proposals

The issues of climate change have been at the core of Intergovernmental Panel on Climate Change's (IPCC's) work for decades. The IPCC is an intergovernmental body (IPCC 2023) responsible for assessing the causes, extent, and consequences of ongoing climate change. Established in 1988 under the aegis of the World Meteorological Organization and the United Nations Environment Programme, following an international political initiative. Open to all United Nations member countries, the IPCC currently brings together 195 states (IPCC 2023). The IPCC's 2023 synthesis report explains the current overview of the scientific understanding of climate change, its widespread impacts and risks, as well as strategies for both mitigation and adaptation. The interdependence of climate, ecosystems and biodiversity, and human societies is highlighted. It emphasises "the value of diverse forms of knowledge and the close linkages between climate change adaptation, mitigation, ecosystem health, human well-being and sustainable development, and reflects the increasing diversity of actors involved in climate action" (IPCC 2023).

The IPCC reports (IPCC 2023) cover a range of topics, including Observed Warming and its Causes; Observed Changes and Impacts; Current Mitigation Progress; Gaps and Challenges; Equity and Inclusion. The IPCC experts provide detailed guidelines for policymakers on specific issues and actions.

1.2.3 Environmental science experts and assessment methods

The challenge of assessing the environmental impacts of human activities has been continuously pursued by SETAC since its founding in the 1970s. Today, the Society of Environmental Toxicology and Chemistry (SETAC) is a global, non-profit professional organization with nearly 20,000 members representing over 3,400 organizations from across 90 countries (SETAC 2024). SETAC is committed to the advancement of environmental science and management through collaborative projects with various stakeholders, other societies and intergovernmental agencies. It organizes workshops to address shared environmental challenges and to identify solutions. For example, a session at the SETAC Europe 2016 congress focused on the development, standardization and implementation of life cycle assessment (LCA) and its integration with economics for transportation infrastructure and operations (Jullien et al. 2016). SETAC also disseminates environmental science through peer-reviewed publications and knowledge bases to promote understanding of emerging environmental topics. Among other scientific issues, the SETAC framed several decades ago the LCA method. This method was adopted in Europe during the early 2000s, leading to the creation of a set of ISO standards to provide a methodological framework for LCA, including ISO 14040 (ISO 14040 2006) and ISO 14043, which was replaced by ISO 14044 in 2006 (ISO 14044 2006). These standards outline principles and definitions but do not provide a list of pollutants or methods to determine impacts. In line with this approach, Environmental Product Declarations (EPDs) were introduced within Europe through the EN 15804 created by CEN, which was in 2012 (NF EN 15804 2014). This standard aimed to establish a homogeneous practice in Europe when applied to construction products used in buildings and, later, infrastructure project. Methods to determine potential impacts are given in the standard and life cycle inventories (LCI) databases of available software are considered in practical cases, assuming hypotheses for calculation. Thus, industrial products may or may not be associated with an EPD, which can be specific to the country or process. Hence, EPDs played a significant role in Europe in improving practices related to the environmental impacts of material use.

1.2.4 Transport experts meeting climate and environment issues

The incorporation of an alternative material, particularly a waste or by-product material, in a pavement structure, embankment, or backfill, requires not only an engineering evaluation but also an assessment of its environmental loads or impacts. Compliance with European laws and regulations is mandatory.

An environmental evaluation of alternative materials used in infrastructure can lead to either a global assessment during its life cycle or a more localized evaluation, depending on the targets considered (resource exploitation, impact to air, water, soil, and waste). Focusing on groundwater and surface water transfer and thus leaching characterization, hydraulic modelling, and impacts on fauna and flora have been under scrutiny for decades. Several European directives have been revised; for example, in October 2022, the Commission adopted a proposal to revise the water legislation, including updated lists of pollutants in surface water and groundwater (European Commission 2023). Leaching is characterized based on EN 12457 since 2001 (NF EN 12457-2 2002).

Regarding recycling, the EC directive 2008/98/EC of the European Parliament and the Council, dated 19 November 2008, established experiments before the use of waste. The application of recycling in road infrastructures raised numerous questions that led to the development of the EN 12920 "Characterization of waste" (EN 12920 2008). Scientific literature across Europe provides many articles dedicated to waste and its recycling or use in roads.

In France, application guides for European standards (EN 12920 2008) have been developed for several categories of alternative materials for road infrastructures. Prior to the publication of these guidelines, communities cooperated to create a national database and state-of-the-art reference (OFRIR 2014) developed and updated between 2002 and 2014. The objective was to provide classified information aligned with various regulations and standards, focusing on road materials as well as the road code of practice. Four categories of alternative materials were investigated: (i) secondary materials and waste resulting from natural resource extraction, (ii) road waste produced in the civil engineering sector, either from the maintenance of road surface courses and dependencies or from building demolition, (iii) residues from industrial thermal processes, and (iv) by-products generated by other industrial activities. The available online information remains relevant, drawing on past experience. Subsequently, at an EU level, under the EC 7th Framework Programme "Sustainable Surface Transport", the DIRECT-MAT project (2009–2011) was launched, involving 15 participating countries. Aiming to bring together the best practices across Europe for recycling and dismantling road materials (DIRECT-MAT 2011).

Today, environmental criteria are increasingly being incorporated into European calls for tenders in the road and construction sectors to meet the objectives of the European Green Deal, which aims for Europe to be carbon neutral by 2050. Some of the main environmental criteria often required in these calls for tenders for road infrastructure projects include:

- Support for climate objectives: Projects must be aligned with EU climate objectives, such as reducing greenhouse gas emissions by 55% by 2030 compared to 1990 levels, requiring a 90% reduction in transport emissions by 2050 to achieve climate neutrality.
- LCA: Invitations to tender require an environmental impact assessment throughout the project lifecycle, from material production to the deconstruction or end-of-life phase of the infrastructure.
- Reduced impact on biodiversity: Projects must demonstrate efforts to minimize the impact on surrounding ecosystems, particularly by implementing measures to protect local flora and fauna.
- Use of recycled materials: The use of recycled materials, such as recycled asphalt, is encouraged in road projects to reduce the reliance on natural resources.
- Reduced emissions during construction: Tenders often include requirements for the use of low-emission technologies and machinery, such as electric or hybrid vehicles on worksites.
- Rainwater management: Road infrastructure must incorporate solutions for stormwater management, aimed at preventing runoff pollution, reusing water, and minimize flooding.
- Innovation for sustainable roads: Innovative solutions, such as CO_2-absorbing road surfaces or construction techniques that limit noise and emissions, are often encouraged.
- The transport infrastructures community regularly reviews research results to update on developments and future work. Dedicated conferences have been held in the transport sector, and several EU networks cooperated with their members on different aspects of transport.

By April 2024, the TRA strategic session covered current objectives such as Sustainable Mobility of People and Goods in line with the European Green Deal. It addressed the implementation of multidisciplinary R&I across all sectors of society, providing a platform for EU networks and significant projects. Matters such as fostering cohesion, reducing regional disparities, improving connectivity, and ensuring access to the internal market for all were discussed, alongside the environmental impacts and loss of biodiversity. It also focused on supporting greener mobility for people and goods via zero-emissions vehicles and systems.

1.3 Known actions of communities targeting the Green Deal objectives

1.3.1 Actions of the EU Commission towards research and innovation communities' projects

Within the framework of the H2020 programme (H2020 2013), specifically SOCIETAL CHALLENGES – Smart, Green and Integrated Transport, the objective was to create a European transport system that is resource-efficient, climate- and environmentally friendly, safe and seamless for the benefit of all citizens, the economy and society. In 2014, the European Commission line of thinking indicated that

> despite its growth, the transport sector must achieve a substantial reduction in greenhouse gases and other adverse environmental impacts, and must break its dependency on oil and other fossil fuels, while maintaining high levels of efficiency and mobility and promoting territorial cohesion.

A substantial amount of work was therefore dedicated to developing sustainable mobility, envisioning technological and societal rupture innovations in transportation solutions, including today's evolution of infrastructures and integration of new technologies, as well as changes in construction practices.

Projects under Horizon Europe, aligned with the Green Deal and focused on transportation, address environmental impacts through several core strategies aimed at sustainability. These strategies include:

Decarbonization and clean energy: Projects focus on reducing greenhouse gas emissions by supporting the development and deployment of clean energy technologies such as renewable energy and hydrogen technologies. These innovations are critical to transitioning energy systems towards sustainability.

Circular economy: Horizon Europe fosters research into technologies that promote resource efficiency and material recycling. This includes projects that contribute to the circular economy, ensuring the reduction of waste and reducing material, thus mitigating pollution and resource depletion.

Sustainable mobility and industry: Projects also target sustainable industrial practices and mobility solutions, aiming to reduce the carbon footprint of industries, enhance energy efficiency, and develop sustainable transport systems. These initiatives are critical in addressing emissions from heavy industries and the transportation sector, both of which are significant contributors to environmental degradation.

Pollution reduction and biodiversity preservation: Environmental impacts are addressed through projects focused on reducing pollution and preserving ecosystems. Horizon Europe supports innovations in sustainable food systems, water conservation, and biodiversity protection, ensuring a holistic approach to environmental sustainability.

In 2021, the Commission launched missions such as the "100 Climate Neutral Cities" with the goal of delivering 100+ climate-neutral and smart cities by 2030. Horizon Europe provided funding to help deliver this mission (100 Climate Neutral Cities 2024). Cities were expected to prepare and implement Climate City Contracts, co-created with citizens and local stakeholders. A portfolio of R&I projects was designed to establish a Global Knowledge base. The EU ministries gather feedback from this mission.

1.3.2 Actions to link academic research output and policy targets

1.3.2.1 Providing a framework

The Joint Research Centre (JRC), as an integral part of the European Commission, plays a key role in providing scientific support to the European Union. According to its Science for Policy handbook, the distinction between Science for Policy and Academic Science is based upon ten criteria, which are perceived or investigated differently by regulatory scientists and academic scientists Sucha and Sienkiewicz (2020). Among these criteria, two stood out in terms of the complementarity approaches between policy and academic perspectives. Firstly, when addressing community goals, the government and industries need to meet specific requirements and provide data, whereas universities are driven by knowledge enhancement. Secondly, the concerns of the role of values cited by Sucha and Sienkiewicz (2020) are collective values versus individual ones. However, when assessing practices aligned with Green Deal requirements, the frontier of single approaches is enlarging in order to include wider boundaries as explained below in Section 1.3.

In the special series addressing UN SDGs, Sanyé-Mengual and Sala from the Joint Research Centre (2022) published a paper investigating the extent to which LCA can address the EU Green Deal policy and monitoring, with reference to specific SDGs. In connection with EU policy, the relations are shown in a recap table between (i) the biodiversity strategy, (ii) the circular economy plan, and (iii) the chemicals strategy for sustainability. Evidence of the relevance of LCA is stipulated respectively with SDG 2/6/14, SDG 6/7/9/12/13, and SDG 2/3/6/9/12/14/15. Unfortunately, SDG11 is not considered in this analysis, which targets

> By 2030, provide access to safe, affordable, accessible and sustainable transport systems for all, improving road safety, notably by expanding public transport, with special attention to the needs of those in vulnerable situations, women, children, persons with disabilities and older persons.

Hence, transportation should still be considered within the frameworks of LCA, the Green Deal, and monitoring as referenced by UN SDGs, at both the EU and national levels.

1.3.2.2 Collecting and sharing data

Climate change is now addressed globally at an international scale. Much of the scientific literature covers countries outside Europe, offering country-specific data regarding energy supply for a wide range of industrial activities. In the context of civil engineering and transport infrastructures, material supply is often more localized.

The International Energy Agency (IEA 2024) states its activities as "the heart of the global dialogue on energy security and clean energy transitions; the agency is the world's leading energy authority". On its website, IEA provides comprehensive data, analysis, and policy recommendations aimed at shaping a secure, sustainable, and affordable energy future for all, while also meeting the climate change objectives of the 2015 Paris Agreement. Thus, for the Green Deal implementation, and according to the items shown in Figure 1.1, it is important to take into account relevant country data and hypotheses regarding the energy mix when conducting research and assessments. Actions by EU networks and platforms to connect research, innovation and policy targets are also important.

The European Environment Agency (EEA 2024) states on its website that it "provides valuable insights on the **state of Europe's environment,** thanks to **reliable data** collected from our extensive network". The EEA offers information on the EU's materials footprint and produces European environmental assessments, including the risks posed by climatic change.

1.3.2.3 Connecting the communities and sharing vision

As illustrated in Figure 1.2, European networks play a central role in connections between the policy and academic communities. These networks support European programmes and contribute to calls for proposals, engaging in a co-construction process of their members.

Such a need to connect has been emphasised by the EUA, which represents universities and national university presidents' conferences from 47 European countries (850 members). The EUA plays an important role in defining the European Union's policies on higher education, R&I. According to the European University Association's Green Deal roadmap for universities (EUA 2023), universities are viewed not only as centres for knowledge generation but also as hubs for partnerships with industries, public sectors, and local communities to drive sustainable innovation. This comprehensive approach allows higher education institutions to support the EU's ambitious targets for reducing emissions and resource use, while also contributing to societal change. Therefore, it is necessary for universities to develop a common roadmap with stakeholders. The key objective would be to incorporate more intersectoral research with industry, taking into consideration the Net-Zero Industry Act and its specific practices.

Other research networks dedicated to transport infrastructures also provide strategic documents to support research. Tables 1.1a and b provide an overview, using key words, of how they address the Green Deal, SDG11, decarbonization, LCA, and the circular economy.

1.4 Lessons learnt addressing the Green Deal in the projects and R&I activities, and scientific literature with environmental assessment

1.4.1 Pathway to address transport infrastructure impacts with LCA

In the field of transport, LCAs were developed after the 2000s, focusing on system boundaries related to materials assessment used during the construction phase of infrastructures. For this purpose, LCA tools and databases were adapted to incorporate, as best as possible, road

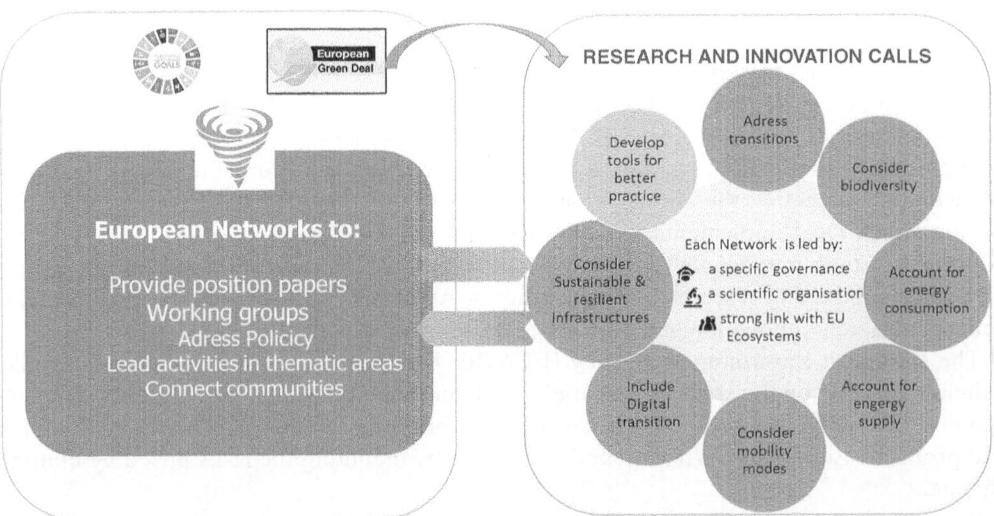

Figure 1.2 Schematic diagram of networks activities to support research and innovation in line with transport infrastructures.

Table 1.1a Overview of how the main transport infrastructures and building European Research Networks are dealing with Green Deal, SDG 11, decarbonization, and LCA in their strategic documents

ETP-networks	*Key words*			
	Green Deal	*SDG 11*	*Decarbonization*	*LCA-circular economy*
ERRAC				
Chéron et al. (2020): Rail Strategic Research & Innovation Agenda, 2020	Present for the context, challenges and Vision, impact and commitment of the Rail European Partnership	Part of the context	Part of the context, impact assessment, transforming project using decarbonization to define environmentally friendly and attractive sustainable mobility	Development of environmental footprint methods and tools that allow public information and monetarization, improvement in recycling potential and practices and development of life cycle assessment databases, incentives circular economy solutions for innovative rail systems
ECTP				
ECTP (2023): Strategic Research & Innovation Agenda 2024–2030, Oct 2023	Basis for EU regulation framework	Not present	One committee on Built Environment Decarbonization, Objective 1: Resilient, adaptive, decarbonized and regenerative built environment; R&I priority 1.2: Make solutions for building and infrastructure renovation and decarbonization more affordable and easier to implement R&I priority 1.4: Demonstrate regenerative and frugal designs, integrating NBS and considering adaptability and life cycle	Topics: · Develop Digital Twins/advanced BIM approaches and tools integrating user data, BMS and LCA & LCC to support the selection of decarbonization pathways with simulation and optimization, considering the construction/renovation process, in particular logistics. · Propose a **new set of indicators on social impact** and standardize the use of S-LCA approaches to support decision-making and raise awareness. · Develop a European common referential for Technical Data and LCA data to foster the use of these data.
FERHL				
FEHRL (2024) FEHRL Position Paper – Horizon Europe Topics for the Third Work Programme (2025–2027)	Linked with a proposal on Mainstreaming biodiversity	Not present	Proposals to outline a process for decarbonization of the road sector including construction, maintenance and operation, new materials, circular materials and models, Mobility network infrastructure,	Circular economy linked with proposals on Mobility network infrastructure

Table 1.1b Overview of how the main transport European research networks are dealing with Green Deal, SDG 11, decarbonization and LCA in their strategic documents

ETP-networks	*Key words*			
	Green Deal	*SDG 11*	*Decarbonization*	*LCA-circular economy*
ERTRAC				
ERTRAC (2024): Integrated Urban Mobility Roadmap, Version: 1.3, 2024, ERTRAC-ALICE-ERRAC Working Group on Urban Mobility	Part of the policy background and high-level targets, seen as a driver for a fair twin transition, urban freight and logistics	Only present for improving road safety policies (*SDG 11.2*)	Decarbonization of mobility is the objective, which is applied to research for cities and districts as living labs, mobility infrastructures, the next generation of MAAS and active travel modes.	For light electric vehicles – new vehicle concepts for urban use, in addition to research on life cycle assessment, increasing lifetime performance and similar issues – research on environmental impacts must also focus on implementation issues.
ERTRAC (2021): New Mobility Services Roadmap, Version: 4, 2021, ERTRAC Working Group: Urban Mobility	Seen as a driver for governance of individual mobility services operating in a city through the long-term sustainability targets envisaged	Not present	Not present	In the policy goals and regulatory frameworks, impact assessment should be multidimensional: *the life cycle assessment* of vehicles should be conducted under different MAAS configurations, rather than specifically for each service.
ERTRAC (2018): Strategic Research Agenda – Input to Ninth EU Framework Programme, Version: 2.8, 2018	Not present	Not present	Decarbonization is very present as one of the main challenges for road transport: ensure mobility in urban areas environmental sustainability: energy and resource efficiency, decarbonization and air quality towards decarbonization: CO_2 evaluation.	Improved vehicle efficiency through optimised design and weight reduction consistent with the circular economy: Develop and apply multi-material approaches that allow cost-effective material separation, recycling and recovery, taking into account environmental impact through life cycle assessment.
ECTRI				
ECTRI (2022): Greening and Digitalization of Transport, March 2022	The Green Deal is the global framework that gives the orientation for future transport systems	Not present	Build knowledge to reach decarbonization goals by applying a system approach, with the digitalization evolution in transport.	Not present
ECTRI (2020): An innovative and socially balanced European transport policy and a consistent strategy for sustainable and smart mobility, Sept 2020	The Green Deal is the global framework which gives the orientation for future transport system target to achieve	Not present	Not present	Not present

data that were previously lacking. Much of this data has been collected from infrastructure owners and from the industrial sites of material providers. In some cases, measurements were performed, but documents from factories and companies were also analysed. Subsequently, dedicated LCA tools have been developed, incorporating specific data collected from industrial sites. Such data are often country-dependent, as civil engineering materials are typically provided locally. The transport distance of materials from extraction and processing plant sites to roadworks locations is both expensive and impactful and consequently minimised. This is the case for both roads and rail.

In the 2010s, *benchmarking* in transportation was carried out in Europe based on case study applications and scientific activities related to LCA. An international symposium was held in Davis, California, in 2014, focused on the implementation of LCA for pavements. It was hosted by the University of California (Davis and Berkeley) and IFSTTAR (France) and sponsored by TRB, FHWA, ISAP, and ISCP. This symposium was the third in a series, which included a similar event in California in 2010 and a follow-up in France in 2012. The symposium featured plenary presentations by experts, who discussed the challenges of implementing LCA for pavements. These were followed by breakout sessions to further discuss the issues, as well a summary of areas of consensus and differences, with suggestions for future actions. Participants included academics (specialising in pavements and LCA), government representatives (national, state, and local), and industry professionals (materials producers, contractors, and national and state trade organizations). One key conclusion of the symposium stipulated that "When applied correctly, LCA can anticipate unintended consequences of a policy or practice that intends to reduce impacts but instead increases them by not considering all life cycle phases or the interaction of pavement systems with road users" (Harvey et al. 2015).

Later, the scope and boundaries of infrastructure LCAs were expanded to consider the whole life cycle covering maintenance, exploitation, end of life, and the application of recycled materials. Questions arose regarding the relative importance of each phase, particularly the use phase. The whole infrastructure was assessed, including earthworks, pavements, and infrastructure structural elements. The biggest challenge in assessing infrastructure remained the availability of the data and the project's state of development. Hence, LCA was even applied at the pre-project level, provided the infrastructure design was sufficiently detailed to describe the lifecycle phases and select appropriate data for assessment. Defining the functional unit is generally easy. However, some papers in the LCA literature present only a reduced number of indicators – namely energy consumption – especially when LCA is performed at the project level or during the design phase.

Besides LCA, various methodologies of assessment for infrastructure certification in civil engineering have been developed, each with different scopes, such as:

- Assessment and certification of buildings/HQE Building, France (HQE 2001)
- Assessment and certification of roads/HQE Sustainable Road, France (Certivéa et al. 2010), and European from an FP7 project (LCE4roads 2014)
- Management and territorial diagnosis approach/HQE Aménagement, France (Carfantan et al. 2005), and the HQE²R European project (Charlot-Valdieu and Outrequin 2002)

Such methods focus on the objectives to be achieved and include environmental evaluation. Some refer to LCA for impact calculations, as seen in LCE4roads, Europe (Zukowska et al. 2014), whereas others consider the connection with the territory (Carfantan et al. 2005). HQE has thus proposed a geographical zoning for the envisaged objectives. Five of the 11 objectives are global, as they represent national issues, whereas the remaining six are based on a more local approach. A participatory approach, involving all project stakeholders, was also proposed. This approach encompasses economic and social development, as well as quality of life, covering a wide range of sustainable development dimensions. The European HQE²R project (Charlot-Valdieu and Outrequin 2002) has developed methods and tools for evaluating urban complexes. The method includes shared diagnosis of sustainable development (by comparison with an essential state), a system of sustainable development indicators for diagnosis, evaluation, and monitoring, recommendations for improving the participation of inhabitants and users, as well as specifications and recommendations for development projects at the programming stage.

1.4.2 *Feedback on environmental assessment in civil engineering with LCA*

Communities today recognize that LCA can identify opportunities for improving environmental performance at all stages of a product or service life, particularly within complex infrastructures, helping to inform decision-makers for strategic planning purposes. The LCA method is today widely investigated in civil engineering, with a big number of extended scientific articles. Hence, individual papers and review papers covering LCA in areas such as roads, railways, buildings, geotechnical engineering and products are available. Several authors have conducted literature reviews to identify the know-how and what is at stake concerning the method, the data used, the indicators provided and the relevance of applying this method in line with the Green Deal requirements. Table 1.2 presents some features from the literature review (including journal articles and conference proceedings), gathering insights from reviews, as well as from historical papers in the field.

The system studied by practitioners includes the industrial and on-site processes used throughout the whole life cycle of its product(s) or service (construction, maintenance, operation, end of life). The analysis is based on a chosen functional unit, according to the study's objectives. At least two scenarios or project variants should be studied for comparisons, as the results provided by LCA should highlight opportunities for improvement. Different impact assessment methods have been developed since the early 1970s to link each inventory data to its potential environmental pressures or damages. A review of the available methods and their origins was presented at the LCA 2014 Congress in California (Rosenbaum 2014).

Part of the available LCA literature on infrastructures is analysed in Table 1.2 to provide feedback for future practitioners regarding the main hypotheses considered. There are different axes to consider when reviewing LCA literature. One is to focus on the domain, such as transport, the type of infrastructure, and the level of focus related to practitioners (e.g., project, bid, construction/operation). Another approach is to consider the type of LCA practised, including its functional units, systems and objectives attached to the decision context, such as policy decisions or procurement decisions. Hence, historical studies on LCAs of *railway* transport systems dealt with itineraries and functional units expressed for single or double tracks in passenger-kilometres, tonne-kilometres, or vehicle-kilometres, depending on the study objectives. These may include either (i) EPDs or choices (Stripple and Uppenberg 2010), or a comparison between ballast track and concrete track (Lee et al. 2008), or (ii) intermodality (Chester and Horvath 2010). For roads, functional units in kilometres and tonnes have been the most widely used for comparing pavement materials and analysing maintenance strategies (Jullien et al. 2014).

The midpoint indicators assessment, such as the CML 2001 method (Guinée et al. 2001) is commonly used in most publications on transport infrastructures because it is more easily interpreted as an indicator of environmental load and thus directly linked to the potential mitigation options in a given situation.

The review papers in Table 1.2 aim to provide feedback on a wide range of studies related to railways, roads, buildings, construction products, and geotechnical engineering by comparing the hypotheses and results of these studies wherever possible. The authors then focused on various aspects such as the life cycle phase, the functional units, the databases referenced, the methods used for impact calculations, and the systems considered. Overall, all the authors agree that the sector and products studied have high environmental impacts. However, making comparisons is challenging due to the wide variation in system boundaries, impact calculation methods, data sources (such as Ecoinvent, ELCD, or product-specific databases), energy mixes, and functional units.

Focus 1: Several features were raised during this recent literature review (Liljenström et al. 2022). The system boundary studied varies between authors and significantly influences the results obtained. The lack of a precise definition of the system boundary in several publications makes it impossible to compare results obtained by different authors. Additionally, the functional unit is not clearly defined in several scientific publications and differs from one work to another. None of the publications present the raw data. To assess the environmental impact of an HSR itinerary, the functional unit must describe the technical features of the railway track and circulation characteristics of the itinerary, because they are crucial for the track design. For example, in Europe (specifically France), the functional unit for the infrastructure assessment could be defined as "a double track high-speed railway of 400 km with a travel time of 1h50mn and with 40 trains per day for each track during 50 years". This functional unit could consider earthwork units composed of cut and fill; civil engineering structures including a tunnel, sewer drainage systems, road rehabilitation, railway foundation, and railway equipment.

Focus 2: The workshop subject focused on the implementation of LCA in pavement operations, establishing a connection between US and European approaches and needs, both at the network and project levels. During the 2014 LCA pavement conference (Harvey et al. 2015), some of the conclusions still need to be applied:

- Up-to-date regional inventory data is required for improvement and credibility of LCA results.
- The regional applicability of impact indicators needs to be taken into consideration.
- The use of LCA is likely to increase faster through incentives and procurement rules in Design/Build/Maintain and Design/Build contracts.
- The inclusion of uncertainty when presenting LCA results is necessary, although rarely illustrated.

Focus 3: Buildings have always been ahead in terms of LCA and construction material data due to sanitary and health requirements for users. However, road stakeholders have actively been experimenting with recycling, incorporating both alternative road materials and non-dangerous waste (OFRIR 2014).

Regarding the LCA of buildings, the authors (Anand and Amor 2017) indicated that previous research has primarily focused on efficiency and associated emissions during the use phase. Highlighting the barriers and efforts required for the comprehensive life application of LCA in the building industry. Their paper structure follows the four steps of LCA as provided by the ISO 14040 and ISO 14044 standards. In terms of goal and scope definition, they emphasize that the use of different functional units contributes to the complexity in preventing comparisons of buildings in LCAs. Concerning system boundaries, the cradle-to-grave is found to be the most

Table 1.2 Investigations performed according to LCA practice for environmental assessment

Focus	*Numbers of papers*	*Life cycle phase*	*Functional units*	*LCA practice investigated by the authors*		
				Data	*Impact*	*System*
1/LCA of roads and rail (Liljenström et al. 2022) *Surface* *Tunnel* *Bridge*	92 review 2016–2020 84 roads 7 rails	Maintenance	Ton.km Passengers.km or Vehicle.km km Single track/lanes 2 tracks/lanes	NA	Energy GWP ODP POCP AP EP Ecology toxicity Human toxicity	Cradle to use
2/LCA of *Pavements* (Harvey et al. 2015)	22 Conference proc 2014	Construction Operation Maintenance End of life/Recycling	Km 1 lane 2 lanes	Country data Generic data Local data	**methods** CML 2001 EcoIndicator 999 ReCiPe Impact 2002+ (Rosenbaum, 2014)	Cradle to site Cradle to gate Cradle to use Cradle to grave
3/LCA of *buildings* (Anand and Amor 2017)	112 review paper	Construction Operation Maintenance	m^2 m^3 ton	• EPDs • specific data	Energy, GHG	No internationally acceptable methods of performing LCA
4/LCA results *Visualization* (Hollberg et al. 2021)	92 Review paper	Identify the most important phase	Define options for design improvement		Sensitivity analysis	Cradle to grave
5/LCA of construction products *European* (Durãoa et al. 2020)	40 review	Design for Construction and End of Life	tonne m^2	**Data** Specific to the product	PEF EN 15 804 and its impacts	Cradle to gate Cradle to use Cradle to grave
6/LCA and LCCA for geotechnical engineering **Retaining wall/slope* **ground improvement* (Samuelsson et al. 2021)	29 review paper	Construction	NA	**Data** Lacking in many studies Ecoinvent	Energy GHG	Cradle to gate

Note: In bold main objective of the focus.

employed. This review establishes that energy and emissions are the most frequently assessed factors in building LCAs and provides a list of tools specifically designed for this purpose. Thus, the key takeaway is that, given the variety of system boundaries, data, and methods employed, achieving robust comparisons between different studies remains unattainable.

Focus 4: Visualisation of LCA results. In their study, Hollberg et al. (2021) focused on the design process of buildings, concentrating on the interpretation phase and the necessity to adapt graphs to suit the needs of decision-makers. They identified six typical requirements: the identification of hot spots, comparison of design options for improvement, sensitivity analysis, benchmarking, and finally the consideration of spatial and temporal distributions. We retained the interest of the visualisation of hot spots to direct efforts towards mitigation.

Focus 5: Regarding end-of-life integration, the authors (Durãoa et al. 2020) noted in their review that the landscape of standards has evolved over time, but without precise indications on how to implement EOL impacts. After the introduction of the LCA international standards of ISO 14040 in the early 1980s for any product, the changes towards ISO 14044 in the 2000s still did not provide a definitive method for selecting impact calculation approaches. In parallel, the European standard EN 15804 was developed specifically for building materials and described environmental indicators. This standard provided the foundation for EPDs in Europe. The review by Durãoa et al. (2020) focused on comparing EPDs and PEF, both of which calculate the environmental footprints of products, based on life cycle thinking. Due to the abundant international literature dedicated to LCA, as shown in Table 1.2, only the methodological output related to EN 15 804 is extracted here.

Focus 6: Authors (Samuelsson et al. 2021) reviewed published conference papers and journal papers on the topics of LCA and LCCA in geotechnical engineering using Google Scholar. They concluded that this is a relatively new research area with a lack of data and a need for assessments to follow a defined method. According to their review, only nine papers had been published before 2010. Three main topics emerged from the 49 papers dedicated to 88% of the papers: geotechnical engineering in general, retaining walls/slope support, and ground improvement. Topics such as piles and foundations, lightweight materials, and excavation and fill, among other topics, were either not fully analysed or not addressed at all. The authors conclude that "there are a limited number of open databases with data for the infrastructure sector".

In the period 2013–2016, the LCE4ROADS FP7EU project (LCE4ROADS 2014), funded by the European Commission, focused on a "Life Cycle Engineering approach to develop a novel EU-harmonized sustainability certification system for cost-effective, safer and greener road infrastructures". According to the LCE4ROADS project itself, it contributed to the implementation of European policies and strategies, boosting the integration of transport in sustainable development. The project promoted technologies and materials that reduce pollutant emissions and the use of natural and financial resources. After a thorough analysis of the state of art in labelling approaches both within and outside Europe and receiving inputs from several stakeholders, the consortium elaborated a list of 25 key performance indicators aimed at promoting the use of greener, more cost-effective and safer technologies in road construction, applicable to both asphalt and concrete pavements. The project deliverable D 1.1, published in 2014 (Zukowska et al. 2014) investigated the labelling approaches and sustainability assessment methodologies in the US and Europe. This was reviewed and updated to reflect the latest label information. If the EU ECOLABEL still does not cover civil engineering, three methods and labels are currently used to assess the sustainability of road infrastructure projects. Private companies or groups provide experts to deliver support for such certification. The Sustainable Transport Council's Greenroads Project Rating Program, which evaluates the qualitative measures of transportation infrastructure project's based on their environmental, social, and economic performances (Sustainable Transport Council 2024).

The CEEQUAL methodology aims to improve infrastructure sustainability by focusing on objectives such as resource efficiency, waste prevention and reduced climate impacts. Case studies related to this methodology are mentioned (BREEAM 2024).

Beyond labelling practices, case studies are often specific to local contexts; it is hardly possible to reach global feedback and know-how. This highlights the importance of feedback provided from certification organisation, which state that companies and stakeholders deploy better projects by taking into account environmental burdens and setting targets for improvements.

1.4.3 Highlights from case studies of infrastructure LCA

Three case studies of infrastructure LCA practice are presented in Table 1.3 to demonstrate key insights. These case studies are focused on heavy traffic infrastructure, which is typically easier to document and offers opportunities for field data collection. The first case study is dedicated to railways, considering two itinerary variants with different traffic volumes and types of use (passengers and/or freight). The second case study compares the environmental impacts of earthworks and pavement construction during the road-building phase to those during the use phase of heavy traffic infrastructure. The third case study focuses solely on road sections supporting heavy traffic. It investigates two geotechnical optional solutions, with and without treatment, in relation to global warming impacts and resource consumption.

All these case studies were performed on real infrastructures, either before or during their construction, using site-specific data. This data includes information on infrastructure design, local climate conditions, constraints such as thicknesses, local materials, traffic requirements, equipment and materials supply and the transportation distances of materials and equipment assessment. The studies were performed using the same LCA tool and homogeneous data. The infrastructures are framed by the same code of practice of the country (in this case, France). All cases are dedicated to the new section of infrastructure located in two regions with Mediterranean and continental climate trends.

The LCA assessment is based on the length of the infrastructure section, with each section corresponding to a specific construction site. If all the local designs are implemented for assessment, LCA provides the following indicators: resource consumption (in tonnes), global warming potential (GWP), acidification potential (AP), eutrophication index (EI), photochemical ozone potential (POCP), ecotoxicity potential (EP), toxicity potential (TP). Selected results from these case studies are summarized in Table 1.3.

Case 1 is about a new railway section pre-project with two variants from 2012, set in a Mediterranean region with two possible itineraries. High-speed rail lines have been built since the 1980s, covering several hundred kilometres and including tunnels, viaducts, and bridges. To analyse existing projects, all available documents for two French high-speed lines (LGV Est and LGV Rhin-Rhône) in eastern France were reviewed (Bosquet et al. 2014). This protocol provided a description of material supply along a given project alignment and the necessary generic data for the elementary functional units assessing the construction of civil engineering structures, such as bridges, viaducts, and tunnels, and of sewer drainage systems. Even for shorter lines, each individual civil engineering structure along an itinerary requires a dedicated assessment for the project (Bosquet et al. 2014). In conjunction with the railway construction, the road network also needs to be rehabilitated. Total road rehabilitation and construction, based on a per-kilometre ratio along the entire railway line length, were assessed using the materials database of the LCA tool (Jullien et al. 2015), whereas each project variant had its local characteristics.

The outcome of this rail project case study at the design level revealed that it is possible to estimate the energy consumption of the construction, maintenance, and use phases. The ratio between the use phase of the rail and the construction phase of the infrastructure is roughly the same, as detailed in Table 1.3.

Case 2 concerns a new national heavy traffic road section in a region with a continental climate, where low winter temperatures are possible. Considering the assessment of a road project, once all the features are defined, it is based on the construction, maintenance, and use phase as described in Zukowska et al. (2014). The monitoring conditions and environmental modelling of earthwork impacts are described in detail in (Capony et al. 2013). In this example, all the midpoint indicators are estimated using several sets of LCI data: local producer data from UPC (Union des Producteurs de Chaux) and the Ecoinvent database. This comparison was made to assess the environmental impacts of materials, such as the relative importance of treated sub-base layers and treated fills, as well as the ratio of earthworks machinery and quicklime use (Zukowska et al. 2014). The traffic impacts are integrated into the assessment throughout the entire lifecycle, with vehicle fleet modelling hypotheses detailed in source (Zukowska et al. 2014), which helped analyse the ratio between earthworks and traffic. The entire road was also assessed separately in the LCE4ROADS EU project to evaluate each lifecycle phase (Zukowska et al. 2014).

In Case 2, the environmental impacts from earthworks construction correspond approximately to two years of traffic for nearly all indicators. Furthermore, the choice of the database does not change the conclusions regarding the significant impacts of treatment in terms of GWP, nor the amount of aggregates saved due to the treatment solution. As for the overall life cycle, Table 1.3 indicates that the use phase is the most important for all impacts, except for the TP. The fact that the toxicity potential indicator TP exhibited an opposite trend, with the construction phase being the most important, then the other indicators, is noticeable. This highlights the need to assess multiple indicators in LCA to capture various potential impacts, rather than focusing solely on greenhouse gases, which are often linked to energy trends.

Case 3 also deals with the same heavy traffic road but for a different new section. It compares two geotechnical solutions for the subgrade layer (Neel et al. 2012). One solution considers a classical granular (untreated) subgrade layer sourced from a quarry located 7.5 km from the construction site. The other solution (treated) consists of 5% clinker (supplied from 21 km), 1% quicklime (supplied from 316 km), 13.7% water, and 80.3% cut material. When assessing the impacts of materials, it is necessary to consider not only the energy consumption or GWP but also the raw resource consumption in tonnes for practical purposes. For the same mechanical properties objectives (120 MPa for the platform), Table 1.3 presents the GWP assessment for the subgrade layer. In such conditions, a small reduction in the amount of binder mass leads to a significant reduction in impacts. Therefore, if the transport distances for the granular materials had been greater than 30 km, the impact evaluation would have led to an opposite conclusion between the treated solution and the non-treated one, if based on the GWP alone.

Case 3, considered during a bid, focused solely on a greenhouse gas approach for material selection in the local context (Neel et al. 2012). To complement such an approach, the sensitivity of 7 LCA environmental indicators related to a capping layer treatment was investigated with the ECORCE LCA tool and its LCI database (Jullien et al. 2015). The four-layer road structure studied is shown in Figure 1.3 (L1–L4). The solution with treatment saves 15.9% of aggregate consumption (about 900 tonnes).

Table 1.3 Highlights from three case studies in France for a rail and a road project

Case study (France)	*Objective and functional unit*	*Data and Impacts*	*% of total impacts of each life cycle phase or solution improvement with mid-points indicators*
No. 1: Rail HSR –south of France Creation of 25 km · New passenger line · New mix line for passengers and freight (Bosquet et al. 2014)	Assessment at the pre-project level of 2 itineraries in terms of energy consumption over 50 years	Collection of the owner data for the Energy and ECORCE database (Jullien et al. 2015)	**Energy: after 10 years** Use 40%; construction 60% of total consumption **after 50 years** Maintenance: about 5%
No. 2: Road with high traffic – centre of France Creation of 8.9 km national 2 × 2 lanes (Capony et al. 2013)	Assessment during the road construction: *Importance of earthworks and of construction/ pavement maintenance versus traffic for 30 years *Importance of the database	Ecoinvent and the ECORCE database (Jullien et al. 2015) for the pavements Collection of environmental data on the roadworks site during 2 years	Maintenance: small and about some % after 30 years (four maintenance operations) **Energy** and **GWP**: Use phase is about 97% and 95% of total **AP, EI, POCP, EP**: Use phase less than 95% of total **TP** construction phase about 90% of total
No. 3: Road with high traffic – centre of France Creation of 9.5 km national 2 × 2 lanes (Neel et al. 2012)	Assessment during the earthworks study to select treated and non-treated subgrade layers in terms of GHG, without counting the traffic impacts	ECORCE database (Jullien et al. 2015) for the base Collection of geotechnical data on the project	**Resources** 1.2 tonnes/m^2 of aggregates **GWP treated** 1,0E + 07 (kg.eq.CO^2) 17% quicklime and 77% cement **GWP classical** 7,53E + 05 (kg. eq.CO_2) 65% aggregates

Table 1.4 shows the variation trends of LCA indicators in percentages, considering the untreated solution as the reference. Most indicator values show an increase with the treated capping layer, but in different proportions. In this example, and with the LCA model implemented in the tool, global warming, ecotoxicity, and chronic toxicity vary significantly between the two solutions. In addition, due to numerous data uncertainties, LCA sensitivity may be considered to be *significant* if there is a relative variation of over 20% compared to the reference case.

Then, the quarry distance for aggregates, which is of practical interest for comparison, does not appear to be *significant* in this study. However, it is worth noting that this supply distance could be much higher in urban areas. Finally, a case-by-base assessment guarantees better alignment with local practices, whereas the project owner may decide to consider one or two indicators in selecting solutions that best meet local environmental targets.

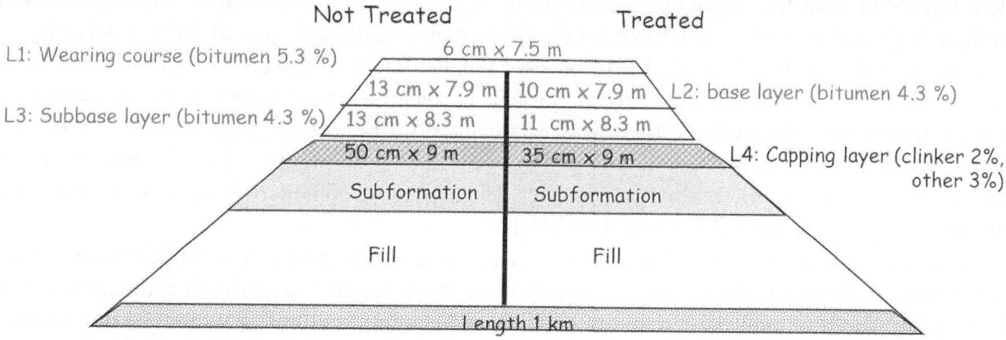

Figure 1.3 Road case design with four layers for comparisons (treatment/quarry distance).

Table 1.4 Indicators comparison for the road structure without (NT) and with (T) treatment

Total road	*Not treated*	*Treated*	*(NT-T)/NT*	*(T20-T30)/T20*
Quarry distance	*20 km*	*20 km*	*20 km*	*30 km*
Energy	$3.3\ 10^6$ MJ/t	$3.58\ 10^6$ MJ/t	−8.5%	−1.7%
Global warming	$1.98\ 10^5$ kg.eq.CO_2	$2.93\ 10^5$ kg.eq.CO_2	−48.0%	−1.7%
Acidification	$2.82\ 10^2$ kg. eq.SO_2/t	$3.46\ 10^2$ kg. eq.SO_2/t	−22.7%	−2.6%
Eutrophication	$5.35\ 10^1$ kg.eq.PO_4^- / t	$6.09\ 10^1$ kg.eq.PO_4^- / t	−13.8%	−3.9%
03 Photochemical	$2.28\ 10^2$ kg.eq.etylene	$2.27\ 10^2$ kg.eq.etylene	0.4%	−2.6%
Ecotoxicity	$6.92\ 10^6$ kg.eq.1.4 D.C.B/t	$1.2\ 10^7$ kg.eq.1.4 D.C.B/t	−73.4%	−1.7%
Chronic toxicity	$1.11\ 10^4$ kg.eq.1.4 D.C.B/t	$1.64\ 10^4$ kg.eq.1.4 D.C.B/t	−47.7%	−3.7%

1.5 Conclusion

This chapter provides an overview of the recent history of environmental impact assessment of transportation infrastructure in relation to the EU Green Deal application.

The collaborative work of various communities over the past decades has resulted in a code of practice and motivations to further environmental assessment of complex infrastructures, aimed at supporting mitigation efforts and identifying the best possible solutions. Engineers need to collaborate with environmental experts, and input from multiple stakeholders and actors is necessary to improve our knowledge and know-how to provide such solutions.

The European Green Deal has significantly changed the research perspective for most of the European communities and research networks. It serves as the policy framework, setting clear targets to limit the impact of transportation systems on climate change. Hence, the LCA methodology and its framework have been more systematically considered for this purpose.

A substantial literature is available today on LCA feedback, although the lack of harmonization across studies makes it difficult to compare its application in civil engineering and transportation infrastructure. However, infrastructures are always different and shaped by local territory. Certification tools have also been developed to encourage good practice. For instance, the HQE Sustainable Territories methodology provides 19 themes to consider, in relation to the local impacts of a project, aligned with the UN SDGs.

A case study-based assessment approach is therefore a sound approach for infrastructure LCA. In this context, feedback from some case studies was investigated and analysed in the final part of this chapter. The way their life cycles are considered for infrastructures provides general answers for practitioners. Although several literature surveys on LCA in transport or civil engineering include several hundreds of references, they do not always address the impacts and outcomes in detail, and practical experience with LCA is needed to fully understand how to apply it. Energy consumption and GWP are the most commonly used indicators in the literature concerning complex infrastructures, mainly due to data availability. These indicators can be considered at the early stage of projects to identify trends. However, other indicators, such as human toxicity, should not be neglected, even though they are not required in the standards EN 15804. This is the case when alternative materials are used, as they are not inert and could release pollutants with water during the use phase. Such materials may need to be considered during maintenance, particularly in some urban areas. The case studies mentioned above relate to heavy traffic infrastructures, which were more thoroughly documented and studied and are necessary to improve territory accessibility. Of course, if one deals with an infrastructure designed for light traffic, especially with low-emission modes, the construction phase impacts will represent a significantly larger proportion of the total life cycle, and the choice of materials should be reconsidered with a focus on the use phase.

In terms of the European Green Deal's target to reduce environmental impacts, namely global warming, while decreasing carbon emissions, LCA models and tools, although not perfect, are valuable for providing a comparative approach, offering significant improvement trends that can support decision-making regarding technical solutions in a local context.

References

100 Climate Neutral Cities. 2024. Horizon Europe support for the European green deal, STUDY panel for the future of science and technology European parliamentary research service scientific foresight unit (STOA) PE 757.823.

Agenda. 2030. https://www.agenda-2030.fr/ (accessed September 23, 2024).

Alibrahim, B.A.I., Al Nabet, N.S.M., Costa, N., Dai, S.Y., Kim, S.R., Liu, H.,Liu,K., Page, H., Risan, J., Shamagin, A., Schrijver, L.,W., Trujillo, J., C., Vaturi, T.,Xu, Z.,You, S. 2023. Voluntary national reviews synthesis report. Office of Intergovernmental Support and Coordination for Sustainable Development – High-Level Political Forum on Sustainable Development: Tonya Vaturi.

Anand, C.K. and Amor, B. 2017. Recent developments, future challenges and new research directions in LCA of buildings: A critical review. *Renewable and Sustainable Energy Reviews* 67: 408–416. https://doi.org/10.1016/j.rser.2016.09.058

Bosquet, R., Jullien, A., Vandanjon, P.O. and Sanchez, F. 2014. Eco-design model of a railway: A method for comparing the energy consumption of two project variants. *Transportation Research Part D: Transport and Environment* 33: 111–124. https://doi.org/10.1016/j.trd.2014.08.003

BREEAM. 2024. https://breeam.com/case-studies (accessed September 30, 2024).

Capony, A., Muresan, B., Dauvergne, M., Auriol, J.C., Ferber, V. and Jullien, A. 2013. Monitoring and environmental modelling of earthwork impacts – A case study. *Resources Conservation and Recycling* 74: 124–133. https://doi.org/10.1016/j.resconrec.2013.03.007

Carfantan, G., Vignes-Rubio, C. and Bonnet, K. 2005. *Méthodologie pour une* démarche de *qualité environnementale sur les opérations d'aménagement dans une perspective de développement durable*. SETUR, Rapport Définitif, France.

Certivéa, LCPC, CSTB, CETE Nord-Picardie et CG 59. 2010. Référentiel technique de certification route durable. Technical report.

Charlot-Valdieu, C. and Outrequin, P. 2002. State of the art review of indicators and systems of indicators. Sustainable renovation of buildings for sustainable neighborhood. Contract N° EV-K4-CT 2000-00025, France.

Chéron, C., Brennan, M., French, R., Furio, N., Gautier, P-E., Goikoetxea, J., Guida, U., Kuzmina, D., Lundgren, M., Mazzariello, C., Morgan, M., Nash, C., Papa, F., Perreal, Y., Schlaht ,J., Stichel, S., Travaini ,G., and Winkler, D. 2020. ERRAC, RAIL Startegic research and innovation agenda. https://rail-research.europa.eu/wp-content/uploads/2020/12/RAIL-Strategic-Research-and-Innovation-Agenda-2020-_FINAL_dec2020.pdf (accessed October 4, 2024).

Chester, M. and Horvath, A. 2010. Life-cycle assessment of high-speed rail: The case of California. *Environmental Research Letters 5*(1): 014003. https://doi.org/10.1088/1748-9326/5/1/014003

DIRECT-MAT. 2011. https://cordis.europa.eu/article/id/90301-infrastructure-recycling-hits-the-road/fr (accessed September 30, 2024).

Durãoa, V., Dinis Silvestre, J., Mateus, R. and de Brito, J. 2020. Assessment and communication of the environmental performance of construction products in Europe: Comparison between PEF and EN 15804 compliant EPD schemes. *Resources Conservation and Recycling* 156: 104703. https://doi.org/10.1016/j.resconrec.2020.104703

ECTP. 2023. Strategic research & innovation agenda, 2024–2030. https://www.ectp.org/fileadmin/user_upload/documents/ECTP/ECTP_Position-papers_and_SRIA/ECTP_SRIA_2030_final_version.pdf (accessed October 4, 2024).

ECTRI Position Paper. 2020. An innovative and socially balanced European transport policy and a consistent strategy for sustainable and smart mobility, September. https://www.ectri.org/wp-content/uploads/2020/09/2020-09_ECTRI-Position-Paper_Sust-Smart-Mobility-Strategy_Final.pdf (accessed October 1, 2024).

ECTRI Position Paper. 2022. Greening and digitalisation of transport, March https://www.ectri.org/wp-content/uploads/2022/03/2022-03-17_ECTRI-PP_Green-Digitalisation-Transport.pdf (accessed October 1, 2024).

EN 12920 Standard. 2008. Characterization of waste – Methodology for the determination of the leaching behaviour of waste under specified conditions (includes Amendment A1:2008).

ERTRAC. 2018. Strategic research agenda – Input to 9th EU framework programme, version: 2.8. https://www.ertrac.org/wp-content/uploads/2022/07/ERTRAC-Strategic-Research-Agenda-SRA-2018.pdf (accessed October 3, 2024).

ERTRAC. 2021. New mobility services roadmap, version: 4, 2021, ERTRAC working group: Urban mobility. https://www.ertrac.org/wp-content/uploads/2022/07/ERTRAC-New-Mobility-Roadmap-V4.pdf (accessed October 3, 2024).

ERTRAC. 2024. Integrated urban mobility roadmap, version: 1.3, 2024, ERTRAC-ALICE-ERRAC working group on urban mobility. https://www.ertrac.org/wp-content/uploads/2024/09/2024-ERTRAC_ALICE_ERRAC_UM_Roadmap.pdf (accessed October 3, 2024).

European Commission. 2019. Communication from the commission to the European parliament, the council, the European economic and social committee of the regions, towards joint programming in research. Brussels.

European Commission. 2023. https://environment.ec.europa.eu/topics/water/water-framework-directive_en (accessed September 29, 2024).

European Environment Agency. 2024. https://www.eea.europa.eu/fr/publications/evaluation-europeenne-des-risques-climatiques-synthese (accessed September 30, 2024).

European University Association. 2023. A green deal roadmap for universities. https://eua.eu/ (accessed October 17, 2024).

FEHRL Position Paper. 2024. Horizon Europe topics for the 3rd work programme (2025–2027) (accessed October 2, 2024).

Guinée, J.B., Gorrée, M., Heijungs, R., Huppes, G., Kleijn, R., de Koning, A., van Oers, L., Sleeswijk, A.W., Such, S., Udo de Haes, H.A., de Bruijn, H., van Duin, R. and Huijbregts, M.A.J. 2001. *Life cycle assessment, an operational guide to the ISO standards*. Ministry of Housing, Spatial Planning and the Environment (VROM) and Center of Environmental Science, Leiden University, The Netherlands.

H2020. 2013. Societal challenges – Smart, green and integrated transport. https://cordis.europa.eu/programme/id/H2020-EU (accessed November 18, 2024).

Harvey, J., Jullien, A., Steyn, W., Kendall, A., Jones, D. and Saboori, Y. 2015. Application of the results of the pavement LCA 2014 symposium to Southern Africa. *Paper presented at the International Southern Africa CAPSA'2015, 11th Conference for Asphalt Pavements for Southern Africa, Sun City*.

Hollberg, A., Kiss, B., Röck, M., Soust-Verdaguer, B., Houlihan, W.A., Lasvaux, S., Galimshina, A. and Habert, G. 2021. Review of visualising LCA results in the design process of buildings Alexander. *Building and Environment* 190: 107530. https://doi.org/10.1016/j.buildenv.2020.107530

HQE. 2001. *Définition explicite de la qualité environnementale, EDQE et SME, Référentiels des caractéristiques HQE*. Association HQE, Paris.

International Energy Agency. 2024. https://www.iea.org/about (accessed September 30, 2024)

IPCC. 2023. Climate change 2023 synthesis report. Summary for policymakers. Contribution of working Groups I, II and III to the Sixth Assessment Report of the Intergovernmental Panel on Climate Change. I Geneva, Switzerland. https://doi.org/10.59327/IPCC/AR6-9789291691647.001

Jullien, A., Dauvergne, M. and Cerezo, V. 2014. Environmental assessment of road construction and maintenance policies using LCA. *Transportation Research Part D: Transport and Environment* 29: 56–65. https://doi.org/10.1016/j.trd.2014.03.006

Jullien, A., Dauvergne, M. and Proust, C. 2015. Road LCA: The dedicated ECORCE tool and database. *International Journal of Life Cycle Assessment* 20(5): 655–670. https://doi.org/10.1007/s11367-015-0858-y

Jullien, A., Horvath, A., Harvey, J. and Proust, C. 2016. Development, standardization and implementation of LCA and integration with economics for transportation infrastructure and operations Session organised at the international SETAC Europe conference, Nantes, France.

LCE4ROADS Project FP7 – Life Cycle Engineering approach to develop a novel EU-harmonized sustainability certification system for cost-effective, safer and greener road infrastructures Grant agreement ID: 605748. https://www.lce4roads.eu/overview (accessed September 30, 2024).

Lee, C.K., Lee, J.Y. and Kim, K. 2008. Comparison of environmental loads with rail track systems using simplified life cycle assessment (LCA). *Presented at the fourteenth International Conference on Urban Transport and the Environment in the 21st Century, WIT Transactions on The Built Environment*, Malta.

Liljenström, C., Björklund, A. and Toller, S. 2022. Including maintenance in life cycle assessment of road and rail infrastructure—A literature review. *The International Journal of Life Cycle Assessment* 27: 316–341. https://doi.org/10.1007/s11367-021-02012-x

Neel, C., Mauduit, C., Carle, Ph., Chorda, J., Denot, A. and Jullien, A. 2012. Evaluation environnementale et ACV: quels usages par les acteurs publics et privés. *Paper presented at the national ECOTECH'S2012 Optimisation de la gestion des déchets de la route: que permettent les outils ACV?* France.

NF EN 12457-2 Standard. 2002. Caractérisation des déchets – Lixiviation – Essai de conformité pour lixiviation des déchets fragmentés et des boues – Partie 2: essai en bâchée unique avec un rapport liquide-solide de 10 l/kg et une granularité inférieure à 4 mm (sans ou avec réduction de la granularité). AFNOR, France.

NF EN 15804+A1 Standard. 2014. Contribution des ouvrages de construction au développement durable – Déclarations environnementales sur les produits – Règles régissant les catégories de produits de construction (Sustainability of construction works – Environmental product declarations – Core rules for the product category of construction products). AFNOR, France.

NF EN ISO 14040 Standard. 2006. Management environnemental – Analyse du cycle de vie – Principes et cadre (Environmental management – Life cycle assessment – Principles and framework). AFNOR, France.

NF EN ISO 14044 Standard. 2006. Management environnemental – Analyse du cycle de vie – Exigences et lignes directrices (Environmental management – Life cycle assessment – Requirements and guidelines). AFNOR, France.

Observatoire Français du recyclage dans les infrastructures routières (OFRIR). 2014. https://Ofrir2.ifsttar.fr (accessed September 23, 2024).

Rosenbaum, R. 2014. Towards the big picture – The path from one-dimensional footprints to complete environmental sustainability assessments. https://www.ucprc.ucdavis.edu/p-lca2014/Presentations.aspx (accessed November 4 2024).

Samuelsson, I., Larsson, S. and Spross, J. 2021. Life cycle assessment and life cycle cost analysis for geotechnical engineering: Review and research gaps. *IOP Conference Series: Earth and Environmental Science* 710: 012031. https://doi.org/10.1088/1755-1315/710/1/012031

Sanyé-Mengual, E. and Sala, S. 2022. Life cycle assessment support to environmental ambitions of EU policies and the sustainable development goals. *Integrated Environmental Assessment and Management* 18(5): 1221–1232.

Society of Environmental Toxicology and Chemistry. 2024. https://www.setac.org/ (accessed September 30, 2024).

Stripple, H. and Uppenberg, S. 2010. Environmental product declaration for railway track foundations on the Bothnia Line. IVL Swedish Environmental Research Institute (accessed November 4 2024).

Sucha, V. and Sienkiewicz, M. 2020. *Joint Research Centre Science for policy handbook*. European Commission, Brussels; Elsevier. https://knowledge4policy.ec.europa.eu/publication/science-policy-handbook_en

Sustainable Transport Council. 2024. https://www.transportcouncil.org/307/projects.html (accessed September 30, 2024).

United Nations – High-Level Political Forum on Sustainable Development. 2024. https://hlpf.un.org/vnrs (accessed September 23, 2024).

Zukowska, E., Jullien, A., Cerezo, V., Lorino, T. and Bouteldja, M. 2014. Deliverable D1.1: Assessment of current labelling approaches applied to roads design, construction maintenance and rehabilitation, as well as the products used to build them (asphalt mixtures and cement based materials). European Project FP7(LCE4ROAD – Life Cycle Engineering approach to develop a novel EU-harmonized sustainability certification system for cost-effective, safer and greener road infrastructures Grant agreement ID: 605748.

Chapter 2

Carbon-friendly construction materials in transportation infrastructure applications in the United States

Current practices and guidelines

Jianxin Huang, Anand Puppala, and Nripojyoti Biswas

2.1 Introduction and background

Development of a robust transportation infrastructure system is vital to every country's economic growth and prosperity (Kumar 2023). Transportation, in particular, bridge and pavement infrastructure systems are essential components, and thus the design and long-term performance of them profoundly affect the overall performance of transportation systems. The United States has approximately 4.2 million miles of roads as of 2022, and almost two-thirds of this road network are paved roads (Bureau of Transportation Statistics 2021, 2024). The American Society of Civil Engineers (ASCE) mentioned in the "2021 Report Card for America's Infrastructure" that over 40% of the road system is in either poor or mediocre condition, and it graded "D" for road infrastructure (ASCE 2021). Each year, transportation agencies spend millions of dollars on road maintenance, which can be reduced by improving and selecting resilient and durable pavement materials. Another important aspect of these materials is that they should be carbon-friendly with low greenhouse gas (GHG) emissions from their manufacturing process.

The U.S. Federal Highway Administration (FHWA) has been working with state and non-state Departments of Transportation (DOTs) in screening and developing construction materials to be used in transportation infrastructure. The utilization of low-carbon materials in civil infrastructure has been a focus in recent years with the aim to reduce GHG emissions and to reach the carbon neutrality goal by 2050 (U.S. Department of State & U.S. Executive Office of the President 2021). Transportation infrastructure, such as federal and state highways, makes up a significant portion of the civil infrastructure in the United States, with highway transportation accounting for 44% of the federal government's 2023 infrastructure spending (USAFacts 2024). Most pavements and roads in the United States use asphalt materials, and these roads are regarded as flexible pavements (FHWA 2018). Rigid pavements, also known as concrete pavements, are a key component in highway transportation systems, and they are mostly used when subsoil conditions are poor and problematic in nature. Stabilization of subsoils or use of resilient base material to support transportation infrastructure is often practised for preparing subsoils to support pavements (Correia et al. 2016, Hoyos et al. 2011, Puppala et al. 2011).

Pavements constructed over weak subgrades undergo distresses such as permanent deformation or rutting under repeated traffic loads over the pavement's design life. Some mitigation measures that have been practised over several decades include replacing the local subgrade soils with better geomaterials, chemical treatment of the problematic soils themselves, and use of innovative infrastructure solutions such as geosynthetics and combinations of them in reinforcing underlying subsoils and bases (Biswas et al. 2024c, Correia et al. 2016, Khan et al. 2020, Little & Nair 2009). However, the subgrade treatment solution depends on several factors,

DOI: 10.1201/9781003530220-2

including the quality of the borrowed material, the initial cost of the ground improvement, and several other socio-economic factors (Kumar et al. 2024).

Chemical stabilization with cement, lime, and fly ash is often considered an economic solution for improving the strength and stiffness material properties of weak soil properties of local soils (Abu-Farsakh et al. 2015, Behnood 2018, Little & Nair 2009). Cement and lime additives are some of the most commonly used materials in construction; however, the production of them is an energy-intensive process and produces a significant amount of GHG, particularly carbon dioxide (CO_2). Cement production contributes to around 8% of the world's total CO_2 emissions. To reduce the carbon footprint of cementitious material production, multiple efforts have been made to develop carbon-friendly stabilizers, incorporate supplementary cementitious materials like fly ash and slags to substitute portion of cement, improve energy efficiency during the cement production process, implement carbon capture and storage technology, and use admixtures to reduce the amount of cement needed in concrete mixes (Adesina 2020, Kazemian & Shafei 2023, Thomas et al. 2012). As an alternative to traditional stabilizers, non-traditional stabilizers such as polymers, enzymes, and others have been utilized for improving material properties in the past few decades, and it should be noted that their energy footprint is not always environment-friendly (Huang et al. 2023a, b, 2024, Kolay et al. 2016, Kumar et al. 2022, 2024, Mekonnen et al. 2020, Sidiq et al. 2023, Tingle & Santoni 2003).

Utilization of low-carbon or environmentally friendly materials, such as recycled materials, in pavement construction could significantly reduce the carbon footprint of the highway infrastructure. However, the effectiveness and performance of these low-carbon materials still need to be evaluated in practice (Althoey et al. 2023). Investigation on the performance and the durability of these low-carbon materials in pavement construction, with certain field validation studies, is needed to enhance the combined resilience and sustainability of the transportation system via utilizing these materials.

This chapter covers a few of the carbon-friendly materials and recycled aggregate materials used in pavement layers in the United States, along with laboratory-based research studies that were used to evaluate and screen these materials. In some cases, durability studies were employed to address the permanency of stabilization methods along with moisture susceptibility issues, and all these results are covered in this chapter. Also, current FHWA practice on the use of low-carbon materials and their evaluation process, along with select practices by various state DOTs, is also presented. This section is followed by various experimental studies that focused on geopolymers, recycled base aggregate, and waste materials, as well as other types of geomaterials. These results are briefly summarized along with salient findings. Final assessments of these studies, as well as the need for performing field validation studies, are mentioned.

2.2 Current practices

2.2.1 *Federal practices*

The U.S. federal government has created a multi-faceted approach and implemented several practices, guidelines, and policies aiming at reducing carbon emissions in the infrastructure construction sector under the guidance of two major acts signed into law recently: the Bipartisan Infrastructure Law (BIL) and Inflation Reduction Act (IRA) (The White House 2022, 2023). These efforts are part of a broader strategy to mitigate GHGs leading to extreme climate hazards and transition to a sustainable and low-carbon economy, focusing on reducing direct emissions from construction practices, promoting the use of sustainable materials, supporting clean energy infrastructure, and setting ambitious targets for carbon neutrality in federal projects. These actions are designed to drive broader adoption of green infrastructure and sustainable practices across the sector, contributing to the nation's overall climate goals.

The BIL, also known as the Infrastructure Investment and Jobs Act, signed into law in November 2021, includes provisions to promote sustainable infrastructure projects. The BIL is the single largest dedicated investment made in American transportation infrastructure since the completion of the Interstate Highway System in 1955 (U.S. DOT 2024a). BIL focuses on reducing carbon emissions by promoting green infrastructure, supporting electrification of transportation infrastructure, and allocating funds for public transit and clean energy projects that reduce emissions (The White House 2022). It encourages the use of sustainable materials, energy-efficient technologies, carbon capture technologies, and climate-resilient designs in construction projects.

The IRA was signed into law in 2022, which aims to address environmental justice and bolster federal involvement in clean energy, including in the transportation infrastructure sector. Apart from BIL and IRA, several executive orders signed over the past few years aimed to address reducing climate hazards and promote the reduction of carbon emissions across different sectors, including infrastructure construction. For example, Executive Order 14008 on tackling the climate crisis at home and abroad aimed at advancing the federal government's role in addressing climate change; it includes directives for the use of low-carbon materials and promoting green construction methods for federal infrastructure projects (The White House 2021b). Executive Order 14057 on catalysing clean energy industries and jobs through federal sustainability (also known as the Federal Sustainability Plan) focuses on driving sustainability in federal operations, which includes the construction and operation of federal buildings and infrastructure. It mandates the adoption of sustainability standards and carbon-neutral approaches in infrastructure development (The White House 2021a). As a result, the federal government has developed specific guidelines and goals for reducing carbon emissions from its own infrastructure projects. One such goal is net-zero buildings and carbon-neutral construction.

The General Services Administration (GSA), which manages federal properties, is working towards making all new federal buildings "net-zero carbon" by 2030. This includes adhering to high standards for energy efficiency and carbon reduction during the construction phase. The GSA also encourages the use of sustainable building materials, such as low-carbon concrete, recycled steel, and renewable and recycled building products and construction materials.

Multiple federal agencies have received federal funds from BIL or IRA. These agencies have been working on developing multiple programmes to tackle the environmental issues, including those that promote carbon reduction in the transportation infrastructure sector. FHWA, under the U.S. Department of Transportation (USDOT), initiated the Low-Carbon Transportation Materials (LCTM) programme to provide incentives and funds to state and non-state DOTs for promoting the use of low-carbon materials that have substantially lower levels of embodied GHG emissions associated with all relevant stages of production, use, and disposal (FHWA 2024b). State DOT recipients will pursue a series of activities that create new processes to qualify LCTMs and then put these processes into practice on federally aided construction projects. Eligible grant activities include identification of eligible materials, collections of material Environmental Product Declaration (EPD), preparation of acceptance and verification procedures for EPD collections, industry outreach activities for training on new procedures, benchmarking global warming potential, identification of thresholds that define LCTM, evaluation and testing of LCTM performance, update and revision of existing specifications to accommodate new LCTMs and their applications, and construction of projects using LCTMs and post-construction monitoring.

In addition to the LCTM programme, FHWA also administers the Carbon Reduction Program (CRP) established under BIL, which provides formula funding over five years for all the states to develop carbon reduction strategies and for projects to reduce transportation carbon dioxide emissions (FHWA 2024a). The CRP outlines various strategies related to carbon reduction in the transportation sector, including traffic management, public transportation, pedestrian facilities, alternative fuels, and port electrification (FHWA 2024a). These strategies further reduce transportation emissions and bolster the adoption of green construction and climate-resilient infrastructure practices.

The University Transportation Centers (UTC) programme administered by FHWA is another programme that provides funds to consortia of colleges and universities across the United States to conduct state-of-the-art research and technology in transportation fields and train the next generation of transportation professionals (U.S. DOT 2024b). The UTC programme has been an on-going programme since 1987 to help address the ever-growing needs for the safe, efficient, and environmentally sound movement of people and goods in the United States (U.S. DOT 2024b). The UTC programme currently supports five national, ten regional, and twenty Tier-1 centres. The UTC programme focuses on multiple areas in transportation research, including the sustainability of transportation infrastructure. The UTC programme works with state, local, and tribal agencies to address the growing environmental concerns and reduce carbon emissions in transportation, which play a key role in conducting advanced, new carbon reduction technologies and using novel low-carbon materials in transportation infrastructure in the United States.

The Environmental Protection Agency (EPA) regulates carbon emissions through rules that affect construction practices, such as the Greenhouse Gas Reporting Program, which mandates large emitters, including certain construction companies, to report their GHG emissions, and the New Source Performance Standards, which apply to certain industrial sectors, including construction equipment that emits carbon pollutants. Under the IRA, two new programmes established by the EPA further boost the identification and standardization of the embodied GHG emissions in transportation in the United States. EPA worked with various agencies to develop the EPD Assistance Program, aiming to help manufacturers of construction materials to develop, verify, and standardize the EPDs – a document that shows the environmental information of the material over its life cycle. Availability of embodied GHG emissions information through EPDs to the public will enable the states, agencies, and stakeholders to choose the desired low-carbon construction materials and products and accurately measure the embodied carbon of the projects during their life cycle. With the disclosure of EPDs and the carbon labels of current and new construction materials, the procurement and utilization of low-carbon materials in construction become feasible, which significantly contribute to the goals of carbon reduction in transportation infrastructures within the country.

Other federal agencies also developed programmes that are relevant to the carbon reduction goals in transportation infrastructure or general civil construction. Federal agencies, like the Federal Emergency Management Agency (FEMA), work to ensure that infrastructure is built to withstand the impacts of climate-related hazards. This includes reducing the carbon impact of constructing resilient infrastructure, such as flood-resistant roads and other infrastructure types.

In general, the involvement of the federal government and agencies to combat climate issues and carbon emissions significantly facilitates the adoption of low-carbon materials and the development and implementation of carbon reduction technologies in transportation infrastructure.

2.2.2 *State DOT practices*

U.S. State DOTs play a critical role in reducing carbon emissions within the infrastructure construction sector, as transportation is a major contributor to GHG emissions. Many state DOTs have implemented practices, guidelines, and policies that align with federal efforts and also reflect state-specific priorities for sustainability. Below are some of the key practices and guidelines that U.S. state DOTs are adopting to reduce carbon emissions in infrastructure construction.

Many state DOTs are actively developing or updating their sustainable transportation plans and GHG reduction plans to meet climate goals. These plans aim to reduce emissions across all areas of transportation, including infrastructure construction. For example, the California Department of Transportation (Caltrans) established the Climate Action Plan for Transportation Infrastructure (CAPTI) in 2021, which outlines strategies to reduce carbon emissions from transportation projects. This includes adopting sustainable construction practices, utilizing low-carbon materials, and increasing the energy efficiency of infrastructure. The Washington State Department of Transportation (WSDOT) has a climate change response strategy, which emphasizes reducing carbon emissions from the construction, maintenance, and operations of infrastructure. WSDOT incorporates GHG accounting and reporting into its project planning process.

State DOTs are increasingly adopting green construction practices to reduce the environmental impact of infrastructure projects. These include the utilization of low-carbon and recycled materials and sustainable pavement design. Many states are encouraging or mandating the use of low-carbon materials like recycled asphalt, recycled concrete, and high-performance, low-carbon concrete. For example, California requires the use of recycled materials and prioritizes the use of lower-carbon alternatives in construction projects. States like Florida and Texas promote the use of sustainable and low-carbon paving materials such as warm-mix asphalt (which requires less energy during production) and recycled asphalt pavement (RAP). This reduces emissions during both construction and maintenance.

Many state DOTs also focus on climate-resilient infrastructure, which considers long-term environmental impacts, including carbon emissions. Some states have adopted policies to ensure that transportation infrastructure is built with an emphasis on sustainability and resilience to climate impacts. In New York State, the New York State Department of Transportation (NYSDOT) uses climate adaptation guidelines to build infrastructure that minimizes emissions over the long term, factoring in energy-efficient roadways, bridges, and maintenance practices. In short, state DOTs across the United States are taking significant steps to reduce carbon emissions in the infrastructure construction sector. These efforts range from using sustainable materials and construction practices to reducing emissions from equipment to promoting electric vehicles and charging infrastructure. As part of their overall sustainability goals, many state DOTs are incorporating carbon footprint tracking, adopting stricter emissions standards, and integrating climate resilience into infrastructure planning and design. Through these guidelines and initiatives, state DOTs are helping to ensure that transportation infrastructure in the country becomes more sustainable and aligned with national climate targets.

2.3 Research and case studies using low-carbon materials

Different carbon-friendly materials have been used in research and practice for transportation infrastructure in the United States over the past decades. These materials include low-carbon concrete, reclaimed asphalt pavement (RAP) aggregates, recycled concrete aggregate (RCA), and other waste materials and by-products. Figure 2.1 shows large stockpiles of RCA and RAP in Texas, United States.

These materials have been used in different construction projects, and many agencies are experimenting with these low-carbon materials in the United States. FHWA characterized 19

(a)

(b)

Figure 2.1 Stockpiles of recycled materials in Texas, USA: (a) RCA material (b) RAP material.

different types of waste and by-product materials that are used in pavement construction in the United States and developed guidelines for the use of these materials (FHWA 2008). Table 2.1 summarizes these materials and their potential uses in transportation infrastructure construction. The use of waste and by-product materials can significantly reduce the carbon footprint of transportation construction and increase the sustainability of the transportation infrastructure.

The focus of this chapter is on LCTMs comprising geomaterials to be used for base/subbase or part of subgrade treatments, and hence the coverage of geomaterial topics is only included here. Among all these materials, concrete or cement is one of the most used construction materials in transportation infrastructure for pavement and bridge infrastructure. In geotechnical engineering, cement is one of the most widely used materials to treat soils for different projects. Due to the high carbon footprint of Ordinary Portland Cement (OPC), different low-carbon stabilizers are investigated to treat different problematic soils for different projects in order to reduce carbon emissions and minimize environmental impacts. In this section, four case studies conducted by the authors over the years using different recycled products or low-carbon materials are discussed. These materials include recycled aggregates like quarry fines (QF) and RAP aggregates, metakaolin (MK)-based geopolymers, recycled concrete aggregate fines (fRCA), Portland limestone cement (PLC), and different silica-based materials.

Table 2.1 The waste and by-product materials used in pavement construction in the United States

Material	*Applications*
Hot mix asphalt (HMA) baghouse fines	Asphalt concrete
Blast furnace slag	Asphalt concrete, cement concrete, granular base, embankment, or fill
Coal bottom ash/boiler slag	Asphalt concrete, granular base, stabilized base
Coal fly ash	Asphalt concrete, cement concrete, embankment or fill, stabilized base, flowable fill
Flue-gas desulfurization (FGD) scrubber material	Stabilized base
Foundry sand	Asphalt concrete, flowable fill
Kiln dusts	Asphalt concrete, stabilized base
Mineral processing wastes	Asphalt concrete, granular base, embankment, or fill
Municipal solid waste (MSW) combustor ash	Asphalt concrete, granular base
Waste glass	Asphalt concrete, granular base
Nonferrous slags	Asphalt concrete, granular base, embankment, or fill
Quarry by-products	Flowable fill
Reclaimed asphalt pavement (RAP) aggregates	Asphalt concrete, granular base, embankment, or fill
Reclaimed concrete aggregate (RCA) material	Cement concrete, granular base, embankment, or fill
Roofing shingle scrap	Asphalt concrete
Scrap tires	Asphalt concrete, embankment, or fill
Sewage sludge ash	Asphalt concrete
Steel slag	Asphalt concrete, granular base
Sulphate wastes	Stabilized base

Source: FHWA (2008)

2.3.1 *Recycled aggregates for pavement bases*

Recycled pavement materials like RAP from asphalt pavements and RCA from concrete pavements are considered alternative construction materials in the United States due to the depletion of natural aggregates, increase in transportation costs, and the growing environmental concerns (Fanijo et al. 2023). Reuse of recycled pavement materials in infrastructure construction is a common practice in many parts of the world, as it has great benefits, such as conservation of natural resources, preservation of the environment, and retention of the existing highway geometrics (Taha et al. 1999). In situ recycling of old pavement layers is becoming increasingly popular, as it offers significant cost savings compared to traditional methods of adding new pavement layers. Additionally, it helps reduce environmental impacts by decreasing the need for natural aggregates and minimizing landfill space that would otherwise be used for demolition waste (Puppala et al. 2012).

In this section, the use of two such materials, including RAP and limestone QF for pavement base/subbase layer construction, is discussed based on the work conducted by the authors. The QF used was a residue deposit produced from limestone quarries, and RAP aggregates were obtained from milling existing hot mix asphalt (HMA) pavement layers. Both recycled materials were characterized in laboratory and field conditions, and the mechanical and geotechnical properties, like shear strength, consolidation, and resilient modulus, were measured. Due to the low strength and low modulus of the QF, cement was added to the fines to enhance the material properties. Both cement-treated limestone quarry fines (CQF) and 100% RAP were used as test pavement base/subbase materials to support new pavement sections constructed on State Highway 360 in Arlington, Texas, USA (Puppala et al. 2012). The performance of these two recycled materials as pavement materials was monitored in the test sections with instrumentation. Figure 2.2 presents a cross-section schematic of a pavement section built with CQF and RAP materials. Pavement surface profiling studies were conducted at different stages to verify the pavement roughness properties.

Laboratory tests conducted to verify the engineering behaviour of RAP and CQF materials show that limestone QF has material characteristics like a sandy material with a small quantity of fines. This material had low unconfined compressive strength (UCS) values without any treatment (around 100 kPa) and exhibited a free vertical swell strain of 6%. After being treated with 2.3% cement, the UCS of CQF improved to 1,200 kPa, and the swelling strain reduced to 0%. The high strength of the CQF is attributed to the cementing reactions of the cement and fine fractions of the QF material (Puppala et al. 2012). The RAP material used showed negligible swelling and compressibility. The UCS value for the RAP material was found to be 340 kPa.

The resilient modulus of CQF increased with an increase in confining stress, as the modulus was around 152 MPa at a low confining pressure of 21 kPa and was around 369 MPa at a high confining pressure of 138 kPa. This trend explains that the cemented material behaves as a strain-hardening material, and the resilient properties of this material depend on the applied confining pressure condition. For RAP material, the resilient moduli varied from 211 to 320 MPa for the same confining pressure values. Overall, these values show that the resilient property magnitudes are sufficient to provide stable and uniform support for pavement sections built in real field conditions. Profilers and field surveys from field test sections built on SH360 showed that the pavement sections built with these recycled aggregates or treated bases performed well, with little distress recorded on both CQF and RAP sections (Puppala et al. 2012).

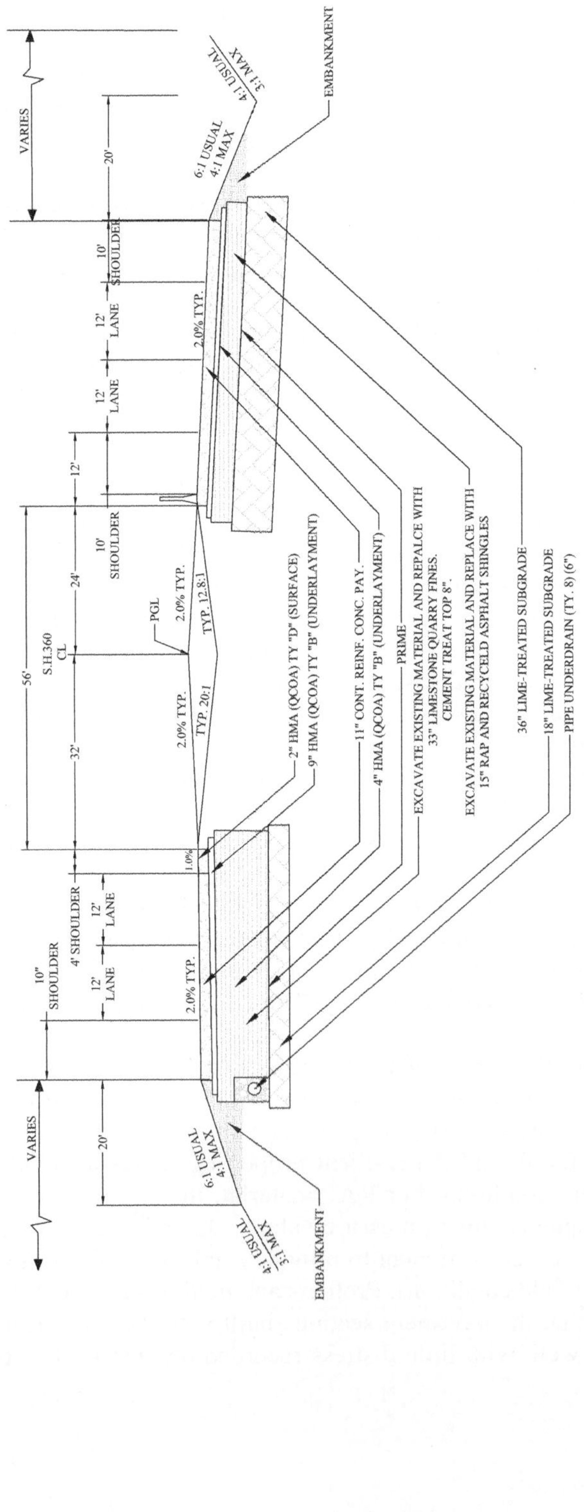

Figure 2.2 Cross-section view of North and Southbound sections showing CQF and RAP as pavement bases built on State Highway 360, Arlington, Texas, USA (Sta. 715+00 to Sta. 737+00).

2.3.2 Stabilization with metakaolin-based geopolymers

Geopolymers are novel binders formed from alkali-aluminosilicate reactions that harden over time. They usually exhibit comparable mechanical properties to conventional cement (OPC) but have lower GHG emissions (Bell et al. 2009). Geopolymers are composed of large three-dimensional networks of covalently bonded aluminosilicates. These covalent bonds ensure that geopolymers normally possess high mechanical strength, good durability, and low shrinking potentials (Guo et al. 2010, Law et al. 2015). The alkali-activated geopolymer materials can be produced from inexpensive, low-carbon aluminosilicate precursors such as MK, calcined clay, and fly ash, which harden rapidly at room temperatures. With these advantages, geopolymers can be considered as an eco-friendly and sustainable alternative to traditional construction materials like cement.

Extensive studies on the production of geopolymers using different source materials have been conducted over the past decades, and no official naming system has been established so far. Other common names for geopolymers are alkali-activated materials, inorganic polymers, or geopolymeric binders (Bernal & Provis 2014, Davidovits 2002). The key components for geopolymer production are the aluminosilicate source material (e.g., precursor) and the alkali activator (Huang et al. 2021). The high pH environment created by the alkali activator dissolves the alumina and silica from the precursor, and the dissolved species undergo a series of physical and chemical reactions (e.g., polymerization) and eventually form the geopolymer. A simplified geopolymerization process can be summarized in five steps: (i) dissolution of source alumina and silica, (ii) speciation equilibrium, (iii) gelation, (iv) reorganization, and (v) polymerization and hardening (Duxson et al. 2007). However, it is important to note that these reactions through the different stages of geopolymerization usually occur simultaneously. During geopolymer preparation, an alkali cation source, water, and an additional silica source are first blended with precalculated proportions to form the alkaline activator solution (AAS). The aluminosilicate precursor is then added to the AAS, and the resulting mixture forms the hardened geopolymer over time.

Jang (2022) used an MK-based, low-carbon geopolymer to stabilize expansive soils with different levels of sulphate content, and this study showed the effectiveness of geopolymers in soil stabilization for the enhancement of physical and engineering properties. In this study, MK was chosen as the precursor to provide source silica and alumina. Alkali activator was prepared by mixing potassium hydroxide, silica fume, and deionized water at designated ratios. The resulting AAS was then mixed with the MK precursor (predetermined to maintain the targeted Si/Al ratio), followed by continuous mixing until a homogeneous solution was obtained.

Due to the infinite number of geopolymer combinations, optimization of the geopolymer composition was conducted and analysed (Jang 2022). Two geopolymer compositions (K331 & K431) were first screened to ensure high initial strength and proper workability. The nomenclature of geopolymer compositions follows a format of KXYZ, where the first letter denotes the source of alkali – K represents KOH – while the three-digit number denotes the SiO_2/Al_2O_3 ratio, the water-to-solid ratio in the geopolymer, and the K/Al ratio, respectively. The two geopolymer compositions were used to treat a natural high-plasticity clay (PI = 33) with a sulphate content of 336 ppm (low sulphate content). To study the effectiveness of MK-based geopolymers on soils rich in sulphate content, laboratory-grade gypsum was added to the natural soil to synthesize a soil with high sulphate (HS) content (10,000 ppm). The natural low-sulphate (LS) clay is designated as LS, while the synthesized HS clay is designated as

HS. The geopolymer contents used in this research work were 8% and 30% of the weight of dry soil. A control treatment comprising 6% lime was used for a comparison study to evaluate the performance of the geopolymers against lime treatment (Jang 2022). Geotechnical properties of geopolymer-stabilized soils, including the UCS, one-dimensional free swelling, and resilient modulus properties, were determined in evaluating the performance of these geopolymers.

The UCS properties of the expansive clays treated with geopolymers under unsoaked or compaction conditions are shown in Figure 2.3. The strength improvement of geopolymer-treated soils behaves similarly to those treated with traditional stabilizers like cement and lime. The strength of the treated soil increases with an increase of geopolymer content and curing time. 8% geopolymer-treated soils (HS and LS) showed similar strength to those treated with 6% lime, while 30% geopolymer significantly improved the strength of the soil.

A one-dimensional free swell test was conducted on untreated, lime-treated, and geopolymer-treated soil specimens after three and 14 days of curing. The mitigation of swelling behaviour of the expansive clay using MK-based geopolymers is investigated, and the results are presented in Figure 2.4. The untreated soil with low and high sulphate contents showed a free swell strain of 12.2% (LS) and 10.8% (HS), respectively. The swelling is due to the expansive nature of the clay mineral present in the soil, and swelling occurred after only one day of moisture intrusion for both soils. Substantial reductions in swelling were observed after lime or geopolymer treatment. For treated soil samples cured for only three days, the swelling strain reduced to nearly 1%. For extended curing of 14 days, the swelling of the LS soil almost diminished. In addition, higher geopolymer content showed better swelling strain reductions. The results in Figure 2.4 show that geopolymers can be used to reduce the swelling potential of expansive soils, as long as sufficient geopolymers are used and proper curing time is guaranteed.

For soil with a high sulphate content (HS soil), considerable swelling occurred one day after the start of the test, most likely due to the formation of ettringite. For soil specimens treated for a short time (e.g., three days), the free swelling strains were near 5%, less than half the swelling

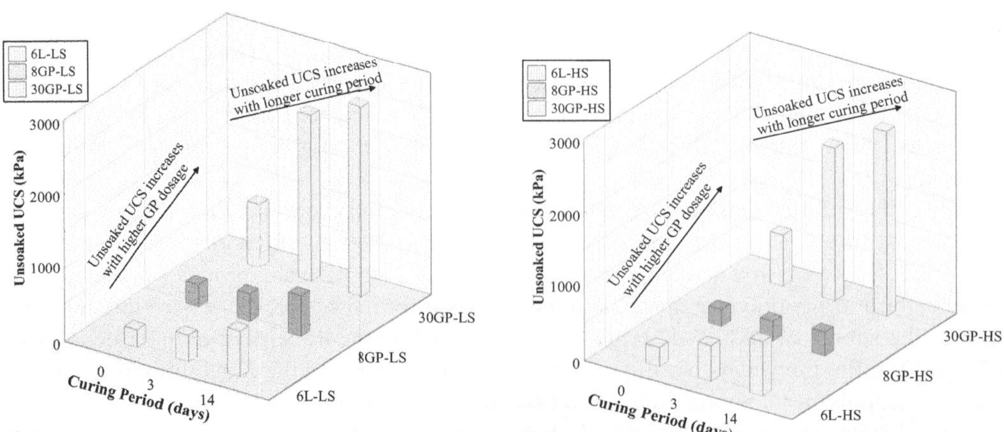

(a) (b)

Figure 2.3 Unsoaked UCS of expansive clays treated with geopolymers: (a) low sulphate soil (LS), (b) high sulphate soil (HS). Notation: GP – Geopolymers (Jang et al. 2022a).

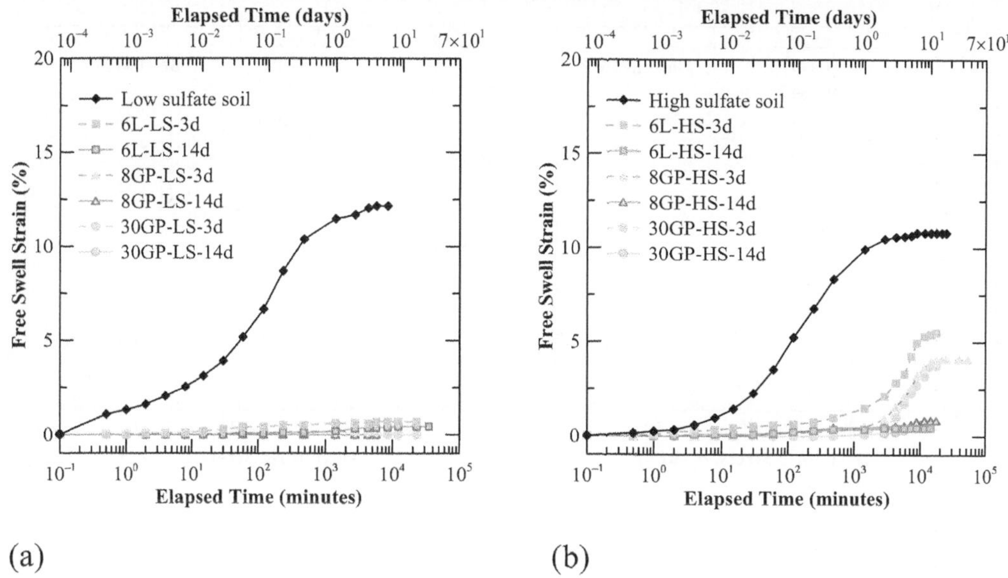

Figure 2.4 Free swell strain of soils treated with geopolymer. (a) low-sulphate soil (LS), and (b) high-sulphate soil (HS). Notation: L – lime (Jang et al. 2022b.)

strain of the untreated HS soil. This applied to all the soil samples treated with 6% lime and 8% and 30% geopolymers, with the geopolymer-treated HS soil showing 1% less swelling. However, when the treated specimens are allowed to cure for 14 days, when a significant degree of reinforcement is reached, the swelling of the treated soils (both lime- and geopolymer-treated soils) becomes negligible, indicating that MK-based geopolymers can efficiently mitigate the swelling potential of expansive soils, even at HS levels.

Resilient modulus (M_R), a key parameter in AASHTO pavement design, of the geopolymer-treated clays was investigated (Jang et al. 2022a). Figure 2.5 presents the M_R of the untreated and geopolymer-treated HS soil specimens. 6L-HS-14d, 8GP-HS-14d, and 30GP-HS-14d denote HS soil treated with 6% lime, 8% geopolymer, and 30% geopolymer treatment, respectively, after 14 days of curing. The untreated soil showed a softening behaviour, typical for natural clay soils, as indicated by the reduction in M_R values with an increase in deviatoric stresses. Geopolymer-treated HS soils exhibited a completely different behaviour, and the M_R values increased with the increase of deviatoric pressures, indicating strain hardening. More specifically, HS soil treated with 8% geopolymer was significantly stiffer than the untreated soil.

The M_R values of the treated specimens were three to four times higher than the untreated at low deviatoric stress levels, while increased by 500% to 700% at high stress levels. The 8%-geopolymer-treated soils are comparable to those treated with 6% of lime, only slightly softer under high confining pressures. This is most likely due to the formed geopolymer network binding partially with the soil particles. With the increase of geopolymer content to 30%, the stiffness of the soil improved substantially, surpassing 6% lime treatment under all stress conditions. The results indicate that sufficient geopolymers are required to guarantee the efficiency and effectiveness of the treatment.

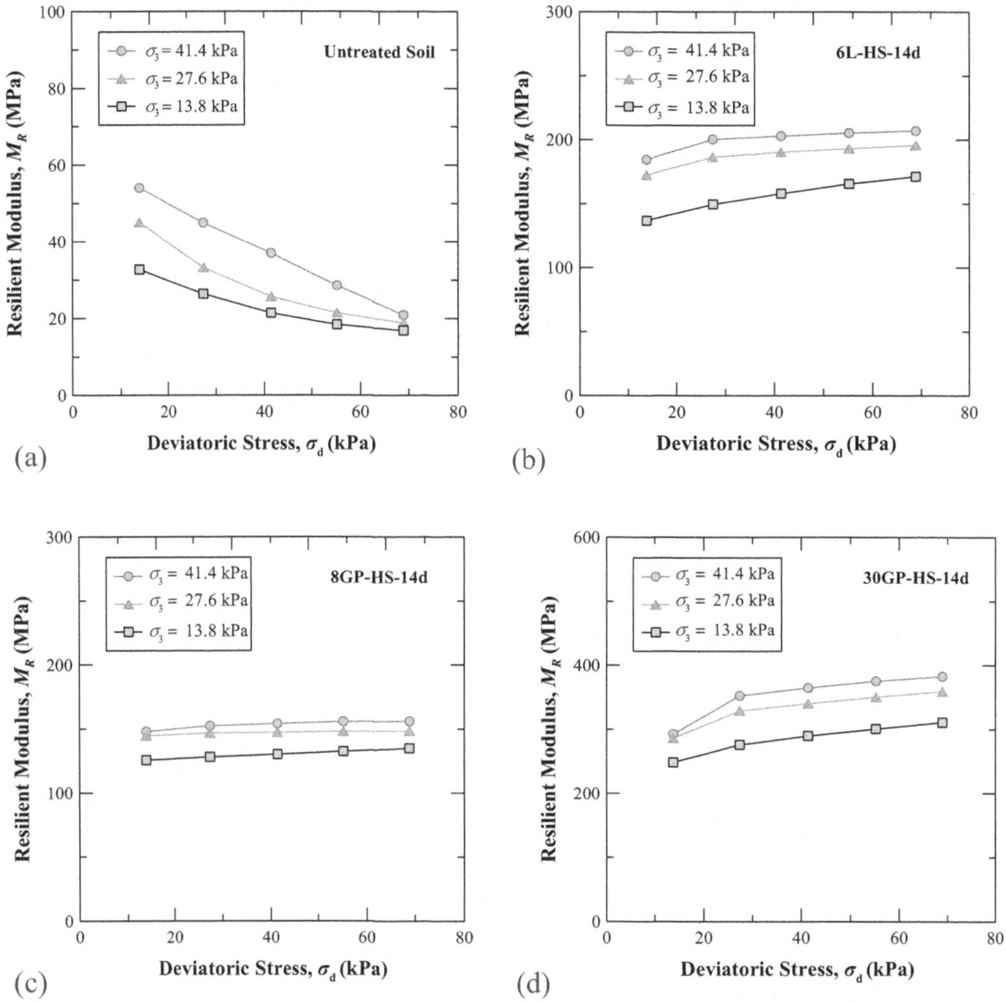

Figure 2.5 Resilient modulus of geopolymer-treated expansive clay as compared to lime (Jang et al. 2022a.)

2.3.3 *Stabilization with waste fines*

Another stabilization area where recycled materials can be effectively used in the field is utilization and construction and demolition (C&D) waste materials. The world generates three billion tonnes of C&D waste materials per year, 60% of which is landfilled, and hence any reuse of these materials in place of cementitious materials of high energy and carbon footprint will reduce the GHGs considerably. This C&D waste is primarily composed of concrete, bricks, plasterboard, and ceramics (Ferdous et al. 2021). Due to such huge and highly variable composition types and quantities, management of C&D waste is a major challenge in the United States and in the world due to limited land.

Studies found that smart recycling of C&D wastes for engineering applications can be an effective strategy in curtailing this problem with immense sustainability benefits (Atta & Bakhoum

2024, Banias et al. 2022). RCA is a major C&D waste, which has recently gained popularity in the transportation sector. RCA, as the name suggests, is waste concrete, which has been stripped clear of debris and steel that comes inherent with the C&D waste. It has already been used extensively as sustainable aggregates in concrete and pavement base layer constructions. RCA has multiple desirable properties, such as enhanced drainage and self-cementing capability, and, depending upon the crushing method, increased angularity, which makes it an ideal substitute for traditional base material (Edil 2018, Panda et al. 2021). Moreover, its utility as an aggregate in the concrete industry has also shown promising results.

As per a survey of state DOTs, RCA is the most commonly used recycled material in the United States (Edil 2018). RCA fines, on the other hand, are often considered as waste materials due to the fine size and are normally disposed of in landfills. However, recent studies show that sand-sized fRCA improved the strength and durability and reduced the plasticity and shrink/swell potential of problematic soils (Kianimehr et al. 2019). In this section, the use of RCA fines in combination with PLC in subgrade soil stabilization is discussed. PLC is a type of blended cement with around 15% of limestone. The inclusion of limestone reduces the carbon footprint in cement while maintain the same performance as the standard cement. PLC is considered a low-carbon cement. Pictures of fRCA and PLC are shown in Figure 2.6.

A subgrade soil with a plasticity index of 25% was treated with 8% PLC with 15% and 30% of fRCA as co-additives, respectively (Sanei 2024). Soil treated with 10% PLC alone was included in the study for comparison purposes. Mechanical properties of the treated soils, including UCS and resilient modulus before and after wet-dry (ASTM D559) and freeze-thaw (ASTM D560) durability cycles, were examined. The untreated and treated soil specimens were cured at room temperature (23°C) and 100% relative humidity for 7 days before testing. Test results showed that the UCS of the untreated soil was 372 kPa, and with 10% PLC treatment, the UCS value improved to 3,080 kPa. The combined use of 8% cement and 15% soil replacement with fRCA (8C_15fRCA) showed an average UCS value of 3,062 kPa, while the UCS of 8% cement and

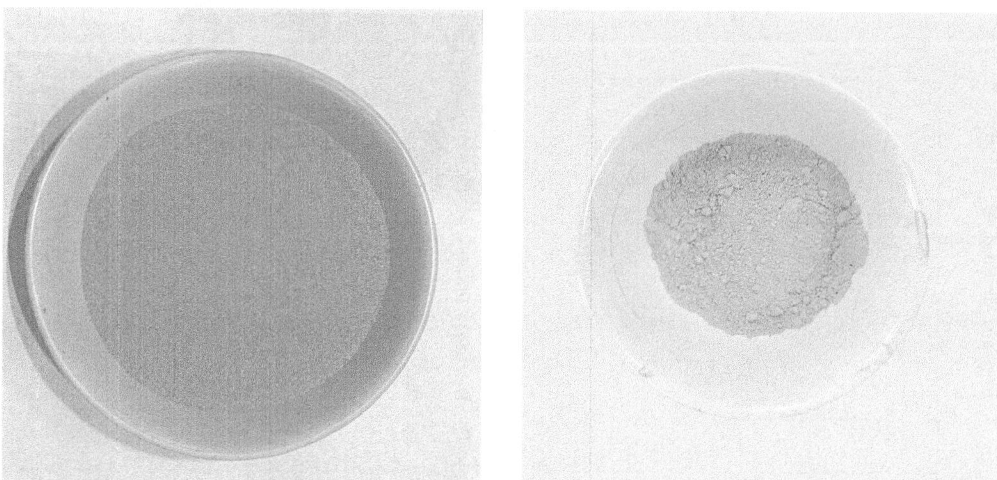

(a) (b)

Figure 2.6 Photos of RCA fines (fRCA) and Portland Limestone Cement (PLC) used for soil stabilization.

30% fRCA treatment (8C_30fRCA) was 2,883 kPa. The UCS values of the fRCA-treated soils were comparable to the subsoils treated with 10% cement alone and are significantly higher than the untreated soil, indicating that inclusion of fRCA in soil stabilization can reduce the demand for cement. This is partly due to the improvement of the soil texture of the weak soil due to the addition of granular RCA fines as well as the possible contributions of fRCA to the cement hydration process if unreacted cement is present.

The resilient moduli of the treated soils are presented in Figure 2.7. The untreated soil showed a softening behaviour as the resilient moduli decreased with an increase in deviatoric stresses, and the stiffness (M_R) was independent of the confining pressure. After treatment with PLC and fRCA, the resilient modulus of the soil improved, and M_R was higher under all the confining pressures and deviatoric stress levels. The inclusion of fRCA in cement-treated soil showed better M_R improvement at both replacement levels (15% and 30%) while using 2% less cement.

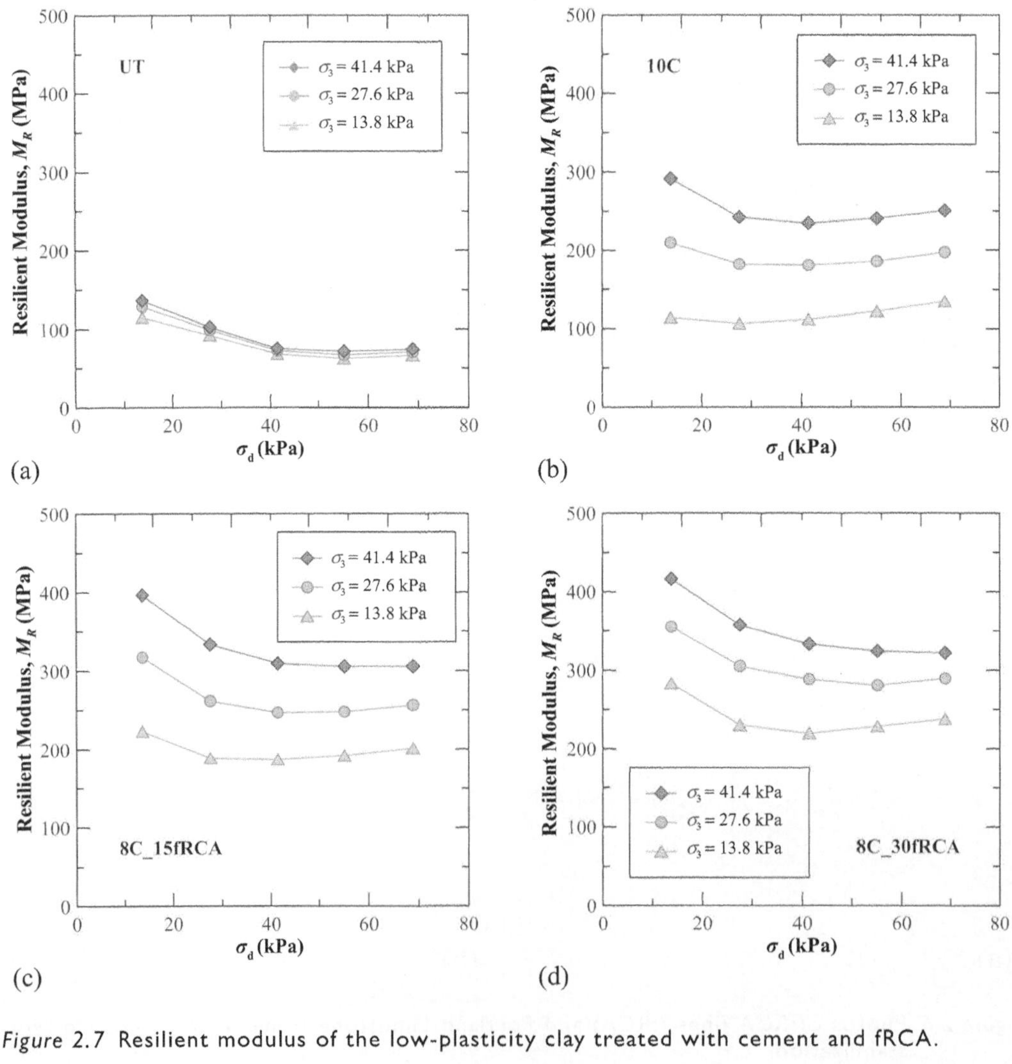

Figure 2.7 Resilient modulus of the low-plasticity clay treated with cement and fRCA.

Table 2.2 Unconfined compressive strength of low-plasticity soil treated with PLC and fRCA after different durability cycles

Soil treatment	*Average UCS of soil after certain numbers of durability cycles, kPa*		
	0 cycles	*6 cycles*	*12 cycles*
Wet-dry cycles (W-D)			
UT	372	NA	NA
10C	3080	5662	6089
8C_15fRCA	3062	5665	5521
8C_30fRCA	2883	6170	6096
Freeze-thaw cycles (F-D)			
UT	372	NA	NA
10C	3080	3304	4215
8C_15fRCA	3062	3916	5220
8C_30fRCA	2883	2674	2714

NA, Results unavailable because samples failed before completion of required cycles.

This shows that fRCA can be used for soil stabilization and slightly reduce the demand for cement, which will bring significant sustainability and environmental benefits.

Durability evaluations of the soil specimens treated with PLC and PLC plus fRCA show that fRCA-reinforced soils showed satisfactory performance, and both soil mixtures withstood 12 cycles of wetting and drying (Sanei 2024). The durability analysis results are documented in Table 2.2. During the first six wet-dry cycles, the UCS of the treated soils continued to increase due to the on-going cement hydration; after 12 cycles, the cement hydration seemed to have completed, and degradation in strength due to the environmental stressors started to appear. Nevertheless, the strength of the treated soil after 12 cycles of wetting and drying was stronger than the original seven-day UCS, indicating the mixtures studied were durable. The freeze-thaw durability test showed similar results; however, the damage resulting from the freezing and thawing cycles was more significant than that from wetting and drying, confirmed by the relatively lower UCS values tabulated in Table 2.2. This is due to the inhibition of freezing temperatures on cement hydration as well as the damage induced by the ice crystal formation during the freezing phase. However, the UCS of the treated soil specimens was all above the required strength value of 1,724 kPa as per the guidelines developed by the U.S. Army Corps of Engineers for pavement design. Similarly, M_R values of the treated specimens showed consistent trends as that of UCS (Sanei 2024). Therefore, the combination of fRCA and reduced cement content is suitable for long-term use in the construction of transportation infrastructure.

2.3.4 *Stabilization with silica products to enhance pozzolanic activity*

To enhance pozzolanic reactions in the treated soils, reduce sulphate-related heave distress from calcium-based stabilizer treatment, and enhance long-term durability of transportation infrastructures, a research study was designed to use novel silica-based co-additives, such as quarry-dust crushed fines or crushed crystalline silica (termed as CS) and laboratory-grade nano-silica (termed as NS) compounds with lime stabilizer to treat high-plasticity soils with different levels of sulphate concentration (Biswas 2022). This study showed that

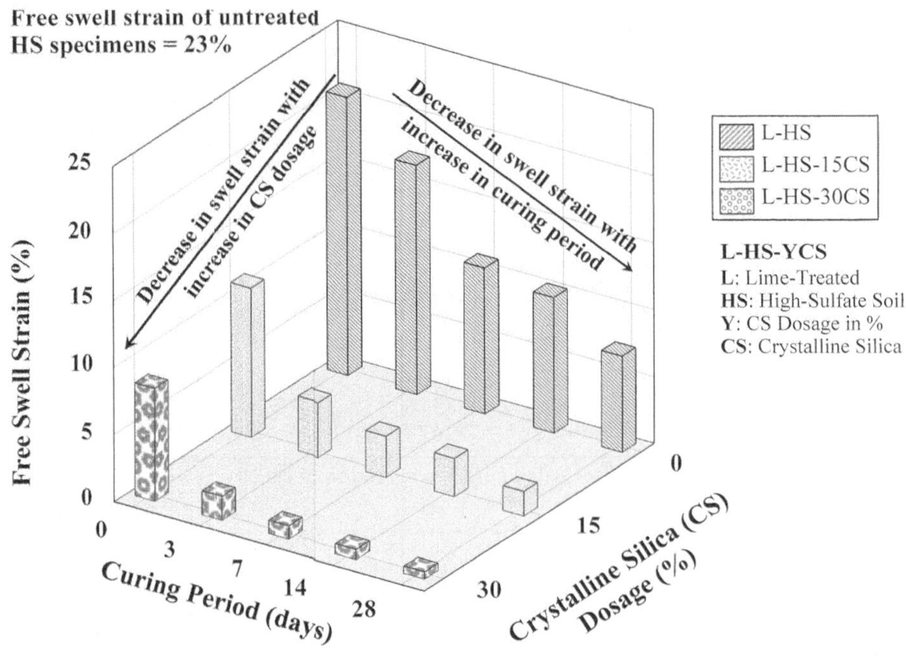

Figure 2.8 Influence of Crystalline Silica (CS) and Lime (L) treatment on High Sulphate (HS) soil.

the application of these co-additives with traditional dolomitic-hydrated lime significantly modified the reaction kinetics. Additionally, when used with lime to treat sulphate-rich soils, the presence of the additional silica phases, such as CS, suppressed the precipitation and subsequent swelling from ettringite crystals. Figure 2.8 presents the influence of CS addition in reducing the swelling in sulphate-rich soils. Other engineering studies also showed that the presence of the silica phases improved the durability and permanency of the chemical stabilizer compared to traditional treatment alone, and the majority of these results can be seen in Biswas et al. (2023, 2024a, b).

Overall, the application of these new treatment techniques will be of immense help for transportation agencies and will help them to use new and more sustainable soil treatment techniques for improving the long-term performance of the transportation infrastructures. The above research studies showcased the potential of various treatments and recycled materials in the inclusion of base or subgrade layers for resilient support of pavements. Still, there are many barriers to translating these laboratory research studies to move forward to field implementation studies. The final section below summarizes a few of these barriers for the field adaptation.

2.4 Barriers to field adaptations and pathways

Authors, particularly the senior author, based on his experience over the past few decades, outlined four barriers below. Addressing these barriers will help bring out the best-performing geomaterials as part of pavement base layer systems.

2.4.1 Budget and economic considerations

The adoption of low-carbon policies and utilization of sustainable materials are highly dependent on the initial costs as well as the available budgets of the agencies. The low-carbon technologies and infrastructure may require high upfront investment, making policymakers and/or stakeholders hesitant to commit to the adaptation of these solutions into field practices (Chegut et al. 2019, Chen et al. 2023). Although certain wastes or recycled materials may have lower initial material costs, the cost of efficient designs using alternative materials may be more expensive than conventional methods due to the investment in initial research and studies required for assurance of performance of these materials (Granoff et al. 2016). These early-stage research studies often last a long duration and will not show immediate outcomes, giving stakeholders low incentives to adopt these policies. In addition, many transportation projects are funded by public budgets. As a result, funding agencies often need to prioritize cost-efficient projects, which makes low-carbon technologies less competitive against the conventional measures.

2.4.2 Availability of sustainable materials

The availability of sustainable materials significantly varies by state, largely due to the differences in regional supply chains, state initiatives, and environmental priorities. For states with high environmental priority and policies consistently in favour of the adoption of low-carbon materials, like California, New York, and Washington, the low-carbon materials are likely available due to the larger number of material suppliers and local supply chains. However, for other states that do not have such policy initiatives, the availability of low-carbon materials may be limited, especially for newer green materials. The availability of low-carbon materials is also restricted in some states due to a lack of certain source materials in the region, which forces these states to rely on imported materials from other states, resulting in high construction cost, which in turn hinder the application of these low-carbon materials. In general, the limited capacity of low-carbon materials is faced in many states in the United States due to the relatively small supply chains. The lack of these sustainable materials can impede the efforts towards the ambitious carbon reduction goals. However, with the growing demand for these low-carbon materials, more states are progressively engaged in incorporation of sustainable construction in different programmes, which incentivizes the local material suppliers to produce more low-carbon materials (FHWA 2024b).

2.4.3 Technology readiness levels

The technical readiness levels of the use of low-carbon materials in different transportation infrastructure constructions vary greatly, depending on the material and its manufacturing potential for large-scale applications and the need for champion agencies to implement them in field studies. Being relatively new, most of the low-carbon materials are waste or recycled products; the uncertainty of the performance of these new materials makes the agencies and stakeholders hesitant to use these materials in the construction projects (Althoey et al. 2023). Additionally, the mega-field construction contracts are often design-build types, which require contractors to be responsible for managing the assets for a longer period, and hence, they often do not consider research solutions for field implementation due to potential liabilities.

Also, the use of many new low-carbon materials in construction is mainly in the research stage, with none or limited large-scale field applications. Transportation infrastructure is built to be durable and should withstand significant wear and tear during its designed life span.

Low-carbon materials, especially the newer ones, may not have sufficient performance data or long-term durability testing to prove their reliability in various climates and in heavy-use conditions. Moreover, the construction of transportation infrastructures should follow established codes and standards (Lopez et al. 2024). These standards are developed based on the conventional construction materials, not for the new, low-carbon materials (Lopez et al. 2024). The lack of standardized protocols for using new low-carbon materials constrains the applications of these materials in real-world infrastructure constructions.

2.4.4 *Policies and priorities*

Policies and priorities with regard to environmental issues and sustainability vary significantly at federal and state levels in the United States. In some cases, the environmental priority may be lowered due to other competing priorities, leading to inconsistent policies regarding the use of low-carbon materials in transportation infrastructure construction. The change in priorities often results in a shift in the use of public funds, which constrains federal and state agencies to prioritize certain materials over others due to the availability of limited funds.

Overall, there are certainly numerous challenges, but they also offer many opportunities for agencies to come up with ways to alleviate these concerns and promote the use of materials that can provide environmentally friendly and resilient solutions. The sustainability focus of many agencies will also lead them to embrace new materials in order to reduce GHG emissions.

2.5 Summary and findings

The chapter gives a brief overview of the latest federal and state policies and programmes in the United States that encourage the use of low-carbon materials and technologies in transportation infrastructure to help combat the environmental challenges and to facilitate the process towards the U.S. carbon neutrality goal by 2050.

Programmes administered by the FHWA under USDOT, including the LCTM programme, CRP, and the UTC programme, have been reviewed, and their contributions to the utilization of low-carbon materials in transportation infrastructure construction on national, state, and local levels have been discussed. Many states in the United States have developed their own carbon reduction programmes, which have been briefly reviewed and discussed in the chapter. It is revealed that the environmental priorities and policies between different states vary greatly, resulting in different levels of utilization of carbon-friendly materials in transportation infrastructure among states.

Four case studies using different recycled or low-carbon materials such as RAP and QF, geopolymers, fRCA and PLC, and quarry dust and silica on different pavement-related projects have been discussed. The laboratory and/or field studies showed promising results of using these low-carbon materials for pavement base or subgrade construction. The use of 100% RAP or cement-treated QF for base and subbase construction showed satisfactory strength and resilient modulus, and field applications indicated good performance of these recycled materials. The laboratory study of using MK-based geopolymers for treatment of expansive clay showed that geopolymers can sufficiently improve the strength and stiffness of weak subgrade soils, even for sulphate-rich soils. fRCA, when used in combination with cement, provides an alternative option for soil stabilization. The inclusion of fRCA in soil stabilization can reduce the demand for cement while maintaining sufficient strength, stiffness, and durability. The addition of silica-based materials to soils treated with lime showed that the presence of silica phase can enhance the durability of the stabilized geomaterials. These materials showcase the potential applications of low-carbon materials in transportation infrastructure construction in the United States.

This chapter also discusses the current barriers to broader utilization and implementation of low-carbon materials in the transportation construction area in the United States. These challenges include the budget and cost considerations, availability of materials, technology readiness levels, and policies and priorities. Addressing these barriers in the future is critical for a broader use of low-carbon materials in construction projects in the country and for achieving the ambitious climate goals of the United States and the world.

Acknowledgements

The authors would like to acknowledge the following individuals for their contributions in different research projects using low-carbon materials during their study at Texas A&M University. The authors would like to thank Dr Jungyeon Jang for PhD dissertation research work regarding MK-based geopolymers for soil stabilization and Mr Muddassir Sanei for master's thesis research work on utilization of recycled concrete fines and low-carbon PLC. Their works added to the knowledge base about using these low-carbon materials in transportation infrastructure construction and generated valuable research data, which made this chapter possible. The authors also appreciate the support from the NSF Industry-University Cooperative Research Center (I/UCRC) programme at the Center for Integration of Composites into Infrastructure (CICI) at Texas A&M University, NSF PD: Dr Prakash Balan; Award #2017796. These supports are crucial for various research studies on the different materials used in transportation infrastructure construction works at TAMU.

Reference list

Abu-Farsakh, M., Dhakal, S. and Chen, Q. 2015. Laboratory Characterization of Cementitiously Treated/ Stabilized Very Weak Subgrade Soil Under Cyclic Loading. *Soils Found.*, *55*(3), 504–516. https://doi.org/10.1016/j.sandf.2015.04.003

Adesina, A. 2020. Recent Advances in the Concrete Industry to Reduce its Carbon Dioxide Emissions. *Environ. Chall.*, *1*, 100004. https://doi.org/10.1016/j.envc.2020.100004

Althoey, F., Ansari, W. S., Sufian, M. and Deifalla, A. F. 2023. Advancements in Low-Carbon Concrete as a Construction Material for the Sustainable Built Environment. *Dev. Built Environ.*, *16*, 100284. https://doi.org/10.1016/j.dibe.2023.100284

ASCE. 2021. *Report Card for America's Infrastructure: Roads*. A. S. o. C. Engineers. https://infrastructurereportcard.org/cat-item/roads-infrastructure/

Atta, I. and Bakhoum, E. S. 2024. Environmental Feasibility of Recycling Construction and Demolition Waste. *Int. J. Environ. Sci. Technol.*, *21*(3), 2675–2694. https://doi.org/10.1007/s13762-023-05036-y

Banias, G. F., Karkanias, C., Batsioula, M., Melas, L. D., Malamakis, A. E., Geroliolios, D., Skoutida, S. and Spiliotis, X. 2022. Environmental Assessment of Alternative Strategies for the Management of Construction and Demolition Waste: A Life Cycle Approach. *Sustainability*, *14*(15), 9674. https://doi.org/10.3390/su14159674

Behnood, A. 2018. Soil and Clay Stabilization with Calcium- and Non-Calcium-Based Additives: A State-of-the-Art Review of Challenges, Approaches and Techniques. *Transp. Geotech.*, *17*, 14–32. https://doi.org/10.1016/j.trgeo.2018.08.002

Bell, J. L., Driemeyer, P. E. and Kriven, W. M. 2009. Formation of Ceramics from Metakaolin-Based Geopolymers. Part II: K-Based Geopolymer. *J. Am. Ceram. Soc.*, *92*(3), 607–615. https://doi.org/10.1111/j.1551-2916.2008.02922.x

Bernal, S. A. and Provis, J. L. 2014. Durability of Alkali-Activated Materials: Progress and Perspectives. *J. Am. Ceram. Soc.*, *97*(4), 997–1008. https://doi.org/10.1111/jace.12831

Biswas, N. 2022. *Role of Silica-Based Co-additives for Effective Treatment of Expansive Soils* [Ph. D. Dissertation, Texas A&M University]. College Station, TX.

Biswas, N., Puppala, A. J. and Chakraborty, S. 2023. Role of Nano- and Crystalline Silica to Accelerate Chemical Treatment of Problematic Soil. *J. Geotech. Geoenviron. Eng.*, *149*(7), 04023044. https://doi.org/10.1061/Jggefk.Gteng-10999

Biswas, N., Puppala, A. J. and Chakraborty, S. 2024a. Experimental Studies and Sustainability Assessments of Quarry Dust for Chemical Treatment of Expansive Soils. *Geotech. Test. J.*, *47*(1). https://doi.org/10.1520/gtj20220243

Biswas, N., Puppala, A. J., Chakraborty, S. and Little, D. N. 2024b. Micro-Mechanical Behavior of Nanosilica-Treated High-Sulfate Soils. *Can. Geotech. J.*, *61*(8), 1606–1621. https://doi.org/10.1139/cgj-2023-0051

Biswas, N., Puppala, A. J., Mirzaiyanrajeh, D. and Lostumbo, J. 2024c. Assessment of Structural Performance of Thin-Asphalt Pavement Reinforced with Wicking Geotextiles over Expansive Subgrades. In A. Carter, K. Vasconcelos and E. Dave, *14th International Conference on Asphalt Pavements ISAP2024* Montreal. Cham.

Bureau of Transportation Statistics. 2021. *Public Road and Street Mileage in the United States by Type of Surface*. United State Department of Transportation. Retrieved Dec. 29, 2024 from https://www.bts.gov/content/public-road-and-street-mileage-united-states-type-surfacea

Bureau of Transportation Statistics. 2024. *System Mileage within the United States*. United States Department of Transportation. Retrieved Dec. 29, 2024 from https://www.bts.gov/content/system-mileage-within-united-states

Chegut, A., Eichholtz, P. and Kok, N. 2019. The Price of Innovation: An Analysis of the Marginal Cost of Green Buildings. *J. Environ. Econ. Manag.*, *98*, 102248. https://doi.org/10.1016/j.jeem.2019.07.003

Chen, L., Huang, L., Hua, J., Chen, Z., Wei, L., Osman, A. I., Fawzy, S., Rooney, D. W., Dong, L. and Yap, P.-S. 2023. Green Construction for Low-Carbon Cities: A Review. *Environ. Chem. Lett.*, *21*(3), 1627–1657. https://doi.org/10.1007/s10311-022-01544-4

Correia, A. G., Winter, M. G. and Puppala, A. J. 2016. A Review of Sustainable Approaches in Transport Infrastructure Geotechnics. *Transp. Geotech.*, *7*, 21–28. https://doi.org/10.1016/j.trgeo.2016.03.003

Davidovits, J. 2002 (October 28–29). 30 Years of Success and Failures in Geopolymer Application, Market Trends and Potential Breakthroughs. In *Geopolymer 2002 Conference*, Melbourne, Australia.

Duxson, P., Fernández-Jiménez, A., Provis, J. L., Lukey, G. C., Palomo, A. and van Deventer, J. S. J. 2007. Geopolymer Technology: The Current State of the Art. *JMatS*, *42*(9), 2917–2933. https://doi.org/10.1007/s10853-006-0637-z

Edil, T. B. 2018. A Review of Recycled Aggregates (RAP and RCA) as Unbound Base Course Material for Sustainable Highway Construction. In H.-H. Tran-Nguyen, H. Wong, F. Ragueneau and C. Ha-Minh, *Proceedings of the 4th Congrès International de Géotechnique – Ouvrages – Structures*. Singapore.

Fanijo, E. O., Kolawole, J. T., Babafemi, A. J. and Liu, J. 2023. A Comprehensive Review on the Use of Recycled Concrete Aggregate for Pavement Construction: Properties, Performance, and Sustainability. *Clean. Mater.*, *9*, 100199. https://doi.org/10.1016/j.clema.2023.100199

Ferdous, W., Manalo, A., Siddique, R., Mendis, P., Zhuge, Y., Wong, H. S., Lokuge, W., Aravinthan, T. and Schubel, P. 2021. Recycling of Landfill Wastes (Tyres, Plastics and Glass) in Construction – A Review on Global Waste Generation, Performance, Application and Future Opportunities. *Resour. Conserv. Recycl.*, *173*, 105745. https://doi.org/10.1016/j.resconrec.2021.105745

FHWA. 2008. *User Guidelines for Waste and Byproduct Materials in Pavement Construction* (FHWA-RD-97-148). Retrieved from https://www.fhwa.dot.gov/publications/research/infrastructure/structures/97148/begin.cfm

FHWA. 2018. *Public Road Length – 2017: Miles by Type of Surface and Ownership/Functional System National Summary*. Retrieved Dec. 29, 2024 from https://www.fhwa.dot.gov/policyinformation/statistics/2017/hm12.cfm#foot1

FHWA. 2024a. *Carbon Reduction Program*. Retrieved Dec. 27, 2024 from https://www.transportation.gov/priorities/climate-and-sustainability/carbon-reduction-program

FHWA. 2024b. *Low-Carbon Transportation Materials Grants Program*. Retrieved Dec. 25, 2024 from https://www.fhwa.dot.gov/lowcarbon/

Granoff, I., Hogarth, J. R. and Miller, A. 2016. Nested Barriers to Low-Carbon Infrastructure Investment. *Nat. Clim. Chang.*, *6*(12), 1065–1071. https://doi.org/10.1038/nclimate3142

Guo, X., Shi, H. and Dick, W. A. 2010. Compressive Strength and Microstructural Characteristics of Class C Fly Ash Geopolymer. *Cem. Concr. Compos.*, *32*(2), 142–147. https://doi.org/10.1016/j.cemconcomp.2009.11.003

Hoyos, L. R., Puppala, A. J. and Ordonez, C. A. 2011. Characterization of Cement-Fiber-Treated Reclaimed Asphalt Pavement Aggregates: Preliminary Investigation. *J. Mater. Civ. Eng.*, *23*(7), 977–989. https://doi.org/10.1061/(Asce)Mt.1943-5533.0000267

Huang, J., Grajales, J. A., Akula, P., Little, D. N., Kim, Y. R. and Rushing, J. F. 2024. Strength and Fracture Properties of Sandy Subgrade Soil Treated with Sodium Polystyrene Sulfonate. *J. Mater. Civ. Eng.*, *36*(5), 04024065. https://doi.org/10.1061/Jmcee7.Mteng-16709

Huang, J., Kogbara, R. B., Hariharan, N., Masad, E. A. and Little, D. N. 2021. A State-of-the-Art Review of Polymers Used in Soil Stabilization. *Constr Build Mater.*, *305*. https://doi.org/10.1016/j.conbuildmat.2021.124685

Huang, J., Makhatova, A., Kogbara, R., Masad, E., Sukhishvili, S. and Little, D. 2023a. Mechanical Properties of Palygorskite Clay Stabilized with Polyelectrolytes. *Transp. Geotech.*, *43*, 101124. https://doi.org/10.1016/j.trgeo.2023.101124

Huang, J., Mohomad, Y., Kogbara, R. B., Masad, E., Sukhishvili, S. and Little, D. 2023b. Stabilization of Calcareous Subgrade Soils with Polyelectrolytes: Mechanisms and Mechanical Properties. *Int. J. Pavement Eng.*, *24*(1), 2190976. https://doi.org/10.1080/10298436.2023.2190976

Jang, J. 2022. *Geopolymer Treatment for Stabilization of Problematic Geomaterials* [Ph. D. Dissertation, Texas A&M University]. College Station, TX.

Jang, J., Biswas, N., Puppala, A. J., Congress, S. S. C., Radovic, M. and Huang, O. 2022a. Evaluation of Geopolymer for Stabilization of Sulfate-Rich Expansive Soils for Supporting Pavement Infrastructure. *Transp. Res. Rec.*, *2676*(9), 230–245. https://doi.org/10.1177/03611981221086650

Jang, J., Puppala, A. J., Biswas, N., Chakraborty, S. and Radovic, M. 2022b. Utilization of Metakaolin-Based Geopolymers for Stabilization of Sulfate-Rich Expansive Soils. *Geo-Congress 2022: Soil Improvement, Geosynthetics, and Innovative Geomaterials*, *331*, 222–231. https://doi.org/10.1061/9780784484012.023

Kazemian, M. and Shafei, B. 2023. Carbon Sequestration and Storage in Concrete: A State-of-the-Art Review of Compositions, Methods, and Developments. *J. CO_2 Util.*, *70*, 102443. https://doi.org/10.1016/j.jcou.2023.102443

Khan, M. A., Biswas, N., Banerjee, A. and Puppala, A. J. 2020. Field Performance of Geocell Reinforced Recycled Asphalt Pavement Base Layer. *Transp. Res. Rec.*, *2674*(3), 69–80. https://doi.org/10.1177/0361198120908861

Kianimehr, M., Shourijeh, P. T., Binesh, S. M., Mohammadinia, A. and Arulrajah, A. 2019. Utilization of Recycled Concrete Aggregates for Light-Stabilization of Clay Soils. *Constr Build Mater.*, *227*, 116792. https://doi.org/10.1016/j.conbuildmat.2019.116792

Kolay, P. K., Dhakal, B., Kumar, S. and Puri, V. K. 2016. Effect of Liquid Acrylic Polymer on Geotechnical Properties of Fine-Grained Soils. *Int. J. Geosynth. Ground Eng.*, *2*(29), Article Unsp 29. https://doi.org/10.1007/s40891-016-0071-5

Kumar, P. 2023. *Resilient Behavior of Chemically Stabilized Geomaterials and Assessment of Moisture-Susceptible Durability* [Ph. D Dissertation, Texas A&M University]. College Station, TX.

Kumar, P., Puppala, A. J., Congress, S. S. C. and Tingle, J. S. 2024. Characterization of Cement-Polymer-Treated Soils under Repeated Loading. *J. Mater. Civ. Eng.*, *36*(4), 04024036. https://doi.org/10.1061/Jmcee7.Mteng-16394

Kumar, P., Puppala, A. J., Tingle, J. S., Chakraborty, S. and Sarat Chandra Congress, S. 2022. Resilient Characteristics of Polymer Emulsion-Treated Sandy Soil. *Transp. Res. Rec.*, *2676*(9), 526–538. https://doi.org/10.1177/03611981221087238

Law, D. W., Adam, A. A., Molyneaux, T. K., Patnaikuni, I. and Wardhono, A. 2015. Long Term Durability Properties of Class F Fly Ash Geopolymer Concrete. *Mater. Struct.*, *48*(3), 721–731. https://doi.org/10.1617/s11527-014-0268-9

Little, D. N. and Nair, S. 2009. *Recommended Practice for Stabilization of Subgrade Soils and Base Materials*. The National Academies Press. https://doi.org/10.17226/22999

Lopez, S., Sutter, L., Hooton, R. D., Van Dam, T., Innis, A. and Senn, K. 2024. Breaking Barriers to Low Carbon Concrete Pavements. *Transportation Research Record*, 03611981241250340. https://doi.org/10.1177/03611981241250340

Mekonnen, E., Kebede, A., Tafesse, T. and Tafesse, M. 2020. Application of Microbial Bioenzymes in Soil Stabilization. *Int. J. Microbiol.*, *2020*(1), 1725482. https://doi.org/10.1155/2020/1725482

Panda, B., Imran, N. T. and Samal, K. 2021. A Study on Replacement of Coarse Aggregate with Recycled Concrete Aggregate (RCA) in Road Construction. In B. B. Das, S. Barbhuiya, R. Gupta and P. Saha, *Recent Developments in Sustainable Infrastructure*. Springer Nature, Singapore.

Puppala, A. J., Hoyos, L. R. and Potturi, A. K. 2011. Resilient Moduli Response of Moderately Cement-Treated Reclaimed Asphalt Pavement Aggregates. *J. Mater. Civ. Eng.*, *23*(7), 990–998. https://doi.org/10.1061/(Asce)Mt.1943-5533.0000268

Puppala, A. J., Saride, S. and Williammee, R. 2012. Sustainable Reuse of Limestone Quarry Fines and RAP in Pavement Base/Subbase Layers. *J. Mater. Civ. Eng.*, *24*(4), 418–429. https://doi.org/10.1061/(Asce)Mt.1943-5533.0000404

Sanei, M. 2024. *Integration of Innovative Materials to Enhance Sustainability and Durability in Transportation Infrastructure* [Unpublished Master's Thesis, Texas A&M University]. College Station, TX.

Sidiq, A., Robert, D. J., O'Donnell, B., Setunge, S., Swarup, A. and Tostvrsnik, F. 2023. Investigation of Enzyme-Based Soil Stabilization in Field Application. *J. Mater. Civ. Eng.*, *35*(5), 04023086. https://doi.org/10.1061/(Asce)Mt.1943-5533.0004742

Taha, R., Ali, G., Basma, A. and Al-Turk, O. 1999. Evaluation of Reclaimed Asphalt Pavement Aggregate in Road Bases and Subbases. *Transportation Research Record*, *1652*(1), 264–269. https://doi.org/10.3141/1652-33

The White House. 2021a. *Executive Order on Catalyzing Clean Energy Industries and Jobs through Federal Sustainability*. Retrieved Dec. 24, 2024 from https://www.whitehouse.gov/briefing-room/presidential-actions/2021/12/08/executive-order-on-catalyzing-clean-energy-industries-and-jobs-through-federal-sustainability/

The White House. 2021b. *Executive Order on Tackling the Climate Crisis at Home and Abroad*. Retrieved Dec. 24, 2024 from https://www.whitehouse.gov/briefing-room/presidential-actions/2021/01/27/executive-order-on-tackling-the-climate-crisis-at-home-and-abroad/

The White House. 2022. *Building a Better America: A Guidebook to the Bipartisan Infrastructure Law for State, Local, Tribal, and Territorial Governments, and Other Partners*. Washington, DC. Retrieved from https://www.whitehouse.gov/wp-content/uploads/2022/05/BUILDING-A-BETTER-AMERICA-V2.pdf

The White House. 2023. *Building a Clean Energy Economy: A Guidebook to The Inflation Reduction Act's Investments in Clean Energy and Climate Action*. Washington, DC. Retrieved from https://www.whitehouse.gov/wp-content/uploads/2022/12/Inflation-Reduction-Act-Guidebook.pdf

Thomas, M., Barcelo, L., Blair, B., Cail, K., Delagrave, A. and Kazanis, K. 2012. Lowering the Carbon Footprint of Concrete by Reducing Clinker Content of Cement. *Transp. Res. Rec.*, *2290*(1), 99–104. https://doi.org/10.3141/2290-13

Tingle, J. S. and Santoni, R. L. 2003. Stabilization of Clay Soils with Nontraditional Additives. *Eighth International Conference on Low-Volume Roads 2003, Vols. 1 and 2*, *1819*(1819), A72–A84. https://doi.org/10.3141/1819b-10

U.S. Department of State & U.S. Executive Office of the President. 2021. *The Long-Term Strategy of the United States: Pathways to Net-Zero Greenhouse Gas Emissions by 2050*. Retrieved from https://www.whitehouse.gov/wp-content/uploads/2021/10/us-long-term-strategy.pdf

U.S. DOT. 2024a. *Investing in America: Biden-Harris Administration Sends $62 Billion to States from the Bipartisan Infrastructure Law for America's Infrastructure*. Retrieved Dec. 24, 2024 from https://highways.dot.gov/newsroom/investing-america-biden-harris-administration-sends-62-billion-states-bipartisan

U.S. DOT. 2024b. *University Transportation Centers*. Retrieved Dec. 24, 2024 from https://www.transportation.gov/content/university-transportation-centers

USAFacts. 2024. *What Does America Spend on Transportation and Infrastructure? Is Infrastructure Improving?* Retrieved Dec. 29, 2024 from https://usafacts.org/state-of-the-union/transportation-infrastructure/

Part I

Green transition for resilience and sustainability

José Neves

Geotechnics plays a crucial role in creating sustainable and resilient transportation infrastructure. This part of the book emphasizes the importance of minimizing environmental impact, promoting durability, and adapting to climate change through various geotechnical techniques. It highlights the use of bio-based, waste-based, and secondary materials such as recycled aggregates and steel slag to reduce environmental footprints. In Europe and the United States, initiatives like the European Green Deal and the Bipartisan Infrastructure Law, respectively, are driving the adoption of sustainable practices. Life Cycle Assessment (LCA) is also presented in this part as a tool for quantifying environmental impacts and guiding material selection in geotechnical projects. Ultimately, integrating sustainable geotechnical practices ensures that transportation infrastructure is environmentally friendly, cost-effective, and resilient for future generations.

Sustainability and resilience are two key principles in transportation geotechnics, ensuring that infrastructure remains functional, safe, and environmentally responsible while withstanding both natural and human-induced challenges. Sustainability in transportation geotechnics focuses on minimizing environmental impacts and reducing resources consumption. It integrates eco-friendly materials, energy-efficient construction techniques, and climate-conscious designs. Resilience, on the other hand, ensures that geotechnical infrastructure can withstand, adapt to, and recover from environmental stresses such as earthquakes, floods, landslides, and the effects of climate change. It emphasizes robust designs and adaptive strategies to enhance long-term performance. Geotechnics plays a critical role in ensuring resilience, promoting sustainability, adapting to climate change, and increasing the long-term durability of transportation infrastructures by addressing the interaction between the ground and engineering structures.

Various geotechnical engineering techniques can contribute to resilient and sustainable green infrastructure. In general, the efficient use of geomaterials, erosion control, soil conservation, ground improvement techniques, resilient foundations, and pavement performance are examples of how geotechnical engineering contributes to the resilience and sustainability of transportation infrastructures. Geotechnics can optimize the use of local and recycled materials, decreasing the carbon footprint of construction and minimizing resource depletion in accordance with the circular economy. Techniques such as bioengineering, geotextiles, and sustainable drainage systems help prevent soil erosion and waterlogging, ensuring long-term functionality. Methods like soil stabilization and the use of geosynthetics enhance weak soils, reducing the need for excessive excavation and the consumption of new materials. Geotechnical principles can also support renewable energy projects integrated into transportation engineering systems, such as solar-powered highways and energy-efficient rail networks. Proper geotechnical investigations ensure that roads, bridges, tunnels, and railways are built on stable ground, minimizing failures due to soil instability or landslides. Geotechnical solutions help mitigate risks from

DOI: 10.1201/9781003530220-3

floods and climate change-induced soil degradation by implementing techniques such as ground reinforcement and retaining structures. Understanding soil behaviour aids in designing road subgrades that resist cracking and rutting under traffic loads and environmental changes.

Bio- and waste-based solutions in transportation infrastructure involve the use of organic materials and biodegradable alternatives to improve sustainability, reduce environmental impact, and enhance durability while promoting a circular economy. Bio-based materials are derived from natural and renewable resources and can be utilized in construction, reinforcement, and erosion control. Waste-based solutions involve recycling industrial, agricultural, and municipal waste to create durable and eco-friendly transportation infrastructure. Cristelo et al. (2025) describe innovative and sustainable soil stabilization techniques as alternatives to traditional methods. The authors examine bio-mediated methods like microbially induced calcite precipitation (MICP) using non-ureolytic pathways and engineering fungi, alongside biofilm and biocrust restoration for enhanced soil stability. The study also investigates alkali-activated cement (AAC) as a substitute for Portland cement, highlighting its potential for waste reutilization and reduced carbon emissions. Furthermore, the impact of these stabilization methods on unsaturated soil behaviour, specifically the soil water characteristic curve (SWCC), is analysed. It stresses the significance of understanding the effects of bio-stabilization on the SWCC to improve climate change resilience and promote vegetation growth. The chapter presents case studies involving both deep and superficial soil stabilization using AAC. The findings emphasize the potential of bio-mediated and AAC techniques for creating more sustainable and effective soil stabilization solutions.

Secondary materials are recycled, reclaimed, or by-product materials used in transportation infrastructure to also promote sustainability, reduce costs, and minimize environmental impacts. These materials can be used in road construction, railway tracks, embankments, and other geotechnical applications without compromising performance. Recycled aggregates and construction waste can replace virgin aggregates in unbound granular mixtures and bituminous mixtures. Recycled concrete aggregate (RCA), resultant from crushed concrete of demolished structures, and reclaimed asphalt pavement (RAP), obtained from milled pavement asphalt, are common construction and demolition waste (CDW) used in road bases, sub-bases, and subgrades and in new pavement layers, respectively. Industrial by-products are materials that come from manufacturing processes and are repurposed for transportation infrastructure. A current by-product used in embankments, bituminous pavements and railway ballast is steel slag (SS) due to its high strength and durability. Other secondary materials include recycled plastics, rubber, and organic and natural waste-based materials. Freire et al. (2025) explore the use of recycled materials, specifically CDW and SS, in transportation infrastructures. The authors emphasize the environmental, economic, and social benefits of using these materials as alternatives to traditional aggregates in road and railway construction. The chapter discusses the processing, characterization, and evaluation of CDW and SS to ensure their quality and suitability for various applications, like unbound and hydraulically bound mixtures. It also reviews existing legislation promoting the use of secondary raw materials in Portugal. Several research projects and case studies are presented, highlighting the performance and durability of these recycled materials in real-world applications. The document concludes that CDW and SS offer a sustainable solution for waste management and resource conservation in civil engineering, with characteristics similar to or superior to natural materials.

LCA is a systematic approach used to evaluate the environmental impacts of a product, process, or infrastructure during its entire lifecycle – from raw material extraction to disposal or recycling. In transportation infrastructure, LCA helps optimize material choices, reduce carbon footprints, and enhance sustainability. Venda Oliveira et al. (2025) analyse the integration of eco-design principles and LCA into geotechnical engineering, particularly for transportation infrastructure. The authors highlight the environmental impact of traditional construction methods and advocate for sustainable alternatives, such as using recycled materials, bio-based methodologies, and innovative design approaches. LCA is presented as a tool to quantify environmental, economic, and social impacts throughout a project's life cycle, emphasizing the importance of considering all phases from material extraction to end of life. A case study demonstrates how LCA can guide material selection in rail embankment repair, showing potential carbon savings through material substitutions. The overall goal is to promote more sustainable practices in the construction industry by incorporating environmental considerations into geotechnical design and decision-making.

Geotechnical engineering is essential for designing, maintaining, and operating transportation infrastructures to withstand environmental stresses while reducing ecological impacts. By integrating sustainable engineering practices, transportation infrastructures can be made more resilient, cost-effective, and environmentally friendly for future generations. Part 1 of this book aims to be an effective support for the practice of geotechnical engineering in the green transition towards the resilience and sustainability of transport infrastructures.

References

Cristelo, N., Salifu, E. and Banerjee, A. 2025. Bio- and Waste-Based Solutions as the Next Generation of Soil Stabilisation Techniques. *in* Gomes Correia, A., Pinto, A., Neves, J., Fortunato, E., (Eds.), *Transportation Geotechnics for Green, Digital, and Modern Infrastructures,* CRC Press (2026).

Freire, A.C., Martins, I., Ferreira, C., Correia, E., Neves, J., Roque, A., Fortunato, E., Paixão, A. and Rios, S. 2025. Use of Construction and Demolition Waste and Steel Slag. *in* Gomes Correia, A., Pinto, A., Neves, J., Fortunato, E., (Eds.), *Transportation Geotechnics for Green, Digital, and Modern Infrastructures,* CRC Press (2026).

Venda Oliveira, P., Correia, A., Gervásio, H. and Heitor, A. 2025. Eco Design and Life Cycle Assessment. *in* Gomes Correia, A., Pinto, A., Neves, J., Fortunato, E., (Eds.), *Transportation Geotechnics for Green, Digital, and Modern Infrastructures,* CRC Press (2026).

Chapter 3

Bio- and waste-based solutions as the next generation of soil stabilization techniques

Nuno Cristelo, Emmanuel Salifu, and Aritra Banerjee

3.1 Introduction

Energy-intensive and consumptive practices that have historically characterized the construction sector are no longer feasible at the required scale due to resource limitations and prohibitive costs (Sizirici et al. 2021). Consequently, construction projects are increasingly assessed based on their carbon footprint, energy efficiency, and life cycle impacts. The construction sector, responsible for a significant amount of global carbon emissions, is expected to play a crucial role in the global objective of reducing greenhouse gas emissions. Until recently, the geotechnical construction industry – a major component of the overall construction sector – had lagged other construction sectors in innovative research and practice towards reducing carbon emissions and transitioning to more holistic and environmentally friendly engineering methods. These methods are increasingly recognized as essential for sustainably addressing complex, interconnected 21st-century challenges such as carbon sequestration, infrastructure rehabilitation, flood and landslide protection, building foundations, waste disposal and hazardous waste management, brownfield remediation, and water resource protection (Chang et al. 2019). Over the last two to three decades, there has been increased awareness, research contributions, and industry interest in innovative, holistic, and sustainable approaches to soil stabilization and management.

3.2 Bio-mediated soil stabilization

Developed countries have demonstrated immense commitment to pioneering the integration of sustainability principles and soil reuse strategies into geotechnical engineering research and practices. A few major research funding initiatives have contributed significantly to the advancement and establishment of bio-mediated and bio-inspired geotechnical engineering approaches as impetus towards the development of sustainable interventions in geotechnics. For example, the EU-funded project TERRE (Training Engineers to Rethink Geotechnical Engineering for a low-carbon future), which ran from 2015 to 2019 (Tarantino et al. 2020), and of greater impact is the Engineering Research Centre for Bio-mediated and Bio-inspired Geotechnics (CBBG), funded by the US National Research Foundation NSF (2015–2025) (DeJong and Kavazanjian 2019). These initiatives, among others across the world, have contributed to the establishment of biogeotechnics as a distinct sub-discipline by targeting innovation in the geotechnical construction industry and exploring novel design concepts for low-carbon geotechnical infrastructure.

DOI: 10.1201/9781003530220-4

The intersection between low-carbon geotechnical engineering and unsaturated soil mechanics, explored in biogeotechnics, aimed to develop novel concepts based on the consideration of soil as a living, active system containing not only mineral components but also living plant roots and a community of microorganisms, interacting with dynamic climatic variations undergoing cyclic wetting and drying, and capable of providing multifunctional sustainable solutions if engineered effectively. Biogeotechnics, which integrates biological processes into geotechnical engineering, offers a promising opportunity to proffer sustainable solutions to typical issues faced by geotechnical engineers, especially in the stabilization and improvement of problematic soils subjected to the increasing frequency of extreme climate events. This research area is advancing rapidly, with some aspects at the stage of scaled-up field trials and potential commercialization.

3.2.1 Current research landscape in biogeotechnics

A variety of nature-inspired bio-based technologies have been extensively explored within biogeotechnics. The bio-mediated technologies focus on engineered biological processes of plant roots and microbial interactions with soil particles in ways that promote the nucleation of minerals and resulting hydromechanical modifications of these particles, secretion of enzymes and/or polymeric substances by microbes or application of extracted enzymes and/or polymeric substances from bio-based sources, changes in pore fluid composition, and the engineered formation of biofilms and surficial soil biocrusts (Abdullah et al. 2018, DeJong et al. 2013, Hamdan and Kavazanjian 2016, van Paassen et al. 2010). Popular technologies in this research area include microbially induced calcite precipitation (MICP), enzyme-induced calcite precipitation (EICP), microbially induced desaturation and precipitation (MIDP), and biopolymers. Numerous review papers, bibliometric studies, state-of-the-art literature, and a dedicated academic journal (Liu et al. 2023) document the diverse expanse of research activities in this field. These documents encompass the rationale, hypotheses, mechanisms, process optimizations, applications, scaled-up trials, challenges, predictions, and prospects of these technologies. Figures 3.1 and 3.2 succinctly capture the trends of research contributions in biogeotechnics over the past two decades (2003–2023), extracted from a comprehensive research landscape and trend mapping conducted using Bibliometrix in R software (Aria and Cuccurullo 2017).

Readers are encouraged to refer to other relevant literature sources to gain a deeper understanding of the mechanisms and latest developments in these research themes.

From Figures 3.1 and 3.2, it is evident that (i) although research activity, in terms of publications, continues to increase with growing interest, the citation-to-publications ratio has declined over the last decade, suggesting that fewer recent studies are as seminal as earlier works; (ii) central research themes have shifted from foundational studies on *Bacillus pasteurii* for ureolytic soil improvement and the mechanical properties of treated soil towards more specialized and organized topics such as MICP and EICP, with a wider range of applications and evaluation methods; (iii) ureolytic MICP and EICP processes have remained the central themes during the review period.

This presents a compelling argument for pursuing novel directions to innovate within this burgeoning field. The subsequent sections of this article will not reiterate concepts widely covered in existing literature but will instead focus on highlighting emerging research directions in bio-mediated soil stabilization that, although promising, have received less attention and offer opportunities for advancing new frontiers within this sub-discipline. These include an overview of techniques such as applying non-ureolytic pathways for biomineralization and soil bio-cementation,

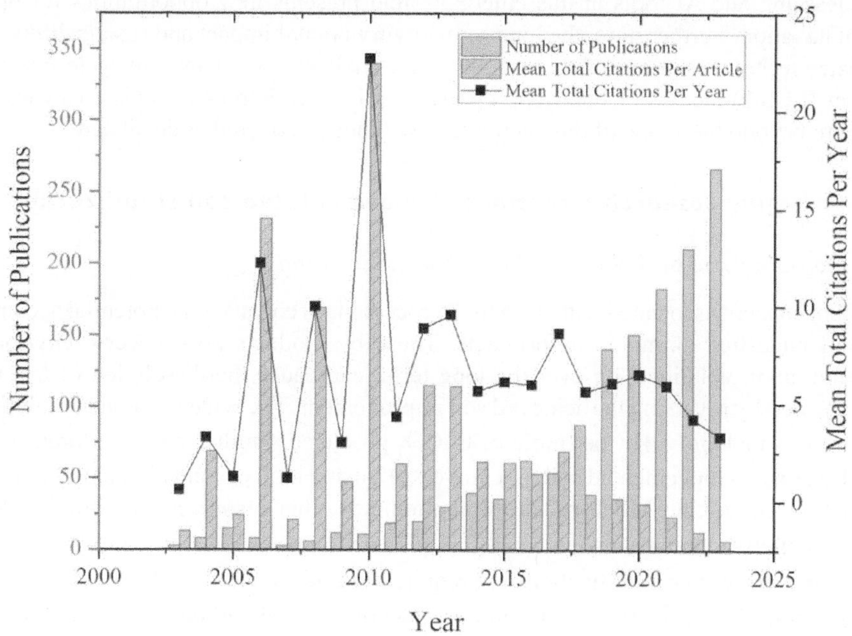

Figure 3.1 Yearly trend in number of publications and mean citations per publication.

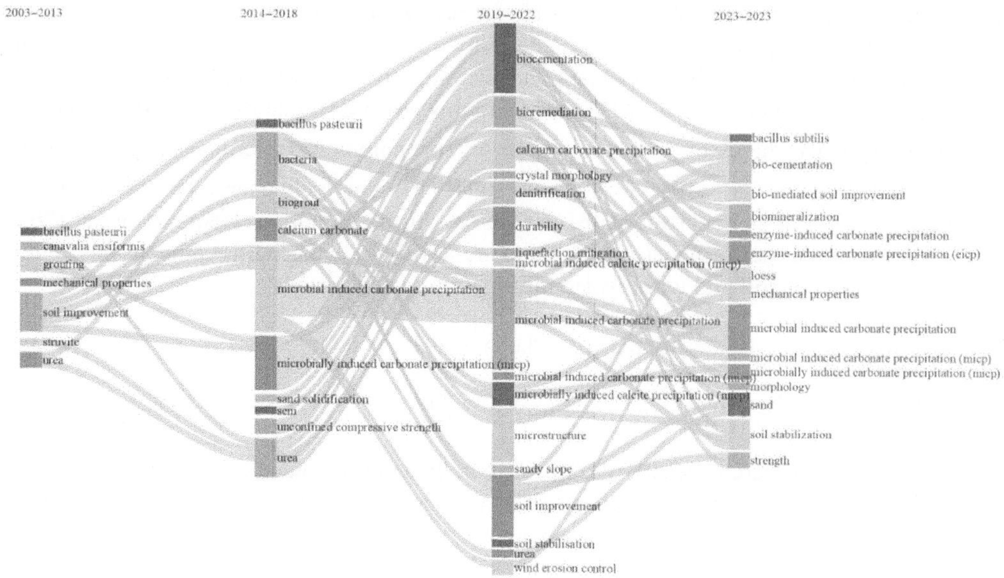

Figure 3.2 Thematic evolution of the research landscape in biogeotechnics from 2003 to 2023.

engineering fungi for soil improvement, promoting the establishment of biological soil crusts as a means for nature-engineered restoration, and the concept of composite or hybrid approaches involving the combination of two or more biogeotechnologies and recycled waste materials for complementary and optimal performance in soil stabilization. In addition, the integration of machine learning and AI tools in this emerging field presents new opportunities for optimization and innovation. Furthermore, the long-term environmental impact and sustainability of these bio-mediated techniques, assessed through life cycle analysis (LCA) and life cycle sustainability assessment (LCSA) studies or other appropriate metrics, are important topics. However, these concepts are beyond the scope of this section and will not be covered in detail here.

3.2.2 *Emerging research directions in bio-mediated soil stabilization*

3.2.2.1 *Non-ureolytic pathways for MICP biocementation*

A significant concern associated with the MICP process via ureolysis is the potential accumulation and fate of undesired ammonia by-products. These by-products pose adverse environmental impacts and, upon volatilization over the long term, can cause the dissolution of $CaCO_3$ (Gat et al. 2017), thereby reversing the achieved soil improvement. The widely studied ureolytic MICP pathway results in 1 mol NH_4^+ per mole of $CaCO_3$ produced. Studies have explored secondary steps such as rinsing biocemented soil to remove NH_4^+ using high pH and high ionic strength (Lee et al. 2019) or converting MICP-produced ammonia to other resources like struvite (Yu et al. 2022). These approaches have demonstrated some success in partially removing ammonium ions $\left(NH_4^+\right)$. However, further investigations are required to understand the influence of site-specific conditions, feasibility at the field scale, and the phenomenon of sorbed or translocated NH_4^+, as well as the additional material costs associated with this post-treatment step.

More recent studies have begun to examine alternative non-ureolytic MICP pathways that exclude the use of urea and urease enzymes, thereby eliminating the production of ammonia by-products and facilitating a direct approach for precipitating $CaCO_3$. In addition to MIDP, another non-ureolytic pathway involves bacteria-mediated oxidation of organic calcium salts or carbonate precursors, such as acetates and lactates, using heterotrophic *Bacillus species* (Fattahi et al. 2020, Hemayati et al. 2023, 2024). In this process (Equations 3.1 and 3.2), aerobic heterotrophs produce energy by metabolizing the organic anions and releasing CO_3^{2-} ions, which react with the available Ca^{2+} to form $CaCO_3$ crystals. The negatively charged bacterial surfaces adsorb Ca^{2+} and serve as nucleation sites for the $CaCO_3$ precipitation. Hemayati et al. (2023, 2024) reported similar performance compared to ureolytic MICP when this pathway was applied for biocementation of soil and wind erosion mitigation of sand dunes.

$$(CH_3COO)_2Ca + O_2 \text{ Heterotrophic Bacteria} \rightarrow CaCO_3 + CO_2 + H_2O \tag{3.1}$$

$$(HCOO)_2Ca + O_2 \text{ Heterotrophic Bacteria} \rightarrow CaCO_3 + CO_2 + H_2O \tag{3.2}$$

It is envisaged that research activity will increase in the coming years, exploring this and other non-ureolytic MICP pathways suitable for soil improvement, with the aim of overcoming the limitations and concerns associated with the popular ureolytic pathway.

3.2.2.2 *Engineering fungi for soil stabilization*

In recent years, several studies have begun to focus on the application of fungi in bio-mediated geotechnical engineering. The vegetative part (hyphal networks or mycelium) of filamentous soil fungi impacts soil in various ways, including inducing biomineralization via ureolytic pathways, similar to bacterial MICP, cellular-based mineralization, precipitation of carbonates and oxalates, and a combination of biophysical, biochemical, and biological actions capable of altering soil behaviour (Gadd 2017, 2007). Fungal growth and mycelial development occur even under sub-optimal environmental conditions and are known to enhance soil stability and structure (Sterflinger 2000). Fungal mycelium networks grow as a mass of fine root-like mesh with a larger surface-to-volume ratio compared to plant roots, typically entangling, aligning, or crosslinking soil particles to form micro- and macro-aggregates, potentially improving soil shear and tensile strengths. These networks interact with available organic and inorganic minerals in the soil matrix, promoting biochemical processes like biomineralization and bioweathering (Bindschedler et al. 2016, El Mountassir et al. 2018, Fang et al. 2018, Fomina et al. 2010, Park et al. 2023). Metabolically, fungi secrete various compounds, including enzymes, proteins, and organic acids (acetic, citric, formic, gluconic, and oxalic), facilitating the formation of metal carbonates and oxalates and the adsorption of nitrate (Bindschedler et al. 2016, Guggiari et al. 2011, Li et al. 2014, Qian et al. 2017). The presence of fungal mycelium in soil can induce changes in soil hydromechanical characteristics and promote the establishment of microbial communities like biofilms and biocrusts, contributing to the formation and stability of soil aggregates and biorestoration (El Mountassir et al. 2018, Rillig and Mummey 2006).

At the laboratory scale, researchers are exploring the engineered growth of fungal mycelium in soil to improve soil shear strength, modify soil hydraulic behaviour, and create lightweight sand-mycelium soil for various applications in geotechnical and geoenvironmental engineering (El Mountassir et al. 2021, Gou et al. 2021, Gou and Li 2024, Lim et al. 2020, Martuscelli et al. 2020, Park et al. 2023, Salifu 2019, Salifu et al. 2022, Salifu and El Mountassir 2021, Salifu and Mountassir 2019). Figure 3.3a–i provides a simplified illustration of the materials and processes involved in creating soil-mycelium material or engineering the growth of mycelium in soil. Figure 3.3i compares the strength improvements recorded by two studies that have explored this method for soil improvement and shows that the variation in soil strength is attributable to differences in fungal strain, treatment conditions, amount of soil used, curing period, and types

(i)

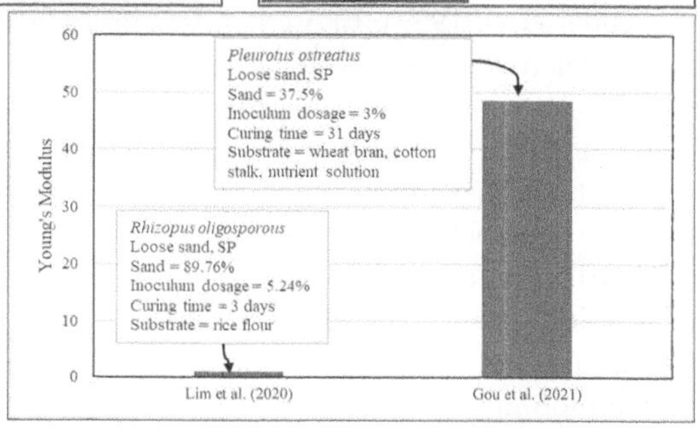

a. Substrates or organic materials (e.g., cotton stalk, wheat bran, wood chips, hemp fiber)
b. Natural granular/reinforcement materials (e.g., silica or calcareous sand, soil)
c. Liquid additives (e.g., water, nutrient broth, nutrient solution)
d. Fungal strain grown on solid media
e & f. Sample incubation in molds under suitable temperature and relative humidity, often in the dark
g & h. Demolded solidified soil-mycelium materials
i. Wide variations in Young's Modulus of mycelium-sand materials due to differences in fungal strain, substrates, amount of sand, and inoculum dosage

Figure 3.3 (a–i) General approach for soil improvement through engineered mycelium growth, forming a lightweight soil-mycelium composite to enhance soil hydromechanical properties.

of substrates. In a pioneering experimental campaign, Salifu (2019) explored the influence of engineered growth of the mycelium of Pleurotus ostreatus on soil hydromechanical behaviour and potential for erosion mitigation. Salifu et al. (2022) and Park et al. (2023) investigated the hydraulic properties of sand treated with *Pleurotus ostreatus* and *Trichoderma virens*, respectively. Their results showed that the growth of these species decreased the infiltration rate and hydraulic conductivity of the treated sand and increased water repellency and air entry suction. Fungal growth has been shown to increase soil water repellency (Salifu and El Mountassir 2021, Zhang et al. 2020) and enhance soil resistance to erosion (El Mountassir et al. 2021, Zhang et al. 2023). Lim et al. (2020, 2023) reported an increase in the strength of sandy soils, with Young's modulus ranging from 0.29 to 1.16 MPa using Pleurotus ostreatus and Rhizopus oligosporus fungi, respectively. Soil improvement achieved by engineered growth of fungal mycelia in soils is attributed to mechanisms such as pore-clogging by mycelial biomass, pore-sealing and induction of water repellency, and improved cohesion due to extracellular polymeric substances and mycelial entrapment of particles.

Fungi-induced calcium carbonate precipitation (FICP) is also gradually gaining research attention. Using ureolytic strains like *Penicillium chrysogenum* and *Neurospora crassa*, Fang et al. (2018) and Martuscelli et al. (2020) have demonstrated that fungi-mediated biomineralization can result in the biocementation of sand, albeit at lower strength compared to ureolytic MICP. Further tests optimizing the treatment process and concentration of cementing solutions are ongoing (Bhurtel et al. 2024). Studies in this emerging area have yet to achieve soil strength improvement values comparable to MICP. While it is arguable that mycelial microfibres could function as nucleation sites and sources of reinforcement in soil, in addition to mediating calcium carbonate precipitation, the overarching advantages or value proposition for FICP over bacteria-mediated MICP are yet to be fully elucidated, particularly concerning challenges related to ammonia by-products, treatment uniformity, and scalability.

3.2.2.3 Biofilms and biocrust restoration

In natural environments, microorganisms rarely exist as free-living entities; they are typically found within communities of intra- or interspecific organisms. Biofilms occur as three-dimensional hydrated matrices (Figure 4a) formed when microorganisms assemble at the interfaces of soil, water, and air media, often aggregated and bound to their substrates using extracellular polymeric substances (EPS) (Flemming and Wingender 2010, Li et al. 2024, Wang et al. 2024). Biofilms are primarily composed of microbial cells and EPS, which facilitates the biofilm attachment to surfaces and serves as a protective encasement for the microbes (Cunningham et al. 1991, Wingender et al. 1999). EPS comprises up to 90% organic material, including polysaccharides, lipids, nucleic acids, and proteins, influencing the biochemical characteristics and structure of biofilms (Nielsen et al. 1997, Wingender et al. 1999). In soil, biofilms create cohesive soil aggregates through the action of EPS and occupy soil pores and void spaces, thereby reducing soil permeability. Studies have shown that engineered biofilm growth in soil can improve the soil's capacity to reduce liquefaction potential (Banagan et al. 2010, Lin et al. 2018) and reduce permeability by up to two orders of magnitude, depending on bacterial composition and concentration, antecedent permeability and soil properties, hydraulic stress conditions, and method of bacterial inoculation (Dennis and Turner 1998, Lin et al. 2018, Mitchell and Nevo 1964, Zamani et al. 2020).

Like biofilms, biocrusts consist of thin, layered assemblages of microorganisms in association with soil particles (a few millimetres thick), typically found in arid and semi-arid regions (Maier et al. 2018, Weber et al. 2022). Biocrusts play an essential role in natural soil stabilization and erosion resistance (Belnap et al. 2011, Eldridge and Leys 2003). They comprise photosynthetic organisms (cyanobacteria, bryophytes, and algae) and heterotrophs (fungi and bacteria), forming patches or mats of soil-microbe aggregates bound by filamentous organisms and polymeric compounds released by the microbes' metabolic processes.

Research is ongoing to determine optimal procedures for rapid biocrust growth. One approach involves cultivating biocrusts ex situ and deploying the inoculum through broad-cast-spreading to promote controlled and uniform distribution of biocrusts for soil stabilization and ecosystem restoration, especially in arid environments (Bethany et al. 2019, Lan and Rossi 2021, Muñoz-Rojas et al. 2021). This engineered approach to biocrust growth aims to overcome limitations posed by natural environmental changes, ensuring effective application for soil stabilization.

3.2.2.4 Composite or hybrid methods

Composite or hybrid methods involve combining biomineralization with various soil amendments such as clayey soils (bentonite/kaolin), fly ash, biopolymers, fungal mycelium, and biofilms (Bhurtel et al. 2024, Figure 3.4). These integrated approaches aim to enhance soil stabilization by leveraging the synergistic effects of multiple materials, agents, and processes.

3.3 Alkali-activated cements as soil stabilizers

Conventional concrete (i.e., based on Portland cement, PC) is currently the second most used worldwide material, in terms of consumption, second only to water. Therefore, to reach the objectives of the United Nations 2030 Agenda for Sustainable Development, it is very clear that the substitution (at least partial) of PC by newer, more sustainable alternatives is one of the most relevant targets to be pursued in the next few decades, if indeed the building and construction sector aims to contribute to the decrease in consumption of natural resources, energy, and, most relevantly, the heavy carbon footprint associated with this industry.

Portland cement, which is the main component of concrete, is obtained through a manufacturing process responsible for approximately 10% of the total CO_2-eq that is released, worldwide, into the atmosphere. To develop, promote (very important!) and implement the consistent use of 'low carbon' cements is thus in order, with significant research efforts being deployed into

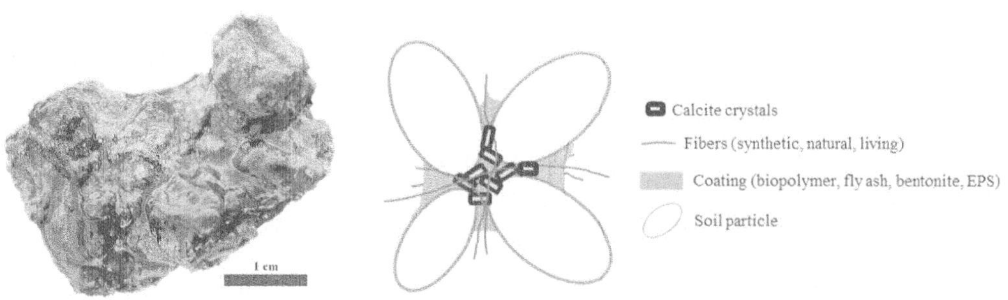

(a) (b)

Figure 3.4 (a) Aggregated mass of soil bound by an assemblage of microorganisms typifying soil-biofilm and biocrust interactions; (b) a schematic illustration of a composite or hybrid approach to soil improvement showing the potential for any two or more multi-agents' contribution to soil strength improvement.

this massive task. The so-called alkaline cements, or alkali-activated cements, aka AAC, are probably the most promising technique with the capacity to actually cause some impact on PC production. This is due to several factors, namely the potential elimination of several of the already mentioned troubling steps related to the production of Portland cement. The most obvious and relevant of these factors is the complete ceasing of the necessity for clinker, as the versatility of the technique (arguably its most interesting feature) allows for most of the raw material to be recovered from industrial wastes and by-products. This eliminates several of the more concerning (from the environmental perspective) steps associated with PC.

However, there is also a significant drawback in the use of AAC, which is the environmental toll related to the other main component of these alkaline cements – the activators. Indeed, these are usually based on common reagents, as discussed further ahead, that can represent, regardless of their reduced content (less than 10% in weight), a significantly hindering factor when facing a live-scale application.

Nevertheless, although AAC still possesses some unresolved issues, associated not with its technical performance but with its environmental context, it has already demonstrated a solid capacity for constituting a valid alternative to PC, especially in less demanding or non-structural applications. The application as a soil stabilizing agent is a good example of such promising applications, mostly due to the lower mechanical performance required (compared with reinforced concrete, for example), which allows very small doses of activator to be used, thus eliminating the 'environmental' disadvantage, but also because geotechnical-related applications are usually very demanding in terms of volume of materials required, which is an advantage when the main part of such materials is an industrial waste or by-product (i.e., geotechnical applications allow the disposal of a very significant volume of waste).

The following is a short presentation of the alkali activation technique and its capacity to be applied in soil stabilization, either at deep or superficial depths.

3.3.1 Activation of aluminosilicates in highly alkaline medium

Based on the research on ancient cements, it seems that the development of binding systems of a different nature to Portland-type cements is not only possible, but it has been around for 2000 years or more. The advantage of applying these (re)discovered cements, from a purely technical perspective, resides in the elimination of the recognized limitations of PC in favour of a more neutral and more stable behaviour, given that calcium is now not indispensable. Furthermore, these systems are based on aluminosilicates, which make up a large part of the inorganic solid compounds in the Earth's crust and are thus more abundant than the raw materials that are used in the manufacture of PC. Clinker constitution shows a predominance of CaO (calcium oxide represents more than 60% of the minerals present in any given type of PC), with a considerable effect on its properties and behaviour, mainly in terms of durability. This material belongs to the group generically known as 'hydraulic binders' because they set in water (even when submerged) as opposed to 'aerial lime binders', which only set in contact with air. The combination of cement with water will lead to hydration reactions, with the consequent formation of another type of mineral. The calcium silicates form the well-known 'calcium silicate hydrates' (C–S–H) and the 'tricalcium aluminate hydrates'.

On the contrary, 'alkaline cements', or 'soil silicates', as initially coined by Glukhovsky in 1952, are very similar, in terms of composition, to many minerals and rocks existing on the surface of the planet. They are formed by a very significant percentage (up to 30%, by weight) of alkali metals (mostly sodium or potassium) and/or alkaline earth metals (mostly calcium).

The formation of any of these two variants can be developed through the activation, with soluble alkali metals, of either aluminosilicate-rich precursors (with no Ca involved) or high-Ca content precursors. Common examples of the former precursors are coal fly ash or metakaolin, while the most relevant example of the latter is blast furnace slag.

Alkaline activation is usually described as a hydration reaction between aluminosilicates and alkali or alkaline earth metals, forming a 3D aluminosilicate mineral polymer constituted by several amorphous to semi-crystalline phases. The phase composition of the products formed is mostly determined by the initial pH of the medium, the type of aluminosilicates in the precursor, and the synthesis conditions. To optimize the reactivity of the precursors, it is essential that these are submitted to a calcination step, involving the loss of constitutive water and changes in the coordination of the aluminium ion with the oxygen. As a direct consequence, the material loses a significant proportion of its crystalline structure, reaching a high state of amorphization, with high entropy and, thus, very eager to chemically combine with other materials. Therefore, and in principle, aluminosilicates such as some types of slag (e.g., from blast furnaces), fly ash (e.g., from burning coal in thermoelectric plants), volcanic ash (with a natural thermal treatment), and tile or brick dust (obtained from industrial ovens) are potential starting materials for alkaline activation, as they have a recognized thermal history.

The alkaline activation process goes through several stages: dissolution – generated by the high pH medium – of the covalent bonds (Si–O–Si, Al–O–Al, Al–O–Si) present in the vitreous phase of the precursor; nucleation and reorganization of the Si–O bonds; and poly-condensation of the dissolved gel, with the Si and Al ions precipitating around the nuclei points. In fact, the ideal AAC consists of aluminium and silicon tetrahedra, alternatively interlinked and sharing all the oxygen atoms. The tetrahedrally coordinated and negatively charged Al ions associate with an alkali metal cation (usually Na^+ or K^+) or an alkaline earth metal (usually Ca^{2+}) to satisfy electrical balance. The sodium or any other activator ion is incorporated into the AAC binders only via a charge-balancing role (with the aluminium ions). The resulting polymeric structure of Al–O–Si bonds is the main structure of the new material.

3.3.2 Soil stabilization with AAC – short review

The following is a comprehensive review of some relevant publications targeting experimental stabilization of granular materials with alkali-activated cements. The works presented were essentially focused on the optimization of the AAC to the characteristics of those specific granular materials, including, but not limited to natural-forming soils. Indeed, several industrial by-products possess significant mechanical, physical, and chemical similarities and have proved their adequacy for the construction of geotechnical-like structures. An increasing number of papers reported the effectiveness of AAC on the enhancement of the geomechanical properties of granular materials, using different types of precursors, like fly ash (Corrêa-Silva et al. 2019, Cristelo et al. 2013, 2015), blast furnace slag (Fasihnikoutalab et al. 2020, Sargent et al. 2013), mixtures of both (Singhi et al. 2016, Yi et al. 2015), and metakaolin (Zhang et al. 2013, 2015), among others. Nonetheless, although several publications can be found covering laboratory-based research on the performance of soil-AAC, mostly focusing on uniaxial compressive strength tests, complemented with chemical and environmental analysis, the use of triaxial tests and the following development of specific constitutive models for this particular combination are still scarce. Understanding the shear behaviour of problematic granular materials after stabilization is very important for future geotechnical engineering projects and field applications, which require strength and deformation analysis at short- and long-term conditions. Therefore, the experimental shear behaviour of soil-AAC is also included in the following review.

3.3.2.1 Stabilization of natural soils

Stabilization of naturally occurring soils with alkaline cements has been developed, with almost no exception, assuming the soil particles to be completely inert, which either facilitates the design of the stabilizer or hinders its potential, depending on the perspective. If, on the one hand, the properties of industrial wastes and by-products can easily interfere with the short- and/or long-term development of the alkaline reactions, it is also possible that such properties somehow increase the reactivity of these precursors, enhancing the quality of the resulting binders. Another main question when designing AAC is the type of final product in itself, which can be, for increased simplicity, divided into two main groups: with or without calcium, producing the so-called N–A–S–H or C–A–S–H gel types, respectively.

Cristelo et al. (2012) analysed the effect of two types of coal fly ash, with very different Ca contents – type F, with <5 wt.%, and type C, with >50 wt.% – on the stabilization of a Portuguese marl (CH, according to ASTM D2487). A combination of sodium silicate and sodium hydroxide (10 molal concentration) was used, in a weight ratio of 2, to activate both precursors, while Type II, 32.4 N PC and calcium hydroxide (lime) were also tested for reference. The highest uniaxial compressive strength (UCS) was obtained with the 20% type F ash (8.5 MPa after 12 weeks), while 20% type C and 20% PC reached maximum values of 3.0 MPa and 2.3 MPa, respectively. However, after only 1 week of curing, the 20% PC compressive strength was higher (0.9 MPa) than that of the 20% type C (0.63 MPa) and substantially higher (0.19 MPa) than that of the 20% type F, which translates to a classical pattern that, since then, has been demonstrated several times: the presence of Ca results in faster reactions and earlier UCS, although their strength development is significantly shorter than that found on low-Ca systems.

3.3.2.2 Stabilization of industrial by-products

Among the industrial by-products with a granular structure, Construction and Demolition Waste (CDW) is clearly the most affirmative possibility, which has been the subject of many research-based developments and, furthermore, is constantly used for live-scale construction, especially as an aggregate for concrete and as a base or sub-base layer in road foundations. Due to the large volumes of CDW generated annually, as well as the high potential for reuse and recycling of some of its components, CDW was identified by the European Commission as a priority stream. On the one hand, the reuse of CDW reduces the consumption of natural, non-renewable resources, while on the other hand, it reduces the landfilling of these inert materials coming from the construction industry. National standards were designed to establish the mandatory conditions for the application of recycled aggregates. These conditions are homogenized across European and national specifications, focusing mostly on chemical and physical properties – density, sulphate content, chloride content, organic material content, constituents, nature and geometry of the fines fraction (quality, content and maximum dimensions) – and mechanical performance (resistance to fragmentation and wear).

The possibility of using AAC to stabilize CDW is, thus, very appealing, in the sense that it will allow for an integrated and sustainable alternative to current practice, which could enhance not only the mechanical but also the environmental capacity of this recycled aggregate. Cristelo et al. (2018) developed a comprehensive study targeting the optimal conditions in which the CDW-AAC combination could function. A high fine content CDW, retrieved from a Portuguese operator, was mixed with type F fly ash, in CDW/FA weight ratios between 100/0 and 60/40, and activated with a combination of 10 molal NaOH and sodium silicate. A parallel study was made with natural residual granitic soil (NGS) as the stabilized aggregate, for reference. After curing for up to 14 days at ambient temperature, the CDW/FA formulations showed a clear advantage in terms of strength gain over the NGS/FA specimens (Figure 3.5). However, at the 28-day curing period, this trend was inverted, and the latter reached higher UCS. Although the ultimate compressive strength was obtained with the natural aggregate, it is important to consider two additional aspects when analysing the data and deciding the best solution. Those aspects are the absolute UCS values reached by the CDW-based solution after 28 days – between 1 and 8 MPa – and the fact that at shorter curing periods (7 and 14 days), its performance was significantly superior, which is relevant in live-scale applications, when time is of the essence.

This trend was somehow confirmed by the BSEM/EDX analysis of the 60/40 pastes, fabricated either with CDW or NGS (Figure 3.6). The CDW allowed the formation of significant contents of C–A–S–H gel, which is similar to the CSH gel formed in PC pastes that develops approximately 90% of its maximum UCS in the first 28 days. By contrast, the paste prepared with natural aggregate (NGS) lacked Ca, resulting in an exclusive N–A–S–H structure, capable of developing ultimate higher compressive strengths than the C–A–S–H gel, but at a much slower rate.

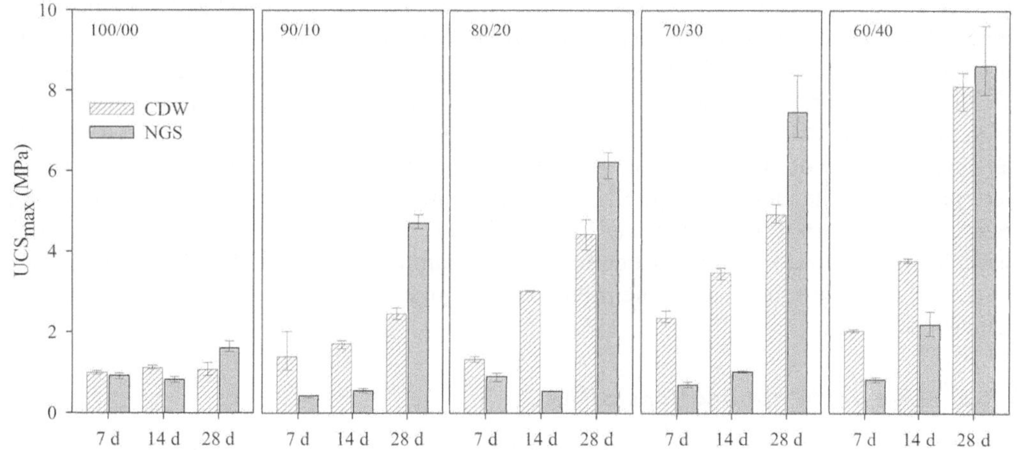

Figure 3.5 Uniaxial compressive strength of the aggregate-ash mixtures. (After Cristelo et al. 2018.)

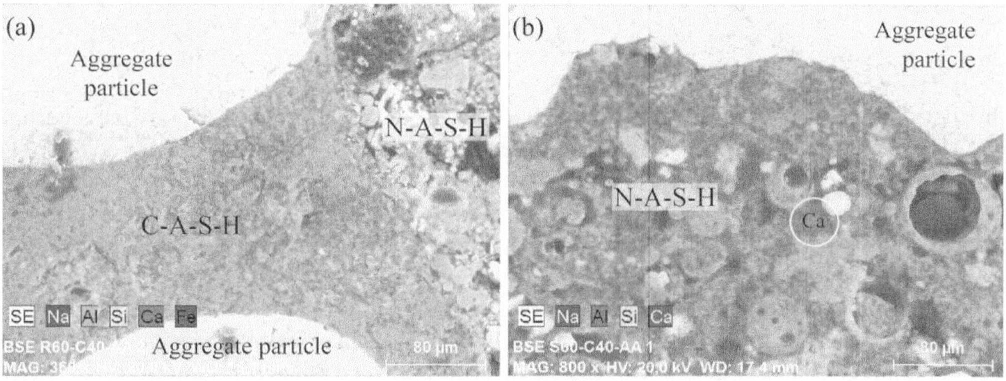

Figure 3.6 BSEM/EDX analyses of the CDW60/40 (a) and NGS60/40 (b) after 28 days curing. (After Cristelo et al. 2018.)

3.3.2.3 Some notes on constitutive modelling

The number of publications on constitutive modelling of soil-AAC is still highly scarce. Several reasons might explain this latency, from the lack of opportunities for field applications (which would require more advanced design knowledge and, thus, justify the development of constitutive laws) to more prosaic concerns, like the aggressive alkaline medium that might damage oedometer or triaxial equipment during the testing stage required for the development of a meaningful constitutive model.

Corrêa-Silva et al. (2021) developed what is believed to be the first serious attempt at establishing a comprehensive constitutive relation for a soil (sandy clay, in this case) stabilized with an alkaline cement. The precursor was a common granulated blast furnace slag, activated with a 3.5 molal NaOH solution – note that the wide range of possible types of alkaline cements will certainly demand that specific models will have to be developed for different AAC compositions. A 'kinematic hardening model' was thus adapted to the behaviour of this particular combination, which includes a 'mobile' elliptic 'bubble' surface defining the elastic region. The position of this 'bubble' is controlled by a kinematic hardening rule, and, in essence, the stiffness of the material falls as this bubble approaches the outside 'structure surface'. Drained and undrained consolidated triaxial tests were performed on soil-AAC specimens with a 7.5% binder content and cured for 28 and 90 days. OCR levels from 1 (normally consolidated) to 12. The results showed that the stress-strain behaviour significantly changed after stabilization, towards a more brittle response, with increased strength and stiffness. The variation of the p'0 and OCR, as well as the (obvious) presence of the binder, factors with a clear influence on the experimental behaviour, were included in the model through the size of the initial surface (pc0). In general, the overall features of the stress-strain response of the stabilized soil were well captured by the kinematic hardening model, demonstrating that current soil constitutive models can be adapted to this new generation of binders, although some modifications might be needed, according to the type of AAC considered.

3.3.3 Live-scale applications of AAC

The following section presents two case studies focused on the two most relevant applications of alkali-activated cements in soil stabilization: deep and superficial treatment/improvement. The works presented comprise only projects related to the in situ treatment of the local soil, as cases addressing the treatment/stabilization of granular materials (not necessarily soils) transported and installed at the site of a particular geotechnical work have yet to be published.

3.3.3.1 Deep soil stabilization

The first documented test of deep stabilization with AAC was published in 2011, when jet grouting was used to create cemented columns in a test site located in Portugal (Cristelo et al. 2011). Prior to that specific test, two preliminary live-scale experiments were conducted in order to develop the necessary knowledge and screen all possible problems related to the application of the technology in the field. The following is a description of those (unpublished) live-scale experimental works.

The first of these preliminary approaches was performed in 2004, during the construction of the Portuguese A13 highway, near the town of Almeirim. The idea was to understand and collate information regarding all factors with the capacity to influence the application of the new type of grout through the jet-grouting method, without a serious concern with the binder formulation or its effect on the soil or on the column formed. The process (Figure 3.7) started with the

Figure 3.7 Preparation of the AAC near the A13 construction, starting with the NaOH activator solution (a,b), followed by the addition of the precursor (c) and 'agitation' prior injection (d).

preparation of the activator and subsequent incorporation of the precursor immediately prior to the injection – both components were already mixed in the 'grout mixer plant' and left in the 'agitator'. The injection was performed through the 'mono fluid jet' system, using common jet parameters (Figure 3.8). No specific problems were encountered, although the recovery of the column was not possible due to the construction schedule of the A13. Regardless, several aspects related to the logistics of the fabrication and injection of the grout were noted.

The following trial occurred in Lisbon, during the construction of the 'Cais do Sodré' subway station, in 2005. This test followed a more comprehensive approach, in the sense that it was not only focused on the construction of a column but also, and more importantly, on its recovery and characterization. The conditions (Figure 3.9), namely the type of soil where the injection occurred, were particularly difficult (the layer injected was a landfill with no specific compaction and formed by a mixture of soil and waste). Nevertheless, a consistent and structurally sound column was recovered (Figure 3.9d).

Figure 3.8 Injection of the AAC in the ground, on the test site near the A13 construction, using the mono-fluid jet system.

Figure 3.9 Injection of the AAC in the ground at the construction site of the 'Cais do Sodré' subway station (a), which was followed by the recovery, after 3 weeks, of the respective column (b,c) clearly very heterogeneous (d).

The samples drilled from the column were submitted to UCS tests (Figure 3.10) and indirect tensile strength tests, yielding average values of 4.2 and 0.93 MPa, respectively.

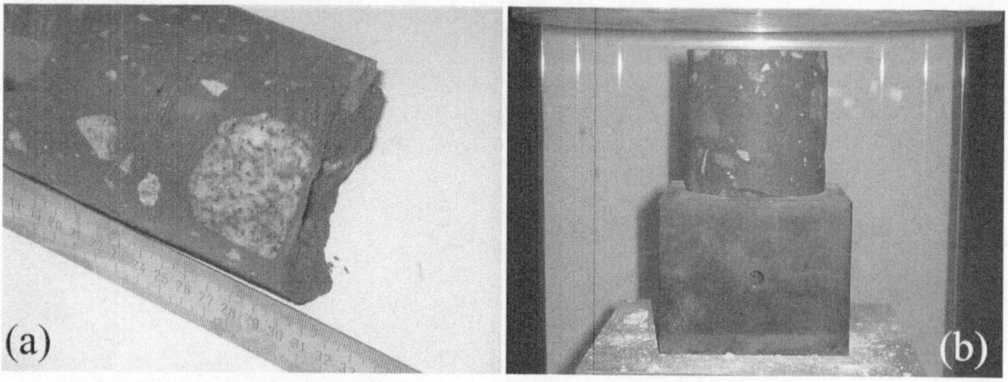

Figure 3.10 Visual inspection (a) and compressive strength testing (b) of one of the samples recovered from the column exhumed at the construction site of the 'Cais do Sodré' station.

3.3.3.2 Superficial soil stabilization

The construction of superficial layers of stabilized material poses a totally different set of problems and challenges than that established by the construction of columns of soil binder. In the former, the application of the binder is probably more complicated, at least in logistical terms, since several components have to be spread and thoroughly homogenized across a significant surface area. By contrast, deep soil mixing applies the stabilizer in one single action, on a shorter area. Furthermore, the curing conditions of the superficial stabilization can vary significantly more, as every particular weather condition will constitute a different problem, from an excessively dry/hot environment to heavy rain. On the other hand, superficial stabilization not only allows for a deep understanding of the soil to be treated, usually resulting in tailor-made binders, but it is also susceptible to a tight quality control of the final product, something which, in deep stabilization, has to rely on a small sample size.

Miranda et al. (2020) produced a full-scale sub(base) layer stabilized with different alkaline cements produced from industrial wastes. This is a fundamental approach to study the viability and impact of the construction process associated with these alternative binders, as well as the effect of the in situ curing conditions on mechanical and durability properties, something which cannot be properly assessed in the laboratory. An 80 m long, 2.5 m wide, 0.2 m thick full-scale pavement foundation was built (Figure 3.11), divided into five sections with different soil-binder combinations, three of which were based on AAC, and the others were prepared with PC or hydrated lime (HL) for comparison purposes. The mechanical properties of all segments were evaluated, on-site and in the laboratory, together with an environmental assessment and a preliminary financial analysis.

The precursor was a class F fly ash (FA), activated with either commercial reagents (sodium hydroxide and sodium silicate, in solution form) or a recycled NaOH solution (CS), originally used to clean aluminium extrusion dies. The preparation of the site started 48 hours before construction and included the removal of the topsoil layer; the development of Falling Weight Deflectometer (FWD) tests along the longitudinal axis (every 2 m); and the measurement of the water content in each segment. All segments were constructed using equipment that is

Figure 3.11 General view of the construction site before stabilization (after mixing with 2% lime to reduce natural water content).

commonly used to build soil layers stabilized with cement or lime. After calibrating the FA flow per square metre, the three segments stabilized with AAC matched the following sequence:

- a Spreading of 1/3 of the pre-determined FA weight (Figure 3.12a), followed by the first passing of the mixer (targeting a thickness of 20 cm).
- b Repetition of the previous step.
- c Spreading of approximately 1/2 of the pre-determined alkaline activators (Figure 3.12b), followed by the second passaging of the mixer (Figure 3.12c).
- d Repetition of step (a).
- e Repetition of step (c).
- f Compaction of the stabilized layer (Figure 3.12d).

The curing occurred without any protective coating layer, and circulation of vehicles and machines was not allowed within the first 36 days, by which time the first in situ testing campaign occurred.

Regarding the mechanical results produced in the field, it is important to note that the highest increase rate of the dynamic deformability modulus, as assessed with the FWD, was observed for the AAC segments, with an increase between 2x and 3x above the modulus of the PC-based segment. A similar trend was observed for the static deformability modulus, assessed through the plate loading test, although, in this case, the PC-based segment showed the highest values, either at 36 or 90 days. Nevertheless, the higher stiffness increase between these two curing periods was observed, as in the case of the dynamic modulus, in the AAC-based segments. Samples collected from the field were tested for UCS after 36 days, with the AAC solutions showing similar values between them. However, after 90 days, the UCS of the best-performing AAC combination almost doubled the values obtained with the PC-based samples.

Figure 3.12 Main steps followed during the layer stabilization with AAC – (a) Mixing the FA with the soil; (b) spreading the activator; (c) mixing the soil, FA and activator; (d) final compaction.

3.4 Unsaturated response of novel forms of soil stabilization

3.4.1 *Overview of unsaturated parameters of soils*

The importance of unsaturated soil mechanics stems from the distresses caused by volume changes induced by fluctuations of moisture regime in expansive or highly plastic clayey soils. Expansive soils are found in many places around the world, including Australia, Canada, Egypt, India, South Africa, Spain, and the USA (Banerjee 2017, Jones and Jones 1987, Puppala et al. 2003, Puppala et al. 2013, Talluri et al. 2013). Due to the presence of expansive soils, various parts of the world have faced a multitude of problems involving the civil and transportation infrastructure (Nelson and Miller 1997, Puppala et al. 2019, Soltani et al. 2017). Civil infrastructure, including embankments and pavements, is highly vulnerable to damage due to the shrink-swell behaviour of expansive soils. It may not be feasible to avoid expansive soil during the construction of such civil infrastructure. Therefore, soil stabilization is frequently employed to mitigate the detrimental effects of shrink-swell behaviour (Petry and Little 2002). Various forms of stabilization have been employed to enhance the mechanical and hydraulic properties of expansive soils to mitigate the distress caused to civil infrastructure such as pavements. These are broadly categorized into three different categories: (i) mechanical stabilization;

(ii) chemical stabilization; and (iii) biological stabilization or biostabilization. Among these, chemical stabilization is the most common.

In the past several decades, conventional calcium-based stabilizers, like lime and cement, have been used (Little and Nair 2009). Cement stabilization rapidly decreases the plasticity of the soil due to the release of calcium ions during the initial hydration reactions of cement. A cation exchange reaction or the accumulation of extra cations onto the clay particles may be involved in the process, or both mechanisms may be in effect (A.J. Puppala et al. 2003). Both processes change the electrical charge density of the clay particles, causing them to be electrically attracted to one another (Bugge and Bartelsmeyer 1961). This causes the clay particles to flocculate or aggregate. When small amounts of cement stabilizer are added to soils, the aggregation process starts quickly (Bugge and Bartelsmeyer 1961). The hydration process of cement stabilization of soil forms calcium-silicate-hydrates (C–S–H), calcium-aluminate-hydrates (C–A–H), and calcium hydroxide ($Ca(OH)_2$). Similar to the lime treatment of soil, strength is developed due to pozzolanic reactions of silica and alumina in clay minerals with calcium ions to produce cementing agents such as C–S–H ($3CaO{\cdot}2SiO_2\ 3H_2O$) and C–A–H ($3CaO{\cdot}Al_2O_3\ 12H_2O$) in cement-stabilized soil (Croft 1967).

Several previous studies have reported increased swelling potential of cement-stabilized soil with high sulphate content, which sometimes leads to failure of transportation infrastructure (Puppala et al. 2004, Yong and Ouhadi 2007). This failure can be attributed to the formation of a detrimental calcium-alumino-sulphate mineral, ettringite, which makes the cement stabilization insufficient or sometimes counterproductive (Little and Nair 2009). In alkaline environments with a pH of more than 10.5, the dissolution of clay releases alumina, which then reacts with Ca^{2+} from Ca-based stabilizers and soluble sulphates to form ettringite ($Ca_6(Al(OH)_6)_2(SO_4)_3{\cdot}26H_2O$) (Biswas et al. 2021, 2023, Chakraborty et al. 2022). The volumetric expansion can reach up to 250% in calcium-based stabilization of sulphate-rich soil due to hydration reaction and crystal growth (Puppala and Cerato 2009). The phenomenon is known as sulphate-induced heave.

While calcium-based stabilizers like cement and lime have many benefits, they also have some disadvantages, especially in terms of their impact on the environment. The biggest concerns are related to carbon dioxide and nitrogen oxide emissions as well as the production of airborne particulate matter (Chang et al. 2015). The cement used for soil treatment stays in place for a long time because cement mixtures degrade slowly. The presence of cement in the soil has the potential to disturb the ecosystem, resulting in problems like increased pH levels and a deterioration of the soil, which can lead to desertification (Bremner 2001). Cement stabilization also has other drawbacks, which include brittle failures and susceptibility to low-temperature cracking (Puppala et al. 2003). Therefore, in recent years, new alternatives for lime and cement stabilization are being investigated. These include the use of novel geotextiles with wicking ability, chemical stabilization by using geopolymers and alkaline cement, and biostabilization using biopolymers and microorganisms such as bacteria that promote stabilization through MICP.

3.4.2 *Effect of wicking geotextiles on the unsaturated response of soils*

The problems of expansive soil and freeze-thaw cycles have been attempted to be resolved by using wicking geotextiles. The wicking geotextiles are a combination of hydrophilic and hygroscopic 4DG fibres with multichannel cross-sections, where DG stands for 'Deep Grove' due to its high surface area, as shown in Figure 3.13. The resilience of the pavement has been enhanced

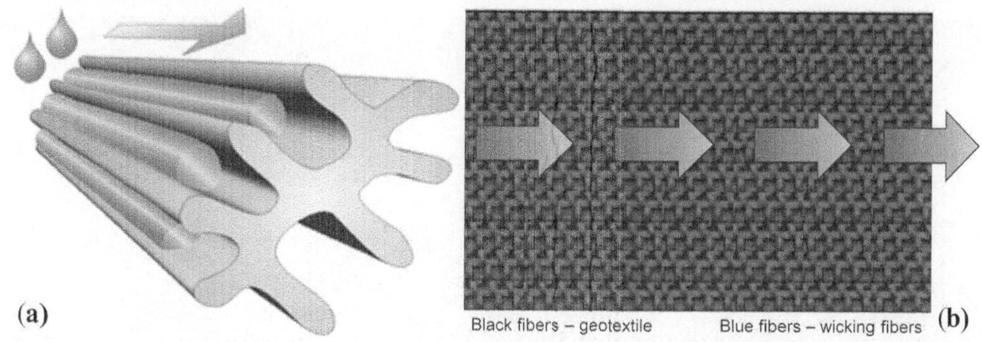

Figure 3.13 (a) A schematic of wicking fabric in wicking geotextiles to represent the higher cross-sectional area as compared to traditional circular fabrics; (b) photograph of wicking geotextile showing the blue-coloured wicking fabric and black-coloured traditional geotextile.

using wicking geotextiles through various studies (Azevedo and Zornberg 2013, Biswas et al. 2021, Guo et al. 2017, Wang et al. 2017, Zhang et al. 2014). The cross-section of the wicking fabric within wicking geotextile has one of the highest available shape factors and a high number of channels per fibre, which enhances the ability of the wicking fabric to maximize capillary action and water transport in an unsaturated condition (Zhang et al. 2014). The main premise of using such a fabric is to expose one side of the wicking geotextile to drier conditions and utilize the capillary action to draw water away from the pavement system. This allows for a higher resilient modulus and stiffness of the pavement system.

The field implementation and performance of the pavement system in Texas, USA, are shown in Figures 3.14–3.16. Previously, the site faced significant distress due to volume changes in the subgrade. The native soil was expansive in nature with high plasticity levels. Two sections were implemented to utilize reclaimed asphalt pavement material and traditional coarse aggregate (also known as flex base), as shown in Figure 3.14. Different forms of instrumentation were used to monitor the performance of the pavement after construction. Figure 3.16a shows the fluctuations of the volumetric water content with rainfall events. It shows the effectiveness of the wicking geotextile in removing water from the pavement layer and maintaining it at conditions that have a higher resilient modulus. The deformations of the pavement system were measured using shape array accelerometers (SAA) and are shown in Figure 3.16b.

3.4.3 *Effect of chemical stabilization of unsaturated response of soils*

The chemical stabilization of soil is employed to modify the properties of soil to enhance the mechanical properties and/or modify the hydraulic properties of soil. Since the stabilizer binds itself to the soil particles and changes the chemical composition of the soil, the physical properties of the soil also change. This can be identified by using the soil water characteristics curve (SWCC) of the soil. SWCC describes the relationship between soil and moisture content or degree of saturation. It helps us understand the distribution of voids in soils and the affinity of the soil to absorb or adsorb water with fluctuations in moisture regime. The effects of soil stabilization on the SWCC of soils have been obtained to be

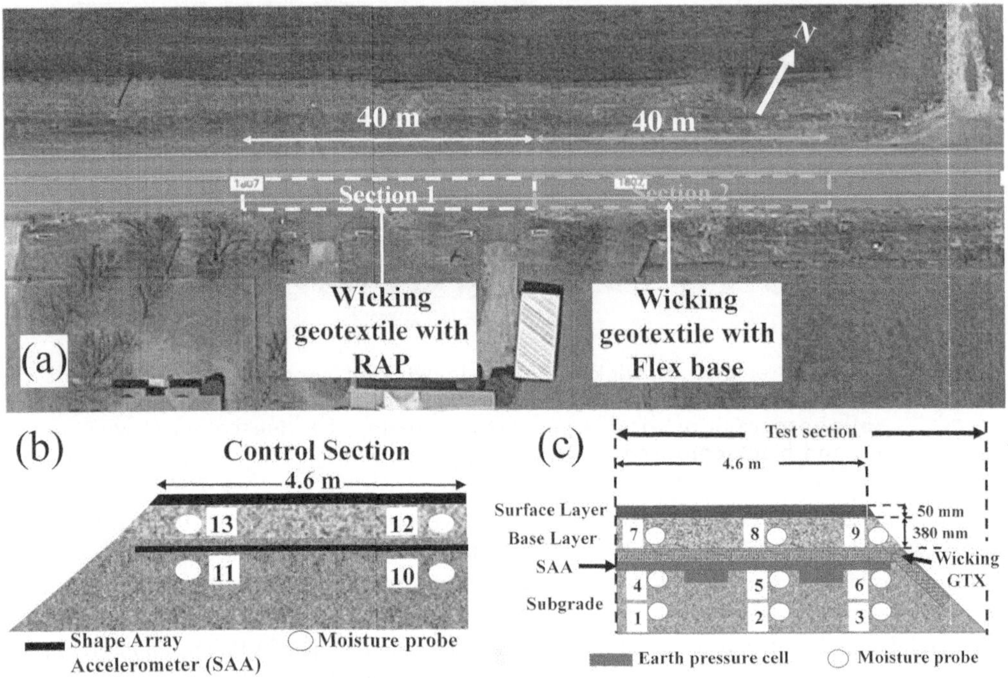

Figure 3.14 The cross-section of the pavement section with and without wicking geotextile: (a) Site location (b) Section details along with instrumentation of control and test sections.

Figure 3.15 Various stages of the field implementation involving wicking geotextile in pavement in Texas, USA: (a) placing the wicking geotextiles above the subgrade; (b) backfilling over the wicking geotextile.

significant. Different forms of stabilizers have varying impacts on the SWCC, and researchers have studied the impact of traditional stabilizers like cement and lime, along with a few studies on the use of non-traditional stabilizers like fly ash, geopolymers, alkaline cement,

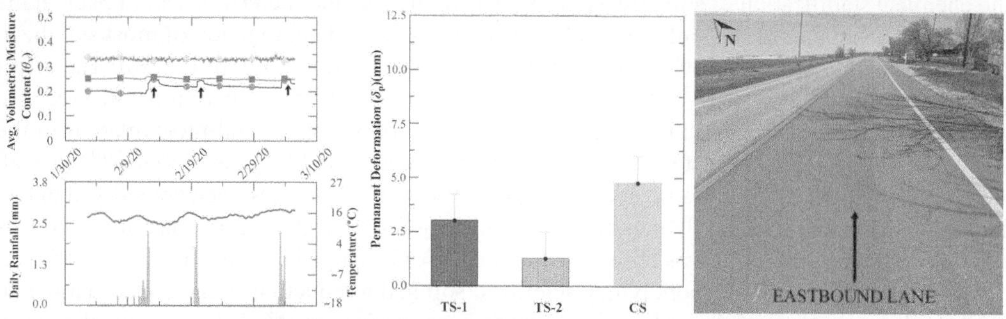

Figure 3.16 Volumetric moisture variation within subgrade for control and test sections; (b) vertical deformation in the pavement; (c) photograph of the site one year after construction.

and others (Araújo et al. 2020, Chen et al. 2023, Puppala et al. 2006). These impacts are categorized as follows.

a Increase in strength and stiffness of soil: The objective of most soil stabilization techniques is to increase the strength of the soil by inducing chemical reactions that promote the formation of pozzolanic products that help bind the soil particles together and form a stiffer geomaterial with higher shear strength. This results in the shift in the SWCC, and for most fine-grained soils it has been observed that the soil suction increases for a particular degree of saturation or volumetric water content (Khattab and Al-Taie 2006, Puppala et al. 2006).

b Changes to soil fabric, pore structure, and permeability: Due to the formation of pozzolan that fills the voids between soil particles, the overall void ratio generally reduces after chemical stabilization, such as by the use of traditional and non-traditional stabilizers like cement, lime, fly ash, and geopolymers (Consoli et al. 2022, Little and Nair 2009, Liu et al. 2016, Mo et al. 2014, Phummiphan et al. 2016, Samuel et al. 2021). Henceforth, typically the SWCC has a higher air-entry value, a steeper SWCC curve, and higher residual suction (Khattab and Al-Taie 2006, Puppala et al. 2006). However, due to the aggregation of soil particles, there have been instances where the size of voids has increased upon the addition of lime or cement, which may result in lower air-entry value for some cases (Barman and Dash 2022, de Brito Galvão et al. 2004). Based on the changes to the pore structure, the permeability of the soil changes with stabilization. In most cases, the permeability reduces with soil stabilization.

c Reduction of soil plasticity and volumetric changes in soil due to moisture fluctuations: The pozzolanic reactions result in reduced soil plasticity in fine-grained soils, which reduces the affinity for water, and thereby the SWCC of treated soils has a lower saturated water content as compared to untreated or natural soils. The fluctuations in the volume of soil due to moisture changes are also reduced by chemical stabilization.

d Changes to SWCC hysteresis: The changes to the soil fabric and reduction of affinity for water lead to a reduction in the SWCC hysteresis, which means that the gap between the drying and wetting curves is smaller. This leads to fewer fluctuations in water content during each drying and wetting cycle and often increases the durability of the soil.

The chemical stabilization of soil affects the SWCC of the soil and thereby has an impact on the unsaturated behaviour of the soil. The higher air-entry value and aggregation of most stabilized soils means higher strength, lower permeability, and increased durability of the soil. The effect of freeze-thaw cycles on the SWCC of stabilized soils has not been studied in detail except in a few studies (Ding et al. 2021); however, the wetting-drying cycles and their relation to the SWCC are well-established. The effect of new forms of stabilizers like geopolymers (Chen et al. 2023) and alkaline cement (Araújo et al. 2020) on fine-grained soils has been considered in this study. The behaviour of geopolymer-stabilized soils consisting of 20%–30% clay and sand was studied (Chen et al. 2023). It was observed that the SWCC changed significantly with stabilization using geopolymer. The geopolymer-stabilized soil had a reduction in the size of large and small voids, as confirmed through mercury intrusive porosimeter (MIP) tests. This change in pore distribution resulted in a higher air-entry value, residual suction, and residual water content (Chen et al. 2023). The slope of the SWCC for geopolymer-treated soil was also flatter as compared to that of untreated soil. This meant that higher energy was required to remove water from geopolymer-stabilized soil.

The SWCC of untreated and treated sandy lean clayey soil from Portugal with alkaline cement was studied and shown in Figure 3.17 (Araújo et al. 2020, Corrêa-Silva et al. 2019). The air-entry value was observed to drop significantly for alkaline cement-treated soil samples, which signifies that the largest void space is larger as compared to that for untreated soil. The slope of the SWCC was also observed to be flatter, which may be a wider distribution of pore spaces from larger ones to smaller ones, as compared to untreated samples. This shows that the introduction of alkaline cements behaves differently as compared to most traditional stabilizers like cement and lime, and further investigation is needed to determine the cause and extent of this phenomenon for other soils.

Overall, different stabilizers have varying effects on the unsaturated response of soil, and it is dependent on the factors discussed earlier. The most important aspect affecting the SWCC of stabilized soil is the changes in soil structure and pore distribution. This dictates the rate of removal of water from the soil and also affects the permeability of saturated and unsaturated soil, which is essential for the design and resilience of civil infrastructure such as embankments and pavements.

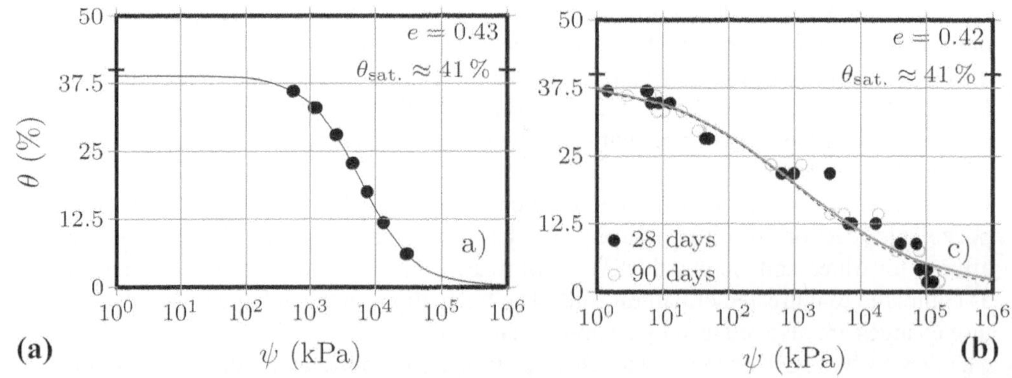

Figure 3.17 The SWCC of (a) untreated soil and (b) alkaline cement-treated soils. (Araújo et al. 2020.)

3.4.4 *Effect of biostabilization using biopolymers and MICP treatment on the unsaturated response of soils*

Researchers have used various forms of biostabilization to enhance the mechanical and hydraulic properties of soils that often promote the growth of vegetation and are carbon-negative. In this section, the relationship between biostabilization and the unsaturated response of soil will be discussed. Limited research has been conducted on determining the variation of SWCC and shear strength of unsaturated bio-stabilized soils by using biopolymers such as guar gum and xanthan gum (Arabani et al. 2024, Chang et al. 2024, Jia et al. 2023). The introduction of biopolymers such as xanthan gum in sandy soil was observed to increase the affinity for water, and the SWCC showed significantly higher saturated volumetric water content, which is due to the formation of hydrogels in biopolymers (Rosenzweig et al. 2012). The slope of SWCC was observed to be slightly flatter, and the residual water content was higher than that for untreated soils. This shows that the pore distribution changes after stabilization with biopolymers (Rosenzweig et al. 2012). Similar observations were made by other researchers when guar gum was to stabilize loess (Jia et al. 2023) and xanthan gum, guar gum, and gellan gum were used independently to stabilize silty soil (Wang et al. 2023).

Biogeotechnical methods, which utilize microbiology for geotechnical applications, have seen significant advancements in recent years (Lai et al. 2021). These methods, including biocementation, bioclogging, and biogas desaturation, have been applied in different areas such as soil strengthening, water leakage sealing, and soil erosion control. They have also been used in environmental protection and restoration, particularly in soil and groundwater remediation (Martin et al. 2024). Enzymes can be utilized to acquire calcite using the identical metabolic pathways employed in MICP. The technique used to enhance this process is referred to as enzyme-induced calcium carbonate precipitation (EICP) (Cui et al. 2024). A clayey sand was stabilized using *Sporosarcina pasteurii*, which is a urease-producing bacteria (Liu et al. 2024). It was observed that the MICP treatment altered the pore distribution of the soil, and the SWCC showed a distinct drying curve (Lee et al. 2022). The air-entry value increased drastically, and the slope of the SWCC was slightly steeper after treatment. The hydraulic conductivity or permeability of the saturated and unsaturated soil was analysed. The saturated hydraulic conductivity was observed to be two magnitudes lower after MICP treatment. This confirms the air-entry value increase in SWCC after treatment. The larger pores are occupied by biocemented products that reduce the permeability of the soil. The slope of the hydraulic conductivity function was observed to be steeper after MICP treatment, which means that the voids are of similar sizes, and this is also confirmed by the steeper slope of the SWCC. Similar observations were made by other researchers who attempted to stabilize clayey sandy soils and sandy soils using MICP treatment with *Sporosarcina pasteurii* as the bacterial culture with calcium chloride and urea as the cementation reagent (Payan et al. 2024).

A similar study was conducted for expansive clayey soil and stabilized using *Bacillus pasteurii*, which is a gram-positive bacterium (Yu et al. 2022). The unsaturated soil characteristics of bio-stabilized expansive soils were observed to be different, probably due to the difference in soil structure. The SWCC was observed to have similar saturated volumetric water contents and slightly higher residual volumetric water contents (Yu et al. 2021). The air entry value was observed to decrease with higher dosages of MICP treatment, and the slope of SWCC was flatter for increased dosages of MICP treatment. This shows that the pore distribution is more varied in treated soil, as both larger voids and smaller voids appear in treated soil. The difference in observation for clayey sand (Lee et al. 2022) and expansive clayey soil (Yu et al. 2021) may be due to the difference in bacterial cultures and soil type. The plate-like structure of clay particles may allow a more dispersed structure after MICP treatment, whereas the larger voids in clayey sand may be occupied by biocemented products and result in higher air-entry value and lower permeability for nearly saturated soil. More detailed studies on different soil types are required to understand the mechanisms involved that impact the unsaturated soil response.

3.5 Conclusions

Two of the most promising and recently developed soil stabilization techniques have been analysed. In both cases, their innovative character demonstrates that more sustainable and equally effective solutions can be developed and applied at an industrial level without compromising the technical requirements or becoming excessively costly. Bio-mediated soil stabilization presents a promising avenue for developing sustainable and effective geotechnical solutions. Emerging techniques, including non-ureolytic MICP pathways, fungal-mediated methods, and biofilm and biocrust restoration, offer innovative ways to enhance soil stability and resilience. Furthermore, composite and hybrid approaches that combine various materials and processes provide opportunities for optimized soil stabilization strategies. Continued research and field trials will be essential to fully realize the potential of these technologies and address the challenges associated with their large-scale implementation. Substitution of Portland cement by alkaline cements, in soil stabilization, is not only highly effective, but it also creates new possibilities for reintroducing significant volumes of wastes in the production chain, given the large volume of material required. Alkaline cements still present some challenges, namely regarding the availability of the precursors in each region or the type and quantity of activator required (not to mention the environmental toll associated with its production), but live-scale trials will continue to help mitigate these concerns, mostly by exposing the most relevant drawbacks and showing at what scale they really need to be tackled. Furthermore, a growing industrial application of these cements will help to develop trust in their capacity and effectiveness, while also helping to create design rules and guidelines. The understanding of the impact of new forms of soil stabilization with regard to the unsaturated soil behaviour and especially the impact of stabilization on SWCC is important. However, very limited studies have been conducted with such an objective. This is essential to understand how climate change resilience may be enhanced through the use of biopolymers. Sustainability in construction and design is essential due to the detrimental impacts of climate change. The presence of higher residual water content and higher organic matter promotes the growth and development of vegetation even in drought conditions. The additional reinforcement obtained from root-soil interaction aids in increasing the shear strength of the treated soil. This is beneficial for extreme conditions where prolonged droughts might be followed by intense precipitation and flooding. Such practices of carbon-negative techniques of stabilization have gained immense attention to enhance the resilience of civil infrastructure and to design effective climate change adaptive structures.

References

Abdullah, A., Hamed, K.T. and Edward, K. 2018. Baseline investigation on enzyme-induced calcium carbonate precipitation. *Journal of Geotechnical and Geoenvironmental Engineering*, 144(11), 04018081. https://doi.org/10.1061/(ASCE)GT.1943-5606.0001973

Arabani, M., Shalchian, M.M. and Haghsheno, H., 2025. The impact of guar gum biopolymer on mechanical characteristics of soil reinforced with palm fiber. *Bulletin of Engineering Geology and the Environment*, 84(1), 30. https://doi.org/10.1007/s10064-024-04064-3

Araújo, N., Corrêa-Silva, M., Miranda, T., Topa Gomes, A., Castro, F., Teixeira, T. and Cristelo, N. 2020. Unsaturated response of clayey soils stabilised with alkaline cements. *Molecules*, 25, 2533. https://doi.org/10.3390/molecules25112533

Aria, M. and Cuccurullo, C. 2017. Bibliometrix: An R-tool for comprehensive science mapping analysis. *Journal of Informetrics*, 11(4), 959–975. https://doi.org/10.1016/j.joi.2017.08.007

Azevedo, M. and Zornberg, J.G. 2013. Capillary barrier dissipation by new wicking geotextile. In: *Panamerican Conference on Unsaturated Soils*, Cartagena, Columbia, pp. 20–22.

Banagan, B.L., Wertheim, B.M., Roth, M.J.S. and Caslake, L.F. 2010. Microbial strengthening of loose sand. *Letters in Applied Microbiology*, 51(2), 138–142. https://doi.org/10.1111/j.1472-765X.2010.02872.x

Banerjee, A. 2017. *Response of Unsaturated Soils Under Monotonic and Dynamic Loading over Moderate Suction States* (Doctoral Dissertation). Department of Civil Engineering, University of Texas at Arlington, Arlington, TX.

Barman, D. and Dash, S.K. 2022. Stabilization of expansive soils using chemical additives: A review. *Journal of Rock Mechanics and Geotechnical Engineering*, 14, 1319–1342. https://doi.org/10.1016/j.jrmge.2022.02.011

Belnap, J., Munson, S.M. and Field, J.P. 2011. Aeolian and fluvial processes in dryland regions: The need for integrated studies. *Ecohydrology*, 4(5), 615–622. https://doi.org/10.1002/eco.258

Bethany, J., Giraldo-Silva, A., Nelson, C., Barger, N.N. and Garcia-Pichel, F. 2019. Optimizing the production of nursery-based biological soil crusts for restoration of Arid land soils. *Applied and Environmental Microbiology*, 85(15). https://doi.org/10.1128/AEM.00735-19

Bhurtel, A., Salifu, E. and Siddiqua, S. 2024. Composite biomediated engineering approaches for improving problematic soils: Potentials and opportunities. *Science of the Total Environment*, 914, 169808. https://doi.org/10.1016/j.scitotenv.2023.169808

Bindschedler, S., Cailleau, G. and Verrecchia, E. 2016. Role of fungi in the biomineralization of calcite. *Minerals*, 6(2), 41. https://doi.org/10.3390/min6020041

Biswas, N., Puppala, A.J. and Chakraborty, S. 2023. Role of nano- and crystalline silica to accelerate chemical treatment of problematic soil. *Journal of Geotechnical and Geoenvironmental Engineering*, 149. https://doi.org/10.1061/JGGEFK.GTENG-10999

Biswas, N., Puppala, A.J., Khan, M.A., Congress, S.S.C., Banerjee, A. and Chakraborty, S. 2021. Evaluating the performance of wicking geotextile in providing drainage for flexible pavements built over expansive soils. *Transportation Research Record: Journal of the Transportation Research Board*, 2675, 208–221. https://doi.org/10.1177/03611981211001381

Bremner, T. 2001. Environmental aspects of concrete: Problems and solutions. In: *Proceedings of the First All Russian Conference on Concrete and Reinforced Concrete*, Moscow, pp. 232–246.

Bugge, W.A. and Bartelsmeyer, R.R. 1961. *Soil Stabilization with Portland Cement*. Highway Research Board.

Chakraborty, S., Puppala, A.J. and Biswas, N. 2022. Role of crystalline silica admixture in mitigating ettringite-induced heave in lime-treated sulfate-rich soils. *Géotechnique*, 72, 438–454. https://doi.org/10.1680/jgeot.20.P.154

Chang, I., Im, J., Prasidhi, A.K. and Cho, G.-C. 2015. Effects of Xanthan gum biopolymer on soil strengthening. *Construction and Building Materials*, 74, 65–72. https://doi.org/10.1016/j.conbuildmat.2014.10.026

Chang, I., Lee, M. and Cho, G.-C. 2019. Global CO_2 emission-related geotechnical engineering hazards and the mission for sustainable geotechnical engineering. *Energies*, 12(13), 2567. https://doi.org/10.3390/en12132567

Chang, I., Cho, G.C. and An Tran, T.P., 2024. Water-retention properties of xanthan-gum-biopolymer-treated soils. *Environmental Geotechnics*, 11(2), 152–163.

Chen, R., Chen, H., Bao, W. and Lai, H. 2023. Soil-water characteristic curve of geopolymer stabilized aeolian sand with clay. *Journal of Traffic and Transportation Engineering*, 23, 128–141. https://doi.org/10.19818/j.cnki.1671-1637.2023.04.009

Consoli, N.C., Silvano, L.W., Lotero, A., Scheuermann Filho, H.C., Moncaleano, C.J. and Cristelo, N. 2022. Key parameters establishing alkali activation effects on stabilized rammed earth. *Construction and Building Materials*, 345, 128299. https://doi.org/10.1016/j.conbuildmat.2022.128299

Corrêa-Silva, M., Araújo, N., Cristelo, N., Miranda, T., Gomes, A.T. and Coelho, J. 2019. Improvement of a clayey soil with alkali activated low-calcium fly ash for transport infrastructures applications. *Road Materials and Pavement Design*, 20(8), 1912–1926.

Corrêa-Silva, M., Rouainia, M., Miranda, T. and Cristelo, N. 2021. Predicting the mechanical behaviour of a sandy clay stabilised with an alkali-activated binder. *Engineering Geology*, 292, 106260.

Cristelo, N., Fernández-Jiménez, A., Vieira, C., Miranda, T. and Palomo, Á. 2018. Stabilisation of construction and demolition waste with a high fines content using alkali activated fly ash. *Construction and Building Materials*, 170, 26–39.

Cristelo, N., Glendinning, S., Fernandes, L. and Pinto, A.T. 2012. Effect of calcium content on soil stabilisation with alkaline activation. *Construction and Building Materials*, 29, 167–174.

Cristelo, N., Glendinning, S., Fernandes, L. and Pinto, A.T. 2013. Effects of alkaline-activated fly ash and Portland cement on soft soil stabilization. *Acta Geotechnica*, 8, 395–405.

Cristelo, N., Glendinning, S. and Pinto, A.T. 2011. Deep soft soil improvement by alkaline activation. *Proceedings of the Institution of Civil Engineers: Ground Improvement*, 164(2), 73–82.

Cristelo, N., Miranda, T., Oliveira, D.V., Rosa, I., Soares, E., Coelho, P. and Fernandes, L. 2015. Assessing the production of jet mix columns using alkali activated waste based on mechanical and financial performance and CO_2(eq) emissions. *Journal of Cleaner Production*, 102, 447–460.

Croft, J.B. 1967. The influence of soil mineralogical composition on cement stabilization. *Géotechnique*, 17, 119–135. https://doi.org/10.1680/geot.1967.17.2.119

Cui, M.J., Chu, J. and Lai, H.J., 2024. Optimization of one-phase-low-pH enzyme-induced carbonate precipitation method for soil improvement. *Acta Geotechnica*, 19(3), 1611–1625. https://doi.org/10.1007/s11440-023-02175-x

Cunningham, A.B., Characklis, W.G., Abedeen, F. and Crawford, D. 1991. Influence of biofilm accumulation on porous media hydrodynamics. *Environmental Science & Technology*, 25(7), 1305–1311. https://doi.org/10.1021/es00019a013

de Brito Galvão, T.C., Elsharief, A. and Simões, G.F., 2004. Effects of lime on permeability and compressibility of two tropical residual soils. *Journal of Environmental Engineering*, 130, 881–885. https://doi.org/10.1061/(ASCE)0733-9372(2004)130:8(881)

DeJong, J.T. and Kavazanjian, E. 2019. Bio-mediated and bio-inspired Geotechnics. In: N. Lu and J.K. Mitchell (Eds.), *Geotechnical Fundamentals for Addressing New World Challenges*. Springer International Publishing (pp. 193–207). https://doi.org/10.1007/978-3-030-06249-1_7

Dejong, J.T., Soga, K., Kavazanjian, E., Burns, S., van Paassen, L.A., Al Qabany, A., Aydilek, A., Bang, S.S., Burbank, M., Caslake, L.F., Chen, C.Y., Cheng, X., Chu, J., Ciurli, S., Esnault-Filet, A., Fauriel, S., Hamdan, N., Hata, T., Inagaki, Y. and Weaver, T. 2013. Biogeochemical processes and geotechnical applications: Progress, opportunities and challenges. *Géotechnique*, 63(4), 287–301. https://doi.org/10.1680/geot.SIP13.P.017

Dennis, M.L. and Turner, J.P. 1998. Hydraulic conductivity of compacted soil treated with biofilm. *Journal of Geotechnical and Geoenvironmental Engineering*, 124(2), 120–127. https://doi.org/10.1061/(ASCE)1090-0241(1998)124:2(120)

Ding, L., Vanapalli, S.K., Zou, W., Han, Z. and Wang, X. 2021. Freeze-thaw and wetting-drying effects on the hydromechanical behaviour of a stabilized expansive soil. *Construction and Building Materials*, 275, 122162. https://doi.org/10.1016/j.conbuildmat.2020.122162

El Mountassir, G., Minto, J.M., van Paassen, L.A., Salifu, E. and Lunn, R.J. 2018. Applications of microbial processes in geotechnical engineering. *Advances in Applied Microbiology*, 104. https://doi.org/10.1016/bs.aambs.2018.05.001

El Mountassir, G., Schellenger, A., Salifu, E. and Lunn, R.J. 2021. Engineering fungal networks for ground improvement. In: R. Maddalena and M. Wright-Syed (Eds.), *Proceedings Resilient Materials 4 Life 2020 (RM4L2020)* (pp. 63–68), *International Conference, Cardiff* University.

Eldridge, D.J. and Leys, J.F. 2003. Exploring some relationships between biological soil crusts, soil aggregation and wind erosion. *Journal of Arid Environments*, 53(4), 457–466. https://doi.org/10.1006/jare.2002.1068

Fang, C., Kumari, D., Zhu, X. and Achal, V. 2018. Role of fungal-mediated mineralization in biocementation of sand and its improved compressive strength. *International Biodeterioration & Biodegradation*, 133, 216–220. https://doi.org/10.1016/j.ibiod.2018.07.013

Fasihnikoutalab, M.H., Pourakbar, S., Ball, R.J., Unluer, C. and Cristelo, N. 2020. Sustainable soil stabilisation with ground granulated blast-furnace slag activated by olivine and sodium hydroxide. *Acta Geotechnica*, 15(7), 1981–1991.

Fattahi, S.M., Soroush, A. and Huang, N. 2020. Biocementation control of sand against wind erosion. *Journal of Geotechnical and Geoenvironmental Engineering*, 146(6). https://doi.org/10.1061/(ASCE)GT.1943-5606.0002268

Flemming, H.-C. and Wingender, J. 2010. The biofilm matrix. *Nature Reviews Microbiology*, 8(9), 623–633. https://doi.org/10.1038/nrmicro2415

Fomina, M., Burford, E.P., Hillier, S., Kierans, M. and Gadd, G.M. 2010. Rock-building fungi. *Geomicrobiology Journal*, 27(6–7), 624–629. https://doi.org/10.1080/01490451003702974

Gadd, G.M. 2007. Geomycology: Biogeochemical transformations of rocks, minerals, metals and radionuclides by fungi, bioweathering and bioremediation. *Mycological Research*, 111(1), 3–49. https://doi.org/10.1016/j.mycres.2006.12.001

Gadd, G.M. 2017. The geomycology of elemental cycling and transformations in the environment. *Microbiology Spectrum*, 5(1). https://doi.org/10.1128/microbiolspec.FUNK-0010-2016

Gat, D., Ronen, Z. and Tsesarsky, M. 2017. Long-term sustainability of microbial-induced $CaCO_3$ precipitation in aqueous media. *Chemosphere*, 184, 524–531. https://doi.org/10.1016/j.chemosphere.2017.06.015

Gou, L. and Li, S. 2024. Dynamic shear modulus and damping of lightweight sand-mycelium soil. *Acta Geotechnica*, 19(1), 131–145. https://doi.org/10.1007/s11440-023-01885-6

Gou, L., Li, S., Yin, J., Li, T. and Liu, X. 2021. Morphological and physico-mechanical properties of mycelium biocomposites with natural reinforcement particles. *Construction and Building Materials*, 304, 124656. https://doi.org/10.1016/j.conbuildmat.2021.124656

Guggiari, M., Bloque, R., Aragno, M., Verrecchia, E., Job, D. and Junier, P. 2011. Experimental calcium-oxalate crystal production and dissolution by selected wood-rot fungi. *International Biodeterioration & Biodegradation*, 65(6), 803–809. https://doi.org/10.1016/j.ibiod.2011.02.012

Guo, J., Wang, F., Zhang, X. and Han, J. 2017. Quantifying water removal rate of a wicking geotextile under controlled temperature and relative humidity. *Journal of Materials in Civil Engineering*, 29, 04016181. https://doi.org/10.1061/(ASCE)MT.1943-5533.0001703

Hamdan, N. and Kavazanjian, E. 2016. Enzyme-induced carbonate mineral precipitation for fugitive dust control. *Géotechnique*, 66(7), 546–555. https://doi.org/10.1680/jgeot.15.P.168

Hemayati, M., Nematollahi, A., Nikooee, E., Habibagahi, G. and Niazi, A. 2024. Non-ureolytic microbially induced carbonate precipitation: Investigating a cleaner biogeotechnical engineering pathway for soil mechanical improvement. *The Journal of Engineering*, 2024(1). https://doi.org/10.1049/tje2.12350

Hemayati, M., Nikooee, E., Habibagahi, G., Niazi, A. and Afzali, S.F. 2023. New non-ureolytic heterotrophic microbial induced carbonate precipitation for suppression of sand dune wind erosion. *Scientific Reports*, 13(1), 5845. https://doi.org/10.1038/s41598-023-33070-w

Jia, Z., Yan, C., Li, B., Bao, H., Lan, H., Liang, Z., Shi, Y. and Ren, J., 2023. Performance test and effect evaluation of guar gum-stabilized loess as a sustainable slope protection material. *Journal of Cleaner Production, 408*, 137085. https://doi.org/10.1016/j.jclepro.2023.137085.

Jones, D.E. and Jones, K.A. 1987. Treating expansive soils. *Civil Engineering*, 57, 62–65.

Khattab, S.A.A., Al-Taie, L.Kh.I. 2006. Soil-water characteristic curves (SWCC) for lime treated expansive soil from Mosul city. In: *Unsaturated Soils 2006. American Society of Civil Engineers*, Reston, VA, pp. 1671–1682. https://doi.org/10.1061/40802(189)140

Lan, S. and Rossi, F. 2021. Combination of chemical and cyanobacterial inoculation promotes biocrust development: A novel perspective for combating desertification. *ACS Sustainable Chemistry & Engineering*, 9(29), 9506–9507. https://doi.org/10.1021/acssuschemeng.1c02856

Lee, M., Gomez, M.G., San Pablo, A.C.M., Kolbus, C.M., Graddy, C.M.R., DeJong, J.T. and Nelson, D.C. 2019. Investigating ammonium by-product removal for ureolytic bio-cementation using meter-scale experiments. *Scientific Reports*, 9(1), 18313. https://doi.org/10.1038/s41598-019-54666-1

Li, Q., Csetenyi, L. and Gadd, G.M. 2014. Biomineralization of metal carbonates by *Neurospora crassa*. *Environmental Science & Technology*, 48(24), 14409–14416. https://doi.org/10.1021/es5042546

Li, Y., Narayanan, M., Shi, X., Chen, X., Li, Z. and Ma, Y. 2024. Biofilms formation in plant growth-promoting bacteria for alleviating agro-environmental stress. *Science of the Total Environment*, 907, 167774. https://doi.org/10.1016/j.scitotenv.2023.167774

Lim, A., Atmaja, P.C. and Rustiani, S. 2020. Bio-mediated soil improvement of loose sand with fungus. *Journal of Rock Mechanics and Geotechnical Engineering*, 12(1), 180–187. https://doi.org/10.1016/j.jrmge.2019.09.004

Lim, A., Henzi, P. and Arvin, O. 2023. Comparison of *Pleurotus ostreatus* and *Rhizopus oligosporus* fungi for loose sand improvement. *Journal of GeoEngineering*, 18(1), 1–10. https://doi.org/10.6310/jog.202303_18(1).1

Lin, H., Suleiman, M.T. and Brown, D.G. 2018. Behavior of biofilm-treated sand. *IFCEE*, 2018, 1–11. https://doi.org/10.1061/9780784481592.001

Little, D.N. and Nair, S. 2009. *Recommended Practice for Stabilization of Sulfate-Rich Subgrade Soils*. Transportation Research Board, Washington, DC. https://doi.org/10.17226/22997

Liu, H., Chu, J. and Kavazanjian, E. 2023. Biogeotechnics: A new frontier in geotechnical engineering for sustainability. *Biogeotechnics*, 1(1), 100001. https://doi.org/10.1016/j.bgtech.2023.100001

Liu, Z., Cai, C.S., Liu, F. and Fan, F. 2016. Feasibility study of loess stabilization with fly ash-based geopolymer. *Journal of Materials in Civil Engineering*, 28, 1–8. https://doi.org/10.1061/(ASCE)MT.1943-5533.0001490

Liu, S., Ding, X., Song, P., Hu, L., Hu, Z., Yu, J. and Chang, F., 2024. Effects of nitrogen sources on MICP-related metabolic mechanism of Sporosarcina pasteurii and soil reinforcement. *Journal of Rock Mechanics and Geotechnical Engineering*, 16(12), 5147-5161. https://doi.org/10.1016/j.jrmge.2024.11.001.

Maier, S., Tamm, A., Wu, D., Caesar, J., Grube, M. and Weber, B. 2018. Photoautotrophic organisms control microbial abundance, diversity, and physiology in different types of biological soil crusts. *The ISME Journal*, 12(4), 1032–1046. https://doi.org/10.1038/s41396-018-0062-8

Martin, P.J., Wen, Y., Woods, A., Fayida, T. and Hobbs, S.R., 2024. Exploring pesticide transport, groundwater, and environmental justice in a changing climate: A community engaged research approach. *Environmental Research Letters*, 19(9), 094009.

Martuscelli, C., Soares, C., Lima, N. and Camões, A. 2020. Potential of fungi to produce biosandstone. *RILEM Technical Letters*, 5, 157–162. https://doi.org/10.21809/rilemtechlett.2020.119

Miranda, T., Leitão, D., Oliveira, J., Corrêa-Silva, M., Araújo, N., Coelho, J., Fernández-Jiménez, A. and Cristelo, N. 2020. Application of alkali-activated industrial wastes for the stabilisation of a full-scale (sub)base layer. *Journal of Cleaner Production*, 242(118427), 1–12.

Mitchell, R. and Nevo, Z. 1964. Effect of bacterial polysaccharide accumulation on infiltration of water through sand. *Applied Microbiology*, 12(3), 219–223. https://doi.org/10.1128/am.12.3.219-223.1964

Mo, B., Zhu, H., Cui, X., He, Y. and Gong, S. 2014. Effect of curing temperature on geopolymerization of metakaolin-based geopolymers. *Applied Clay Science*, 99, 144–148.

Muñoz-Rojas, M., Machado de Lima, N.M., Chamizo, S. and Bowker, M.A. 2021. Restoring post-fire ecosystems with biocrusts: Living, photosynthetic soil surfaces. *Current Opinion in Environmental Science & Health*, 23, 100273. https://doi.org/10.1016/J.COESH.2021.100273

Nelson, J. and Miller, D.J. 1997. *Expansive Soils: Problems and Practice in Foundation and Pavement Engineering*. John Wiley & Sons, Colorado State University.

Nielsen, P.H., Jahn, A. and Palmgren, R. 1997. Conceptual model for production and composition of exopolymers in biofilms. *Water Science and Technology*, 36(1), 11–19. https://doi.org/10.1016/S0273-1223(97)00318-1

Payan, M., Sangdeh, M.K., Salimi, M., Ranjbar, P.Z., Arabani, M. and Hosseinpour, I., 2024. A comprehensive review on the application of microbially induced calcite precipitation (MICP) technique in soil erosion mitigation as a sustainable and environmentally friendly approach. *Results in Engineering*, 24, 103235. https://doi.org/10.1016/j.rineng.2024.103235.

Park, J.S., Lin, H., Moe, W.M. and Salifu, E. 2023. Hydraulic properties of sands treated with fungal mycelium of *Trichoderma virens*. *Journal of Geotechnical and Geoenvironmental Engineering*, 149(11). https://doi.org/10.1061/JGGEFK.GTENG-11111

Petry, T.M. and Little, D.N. 2002. Review of stabilization of clays and expansive soils in pavements and lightly loaded structures—History, practice, and future. *Journal of Materials in Civil Engineering*, 14, 447–460. https://doi.org/10.1061/(ASCE)0899-1561(2002)14:6(447)

Phummiphan, I., Horpibulsuk, S., Sukmak, P., Chinkulkijniwat, A., Arulrajah, A. and Shen, S.-L. 2016. Stabilisation of marginal lateritic soil using high calcium fly ash-based geopolymer. *Road Materials and Pavement Design*, 17, 877–891. https://doi.org/10.1080/14680629.2015.1132632

Puppala, A., Wattanasanticharoen, E. and Hoyos, L. 2003. Ranking of four chemical and mechanical stabilization methods to treat low-volume road subgrades in Texas. *Transportation Research Record: Journal of the Transportation Research Board*, 1819(1), 63–71.

Puppala, A.J. and Cerato, A. 2009. Heave distress problems in chemically-treated sulfate-laden materials. *Geo-Strata*, 10, 28–32.

Puppala, A.J., Congress, S.S.C. and Banerjee, A. 2019. Research advancements in expansive soil characterization, stabilization and geoinfrastructure monitoring. In: G.M. Latha (Ed.), *Frontiers in Geotechnical Engineering*. Springer Nature, Singapore, pp. 15–29. https://doi.org/10.1007/978-981-13-5871-5_2

Puppala, A.J., Griffin, J.A., Hoyos, L.R. and Chomtid, S. 2004. Studies on sulfate-resistant cement stabilization methods to address sulfate-induced soil heave. *Journal of Geotechnical and Geoenvironmental Engineering*, 130, 391–402. https://doi.org/10.1061/(asce)1090-0241(2004)130:4(391)

Puppala, A.J., Manosuthikij, T. and Chittoori, B.C.S. 2013. Swell and shrinkage characterizations of unsaturated expansive clays from Texas. *Engineering Geology*, 164, 187–194. https://doi.org/10.1016/j.enggeo.2013.07.001

Puppala, A.J., Punthutaecha, K. and Vanapalli, S.K. 2006. Soil-water characteristic curves of stabilized expansive soils. *Journal of Geotechnical and Geoenvironmental Engineering*, 132, 736–751. https://doi.org/10.1061/(ASCE)1090-0241(2006)132:6(736)

Puppala, A.J., Wattanasanticharoen, E. and Punthutaecha, K. 2003. Experimental evaluations of stabilisation methods for sulphate-rich expansive soils. *Proceedings of the Institution of Civil Engineers – Ground Improvement*, 7, 25–35. https://doi.org/10.1680/grim.2003.7.1.25

Qian, X., Fang, C., Huang, M. and Achal, V. 2017. Characterization of fungal-mediated carbonate precipitation in the biomineralization of chromate and lead from an aqueous solution and soil. *Journal of Cleaner Production*, 164, 198–208. https://doi.org/10.1016/j.jclepro.2017.06.195

Rillig, M.C. and Mummey, D.L. 2006. Mycorrhizas and soil structure. *New Phytologist*, 171(1), 41–53. https://doi.org/10.1111/j.1469-8137.2006.01750.x

Salifu, E. 2019. *Engineering Fungal-mycelia for Soil Improvement* (PhD Thesis). University of Strathclyde.

Salifu, E. and Mountassir, G.E. 2019. Preliminary observations of the shear behaviour of fungal treated soil. *E3S Web of Conferences*, 92. https://doi.org/10.1051/e3sconf/20199211017

Salifu, E. and El Mountassir, G. 2021. Fungal-induced water repellency in sand. *Géotechnique*, 71(7), 608–615. https://doi.org/10.1680/jgeot.19.P.341

Salifu, E., El Mountassir, G., Minto, J.M. and Tarantino, A. 2022. Hydraulic behaviour of fungal treated sand. *Geomechanics for Energy and the Environment*, 30. https://doi.org/10.1016/j.gete.2021.100258

Samuel, R., Puppala, A.J., Banerjee, A., Huang, O., Radovic, M. and Chakraborty, S. 2021. Improvement of strength and volume-change properties of expansive clays with geopolymer treatment. *Transportation Research Record: Journal of the Transportation Research Board*, 2675, 308–320. https://doi.org/10.1177/03611981211001842

Sargent, P., Hughes, P.N., Rouainia, M. and White, M.L. 2013. The use of alkali activated waste binders in enhancing the mechanical properties and durability of soft alluvial soils. *Engineering Geology*, 152(1), 96–108.

Singhi, B., Laskar, A.I. and Ahmed, M.A. 2016. Investigation on soil–geopolymer with slag, fly ash and their blending. *Arabian Journal of Science and Engineering*, 41, 393–400.

Sizirici, B., Fseha, Y., Cho, C.-S., Yildiz, I. and Byon, Y.-J. 2021. A review of carbon footprint Reduction in construction industry, from design to operation. *Materials*, 14(20), 6094. https://doi.org/10.3390/ma14206094

Sharma, M., Satyam, N. and Reddy, K.R., 2021. State of the art review of emerging and biogeotechnical methods for liquefaction mitigation in sands. *Journal of Hazardous, Toxic, and Radioactive Waste*, 25(1), p.03120002. https://doi.org/10.1061/(ASCE)HZ.2153-5515.0000557

Soltani, A., Taheri, A., Khatibi, M. and Estabragh, A.R. 2017. Swelling potential of a stabilized expansive soil: A comparative experimental study. *Geotechnical and Geological Engineering*, 35, 1717–1744.

Sterflinger, K. 2000. Fungi as geologic agents. *Geomicrobiology Journal*, 17(2), 97–124. https://doi.org/10.1080/01490450050023791

Talluri, N., Puppala, A.J., Chittoori, B.C.S., Gaily, A.H. and Harris, P. 2013. Stabilization of high-sulfate soils by extended mellowing. *Transportation Research Record: Journal of the Transportation Research Board*, 2363, 96–104. https://doi.org/10.3141/2363-11

Tarantino, A., El Mountassir, G., Wheeler, S., Gallipoli, D., Russo, G., Augarde, C., Urciuoli, G., Pirone, M., Stokes, A., Van De Kuilen, J.W., Gard, W., Fourcaud, T., Romero, E., Priegue, A., Smith, C.C., Larrey-Lassalle, P., Becker, P., Ferrari, A., Dainese, R. and Speranza, G. 2020. TERRE project: Interplay between unsaturated soil mechanics and low-carbon geotechnical engineering. *E3S Web of Conferences*, 195. https://doi.org/10.1051/e3sconf/202019501002

van Paassen, L.A., Daza, C.M., Staal, M., Sorokin, D.Y., van der Zon, W. and van Loosdrecht, M.C.M. 2010. Potential soil reinforcement by biological denitrification. *Ecological Engineering*, 36(2), 168–175. https://doi.org/10.1016/j.ecoleng.2009.03.026

Wang, F., Han, J., Zhang, X. and Guo, J. 2017. Laboratory tests to evaluate effectiveness of wicking geotextile in soil moisture reduction. *Geotextiles and Geomembranes*, 45, 8–13. https://doi.org/10.1016/j.geotexmem.2016.08.002

Wang, G., Li, T., Zhou, Q., Zhang, X., Li, R. and Wang, J. 2024. Characterization and environmental applications of soil biofilms: A review. *Environmental Chemistry Letters*, 22(4), 1989–2011. https://doi.org/10.1007/s10311-024-01735-1

Weber, B., Belnap, J., Büdel, B., Antoninka, A.J., Barger, N.N., Chaudhary, V.B., Darrouzet-Nardi, A., Eldridge, D.J., Faist, A.M., Ferrenberg, S., Havrilla, C.A., Huber-Sannwald, E., Malam Issa, O., Maestre, F.T., Reed, S.C., Rodriguez-Caballero, E., Tucker, C., Young, K.E., Zhang, Y. and Bowker, M.A. 2022. What is a biocrust? A refined, contemporary definition for a broadening research community. *Biological Reviews*, 97(5), 1768–1785. https://doi.org/10.1111/brv.12862

Wingender, J., Neu, T.R. and Flemming, H.-C. 1999. What are bacterial extracellular polymeric substances? In: *Microbial Extracellular Polymeric Substances*. Springer, Berlin, Heidelberg, pp. 1–19. https://doi.org/10.1007/978-3-642-60147-7_1

Yi, Y. and Li, C. 2015. Alkali-activated ground-granulated blast furnace slag for stabilization of marine soft clay. *Journal of Materials in Civil Engineering*, 27(4).

Yong, R.N. and Ouhadi, V.R. 2007. Experimental study on instability of bases on natural and lime/cement-stabilized clayey soils. *Applied Clay Science*, 35, 238–249. https://doi.org/10.1016/j.clay.2006.08.009

Yu, X., Yang, H. and Wang, H. 2022. A cleaner biocementation method of soil via microbially induced struvite precipitation: A experimental and numerical analysis. *Journal of Environmental Management*, 316, 115280. https://doi.org/10.1016/j.jenvman.2022.115280

Zamani, A., Phradichith, C., DeJong, J., Nelson, D. and Parales, R. 2020. Effect of biofilm treatment strategy on the permeability reduction of sands. *Geo-Congress*, 2020, 150–158. https://doi.org/10.1061/9780784482834.017

Zhang, M., Zhao, M., Zhang, G., Nowak, P., Coen, A. and Tao, M. 2015. Calcium-free geopolymer as a stabilizer for sulfate-rich soils. *Applied Clay Science*, 108, 199–207.

Zhang, X., Fan, X., Han, C., Wang, C. and Yu, X. 2020. Improving soil surface erosion resistance by fungal mycelium. *Geo-Congress*, 2020, 523–531. https://doi.org/10.1061/9780784482780.051

Zhang, X., Fan, X., Wang, C. and Yu, X. 2023. A novel method to improve the soil erosion resistance with fungi. *Acta Geotechnica*, 18(5), 2827–2845. https://doi.org/10.1007/s11440-022-01673-8

Zhang, X., Presler, W., Li, L., Jones, D. and Odgers, B. 2014. Use of wicking fabric to help prevent frost boils in Alaskan pavements. *Journal of Materials in Civil Engineering*, 26, 728–740. https://doi.org/10.1061/(ASCE)MT.1943-5533.0000828

Zhang, Z., Provis, J.L., Wang, H., Bullen, F. and Reid, A. 2013. Quantitative kinetic and structural analysis of geopolymers. Part 2. Thermodynamics of sodium silicate activation of metakaolin. *Thermochimica Acta*, 565, 163–171.

Chapter 4

Use of construction and demolition waste and steel slag

Ana Cristina Freire, Isabel Martins, Cláudia Ferreira, Eugénia Correia, José Neves, António José Roque, Eduardo Fortunato, André Paixão, and Sara Rios

4.1 Introduction

In Europe, the aggregate industry is the largest non-energy extraction industry (UEPG 2024). Civil engineering works, particularly the construction of transport infrastructures like roads and railways, drive significant demand. Due to the ineffective use of resources, exploiting the potential of producing aggregates from alternative sources has been increasingly crucial. Annually, the significant amount of waste generated poses major economic, social, and environmental problems. Construction waste disposal in landfills has high environmental impacts, depleting the planet's non-renewable resources. Awareness of the need to reuse materials and recycle waste, reducing landfill volumes, requires the implementation of effective material management practices, in line with the 2030 Agenda and Circular Economy principles.

The 2030 Agenda, defined in 2015 by the United Nations (UN), aims to transform the world through 17 Sustainable Development Goals (SDGs) and 169 targets, organized into five guiding principles: People, Planet, Prosperity, Peace, and Partnerships (UN 2015). At the European level, the European Commission (EC) is committed to the 2030 Agenda and shares the UN's concerns about sustainability, having presented a political agenda aiming to achieve sustainability in the European Union (EU) and beyond. The SDGs are an inherent aspect of the political guidelines of the EC and lie at the heart of the policymaking on internal and external actions across all sectors (EU 2024). The EC has proposed the European Green Deal to address climate change and environmental degradation, aiming to reduce net greenhouse gas emissions by at least 55% by 2030, compared to 1990 levels. The European Green Deal seeks to transform the EU into a modern, resource-efficient, and competitive economy with no net greenhouse gas emissions by 2050 (EU 2019).

Portugal is also committed to implementing the 2030 Agenda. Respecting national priorities and strategic guidelines, Portugal's implementation of the 2030 Agenda promotes sustainable development and identifies priority SDGs in line with the country's strategic development vision, mainly through the National Reform Programs (Cabaço 2017).

Using recycled and alternative aggregates aligns with several SDGs, including SDG 12: Responsible Consumption and Production, SDG 11: Sustainable Cities and Communities, and SDG 13: Climate Action, as well as European and Portuguese Action Plans for the Circular Economy. These materials, such as crushed concrete, mixed concrete, and Reclaimed Asphalt Pavement (RAP), offer economic and environmental alternatives for transport infrastructures. Artificial aggregates, such as steel slag and municipal solid waste (MSW) slag, must meet quality and durability requirements for their intended applications.

DOI: 10.1201/9781003530220-5

This chapter discusses the application of recycled materials in transport infrastructures, focusing on unbound and hydraulically bound mixtures (HBM), as well as the use of steel slag aggregates in unbound mixtures, from a circular economy perspective.

4.2 Framework

4.2.1 *Waste processing*

Acting towards a resilient, healthier, and zero-carbon world is essential for transitioning to a circular economy, particularly where high resource consumption and the potential to close material loops exist. Transportation infrastructures meet these criteria, requiring consideration of resource availability, local conditions, construction processes, and maintenance practices to ensure sustainability (Van Dam et al. 2015).

At the end of their life, transportation infrastructures offer the opportunity to recover waste materials, returning them to the life cycle, providing economic value, improving resource efficiency, developing markets for secondary materials, and reducing greenhouse gas emissions.

In Europe, more than 90% of road infrastructures are paved with asphalt, and their rehabilitation involves establishing the most adequate destination for the waste produced. This waste, known as RAP, comprises both the asphalt milled or excavated from existing pavement layers and the asphalt returned from reject, surplus, or failing production. To assess the feasibility of using RAP, it is necessary to characterize the material and to evaluate its potential recycling rate. Subsequently, RAP is processed through a set of operations, namely, by crushing, milling, or screening, to generate the reclaimed asphalt (RA), a product suitable to be used as a constituent for the rehabilitation of roads, after being properly tested and classified. If all the constituents of RA could perform the same function as in their original application, then the most appropriate option would be reusing RA. On the other hand, if the binder of RA could not perform the same function as in its original application, then RA would be recycled.

For RA reuse or recycling, it is also necessary to define whether it has been performed in situ or in a mixing plant. This choice must consider factors like the type of rehabilitation (surface or structural), the distance to the mixing plant, the quality of RAP, the specifications for the rehabilitated road, and even the need for a fast reopening of the road infrastructure. Additionally, the adequate range of mixing temperature – cold mix (CM at ambient temperature), half-warm mix asphalt (half-WMA ≅ 70°C–100°C), warm mix asphalt (WMA ≅ 100°C–150°C), and hot mix asphalt (HMA ≅ 120°C–190°C) must also be set (EAPA 2014).

Potential limitations to the continued valorization include the decrease in aggregate size due to the removal of RAP and comminution operations, as well as the decrease in binder performance due to ageing during production, use, and processing. It should also be noted that the data concerning the use of RA in Europe indicates a growing trend, which highlights that asphalt reuse is increasingly adopted in relation to recycling, approximately at a ratio of 3:1 (EAPA 2022).

Along with RA, other different types of waste and industrial by-products have also been successfully used in transportation infrastructures, namely, concrete from construction and demolition waste (CDW), as detailed in Section 4.2.2. These materials often go through processing (namely, sorting, comminution, and sieving) before being applied as substitutes for traditional virgin raw materials and can even lead to a superior performance of road networks (Poulikakos et al. 2017).

Nevertheless, in this whole process of waste recovery, special attention must be paid to the presence of hazardous substances. The use of tar in road infrastructure was well known until the 1980s, when it was banned. Therefore, the presence of relatively high concentrations of polycyclic aromatic hydrocarbons (PAHs), when there is suspicion of tar existence, must be analysed by carrying out specific tests to minimize future contamination. The same applies to other substances, particularly heavy metals and their compounds, whose release over time can be determined by performing leaching tests (Roque et al. 2016).

As regards the sustainability of railway infrastructures, waste management also represents a relevant area to improve resource efficiency. Rail transport is recognized as a low-carbon mode, but the total freight accounted for only 5.5% in the EU in 2022, approximately five times less than by road (Eurostat 2023). Therefore, adequate practices to manage the used resources of railway infrastructures are highly valuable. Replacing ballast aggregates with natural aggregates has been a common practice. However, different studies have shown the possibility of reusing these aggregates, after cleaning, if factors like the condition of the tracks, the type of technology used, and the specific regulations governing rail infrastructure in each region do not hinder it. Some of the approaches throughout the whole value chain are presented: (i) increasing the percentage of reuse of concrete sleepers and ballast (DTMR 2018); (ii) replacing traditional concrete with sustainable concrete options; (iii) using sleepers from alternative recycled materials, namely, from recycled plastic (UIC 2021).

A more substantiated decision on the environmental and economic advantages of processing waste and incorporating recycled and reused materials in sustainable rehabilitation of transportation infrastructures must make use of life cycle assessment (LCA) and life cycle cost analysis (LCCA) tools, enabling the appropriate allocation of the different impacts (Zhao, Goulias, and Peterson 2021).

The rehabilitation of transportation infrastructures should estimate the amount of waste the project will generate and provide an adequate characterization of these materials. Tracking the waste quantities generated during a project, recording, and reporting the final volumes of waste quantities generated, and the use of recycled materials in a project are essential for implementing the principles of the circular economy. After waste processing, the feasibility of using the recycled materials in transportation infrastructures will be determined by their compliance with the technical requirements, by their cost compared to natural materials, and by the quantities available for the project. Waste processing and reuse of materials is an important investment in transportation infrastructures, as it contributes to preventing the depletion of resources and avoiding the disposal of materials.

4.2.2 *Different types of recycled and alternative materials and some applications*

A holistic view throughout the whole life cycle of construction materials and over the entire value chain must be adopted to minimize economic, environmental, and climate change-related impacts. This applies to a broad range of construction materials used in different applications and provides opportunities for innovative approaches.

Existing circular economy and material recovery models that prioritize recycling and reuse efforts demonstrate a sustained commitment to support circular practices in the construction and demolition sector. In the EU, CDW accounts for more than a third of all generated waste, being a priority waste stream to deal with (ETPC 2019).

Finding solutions to increase energy efficiency and circularity of these materials would have a high impact on the sustainability of the construction value chain, and hence, materials with an increased sustainability and circularity, such as CDW and steel slag, can be the key for achieving SDG goals. The use of recycled and alternative aggregates is expected to grow as the construction industry becomes more aware of the environmental and economic benefits of using these materials (Roadmap 2022). CDW is a broad term that covers various materials generated during the construction, renovation, and demolition of buildings and transport infrastructure projects, among others (Figure 4.1).

Figure 4.1 CDW types: (a) Reclaimed Asphalt Pavement (RAP); (b) mixed CDW.

As the construction industry strives for sustainability, the use of recycled CDW is becoming increasingly important for transport infrastructures. By incorporating recycled and alternative materials, it is possible to minimize the environmental impact of transport infrastructure projects and promote circularity in the construction sector.

Different types of waste or by-products can be recycled and used as alternative materials in transport infrastructures, namely:

- Recycled concrete aggregate (RCA): produced by crushing and processing of old concrete pavement or demolition debris. It can be used as a substitute for virgin gravel or crushed stone in various applications, including pavement bases and sub-bases, and capping layers, and as concrete aggregate (FHWA 2016).
- RA: the processed site-won asphalt, by milling, crushing, sieving or screening, and blending, is suitable and ready to be used as constituent material for asphalt after being tested, assessed, and classified (Mohajerani et al. 2017). It can be used primarily as a substitute for virgin asphalt binder and virgin gravel in pavement construction, namely, for pavement bases, sub-bases, and capping layers.
- Steel slag: a byproduct of steel making, being a manufactured aggregate produced during the separation of the molten steel from impurities in steel-making furnaces (FHWA 2016). The primary applications of steel slag are its use as a granular base or sub-base layer and as an aggregate material in construction applications. However, its application as aggregate in bituminous mixtures has proven to have good results (Freire and Maia 2019).
- Crushed glass: this waste can be used as a substitute for sand or gravel in a variety of applications. Research has covered various applications, namely, as aggregate in asphalt, sub-bases, and unbound bases; as lightweight engineering material; cementitious material; and as aggregate in concrete (Mohajerani et al. 2017).
- Mixed concrete and masonry: the mixture of concrete, bricks, and blocks constitutes a significant portion of CDW. These materials can be crushed and recycled into aggregates for road base layers and embankments.
- Plastics: plastic waste can be recycled into various products, including plastic lumber, drainage pipes, and packaging materials. Studies regarding its incorporation into asphalt have been developed, considering the wet or dry process incorporation (Heydari et al. 2021).

4.2.3 Legislative framework in Portugal

There are several international and Portuguese technical and legislative support documents on the use of recycled and alternative aggregates in civil engineering, particularly in transport infrastructures. At an international level, different countries apply specific criteria developed according to national realities.

In Portugal, the management of CDW was initially regulated by the General Waste Management Regime (GWMR) approved by Decree-Law (DL) no. 178/2006, 5 September, as well as by legislation referring to the special flows often mixed in CDW, such as packaging waste, waste from electrical and electronic equipment, or even used oils. CDW has a high potential for recycling, and a considerable part of its constituents has a high resource value. These aspects justified the need to create a specific legal system for this stream – DL no. 73/2011, 17 June – which established technical standards for CDW management operations, following the provisions of Article 20 of DL no. 178/2006, by ensuring the application to the CDW stream of the waste reduction, reuse, and recycling policies advocated in the Program of the XVII Portuguese Constitutional Government. Therefore, the GWMR was regulated by DL no. 46/2008, 12 March, which fulfilled these needs. Also noteworthy is that DL no. 73/2011 made the third amendment to DL no. 178/2006 by republishing it and transposing Directive no. 2008/98/EC of the European Parliament and of the Council of November 19 on waste and amending various legal systems on waste.

The management of CDW and the use of recycled aggregates and other alternative aggregates at the national level are currently regulated by the New General Waste Management Regime (NGWMR), as set out in DL no. 102-D/2020, dated 10 December. This law approved the GWMR, the legal regime for the landfill of waste, and amended the regime for the management of specific waste streams by transposing Directives (EU) 2018/849, 2018/850, 2018/851, and 2018/852. This DL was ratified by the Declaration of Rectification no. 3/2021, 21 January. Law no. 52/2021, 10 August, has amended DL no. 102-D/2020 by parliamentary consideration. Article 21 of DL no. 102-D/2020 establishes objectives and targets for reducing waste production, particularly in the construction and public works sector.

Aiming to promote recovery and incorporation into works, the Portuguese Environment Agency (APA), as the entity responsible for implementing environmental policies in Portugal, also put forward, for the use of CDW on site, after due treatment, several documents (General Rules), referring to milling and crushing of RA, railway ballast waste and incorporation of mixed concrete and CDW waste.

According to Article 28 of the NGWMR, in Portugal, it is compulsory to use at least 10% of recycled materials or materials incorporating recycled materials in relation to the total quantity of raw materials used on site when contracting works under the Public Contracts Code (CCP). In addition, paragraph 7 of the same article stipulates that these materials must be certified by the competent national or European bodies, in accordance with the applicable legislation in force. Recycled and/or alternative aggregates contribute to achieving the target required by law and can be used in geotechnical, drainage, and paving works.

The Portuguese National Laboratory for Civil Engineering (LNEC) has developed and published a set of relevant LNEC specifications, providing recommendations and defining minimum requirements for the use of recycled aggregates and recycled materials from CDW concerning different applications:

- LNEC E 471 (2009) – Guide for the use of coarse recycled aggregates in concrete
- LNEC E 472 (2009) – Guide for the production of recycled hot mix asphalt
- LNEC E 473 (2009) – Guide for the use of recycled aggregates in unbound pavement layers
- LNEC E 474 (2009) – Guide for the use of recycled materials coming from CDW in embankment and capping layer of transport infrastructures
- LNEC E 483 (2016) – Guide for the use of recycled aggregates from RA for unbound road pavement layers
- LNEC E484 (2016) – Guide for the use of materials resulting from CDW in rural and forest roads
- LNEC E485 (2016) – Guide for the use of materials resulting from CDW in backfill of trenches

At a national level, in Portugal, the construction owners can also publish construction specifications promoting the incorporation of recycled and/or alternative aggregates. Those specifications must comply with the applicable law and should respect the recommendations set by LNEC specifications (IP 2014).

4.3 Recycled and alternative aggregates for unbound and hydraulically bound mixtures

4.3.1 *Recycling CDW for unbound granular road layers*

4.3.1.1 Advances on the use

The use of CDW in unbound granular (UG) pavement layers allows promoting the reuse, recovery, or recycling of applied materials in future maintenance, conservation, and/or rehabilitation operations by considering the concept of the life cycle of transport infrastructures and materials applied to the circular economy.

Research based on experimental work is essential to overcome empirical approaches in the design and construction using recycled and alternative aggregates for transportation infrastructures. Experimental investigation involving laboratory and in situ tests makes it possible to validate the adequate mechanical behaviour of both the recycled and the alternative materials (Gomes-Correia, Winter, and Puppala 2016). Considering the enormous variety of recycled materials and conditions of their application, it is advisable to conduct thorough research based on experimental sections, instrumented and observed over the life cycle of the materials.

In recent years, several studies, as well as research and innovation projects, have been developed to identify, characterize, and validate the characteristics, evaluation methods, and application technologies regarding recycled and alternative aggregates in UG layers, namely, for roads and railways. The ALT-MAT (ALTerials ALTernative in road Construction project) and SAMARIS (Sustainable and Advanced MAterials for Road InfraStructure) projects are some examples of initiatives developed in this context (EC 1999, SAMARIS 2004). Also noteworthy is the DIRECT_MAT project (DISmantling and RECycling Techniques for road MATerials – Sharing Knowledge and Practices). The main objectives of this project were to share and disseminate, on a European scale, the national knowledge of the countries involved and the practices verified in situ as they refer to the demolition and recycling of pavements and pavement-related materials to promote and improve recycling practices for the construction and maintenance of road infrastructures (EC 2011). With the RE-ROAD project (End-of-Life Strategies for Asphalt Pavements), the aim was to develop existing knowledge and innovative technologies identified for end-of-life strategies for bituminous road infrastructures. These strategies have a significant impact on the energy efficiency and environmental footprint of the European transport system from the viewpoint of the infrastructure life cycle (Kalman et al. 2013). More recently, the EU-funded project REcircularPAV – ReEngineering asphalt mixture for circular road PAVements evaluated the feasibility of using biomaterials along with asphalt pavement recycling to minimize the use of virgin aggregates and petroleum bitumen for asphalt pavements (EC 2023). Another recent initiative is the ReBuild17 – Building Circularity project, developed in the Azores, Portugal. This project aims to create alternatives to reduce the production of CDW and focuses on raising awareness among companies and entities for the valorization of this waste by suggesting alternatives and forms of reincorporation into construction products and ultimately generating a chain of circularity with this process (EEAGrants 2021).

4.3.1.2 Application in road pavement layers

At a national level, reference is made to the SUPREMA project – Sustainable Application of Recycled Construction and Demolition Materials in Road Infrastructures – developed by LNEC from 2010 to 2014, in collaboration with the University of Lisbon. The main goal of the project was to promote the sustainable and resilient application of CDW in road bases, sub-bases, and subgrades (Freire et al. 2021). The project consisted of an extensive experimental programme comprising the assessment of the geoenvironmental and geomechanical characteristics of selected recycled materials, from different origins, by screening methodology and final composition; the behaviour comparison of UG layers with or without recycled aggregates; the definition of the pavement design parameters with the use of recycled materials; as well as the evaluation and validation of construction methodologies to be applied, depending on the type of recycled materials, to UG layers (base, sub-base, and subgrade).

Four types of recycled (crushed mixed concrete – CMC, crushed reclaimed asphalt – CRA, and a mix of 70% natural aggregate and 30% milled RA – 70% CL+30% MRA) and virgin aggregates (crushed limestone – CL) were identified, selected, and characterized in the laboratory to evaluate the chemical, geometrical, physical, and mechanical properties (Freire et al. 2021). Leaching tests were done on the selected materials to validate the compliance with the legal limits of acceptance of waste for landfill disposal (Roque et al. 2016).

A full-scale experimental pavement section was built, which was instrumented with strain gauges and pressure cells during construction (Figure 4.2). The pavement of the experimental sections was composed of aUG base layer, 0.30 m thick, containing recycled materials. The construction procedures adopted were similar for all the sections. To ensure a proper compaction in the total depth of the base layer, this operation was performed in two sub-layers with 0.18 m and 0.12 m thicknesses, respectively. The HMA layer was built with a thickness of 0.07 m. The quality control of the compaction of the UG layer was performed relative to the modified Proctor test.

The Falling Weight Deflectometer (FWD) tests were carried out during and after construction of the experimental test sections (Figure 4.3), which made it possible to assess the mechanical behaviour of the pavement layers.

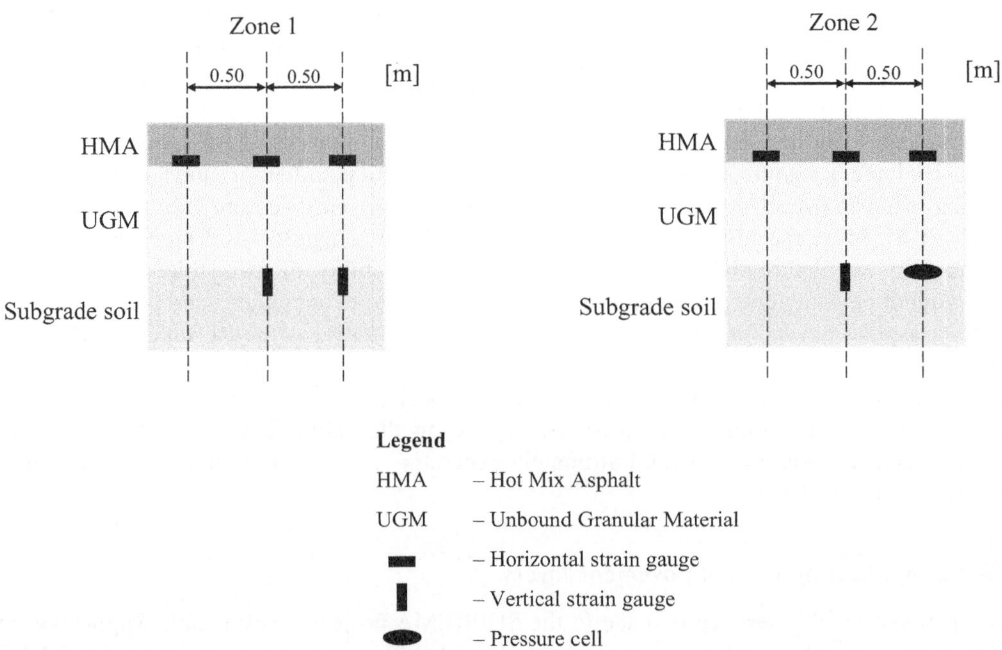

Figure 4.2 Test section instrumentation plan for Zone 1 (left) and Zone 2 (right).

Figure 4.3 Construction of the experimental test sections and FWD tests.

The analysis of the behaviour of the pavement of the experimental test section, together with the reading of the applied instrumentation, makes it possible to obtain the deformability modulus of the constituent layers and the consequent validation of the load capacity of the recycled materials applied on the UG layer. The test section with the mix of 70% natural aggregate and 30% milled RA (70% CL + 30% MRA) presented a deformability modulus of the UG layer between 323 and 355 MPa, with reduced variability. These results are of the same order of magnitude as those presented in the section constructed with the reference material (ABGE), between 253 and 372 MPa. The test section with CMC presented a deformability modulus of the UG layer between 142 and 285 MPa. These results confirmed that, in general, recycled materials from CDW demonstrated an acceptable behaviour in terms of deformability modulus, supporting the application of recycled materials from this type of waste. In addition, the environmental impact was evaluated to control potential contamination of surface water, groundwater, and soil during the use of recycled materials, which presented satisfactory results (Neves et al. 2023).

4.3.1.3 Application in road and railway drainage structures

Following the results obtained in the SUPREMA project, other applications have been evaluated, such as drainage structures, by the CDW (concrete aggregate) characterization and assessment, namely, the macropetrographic description and micropetrographic examination of the RCAs, the physical properties with the slake durability test (SDT), and the performance characterization with the permeability test (Roque, da Silva, and Almeida 2022).

4.3.1.3.1 PETROGRAPHIC CHARACTERISTICS

The macropetrographic description and micropetrographic examination of RCAs from construction and demolition waste samples (CACDW) were performed according to ASTM C856-11:2012. The micropetrographic examination was carried out on thin sections observed with an optical microscope. The macropetrographic description and micropetrographic examination were executed before and after the SDT (ISRM 1977).

Macroscopic observation samples evidenced that CACDW is a heterogeneous and multiphase material, composed of natural aggregates of coarse-grained and fine-grained sizes, with an adhered cementitious matrix, porous (presence of some pores $\leq$3 mm, dispersed heterogeneously) and light grey in colour. After the SDT, no relevant signals of weathering were identified. From a microscopic point of view, CACDW has a complex structure composed of (i) a cementitious matrix with several black points, heterogeneously distributed, which may correspond to the existence of pores; (ii) coarse-grained and fine-grained (quartzite sand) aggregates scattered across the cementitious matrix; and (iii) a transition zone, corresponding to the matrix-aggregate interface (Roque, Rodrigues, and da Silva 2020).

4.3.1.3.2 PHYSICAL PROPERTIES (SLAKE DURABILITY TEST)

The durability of CACDW was also evaluated considering the results obtained in the SDT and Gamble's classification (Gamble 1971), compared with the durability of two reference natural aggregates, a basalt (BAS) and a limestone (LIM). The specimens of CACDW, BAS, and LIM were subjected to two rotation cycles as recommended by the ISRM suggested method (ISRM 1977) (200 rotations + 200 rotations). To extend the period of abrasion of the materials, the specimens were subjected to 600 additional rotations (1,000 rotations in total). The slake durability index (Id), which corresponds to the percentage ratio of final and initial dry masses of the materials in the drums at the end of each cycle, was calculated in accordance with the ISRM suggested method, that is, for the second cycle of wetting and drying (Id2, 400 rotations). The results of Id1 (200 rotations) and Id5 (1,000 rotations), the first and third cycles of wetting and drying, respectively,were also calculated. The SDTs were carried out with tap water (W), using specimens without prior immersion before the test and with prior immersion for 15 days before the test.

The Id2 results compared with Gamble's classification lead to the conclusion that all materials belong to the class of very high durability (Id2 $\geq$ 98.0%). The values obtained for Id5 also show that the increase in the number of rotations did not contribute to changing the classification of the durability of the materials in accordance with Gamble's classification (Roque, Rodrigues, and da Silva 2020). Previously immersing the specimens in water for 15 days also did not influence the durability of the materials.

4.3.1.3.3 PERFORMANCE CHARACTERIZATION (PERMEABILITY TEST)

Specimens used in the permeability tests were prepared from two grain size ranges: 0.25–2 mm (sand) and 2–20 mm (fine gravel). The dimensions of the tested specimens were 9 cm in diameter and 25 cm in height, which correspond to the dimensions of the rigid-wall column of the permeameter. Permeability tests were carried out at constant head at a controlled room temperature of about 20°C. In Phase 1 of the test, a minimum volume of about 250 L was expected to

Table 4.1 Results of the permeability tests obtained with CA_{CDW}, BAS, and LIM

Material	*Grain size range*	*Time (day)*		*Effluent volume (l)*		*K × 10⁻³ (m/s)*	
		Phase 1	*Phase 2*	*Phase 1*	*Phase 2*	*Phase 1*	*Phase 2*
CACDW	Sand (0.25–2 mm)	29	12	257	53	0.4	0.3
	Fine gravel (2–20 mm)	17	11	263	37	2.5	2.5
BAS	Sand (0.25–2 mm)	29	19	265	50	0.9	1.8
	Fine gravel (2–20 mm)	17	11	270	40	2.8	2.8
LIM	Sand (0.25–2 mm)	27	19	183	30	0.05	0.02
	Fine gravel (2–20 mm)	17	11	267	40	1.7	1.7

Source: Adapted from Roque, da Silva, and Almeida (2022).

CACDW, concrete aggregate of construction and demolition waste; BAS, Basalt; LIM, Limestone; k, hydraulic conductivity.

be collected. In Phase 2, the permeameter column was removed from the test device and placed in an oven, at a temperature of 45°C, for a minimum period of seven days. After removing the column from the oven, the specimens were given the necessary time to reach room temperature. In Phase 2, a minimum volume of about 30 L was expected to be collected. All permeability tests were carried out with a hydraulic head of 0.30 m until the hydraulic conductivity, *k*, was approximately constant (Roque, da Silva, and Almeida 2022). Table 4.1 presents the values for *k* obtained in both phases for the three materials. As expected, the *k* of fine gravel is higher than the *k* of sand, verifying that the *k* of CACDW, in both grain size ranges, is lower than the *k* of BAS and higher than the *k* of LIM. *k* values typically required for the drainage structures in the road pavements are equal to or higher than 1×10^{-3} m/s (CDOT 2019). So, only the fine gravel of the three aggregates met this requirement. Regarding the inside face of retaining walls, the structures with drainage functions should typically present $k > 1 \times 10^{-5}$ m/s (DTMR 2018). In this case, the minimum requirement was fulfilled in both grain size ranges of the three materials.

Considering the variation of *k* as a function of the collected volume in the permeability tests carried out with CACDW, which was similar to the curves obtained with the natural aggregates (BAS and LIM), it was concluded that neither the formation of tufa, a precipitate of insoluble calcium carbonate, nor the recementation of cementitious materials that could not be completely hydrated were observed (Roque, da Silva, and Almeida 2022). These results agree with previous studies (McCulloch et al. 2017), which also did not observe significant tufa formation and cementation phenomena, especially if the percentage of fines is limited to a small percentage and a proper processing and handling of these recycled aggregates is carried out (Roque, da Silva, and Almeida 2022). However, several studies reported the possibility of the formation of tufa and recementation of cementitious materials (Bruinsma, Peterson, and Snyder 1997, Kim et al. 2014), which, when occurring, can significantly reduce the drainage capacity of structures that incorporate them or are adjacent to them.

4.3.2 *Valorization of construction and demolition waste within hydraulically bound mixtures*

4.3.2.1 Developments on the use of hydraulically bound mixtures

HBM are mixtures in which the hydraulic binder induces a fast setting and hardening process of the material characteristics. HBM is a traditional technology that continues to be applied as a pavement material (Walters and Edwards 2009, Liu et al. 2020). The most current HBM used in road pavements are cement-bound granular mixtures (CBGM), aimed at preserving and enhancing local and natural resources, thus contributing to more sustainable construction practices and promoting the circular economy model (Gomes Correia, Winter, and Puppala 2016, Plati 2019). In addition, CBGM can also incorporate waste materials (Xuan et al. 2010, Pérez et al. 2013, Pasetto and Baldo 2016).

The application of recycled aggregates (RAg) from CDW in CBGM (CBGM-CDW) is still limited. Literature presents studies demonstrating the advantages of CBGM-CDW through laboratory tests (Pasetto 2000, Xuan et al. 2010, Abu El-Maaty 2013) and full-scale experiments (Agrela et al. 2012, Stehlík, Hyzl, and Varaus 2016, Stehlík et al. 2016, Qamhia et al. 2019). Also, other industrial waste materials (e.g., steel slag, waste foundry sand, ladle furnace slag, glass waste, and coal ash) have been studied to be incorporated as aggregate in CBGM (Pasetto 2000, Pasetto and Baldo 2013, 2015).

The mechanical performance of CBGM-CDW can be evaluated by laboratory tests, such as unconfined compression strength (UCS) and indirect tensile strength (ITS). The durability of CBGM-CDW is significant and is currently mostly evaluated by wet/dry or freeze/thaw methods (Neves 2023). CBGM-CDW tends to have higher optimum moisture content, lower maximum dry density, and lower resistance to crushing due to the higher porosity created by the adhered mortar and porous constituents (e.g., crushed brick). Some of the residual, non-hydrated cement stored in the adhered mortar can lead to a recementation process, which will occur when non-hydrated cement (exposed by the recycling process or CBGM-CDW mixing and compaction) is mixed with water and granular materials. High compaction energy can maximize the release of non-hydrated cement and aggregate breakage during the construction phase, mitigating aggregate breakage during service life. The recementation process may reduce the added cement required for the CBGM-CDW.

There is still some uncertainty in the mechanical performance and durability of CBGM-CDW. In mechanical performance, essential contributions would be the characterization of flexural strength and fatigue resistance. Fatigue resistance is significant and associated with reflection cracking in flexible pavements. Regarding durability, the experience with CBGM-CDW is still minimal, including traditional tests, such as wet-dry and freeze–thaw methods.

4.3.2.2 Application in road pavements

A large-scale application of CBGM-CDW in pavement construction can generate demand and add value to recycled materials. The use of CBGM-CDW will also contribute to reducing the CDW volumes discharged in landfills and to saving the consumption of non-renewable natural resources. The main application of CBGM-CDW has been to unbound and bound base and sub-base layers of flexible, semi-rigid, and rigid pavements.

In Portugal, the ADVAMCE project – Development of techniques to add value to recycled materials to promote the circular economy in transport infrastructures – was developed at the University of Lisbon (2018–2022) and in cooperation with Brazil. This work aims to promote the use of waste materials, particularly RCAs derived from CDW and coconut fibres (CFs) derived from waste coconut, in the construction of CBGM-CDW for base and sub-base layers of flexible pavements. The use of CF in the project was related to this waste being an environmental problem in some countries (e.g., Brazil). Incorporating organic fibres into CBMG could be interesting and significantly contribute to the circular economy. Using organic by-products in construction may raise the question of their durability over time and compromise the intended effect. However, the case of CF is particular because they are highly durable, so their use in road pavements is an advantage. To evaluate the feasibility of the materials, an experimental programme was developed in the laboratory consisting of mechanical and durability tests (Crucho, Picado-Santos, and Neves 2021, 2022).

The project studied several CBGM-CDW with varying cement contents and the CF reinforcement. The RCA was almost entirely constituted by concrete or concrete products (81.0%) and unbound aggregate or natural stone (18.3%). The other constituents were bituminous materials and clay masonry units (0.7%). Six HBM were produced: two mixtures using 100% natural aggregate with 2% cement, one with no CF and the other with CF reinforcement; two mixtures produced with 100% RCA and 2% cement, without and with CF; and two mixtures using 100% RCA and 3% cement, without and with CF. The cement was the Portland cement CEM II/A-L 42.5 R. The dosage of cement (2% and 3%) was by mass of dry aggregates. The dosage of CF was 0.1% by mass of dry aggregates and cement. The aggregates and cement were mixed firstly, followed by water, and finally the CF. It should be noted that low cement dosages were adopted in this project with two main objectives: on the one hand, to avoid the appearance of cracking that is typical of hydraulic materials, and which can have negative consequences for the medium- and long-term performance of pavements; and on the other hand, to further enhance the benefits of using recycled aggregates and fibre reinforcement.

The materials were tested in various strength tests: immediate bearing index, unconfined compressive strength, ITS, and flexural strength. In general, the mechanical performance tests highlighted the suitability of RCA to be used in CBGM and the effective fibre reinforcement due to the presence of CF (Crucho, Picado-Santos, and Neves 2022). The CBGM-CDW, which incorporated RCA, showed a similar strength to that of the natural material. The cement significantly increased the material strength. It was proved that using RCA did not compromise the overall performance of the CBGM-CDW according to the technical requirements. Concerning fibre reinforcement, the main effect was not so much to increase strength as to guarantee the integrity of the material after cracking. Figure 4.4 clearly shows the impact of fibre reinforcement on the flexural strength test case. Including fibre elements in the material can increase its durability and mitigate pavement degradation. The project also included the assessment of the durability of CBGM-CDW through wetting-and-drying, freezing-and-thawing, and ITS tests.

During this project, a full-scale pilot section was built in 2022, which consisted of CBGM-CDW with 2% of cement and 0.1% of CF in the sub-base layer of a semi-rigid pavement. The monitoring of this stretch has made it possible to demonstrate the conclusions reached in the laboratory study.

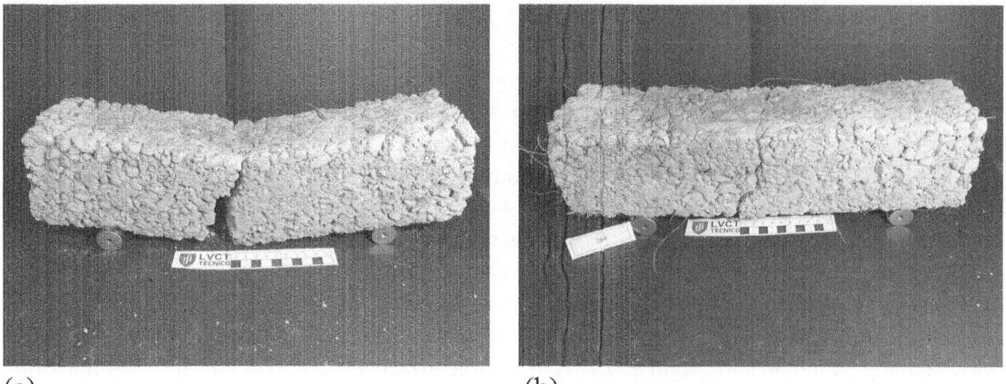

(a) (b)

Figure 4.4 Post-flexural strength of CBGM-CDW: without fibre (left); with fibre (right).

4.4 Steel slag aggregates in unbound layers

4.4.1 Production of steel slags

The use of steel slags in transport infrastructures offers a solution for managing these industrial by-products. Safe disposal of steel slags is increasingly problematic due to a 50% rise in steel production over the last decade (World Steel Association 2020). Various types of slags are produced depending on the industrial process. In the integrated steel industry, steel is produced directly from iron ores, generating granulated blast furnace slags. In the semi-integrated steel industry, steel is produced from scrap melting using methods like the Siemens-Martin (Open Hearth OH), the Oxygen Converter (Linz-Donawitz LD or Basic Oxygen Furnace BOF), or the Electric Arc Furnace (EAF). The production method affects the chemical composition and properties of the steel slag.

In Portugal, two EAF plants produce two types of steel slags: oxidizing or black slags (EAF slags) and reducing or white slags (ladle slags, LFS). Portuguese EAF, with an annual production of 240,000 tonnes, has been certified as an Inert Steel Aggregate for Construction (ISAC), increasing its commercial value (Roque et al. 2010). This certification clearly defines the ISAC production process and chemical properties. The process comprises three phases: (i) flow and cooling – ISAC is transferred to an impervious zone and cooled with water before storage; (ii) separation and recycling of metallic components; (iii) sieving and grading into various grain sizes, and maturation via air hydration to neutralize remaining lime (Gomes Correia et al. 2012). This section focuses on ISAC properties and performance for transport infrastructures.

4.4.2 Steel slag characterization

The macropetrographic description of ISAC was performed according to EN 932-3:2010, and the micropetrography examination was carried out on thin sections with an optical microscope, according to EN 12407:2007. Both were considered before and after the SDT, using procedures to characterize CACDW, BAS, and LIM.

Before the SDT, ISAC exhibited no signs of weathering, which were identified after the SDT with the oxidation of iron particles (Roque, Rodrigues, and da Silva 2020). The micropetrographic examination before the SDT showed that ISAC is composed of opaque minerals and a mineral species that could not be identified with an optical microscope due to its intrinsic characteristics. After SDT, weathering was observed, resulting in an increase in the concentration of opaque minerals and iron oxides (Roque, Rodrigues, and da Silva 2020).

The variability of the chemical composition depends on many factors, particularly the production processes (Mica et al. 2024). The main chemical elements of ISAC are iron oxide, calcium oxide, and silicon oxide. A small amount of magnesium oxide is also common, which, together with calcium oxide, can lead to some swelling potential due to the hydration of these components when in contact with water. This is why the maturation period is crucial, and most guidelines specify minimum maturation times (Mica et al. 2024). Leaching tests have shown that ISAC poses no health or environmental risks (Gomes-Correia et al. 2012).

ISAC exhibits distinct physical and morphological properties compared to natural crushed rocks. Usually granular and angular, ISAC is characterized by a dense and hard composition, typical of EAF steel slag. For example, while particles of natural rock (i.e., limestone and granite) have a specific gravity of around 2.7, ISAC particles show a specific gravity between 3.2 and 3.6 (Gomes-Correia et al. 2012, Delgado et al. 2022, Alves et al. 2023). The colour ranges from dark grey to black. Its particle size distribution can be adjusted, making it versatile for many applications as an alternative to hard natural crushed rocks, including as a monogranular aggregate for railway ballast or as a well-graded crushed aggregate in road construction (Mica et al. 2024).

Regarding the common parameters considered in the physical and mechanical characterization of well-graded crushed aggregates (0/31.5 mm grain size distribution), ISAC typically meets the same requirements as natural crushed hard rocks. ISAC's Los Angeles and micro-Deval indices – two parameters required in road and railway construction – typically fall within the ranges of 23–28 and 11–12, respectively (Gomes-Correia et al. 2012, Paixão and Fortunato 2021, Delgado et al. 2022).

To highlight ISAC's distinct morphological properties, some studies have employed advanced characterization methods as an alternative to conventional particle characterization involving sieving and manual measurement with callipers. Delgado et al. (2022) used a 2D characterization method with AIMS equipment (Gates et al. 2011). They compared granite particles with ISAC (4–30 mm) and found no significant differences in sphericity and surface texture compared to granite particles commonly accepted for railway ballast applications. However, ISAC had a higher angularity index (AI). Paixão and Fortunato (2021) used close-range photogrammetry for a 3D analysis, also comparing granite and ISAC railway ballast particles. The study included a micro-Deval abrasion test, evaluating three morphological parameters (Masad et al. 2005): 3D Form Index (FI_{3DN}); 3D AI (AI_{3DN}); and 3D Micro-Texture index (MT_{3DN}). The study revealed that slag particles were more complex and irregular (higher 3D index values), retaining these characteristics longer than granite during the micro-Deval abrasion test, as shown in Figure 4.5, corresponding to different test stages (0, 2,000, and 14,000 revolutions). The 3D analysis further indicated lower surface wear in slag aggregate than in granite, across different ranges of surface curvatures.

Results from the SDT, following the methodology for testing CACDW, BAS, and LIM, were presented by Roque, Rodrigues, and da Silva (2020). Comparing the Id2 values with Gamble's classification, ISAC belongs to the class of very high durability (Id2 ≥ 98.9%). The Id values show that increasing the number of rotations (from 400 to 1,000) and extending the contact with wetting solutions (15-day immersion before SDT) did not change ISAC's durability classification

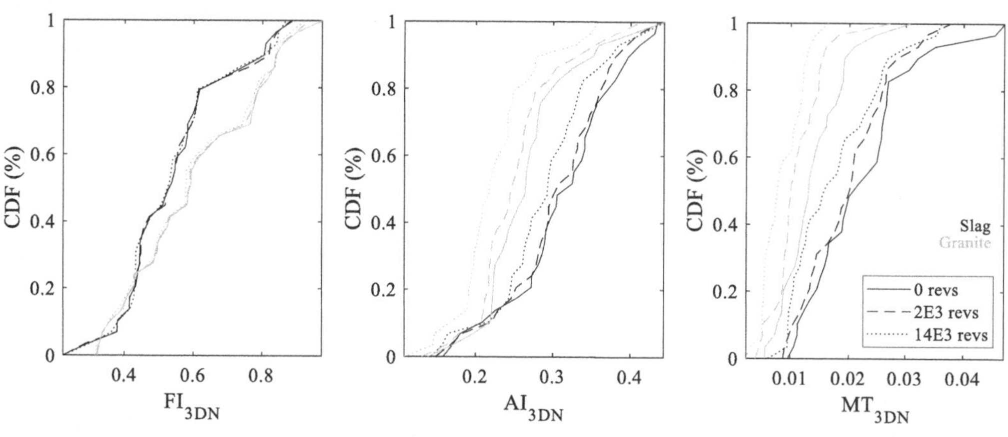

Figure 4.5 Cumulative distributions (CDF) of indices FI_{3DN}, AI_{3DN} and MT_{3DN} determined for two sets of railway ballast particles: granite and EAF slag (Paixão and Fortunato 2021).

according to Gamble's classification. The comparative analysis of the durability of CACDW and ISAC shows that, under similar testing conditions, Id values were slightly higher for ISAC.

4.4.3 *Mechanical and hydraulic performance*

4.4.3.1 *Laboratorial characterization*

The mechanical performance of aggregates for granular layers of transport infrastructures is often analysed in terms of resilient behaviour and permanent deformation to prevent excessive pavement deformation or cracks due to fatigue. ISAC has been compared with natural (igneous or sedimentary rocks) or artificial aggregates (such as recycled concrete) using EN 13286-7 for cyclic load triaxial testing. Figure 4.6a presents this comparison in terms of the permanent axial strain (ε_1^c, calculated as the difference between the permanent axial strain at the end of the conditioning and after the first 100 cycles) and the resilient modulus (E_c, defined for a stress state of 250 kPa mean effective stress and 500 kPa deviatoric stress).

Delgado et al. (2021) extended this study to assess the permanent deformation performance of ISAC aggregates after one and two million load cycles, using a 4/30 mm grain size distribution corresponding to scaled-down railway ballast. The authors investigated the shakedown behaviour and compared ISAC and granite aggregates with the same grain size distribution under cyclic triaxial loading. Figure 4.6b depicts the permanent vertical strain rate plotted against the permanent vertical strain, highlighting that for higher stress levels (II) a higher shakedown tendency was observed for ISAC under a relatively high number of cycles (approximately 2E6), compared to the granite aggregate, which tends to remain in a plastic creep regime.

Marsal's breakage index (B_g) was also analysed after long-term tests, comparing ISAC with granite aggregate. Notably, granite aggregates exhibit a slightly higher particle breakage index

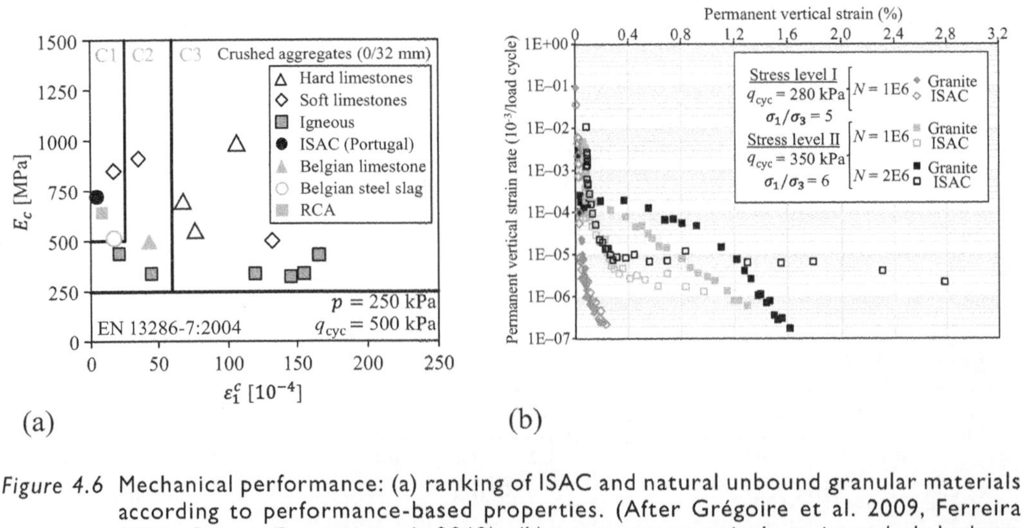

Figure 4.6 Mechanical performance: (a) ranking of ISAC and natural unbound granular materials according to performance-based properties. (After Grégoire et al. 2009, Ferreira 2010, Gomes-Correia et al. 2012); (b) permanent vertical strain and shakedown occurrence (after Delgado et al. 2021).

compared to ISAC at equivalent stress levels. While ISAC presented a maximum B_g of 6%, the granite aggregate showed a slightly higher B_g index value of approximately 8%. This aspect helps explain the plastic creep regime identified in the granite aggregate during the permanent deformation tests (Figure 4.6b). Figure 4.7 shows the resilient moduli obtained by Delgado et al. (2021) as a function of the confining pressure (σ_3) before and after the long-term permanent deformation tests. For all stress paths applied before the permanent deformation tests, ISAC presented higher values of the resilient modulus compared to the granite aggregate.

In terms of shear strength, monotonic triaxial tests were also performed on ISAC, demonstrating very high strength parameters in agreement with results from other EAF slags (Table 4.2). This is due to high grain size, particle angularity, and particle strength leading to significant interlocking and dilatancy. Although granular materials typically present a nonlinear strength envelope over a large range of effective stresses, a linear envelope was assumed for the range of confining stresses analysed to obtain the results in Table 4.2. An artificial cohesion intercept is included, though the materials lack cohesion due to the absence of fine particles or cementation.

Permeability is another important characteristic for the application of these materials in transport infrastructures. Permeability tests were conducted on ISAC specimens with the same grain size distribution as the other materials presented before (CACDW and natural aggregates: BAS and LIM) for a proper comparison. The preparation of the specimens and the permeability tests were carried out according to the procedures used with the other three aggregates. Table 4.3 presents the

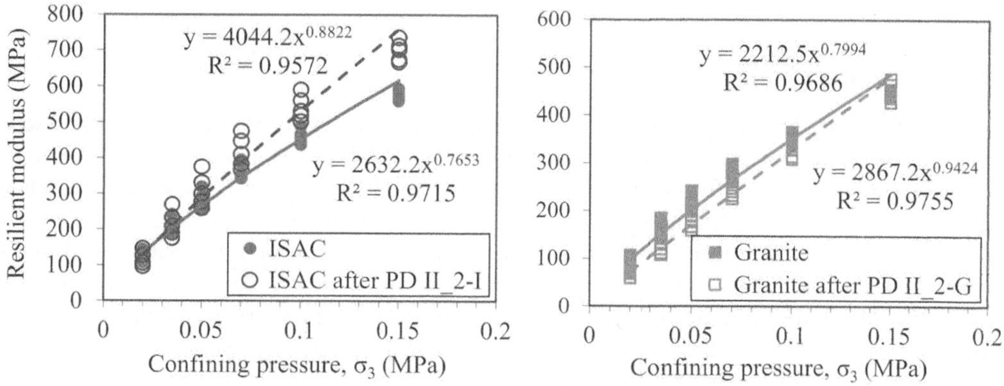

Figure 4.7 Influence of the confining pressure on the resilient moduli before and after 2E6 load cycles (Delgado et al. 2021).

Table 4.2 Strength parameters obtained in EAF slags, including ISAC

Reference	*Grain size range [mm]*	*Isotropic consolidation effective stress [kPa]*	*Triaxial testing*	
			ϕ'_{peak}[°]	c'[kPa]
Maghool et al. (2019)	0.4–20	15–60	55	1
		30–120	50	17
		60–240	42	69
Delgado et al. (2019)	4.0–30	20–85	44	11
Jia, Markine, and Jing (2021)	0.1–25	50–200	47	31
Alves et al. (2023)	0–30 (8% fines)	20–70	64	85

Table 4.3 Results of the permeability tests obtained with ISAC

Grain size range	*Time (day)*		*Effluent volume (l)*		*k × 10^{-3} (m/s)*	
	Phase 1	*Phase 2*	*Phase 1*	*Phase 2*	*Phase 1*	*Phase 2*
Sand (0.25–2 mm)	28	12	255	30	0.2	0.05
Fine gravel (2–20 mm)	17	11	270	40	1.9	1.9

Source: Adapted from Roque, da Silva, and Almeida (2022)

values for *k*, as well as the total test time and the total volume of effluent solution collected during Phase 1 and Phase 2. Considering the *k* of sand, the fine gravel of ISAC and the results in Table 4.1 for the *k* of CACDW, BAS, and LIM, the sequence is $k_{BAS} > k_{CACDW} > k_{ISAC} > k_{LIM}$.

Considering the previously addressed values of *k* typically required for the drainage layers of road pavements, only ISAC fine gravel, or coarser grain size, could be used. For the inside face of retaining walls, which typically requires $k > 1 \times 10^{-3}$ m/s, the minimum requirement was fulfilled with both grain size ranges of ISAC. As with CACDW, there was no evidence of tufa formation or recementation phenomena with ISAC. However, Boyer (1994) reports the possibility of this recycling reducing the drainage capacity of UG layers and associated drainage structures due to tufa formation. It is important to control the percentage of fines and reactive elements responsible for tufa formation and recementation and to adequately process and handle these recycled aggregates (Roque, da Silva, and Almeida 2022).

4.4.3.2 *In situ evaluation*

The construction of test sections and the monitoring of their behaviour play an essential role in evaluating the in-service performance of structures using non-traditional materials. In a research project involving ISAC, a full-scale field trial embankment measuring 60 m in length and up to 3.5 m in height was built, characterized, instrumented, and monitored (Gomes Correia et al. 2012). This trial included three distinct sections: (i) a section with granite residual soil in the embankment and capping layer and crushed granite aggregate in the base layer; (ii) a section with granite residual soil in the embankment and capping layer and ISAC in the base layer; and (iii) a section with ISAC in the embankment, capping layer and base layer.

Tests conducted during construction, for approximately two and a half years post-completion, revealed that layers incorporating ISAC exhibited superior mechanical performance compared to layers with natural materials (soil and crushed granite aggregate). Deflections measured with the FWD on the surface of areas built with ISAC were about five times less than those recorded on areas constructed with natural materials.

Approximately ten years after the field trial construction, a visual inspection and testing campaign was conducted to assess the pavement's mechanical performance (Fortunato, Roque, and Correia 2023). FWD tests performed on various pavement areas using two load levels (30 kN and 47 kN) on a 0.45 m diameter plate indicated lower deflections in the zones with ISAC, signifying superior pavement performance. This effect was more pronounced at higher load values, underscoring ISAC's low deformability, as higher loads engage deeper pavement layers (base and capping layers). Deflections in ISAC zones were significantly lower than those measured ten years prior (during construction), whereas deflections in other zones increased over time. This suggests that the pavement structure in ISAC zones became more rigid over time, while it became more flexible in other zones (Gomes Correia et al. 2012). Maximum deformations of 0.25% were recorded ten years after embankment construction in both the ISAC and the granite residual soil sections, indicating good embankment performance (Fortunato and Veiga Pinto 1998). Pavement cracks, mainly longitudinal cracks and alligator cracking, were less severe and less frequent in ISAC zones. It can be concluded that ISAC possesses similar or superior characteristics compared to natural materials.

4.5 Concluding remarks

Civil engineering works, such as construction, reconstruction, rehabilitation, conservation, and demolition, often generate significant quantities of waste referred to as CDW. The large amounts of waste resulting from these activities are a problem that raises concerns at national and international level. The disposal of this waste, either illegally or in uncontrolled landfills, presents high environmental impacts, in addition to promoting the reduction in our planet's limited natural resources. Then, there is the awareness of the need to reuse and recycle this waste, with the final objective of eliminating landfilling. The use of recycled and alternative aggregates in geotechnical works and transport infrastructure construction, such as roads and railways, is a solution with environmental, economic, and social benefits. Proper identification, characterization, and evaluation of these materials and the presence of legislation are essential to support and promote confidence in the application of these secondary raw materials.

Taking account of current challenges and possible contributions regarding the promotion of CDW use, this chapter presented and discussed several applications of CDW for unbound and HBM and drainage structures, as well as of steel slag aggregates for unbound mixtures, within a circular economy approach. Generally, it was concluded that recycled materials from CDW demonstrated suitable behaviour in terms of deformability modulus, supporting the application of recycled materials from this type of waste. The environmental impact regarding the control of potential contamination of surface water, groundwater, and soil during the use of recycled materials also presented satisfactory results. Studies on HBM using organic by-products were also referred to, leading to the conclusion that they are highly durable, and so their use in road pavements is an advantage. Regarding industrial by-products, such as steel slags, their application in transport infrastructures is a solution for managing these materials. An extended study with certified ISAC was presented, which led to the conclusion that this material possesses similar or superior characteristics compared to natural materials.

Acknowledgements

José Neves acknowledges "Fundação para a Ciência e a Tecnologia" (FCT), in Portugal, because part of the research activity carried out at Civil Engineering Research and Innovation for Sustainability (CERIS) has been funded by FCT in the framework of project UIDB/04625/2020.

Reference list

Abu El-Maaty, A. E. 2013. Utilization of cement treated recycled concrete aggregates as base or subbase layer in Egypt. *Ain Shams Eng J* 4, no. 4: 661–73. https://doi.org/10.1016/j.asej.2013.02.005.

Agrela, F., A. Barbudo, A. Ramírez, J. Ayuso, M. D. Carvajal, and J. R. Jiménez. 2012. Construction of road sections using mixed recycled aggregates treated with cement in Malaga, Spain. *Resour Conserv Recy* 58 (January): 98–106. https://doi.org/10.1016/j.resconrec.2011.11.003.

Alves, R., S. Rios, E. Fortunato, A. V. da Fonseca, and B. G. Delgado. 2023. Mechanical behaviour of steel slag–rubber mixtures: Laboratory assessment. *Sustainability* 15, no. 2: 1563. https://doi.org/10.3390/su15021563.

Boyer, B. W. 1994. Alkaline leachate and calcareous tufa originating from slag in a highway embankment near Baltimore, Maryland. *Transport Res Rec* 1434: 3–7.

Bruinsma, J. E., K. R. Peterson, and M. B. Snyder. 1997. Chemical approach to formation of calcite precipitate from recycled concrete aggregate base layers. *Transpor Res Rec* 1577, no. 1: 10–17. https://doi.org/10.3141/1577-02.

Cabaço, L. 2017. *Relatório nacional sobre a implementação da Agenda 2030 para o Desenvolvimento Sustentável Portugal*. Ministério dos Negócios Estrangeiros, República Portuguesa. Relatório Nacional Voluntário (in Portuguese).

CDOT. 2019. Seepage & groundwater. In *Drainage Design Manual*. Colorado Department of Transportation (CDOT), Division of Engineering and Maintenance, Denver, CO.

Crucho, J., L. Picado-Santos, and J. Neves. 2021. Assessment of a cement bound granular mixture using recycled concrete aggregate and coconut fiber. In *Airfield and Highway Pavements 2021*, ed. H. Ozer, J. F. Rushing, and Z. Leng, 246–57. ASCE, Reston, VA.

Crucho, J., L. Picado-Santos, and J. Neves. 2022. Mechanical performance of cement bound granular mixtures using recycled aggregate and coconut fiber. *Appl Sci* 12, no. 4: 1936. https://doi.org/10.3390/app12041936.

Delgado, B. G., A. V. da Fonseca, E. Fortunato, and L. M. G. da Motta. 2022. Particle morphology's influence on the rail ballast behaviour of a steel slag aggregate. *Environ Geotech* 9, no. 6: 373–82. https://doi.org/10.1680/jenge.18.00203.

Delgado, B. G., A. V. da Fonseca, E. Fortunato, and P. Maia. 2019. Mechanical behavior of inert steel slag ballast for heavy haul rail track: Laboratory evaluation. *Transp Geotech* 20 (September): 100243. https://doi.org/10.1016/j.trgeo.2019.100243.

Delgado, B. G., A. V. da Fonseca, E. Fortunato, A. Paixão, and R. Alves. 2021. Geomechanical assessment of an inert steel slag aggregate as an alternative ballast material for heavy haul rail tracks. *Constr Build Mater* 279 (April): 122438. https://doi.org/10.1016/j.conbuildmat.2021.122438.

DTMR. 2018. *Transport and main roads specifications. MRTS06 reinforced soil structures*. Department of Transport and Main Roads (DTMR), Queensland, Australia.

EAPA. 2014. *The use of warm mix asphalt*. EAPA Position Paper. EAPA – European Asphalt Pavement Association.

EAPA. 2022. *Asphalt in* figures 2022. European Asphalt Pavement Association. https://eapa.org/asphalt-in-figures-2022-view/

EC. 1999. *ALT-MAT: Alternative materials in road construction*. Final Report for Publication, ALT-MAT Contract No.: RO-97-SC.2238. https://cordis.europa.eu/project/id/RO-97-SC.2238 (accessed June 20, 2024).

EC. 2011. *Final report summary – DIRECT_MAT (DISmantling and RECycling Techniques for road MATerials – Sharing knowledge and practices)*.

EC. 2023. *ReEngineering asphalt mixture for circular road PAVements project*. https://cordis.europa.eu/project/id/101033561/es (accessed June 20, 2024).

ECTP. 2019. *ETPC materials & sustainability*. Horizon Europe (2021–2027) Position Paper.

EEAGrants. 2021. *ReBuild17 – Projeto para a promoção da economia circular no setor da construção*. https://www.eeagrants.gov.pt/pt/programas-mfeee-2014-2021/ambiente/noticias/projeto-rebuild-17/ (accessed June 18, 2024).

EU. 2019. European Commission. *Communication from the Commission to the European Parliament, the European Council, the Council, the European and Economic and Social Committee and the Committee of the Regions*. The European Green Deal.

EU. 2024. *European Commission EU 'whole-of-government' approach (europa.eu)*. https://commission.europa.eu/strategy-and-policy/sustainable-development-goals/eu-whole-government-approach_en (accessed June 20, 2024).

Eurostat. 2023. *Freight transport statistics – modal split*. https://ec.europa.eu/eurostat/statistics-explained/index.php?title=Freight_transport_statistics_-_modal_split (accessed May 23, 2024).

Ferreira, S. 2010. *Comportamento mecânico e ambiental de materiais granulares: aplicação às escórias de aciaria nacionais*. PhD diss., Universidade do Minho (in Portuguese).

FHWA. 2016. *User guidelines for waste and byproduct materials in pavement construction.* Report FHWA-RD-97-148. U.S. Department of Transportation. Federal Highway Administration.

Fortunato, E., and A. V. Pinto. 1998. Structural behaviour of soil-rockfill mixtures embankments. In *The Geotechnics of Hard Soils-Soft Rocks,* Vol. 1, ed. A. Evangelista and L. Picarelli, 1101–7. CRC Press, Rotterdam.

Fortunato, E., A. J. Roque, and A. Gomes-Correia. 2023. Case study of an inert steel aggregate in road construction: Characterization and monitoring of the structural behaviour. In *Advances in Geoengineering, Geotechnologies, and Geoenvironment for Earth Systems and Sustainable Georesources Management,* ed. H. I. Chaminé and J. A. Fernandes, 57–61. Springer, Cham. https://doi.org/10.1007/978-3-031-25986-9_9.

Freire, A. C. and A. Maia. 2019. Estado da arte ibero-brasileira sobre a aplicação de agregados de escórias de aciaria em misturas betuminosas. *Geotecnia* 146 (July): 19–50. https://doi.org/10.24849/j.geot.2019.146.02.

Freire, A. C., J. Neves, A. J. Roque, I. M. Martins, and M. L. Antunes. 2021. Feasibility study of milled and crushed reclaimed asphalt pavement for application in unbound granular layers. *Road Mater Pavement* 22, no. 7: 1500–20. https://doi.org/10.1080/14680629.2019.1701539.

Gamble, J. C. 1971. *Durability-plasticity classification of shales and other argillaceous rocks.* PhD diss., University of Illinois at Urbana-Champaign.

Gates, L., E. Masad, R. Pyle, and D. Bushee. 2011. *Aggregate Imaging Measurement System 2 (AIMS2): Final report.* Federal Highway Administration (FHWA), Washington, DC.

Gomes-Correia, A., A. J. Roque, S. M. Reis Ferreira, and E. Fortunato. 2012. Case study to promote the use of industrial byproducts: The relevance of performance tests. *J ASTM Int* 9, no. 2: 1–18. https://doi.org/10.1520/JAI103705.

Gomes-Correia, A., M. G. Winter, and A. J. Puppala. 2016. A review of sustainable approaches in transport infrastructure geotechnics. *Transp Geotech* 7 (June): 21–28. https://doi.org/10.1016/j.trgeo.2016.03.003.

Grégoire, C, B. Dethy, J. Detry, and A. Gomes-Correia. 2009. Characterizing natural and recycled granular materials for (sub) base layers of roads by cyclic triaxial testing. In Tutumluer, Erol, AL-Qadi, Imad L., (Eds.), *8th Int Conf Bearing Capacity of Roads, Railways, and Airfields,* 215–23. University of Illinois at Urbana-Champaign, Taylor & Francis, London.

Heydari, S., A. Hajimohammadi, N. H. Seyed Javadi, and N. Khalili. 2021. The use of plastic waste in asphalt: A critical review on asphalt mix design and Marshall properties. *Constr Build Mater* 309 (November): 125185. https://doi.org/10.1016/j.conbuildmat.2021.125185.

IP. 2014. *Caderno de Encargos Tipo de Obra.* Infraestruturas de Portugal (in Portuguese), Almada, Portugal.

ISRM. 1977. *Suggested Method for Determination of the Slake Durability Index - Part 2. Commission on Testing Methods,* 92–94. Pergamon Press, London.

Jia, W., V.L. Markine, and G. Jing. 2021. Analysis of furnace slag in railway sub-ballast based on experimental tests and DEM simulations. *Constr Build Mater* 288 (June): 123114. https://doi.org/10.1016/j.conbuildmat.2021.123114.

Kalman, B, V. Mouillet, K. Mollenhauer, A. Enell, E. Nielsen, and S. Werkmeister. 2013. *Re-road: End of life strategies of asphalt pavements.* Reroad. Fehrl. Org, Sweden.

Kim, J., B. Nam, Z. Behring, and B. Al Muhit. 2014. Evaluation of recementation reactivity of recycled concrete aggregate fines. *Transport Res Rec* 2401, no. 1: 44–51. https://doi.org/10.3141/2401-05.

Liu, Y., P. Su, M. Li, Z. You, and M. Zhao. 2020. Review on evolution and evaluation of asphalt pavement structures and materials. *J Traffic Transp Eng* 7, no. 5: 573–99. https://doi.org/10.1016/j.jtte.2020.05.003.

Maghool, F., A. Arulrajah, C. Suksiripattanapong, S. Horpibulsuk, and A. Mohajerani. 2019. Geotechnical properties of steel slag aggregates: Shear strength and stiffness. *Soils Found* 59, no. 5: 1591–601. https://doi.org/10.1016/j.sandf.2019.03.016.

Masad, E., S. Saadeh, T. Al-Rousan, E. Garboczi, and D. Little. 2005. Computations of particle surface characteristics using optical and X-ray CT images. *Comp Mater Sci* 34, no. 4: 406–24. https://doi.org/10.1016/j.commatsci.2005.01.010.

McCulloch, T., D. Kang, R. Shamet, S. J. Lee, and B. H. Nam. 2017. Long-term performance of recycled concrete aggregate for subsurface drainage. *J Perform Constr Fac* 31, no. 4: 04017015. https://doi.org/10.1061/(ASCE)CF.1943-5509.0000994.

Mica, N., S. Rios, A. V. da Fonseca, and E. Fortunato. 2024. The use of steel slags in transport infrastructures: A critical review. *Geotech Test J* 47, no. 1: GTJ20230297. https://doi.org/10.1520/GTJ20230297.

Mohajerani, A., J. Vajna, T. H. H. Cheung, H. Kurmus, A. Arulrajah, and S. Horpibulsuk. 2017. Practical recycling applications of crushed waste glass in construction materials: A review. *Constr Build Mater* 156 (December): 443–67. https://doi.org/10.1016/j.conbuildmat.2017.09.005.

Neves, J. 2023. Review on laboratory testing for hydraulically bound mixtures used in road applications. In *Ad Geoengineering, Geotechnologies, and Geoenvironment for Earth Systems and Sustainable Georesources Management*, ed. H. I. Chaminé and J. A. Fernandes, 63–67. Springer, Cham. https://doi.org/10.1007/978-3-031-25986-9_10.

Neves, J., A.C. Freire, I. Qamhia, I.L. Al-Qadi, and E. Tutumluer. 2023. Full-scale accelerated pavement testing and instrumentation. In *Ad Testing and Experimentation in Civil Engineering*, ed. C. Chastre, J. Neves, D. Ribeiro, M. G. Neves, and P. Faria, 163–86. Springer, Cham. https://doi.org/10.1007/978-3-031-05875-2_7.

Paixão, A., and E. Fortunato. 2021. Abrasion evolution of steel furnace slag aggregate for railway ballast: 3d morphology analysis of scanned particles by close-range photogrammetry. *Constr Build Mater* 267 (January): 121225. https://doi.org/10.1016/j.conbuildmat.2020.121225.

Pasetto, M. 2000. The re-utilisation of discarded building materials in cement-stabilised layers of road and airfield pavements. In *Waste Management Series*, Vol. 1: 548–66. Elsevier. https://doi.org/10.1016/S0713-2743(00)80066-5.

Pasetto, M., and N. Baldo. 2013. Cement bound mixtures with metallurgical slags for road constructions: Mix design and mechanical characterization. *Inżynieria Mineralna* 14, no. 2: 15–20.

Pasetto, M., and N. Baldo. 2015. Experimental analysis of hydraulically bound mixtures made with waste foundry sand and steel slag. *Mater Struct* 48, no. 8: 2489–503. https://doi.org/10.1617/s11527-014-0333-4.

Pasetto, M., and N. Baldo. 2016. Recycling of waste aggregate in cement bound mixtures for road pavement bases and sub-bases. *Constr Build Mater* 108: 112–18. https://doi.org/10.1016/j.conbuildmat.2016.01.023.

Pérez, P., F. Agrela, R. Herrador, and J. Ordoñez. 2013. Application of cement-treated recycled materials in the construction of a section of road in Malaga, Spain. *Constr Build Mater* 44 (July): 593–99. https://doi.org/10.1016/j.conbuildmat.2013.02.034.

Plati, C. 2019. Sustainability factors in pavement materials, design, and preservation strategies: A literature review. *Constr Build Mater* 211 (June): 539–55. https://doi.org/10.1016/j.conbuildmat.2019.03.242.

Poulikakos, L. D., Papadaskalopoulou, C., Hofko, B., Gschösser, F., Falchetto, A.C., Bueno, M., Arraigada, M., Sousa, J., Ruiz, R., Petit, C. and Loizidou, M.. 2017. Harvesting the unexplored potential of European waste materials for road construction. *Resour Conserv Recycl* 116 (January): 32–44. https://doi.org/10.1016/j.resconrec.2016.09.008.

Qamhia, I., E. Tutumluer, H. Ozer, H. Shoup, S. Beshears, and J. Trepanier. 2019. Evaluation of chemically stabilized quarry byproduct applications in base and subbase layers through accelerated pavement testing. *Transport Res Rec* 2673, no. 3: 259–70. https://doi.org/10.1177/0361198118821099.

Roadmap. 2022. *Materials 2030 roadmap*. https://www.ami2030.eu/wp-content/uploads/2022/12/2022-12-09_Materials_2030_RoadMap_VF4.pdf (accessed May 23, 2024).

Roque, A. J., P. F. da Silva, and R. P. M. Almeida. 2022. Recycling of crushed concrete and steel slag in drainage structures of geotechnical works and road pavements. *J Mater Cycles Waste* 24, no. 6: 2385–400. https://doi.org/10.1007/s10163-022-01486-7.

Roque, A. J., F. Castro, A. Gomes-Correia, S. Silva, and A. Cavalheiro. 2010. Laboratory and field leaching tests for predicting the environmental impact of Portuguese steel slag. In *Proc 6th Int Cong Environmental Geotechnics*, New Delhi, India, 1166–71.

Roque, A. J., I. M. Martins, A. C. Freire, J. M. Neves, and M. L. Antunes. 2016. Assessment of environmental hazardous of Construction and Demolition Recycled Materials (C&DRM) from laboratory and field leaching tests application in road pavement layers. *Procedia Engineering* 143: 204–11. https://doi.org/10.1016/j.proeng.2016.06.026.

Roque, A. J., G. M. Rodrigues, and P. F. da Silva. 2020. Re-cycling of construction and demolition waste and steel slag: Characterization of the durability. *J Mater Cycles Waste* 22, no. 6: 1699–711. https://doi.org/10.1007/s10163-020-01061-y.

SAMARIS. 2004. *Literature review of recycling of by products in road construction in Europe. Deliverable 5* (SAM-06-DE05). Sustainable and Advanced Materials for Road Infrastructure.

Stehlík, D., P. Hyzl, and M. Varaus. 2016. Waste building materials to roads in Czech Republic. *App Mech Mater* 832 (April): 10–17. https://doi.org/10.4028/www.scientific.net/AMM.832.10.

UEPG. 2024. *European Aggregates Association (UEPG)*. https://www.aggregates-europe.eu/ (accessed June 20, 2024).

UIC. 2021. *UIC sustainability. Circular practices in the railway and ways forward.* REUSE. Project final Report, June 2021. International Union of Railways (UIC).

UN. 2015. United Nations. *Transforming our world: The 2023 Agenda for sustainable development.* A/RES/70/1. United Nations. sustainabledevelopment.un.org.

Van Dam, T. J., Harvey, J., Muench, S.T., Smith, K.D., Snyder, M.B., Al-Qadi, I.L., Ozer, H., Meijer, J., Ram, P., Roesler, J.R. and Kendall, A.. 2015. *Towards sustainable pavement systems: A reference document.* United States. Federal Highway Administration.

Walters, C.E., and J. Edwards. 2009. *Increasing the use of hydraulically bound mixtures in construction.* Final Report. WRAP, Banbury.

World Steel Association. 2020. *2020 World Steel in Figures*. Worldsteel, Beijing.

Xuan, D., L. J. M. Houben, A. A. A. Molenaar, and Z. Shui. 2010. Cement treated recycled demolition waste as a road base material. *J Wuhan Univ Technol* 25, no. 4: 696–99. https://doi.org/10.1007/s11595-010-0073-4.

Zhao, Y, D. Goulias, and D. Peterson. 2021. Recycled asphalt pavement materials in transport pavement infrastructure: Sustainability analysis & metrics. *Sustainability* 13, no. 14: 8071. https://doi.org/10.3390/su13148071.

Chapter 5

Eco-design and life-cycle assessment

Paulo J. Venda Oliveira, António A. Santos Correia, Helena Gervásio, and Ana Heitor

5.1 Introduction

The construction sector is responsible for the mining and extraction of close to 50% of all resources worldwide. This contributes to about 40% of greenhouse gas emissions and about 1/3 of the total waste being produced in the EU (COM(2011)571). For these reasons, it is one of the key sectors in the pursuit of decarbonization targets and other sustainability goals.

Geotechnical works often involve significant disturbance to natural landscapes, including excavation, land clearing, and soil changes and waterways. Moreover, the potential environmental impacts of a geotechnical asset, or any other construction asset, occur over their complete life cycle, e.g., from construction, service, and maintenance, and until it reaches its end of life. Hence, the evaluation of all potential impacts requires approaches such as eco-design and life-cycle analysis (LCA). Eco-design refers to the process of incorporating environmental considerations and principles into the design and development of products, processes, and systems, being the primary goal to minimize environmental impacts throughout the complete life cycle of the product or system. Therefore, one key aspect of eco-design is consideration of LCA, which is a systematic procedure to quantify the potential impacts of a product or system throughout its entire life cycle, e.g., from extraction of raw materials to disposal. LCA may include not only environmental indicators but also social and economic indicators, thus fulfilling the holistic concept of sustainability.

LCA is becoming popular among the scientific community; however, in practice, the evaluation of the sustainability indices of construction works usually relies on rating systems such as CEEQUAL or the EGIs system. These types of tools are voluntary certification schemes, usually based on the evaluation of selected criteria by comparing the performance of the construction work with predefined thresholds or reference values. Quantitative and qualitative indicators are then translated into grades that are further aggregated into a final score. The main drawback of these systems is that they are not comparable due to several disparities in terms of system boundaries, indicators reference values and calculation methods.

On the other hand, CEN/TC350 is the technical committee mandated to develop horizontal standards for the sustainability assessment of construction works (buildings and civil engineering works), focusing on LCA. In this case, the most relevant standards that may apply to geotechnical works are EN 15643:2021 and EN 17472:2022. The former establishes the framework and the general requirements for sustainability assessments of civil engineering works, which includes geotechnical works. While the latter sets the requirements and provides specific methods for the assessment of environmental, economic and social performances of a civil engineering work, considering simultaneously its functionality and technical characteristics.

DOI: 10.1201/9781003530220-6

The above standards provide a standardized method, enabling comparability of alternative options and supporting decision-making.

In summary, eco-design and LCA are essential for geotechnical works to minimize environmental, economic and social impacts, conserve natural resources, and promote sustainable development practices, ultimately contributing to the resilience and sustainability of infrastructure projects in the face of environmental challenges.

Hence, the main goal of this chapter is to discuss the application of eco-design and life-cycle analysis to transportation geotechnics. The chapter is organized as follows: after this introductory section, a comprehensive state of the art is provided on the use of eco-design and LCA in geotechnical works, focusing on transportation geotechnics; then, a case study is provided to illustrate the application of LCA on a geotechnical work, and finally, conclusions are drawn in the last section of the chapter.

5.2 State of the art

The world is facing unprecedented global crises and challenges. Recent challenges drive the need to urgently find socially acceptable solutions that protect the environment without compromising economic development and social wellbeing, some examples are:

a Climate change (e.g., extreme weather events from hurricanes to flooding and heat waves and wild fires are becoming more frequent and with devastating effects),
b Resource depletion (due to population growth, consuming habits, and industrialization),
c Growing energy demand (driven by economic development, rising population and social wellbeing, and technological developments),
d Waste management (excessive consumption, product packaging, food waste, lack of environmental education/laws, lack of suitable places for waste disposal, and pollution of soil, water, and air),
e Terrorism and cyberattacks (due to instability and conflicts due to poverty and inequality, lack of ethics in corporate management),
f Pandemics (driven by lack of hygiene and health care, large-scale trade, massive traveling of people).

The path toward finding global solutions for such world challenges and crises began in 1987 with the publication of the "Our Common Future" report (Gro Harlem Brundtland, at the time Prime Minister of Norway), where a global agenda for change was proposed for the first time. In the Brundtland Report, the sustainable development was defined as the development that fulfills the needs of the present without compromising the ability of future generations to meet their own needs, ensuring a balance between economic growth, care for the environment, and social wellbeing (Brundtland 1987).

One of the human activities that contribute most to the depletion of natural resources, energy consumption, and greenhouse gas emissions (GHG) is the construction sector, which includes current geotechnical practices (Raymond et al. 2021b). In spite of the global construction sector being a driving force for the world economy (e.g., 13% of the global GDP and employing more than 100 million people, WEC 2024), it is responsible for about 40% of annual global CO_2 emissions. This includes about 7.3% of annual global CO_2 emissions related to the construction of infrastructure (including just the embodied carbon emissions from cement, iron, steel, and aluminum) (Architecture2030 2024). It is also known that additional emissions from materials used in infrastructure, such as soils and earthworks, asphalt, aggregate, PVC, and HDPE piping, among others, are still yet to be fully accounted for. Additionally, the next 25 years are expected to see the largest wave of infrastructure growth in human history: 60%–75% of the infrastructure that will exist in 2050 has yet to be built (Architecture2030 2024; ECESP 2021). This is further exacerbated by the fact that many of the key infrastructure built in the 1950s and 1960s are aging and now approached its end of life with the need for upgrade or replacement (ECESP 2021). It is therefore crucial to find sustainable solutions that may disrupt our current embodied carbon emissions trend (associated with infrastructure construction, including products and materials, from raw material extraction to manufacturing, transportation, and end-of-life disposal or recycling). This is because the embodied carbon emissions of infrastructure projects are locked in once the project is constructed and cannot be taken back or reduced. However, infrastructures also have another important source of carbon emissions, known as operational carbon emissions, related to the use of infrastructure (e.g., energy consumed by vehicles on road). Taking into consideration the very long lifespan (80–100 years) of infrastructure, the operational carbon emissions are normally higher than the embodied carbon emissions of an infrastructure (GIH 2021).

In 2018, the European Union presented a climate-neutral vision for achieving net-zero greenhouse gas (GHG) emissions economy by 2050, in line with the COP21 Paris Agreement (EUCO-29/19 2019). For such ambitions to be fully realized, hundreds of billions of euros of investment and a solid regulatory framework underpinning sustainable financing (EU Taxonomy, European Commission, 2020) are required. This means that sustainability principles should be at the core of all economic activities, including the construction sector, as a general criterion for accessing funding and financial leverage (green finance). Thus, reaching net-zero GHG emissions, economy will require a contribution from the construction sector, which is one of the largest users of global resources and largest GHG emission contributors. While a path has been proposed with regard to buildings (Saieg et al. 2018; Wilberforce et al. 2023; Wu and Skye 2021), very little has been done regarding infrastructure and even less in the geotechnical engineering sector.

Infrastructure comprises the basic physical structures, facilities, equipment, and services needed for the operation of a society or organization, i.e., the facilities and services necessary for an economy to function (PAS-2080 2023). Nevertheless, infrastructure plays a crucial role in any economy, it has a number of specific features that justify the delay in adopting sustainability principles. For instance, they are high capital and cost intensive, requiring significant investments have a very long lifespan (80–100 years). In addition, the sector is highly government dominated, so it deals with public tenders in the form of long-term investments and a small number of stakeholders.

As the geotechnical engineering activities (which includes ground investigation, design, and construction works) occur at an early stage of any construction project, it can influence the sustainability of infrastructure projects positively or negatively by setting a precedent for later project stages (Basu et al. 2014). Thus, it is required for the geotechnical engineers to evaluate the sustainability of current practices and to begin providing the data, tools, and expertise necessary to realize effective gains with respect to sustainability considerations in construction projects (Raymond et al. 2021b). However, the current project decision-making in geotechnical engineering prioritizes minimizing cost. This is mainly due to the absence of sustainability thresholds or benchmarks for routine geotechnical practice (Raymond et al. 2021b). Such benchmarks are essential to mitigate environmental impacts and improve sustainability practices (Das et al. 2018) of any geotechnical engineering activity. Sustainability benchmarks can be developed using eco-design and LCA.

The integration of sustainability principles into the planning, design and construction of transportation infrastructure is one of the emerging trends in contemporary geotechnical engineering. It has become imperative to develop approaches that integrate environmental considerations from the initial planning and design stages to the end of the life cycle of these infrastructures. In this context, eco-design and life-cycle assessment (LCA) are essential tools to guide the development of more sustainable solutions in transportation geotechnics.

The concept of eco-design in transportation geotechnics aims to integrate environmental considerations into the design process not only of materials and products but also of systems and infrastructures. It can be applied in urban planning and finance, infrastructure design, and material selection, as well as construction, operation, and end-of-life stages of infrastructures. The life-cycle assessment (LCA) is a comprehensive and systematic approach for evaluating the environmental impacts of a product or system over its entire life cycle (i.e., from cradle to grave). However, for a complete LCA, it is also necessary to consider economic and social impacts, for instance, by considering the life-cycle cost assessment (LCCA) and life-cycle sustainability assessment (LCSA). LCA approaches are employed to compare alternative products or systems by weighing their trade-offs while minimizing the risk of shifting burdens to other types of impacts, life-cycle stages, geographic regions, or generations. These two concepts are closely interlinked and can be taken as intrinsic parts of a novel methodology based on the life-cycle thinking approach. This novel methodology incorporates the concepts of EU Green Deal and circular economy, and it aims to provide a framework to support decision-makers toward a more sustainable, resilient, and equitable infrastructure design (Pettinaroli et al. 2023a).

In the next subsections, the main findings from previous studies is reviewed in relation to the adoption of eco-design and LCA in geotechnics, particularly in transportation infrastructure.

5.2.1 *Eco-design in transportation geotechnics*

The concept of eco-design in transportation geotechnics is something relatively new. Indeed, a search in the Web of Science™ (papers from 2010 onward) using the words "eco-design" and "transportation geotechnics" revealed only a single paper focusing on such words or topics (Pettinaroli et al. 2023a). When the word "eco-design" is replaced by a broader word such as "sustainability", ten papers appear (i.e., Basu et al. 2014; Correia et al. 2016; Cristelo et al. 2016; Indraratna et al. 2024; Parente et al. 2016; Raza et al. 2021a, b; Soga 2010; Steedman 2010; Vanícek et al. 2016), even though none of them are directly related with eco-design. In a new search using the word "eco-geotechnics," two additional papers are found (Ng et al. 2022; Olinic et al. 2023). This small number of papers demonstrates that the concept of eco-design is new in the field of transportation geotechnics. A brief inspection of the papers reveals that eight are published in international peer-review journals, while the remaining five papers refer to international conferences. A closer inspection of the papers reveals that eleven papers focus on sustainability applied to geotechnics and/or transportation infrastructure. For instance, Basu et al. (2014) review the philosophies and definitions of sustainability in geotechnical engineering, while Steedman (2010) introduces a definition of sustainability for policy makers and engineers, highlighting the importance of technological innovation in the use of resources, energy, and human capital, as well as associated carbon emissions in geotechnical projects. Some examples of sustainability concepts applied to transportation infrastructure are introduced in Vanícek et al. (2016) and Parente et al. (2016). The former paper discusses the benefits of applying a sustainability approach to facilitate the transition to building green, smart, and low-cost infrastructure that is robust enough to guarantee the functionality even under extreme situations. The latter paper explores the advantages of integrating metaheuristics, data mining, and geographic information systems for a sustainable allocation/management of equipment in infrastructure earthworks. Other examples of sustainability implemented through material reuse or recycling in transportation infrastructure are presented in the papers by Correia et al. (2016), focusing on natural geomaterials and recycled aggregates; Indraratna et al. (2024), focusing on coal wash, steel furnace slag, fly ash, and rubber elements from recycled tires; and Cristelo et al. (2016), focusing on construction and demolition waste. Four papers showcase new approaches to assess the sustainability of geotechnical infrastructure. For instance, Soga (2010) developed a methodology using embodied energy and carbon emissions and illustrated those concepts in a case study of a high-speed railway infrastructure in United Kingdom. Raza et al. (2021a) and Raza et al. (2021b) reviewed the available sustainability assessment tools, highlighting their strengths and weaknesses, and introduced a specific tool for assessing sustainability in geotechnical projects—the geotechnical sustainability assessment tool (Geo-SAT). The proposed tool incorporates not only the environmental, economic, and social aspects but also the engineering aspects. Finally, Pettinaroli et al. (2023a) proposed a new holistic sustainability assessment tool based on the life-cycle thinking approach, which incorporates the concept of the EU Taxonomy, the EU Green Deal, and Envision framework to guide decision-makers toward a more sustainable, resilient, and equitable geotechnical infrastructure design of ground improvement work. The last two papers deal with eco-geotechnics, which is an emerging discipline that integrates knowledge from soil mechanics, rock mechanics, engineering geology, ecology, botany, and atmospheric science to achieve overall sustainability. The paper Ng et al. (2022) emphasizes the vital importance of atmosphere–plant–soil interactions, especially in the context of climate change. The paper reviews theoretical frameworks for modeling these interactions, as well as the mechanical and hydrological effects of plants on soil. It also explores the optimization of plant growth and the use of vegetation to control sandstorms and reduce wind erosion, to improve stability of slopes and embankments, and to protect landfill from infiltration and percolation. Finally, the paper by Olinic et al. (2023) reviews the benefits of using different types of plants to control erosion and improve the stability of slopes.

The literature search identified also other documents related to eco-design. This included a book focusing on 'Eco-design of buildings and infrastructure - Developments in the Period 2016–2020' (Peuportier et al. 2020), and some EU initiatives developed under the 'Eco-design for Sustainable Products Regulation framework'. Some examples are the current Eco-design Directive 2009/125/EC (Directive-125/EC 2009), which will be replaced by the proposal COM(2022) 142 Final (COM/2022/142-final 2022), as well as the new 'Digital Product Passport' under preparation.

Peuportier et al. (2020) suggest that eco-design must consider all the stages of a product's life, e.g., manufacturing and distribution, use, maintenance, end of life, and recycling. In this way, it is possible to make the product intrinsically sustainable. For geotechnical assets, it is critical to embed such considerations throughout the asset service life, which is often 100 years or more. Thus, the eco-design of any infrastructure must include not only the construction stage but also the service and refurbishing or deconstruction stages. The book also emphasizes the importance of considering the challenges of urban and territorial planning in relation to strategic infrastructure due to the need to conciliate the protection of the environment and biodiversity with the functions of the infrastructure.

Regarding EU initiatives on eco-design, it is apparent that the current product design does not sufficiently consider the environmental impacts over the life cycle, including circularity aspects. In addition, it is still too difficult for economic players and citizens to make sustainable choices in relation to products they procure due to a lack of information and affordable options, and the current product legislative framework is not too ambitious (COM/2022/142-final 2022). Thus, new requirements must be applied to (eco-)design of products, aiming to significantly strengthen its sustainability. Examples of this could include the use of the digital product passport and ensure high-quality recycling of products as well as increase their recycled content. This in turn can ensure value retention, foster circular economy practices, and improve availability of information on the sustainability footprint of products. While the roll out of these measures likely imply increased costs for both product manufactures and consumers, they can be offset by financial savings in the long term.

In summary, the concept of eco-design in transportation geotechnics is a broad approach that aims to integrate environmental considerations into the design process not only of materials and products but also of systems and infrastructure. When it is applied to transportation infrastructure, five actions should be considered:

1 Use smart urban and territorial planning by promoting a conciliation between the protection of the environment and biodiversity, people, and with the functions of the infrastructure;
2 Prioritize existing infrastructure and materials by reusing and upgrading, including the materials incorporated in old legacy infrastructure (e.g., urban/infrastructure mines);
3 Design the infrastructure for the future by prioritizing adaptable and flexible infrastructure uses while embedding climate resilience and nature-based solutions;
4 Implement and scale up low-carbon construction techniques by optimizing construction machinery and optimize logistics and transportation to and from construction sites;
5 Maximize infrastructure and materials efficiency and emissions reductions by optimizing all infrastructure systems, specifying low-carbon materials and promote a greater use of green energy sources.

These five actions are supported on three pathways that must be facilitated by stakeholders across the life-cycle of the transportation infrastructure sector in order to decarbonize (UN-Environment-Programme 2023):

a AVOID the extraction and production of raw materials, by avoiding waste and build with less and improve circularity;
b SHIFT to regenerative and/or nature-based materials wherever possible, by promoting innovation and research;
c IMPROVE nonrenewable infrastructure materials and processes, by establishing reused and recycled materials supply chains and promoting decarbonization.

Since transportation infrastructure is high capital and cost intensive, requiring significant investments, the financial institutions that support such investments should adopt sustainability criteria in their economic and financial evaluations, i.e., sustainable financing should be one of the driving forces toward sustainability of transportation infrastructure.

Thus, the integration of eco-design in transportation geotechnics could be a powerful tool to build a more sustainable, resilient, and equitable infrastructure. For such, the eco-design should embrace the following stages:

- Use smart urban and territorial planning to carry out a sustainable cost–benefit analysis of the effective need for the infrastructure, evaluating the pros and cons of a new construction against the rehabilitation of an existing one;

- Conduct a preliminary design, considering engineering and sustainable aspects while prioritizing solutions that maximize infrastructure and materials efficiency and emissions reduction. For instance, the route layout of the transportation infrastructure must be optimized, prioritizing resource efficiency, energy use, and traffic management;
- Design the transportation infrastructure assets for the future by prioritizing adaptable and flexible infrastructure uses, considering the need for climate resilience, and choosing nature-based solutions. Preference should be given to choosing materials and products that are recycled, have low carbon emissions, and have a digital passport. The design should be supported by an effective LCA;
- Implement and scale up low-carbon construction techniques by optimizing construction machinery and optimize logistics and transportation to and from construction sites. Tenders and funding should favor sustainable companies and the adoption of sustainable construction solutions;
- Accounting for the very long lifespan (80–100 years) of the transportation geotechnics infrastructure, optimize maintenance costs, and maximize uses and carbon emission reductions;
- Consider the end of life of the transportation infrastructure assets and its deconstruction. The materials incorporated into the existing infrastructure must be able to be mobilized/valued in recycling/reuse operations (infrastructure mines), i.e., promoting circular economy.

More recently, some examples of sustainable practices associated with the materials, products, or systems applied in the transportation geotechnical sector have been proposed, as follows:

- The use of recycled materials, such as industrial by-products (e.g., blast furnace slag and fly ash), recycled aggregates from mining old infrastructure and construction and demolition waste, to replace conventional materials (e.g., natural aggregates) (Correia et al. 2016; Cristelo et al. 2016; Indraratna et al. 2024; Puppala 2016; Vanícek et al. 2016; Venda Oliveira et al. 2013; Venda Oliveira et al. 2022). For example, in road construction, the use of recycled aggregates can significantly reduce the need to extract virgin materials and the generation of waste;
- Soil stabilization using biogeotechnical methods, which involve the use of plants, biopolymers, and microorganisms to improve soil properties, is another promising approach (Armistead et al. 2020; Chang et al. 2020; Correia et al. 2023; Mendonça et al. 2021; Ng et al. 2022; Olinic et al. 2023). This technique not only strengthens the soil and improves infrastructure resilience but also contributes to carbon capture and local biodiversity;
- The development and implementation of sustainable drainage systems, such as permeable pavements, has gained prominence recently (Gu et al. 2024; Husain et al. 2024). These systems allow rainwater to infiltrate the soil, reducing surface run-off and erosion and improving groundwater quality because water is filtrated.
- The use of geothermal energy in pavements of transportation infrastructure to minimize temperature differential in winter and summer, thus reducing maintenance costs (Gu et al. 2022).
- Geosynthetics, widely used in transportation infrastructure, are evolving to incorporate more sustainable materials (Dąbrowska et al. 2023; Palmeira et al. 2021; Vieira 2022; Yoo 2023). Geosynthetics made from recycled plastics or natural fibers are examples of how the industry is adapting traditional products to the needs of eco-design.
- The use of sensors, unmanned aerial vehicles, satellite data, machine learning and artificial intelligence, digital technologies to monitor the structural health, and environmental behavior of geostructures that incorporate transportation infrastructure assets offers a way of optimizing maintenance and ensuring that environmental impacts are minimized over the lifetime of the infrastructure (Bednarski et al. 2021; Du et al. 2020; Flammini et al. 2016;

Maghazei and Steinmann 2020; Parente et al. 2016; Scoular et al. 2020; Suhartomo et al. 2019; Tinoco et al. 2016; Wu et al. 2022).

- The use of renewable energy solutions, such as solar panels and wind turbines, in transportation infrastructure is a trend that is set to grow. This practice not only reduces the environmental impact of infrastructure operation but also contributes to the generation of green energy.
- There are also important research advances in the area of the development of self-healing geotechnical materials that can repair themselves in response to damage, similar to what is already observed in mortars and concretes, which could extend the service life of infrastructures and reduce the need for frequent maintenance (De Belie and Wang 2016; Javeed et al. 2024; Pacheco et al. 2022; Spencer et al. 2020).
- Another future trend is the use of 3D printing technology, which has significant potential to revolutionize the way geostructures incorporated in transportation infrastructure are built, by enabling the use of local materials to manufacture products and systems. This could drastically reduce the transport of materials and, consequently, the associated emissions.

5.2.2 *Life-cycle assessment of geotechnical works*

As stated previously, the concept of life-cycle assessment (LCA) is a comprehensive and systematic approach for evaluating the environmental impacts of products, systems, or infrastructures, over their entire life cycle (e.g., from cradle to grave). However, an LCA is only complete when economic and social impacts are also considered. While the assessment becomes more complex, it is more robust, because it relies on more comprehensive impact categories, albeit some are more subjective. The incorporation of economic impacts into LCA can be incorporated through LCCA (Balasbaneh et al. 2021; Ferrazo et al. 2024b; Miguel et al. 2023; Purdy et al. 2022; Samuelsson et al. 2024). The incorporation of social impacts (Ferrazo et al. 2024a; Rocha et al. 2022) can be done through LCSA (Raymond et al. 2021b). There are other more specific approaches where, for example, dynamic effects are considered instead of a static assessment (dynamic life-cycle assessment, DLCA, Pehnt. 2006), or the environmental impacts of changes in an economic sector usually neglected are analyzed (economic input-output life-cycle assessment, EIO-LCA, Henrickson et al. 1998; Roux 2020). In summary, LCA approaches are used to compare alternative products, services, systems, or infrastructures by weighing their trade-offs while minimizing the risk of shifting burdens to other types of impacts, life-cycle stages, geographic regions, or generations.

As far as infrastructure is concerned, the challenge is to reach a compromise between an adequate and comprehensive approach to sustainability principles and providing a scheme that is understandable and accessible to society, including designers and professionals, decision-makers, and users (Pettinaroli et al. 2023a). Thus, combining and/or integrating economic and social aspects with LCA methodology can be an effective and powerful way to successfully implement infrastructure projects.

Despite the benefits that can be derived from incorporating sustainability principles into the life-cycle assessment of geotechnical assets in transportation infrastructure, this is still an unusual reality nowadays within the geotechnical community. Nonetheless, in recent years, the application of different LCA approaches (or LCA-based methods) in various areas of geotechnical engineering (Table 5.1) has been reported in a several studies. For instance, these include deep foundations, dams, ground improvement, soil stabilization, earth-retaining walls, slope stability, tunnels, embankments, excavations, bio-based methods (e.g., the use of bacteria, enzymes, biopolymers, or similar), geotechnical investigation (field testing), the use of geosynthetics, among other applications. Indeed, an expanded search including the Web of Science™ revealed that 101 works (journal and conference papers, reports, and theses) were published between 2007 and 2024 on the application of environmental sustainability methods to geotechnical works, as described in Table 5.1. Figure 5.1 shows the distribution of these works among the different areas of geotechnical engineering. The data indicate that the majority of published works on the use of LCA-based methods are related to ground improvement and/or soil stabilization works (≈ 27%), followed by earth-retaining walls (≈ 15%) and deep foundations/earthworks (≈ 10%). In the 5%–9% range of published works, applications for slope stabilization, embankments/excavations, tunnels, and bio-based methods are showcased. The data also indicates that LCA has rarely been employed in important areas of geotechnical engineering, such as dams, the use of geosynthetics, and geotechnical investigation (both in the laboratory and in the field).

For incorporating LCA approach into geotechnical design, it is important to identify the major four phases of any life cycle assessment. ISO 14040 (2006) and ISO 14044 (2006) suggests that an LCA is carried out in four distinct phases:

a Goal and scope definition
b Life cycle inventory or inventory analysis
c Life cycle impact assessment or impact assessment
d Life cycle interpretation

In this way, the LCA methodology starts with a definition of the goal and scope of the study. The goal of an LCA study is an explicit statement, which sets out the context of the study and explains how and to whom the results are to be communicated. According to ISO standards, a robust goal statement must address several key components. For instance, the intended application (e.g., internal decision-making, public disclosure, or regulatory compliance), reasons for conducting the study (e.g., compare different alternatives, search for improvements), target audience (e.g., users or consumers, regulatory bodies, decision-makers), and public

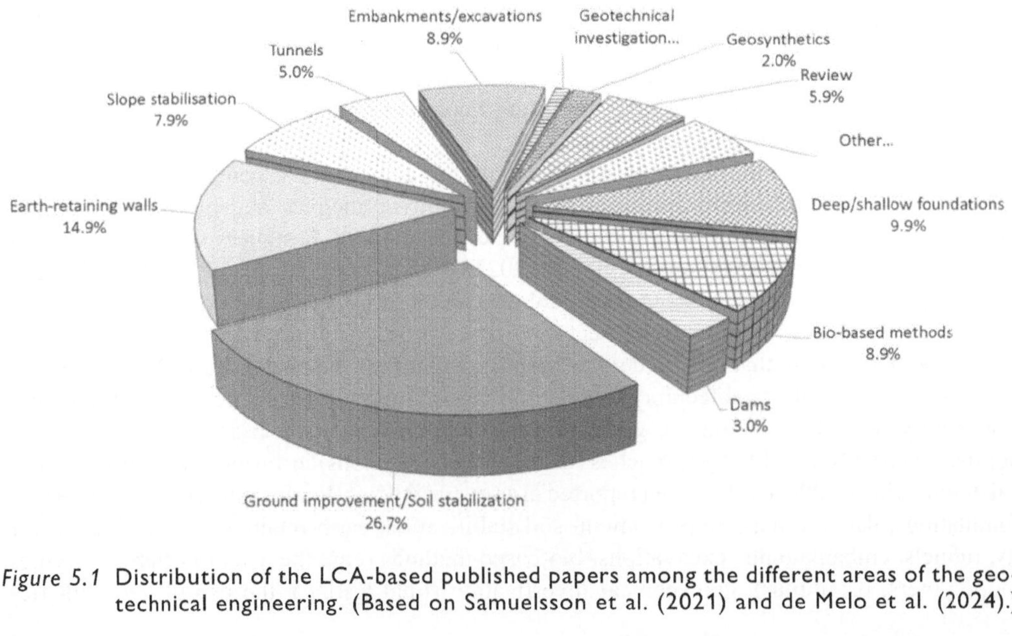

Figure 5.1 Distribution of the LCA-based published papers among the different areas of the geotechnical engineering. (Based on Samuelsson et al. (2021) and de Melo et al. (2024).)

Table 5.1 List of papers focusing on LCA-based works applied to geotechnical engineering

Topic	*References*	*No. papers*
Ground improvement/Soil stabilization	Harbottle et al. (2007); Egan and Slocombe (2010); Salemans and Blauw (2010); Pinske (2011); Spaulding et al. (2012); Craig et al. (2015a); Craig et al. (2015b); Rocha et al. (2016); Shillaber et al. (2016); Vijayakumar et al. (2016); Raymond et al. (2017); Shillaber et al. (2017); Ashfaq et al. (2021); Zahmak et al. (2021); Ghadir et al. (2021); Ashfaq et al. (2021); Rocha et al. (2020); Raymond et al. (2021b); Sánchez-Garrido et al. (2022); Rocha et al. (2022); Pettinaroli et al. (2023b); Chica et al. (2023); Miguel et al. (2023); Saldanha et al. (2023); Ferrazo et al. (2024b); Araújo et al. (2024)	27
Earth-retaining walls	Chau et al. (2008); Rafalko et al. (2010); Inui et al. (2011); Lee and Basu (2015); Giri and Reddy (2015); Phillips et al. (2016); Damians et al. (2017); Zastrow et al. (2017); Pons et al. (2018); Balasbaneh and Marsono (2020); Seol et al. (2021); Balasbaneh et al. (2021); Speranza et al. (2022); Ongpeng and Ginga (2022); Kumar and Parihar (2023).	15
Deep/shallow foundations	Misra (2010); Misra and Basu (2011); Giri and Reddy (2014); Lee and Basu (2016); Li and Chen (2017); Lee and Basu (2018); Pujadas-Gispert et al. (2018); Li and Zheng (2020); Pujadas-Gispert et al. (2021); Lee and Basu (2022).	10
Bio-based methods	Suer et al. (2009); Storesund et al. (2012); Chang et al. (2016); Martin et al. (2020); Raymond et al. (2021a); Deng et al. (2021); Saldanha et al. (2021); Alotaibi et al. (2022); Faruqi et al. (2023).	9
Embankments/ excavations	Blanck et al. (2016); von der Tann et al. (2022); Ashfaq and Moghal (2022); Samuelsson et al. (2022); Curpen et al. (2023); Lee and Basu (2023); Perri et al. (2023); Samuelsson et al. (2024); Bai et al. (2024).	9
Slope stabilization	Cordi et al. (2013); Hoerbinger and Obriejetan (2015); Das et al. (2018); Leal et al. (2018); Zhang et al. (2019); Leal et al. (2020); Raymond et al. (2021a); Yao et al. (2024).	8
Tunnels	Chau et al. (2012); Huang et al. (2014); Arena (2019); Hausberger et al. (2023); Shi et al. (2024).	5
Dams	Liu et al. (2013); Noh et al. (2018); Yuguda et al. (2020)	3
Geosynthetics	Elsing et al. (2012); Werth and Höhny (2012)	2
Geotechnical investigation	Purdy et al. (2022).	1
Review	Basu et al. (2015); Kendall et al. (2018); Finnveden et al. (2009); Song et al. (2020); Samuelsson et al. (2021); de Melo et al. (2024).	6
Other	Lubrecht (2012); Frischknecht et al. (2013); Goldenberg and Reddy (2017); Balaguera et al. (2018); Raymond et al. (2021a); Ding et al. (2023).	6

release of results (determine whether the results will be disclosed publicly). Once the goal is established, the next step is to define the scope of the study. The scope of an LCA study describes the detail and depth of the study to demonstrate that the goal can be achieved within the stated limitations. The ISO standards suggest that the scope should encompass the following elements, including: (i) the functional unit (it is a unit of use for a certain period of time, serves as a reference for comparing alternative products or processes; it must be specify precisely what is being assessed and quantify the service delivered by the system or infrastructure); (ii) system boundaries (the limits or boundaries to be included in the assessment, describing what is taken into the assessment and what is left out); (iii) reference flows (the flows of materials or energy required to fulfil the functional unit); (iv) assumptions and limitations (assumptions or decisions made throughout the study that may influence the results, i.e., all simplifications introduced in the LCA model); (v) allocation procedures (address the issue of multifunctionality by specifying the method used to partition inputs and outputs flows, ensuring consistency and comparability); and (vi) and impact assessment (specifying the impact categories of interest and the methodology used).

In the second phase (life cycle inventory, LCI), a detailed inventory of all flows throughout the life cycle of a product, process, or system is done. The LCI focus is on collecting relevant data and associated modeling through inputs and outputs of a product, service, or system. For instance, an environmental input means something that is taken from the environment (e.g., raw materials, energy) to be introduced into the product's life cycle, while an environmental output means something that your product's life cycle puts into the environment (e.g., emission of pollutants and waste streams). LCI analysis starts with creating a flow model of the system involved in producing the product, process, or system. This model is often depicted as a flow diagram, illustrating the activities within the system boundary. The more detailed the diagram, the more accurate the results. Data collection is then conducted for all activities within the system boundary. This phase is normally very challenging, especially when dealing with proprietary or confidential information. International databases are used, but secondary data sources may be used, such as literature sources. However, it is crucial to ensure the accuracy and reliability of the data collected, especially if procured from other sources.

In the third phase, life cycle impact assessment (LCIA), potential environmental impacts on the environment and human health are assessed. As outlined by the ISO 14040 (2006) and ISO 14044 (2006) standards, the key steps involved in LCIA are: (i) selection of impact categories (such categories should be tailored to the justifiable geographical region of the study); (ii) classification of inventory results (the results from LCI are classified into the chosen impact categories); (iii) characterization (by using characterization factors, the LCI results within each impact category are quantitatively transformed in impact categories indicators; this allows for comparison by converting all classified flows into common units, e.g., CO_2 equivalents). Weighting of impact categories (relative significance of each impact category) can be done but with caution, allowing aggregation of impact scores into a single indicator for comparison.

The last stage is the life cycle interpretation, which is a systematic technique for identifying, quantifying, verifying, and evaluating a set of conclusions and recommendations from the life cycle inventory and/or LCIA results. All conclusions, limitations, and recommendations must be duly justified. The level of confidence in the final results must be determined. This ensures all data are communicated in a fair, complete, and accurate manner. This is done by identifying the data elements that contribute the most to each impact category.

A complete LCA analysis will normally require a great deal of data and as a result will be time consuming and expensive. In extreme cases, a study can take several years and cost millions of euros. In practice, a simplified approach of LCA is often used, tailored to the product, service, system, or infrastructure. The challenge is how to simplify the LCA model without causing a substantial reduction of its accuracy, i.e., the LCA model must be reliable to support its conclusions.

In theory, the LCA methodology refers to the entire life cycle of the product, service, system, or infrastructure, i.e., it refers to a of cradle-to-grave system boundary. However, in the construction sector, it is common to introduce simplifications in the LCA model and to define the system boundary as part of the life cycle, for example, cradle to gate or cradle to site. The usual system boundaries used in a build LCA are illustrated in Figure 5.2. In a cradle-to-grave system boundary the following four stages are considered in an LCA (EN 15804:2012): (i) product stage (activities A1–A3: raw material supply, transport, and processing); (ii) construction stage (activities A4–A5: transport and construction/installation); (iii) use stage (activities B1–B7: use, maintenance, repair/replacement/refurbishment, and energy/water use); and (iv) end-of-life stage (activities C1–C4: deconstruction/demolition, transport, waste processing/disposal). For simplicity, other system boundaries are commonly used, which perform only a partial LCA analysis, such as the cradle-to-gate system boundary, which only analyses the product stage, and the cradle-to-site system boundary, which combines the product and construction stages.

Despite the current stage of development of knowledge on LCA and the numerous successful cases of its application to various areas, there is a limited number of cases of the LCA methodology being applied to geotechnical engineering and even fewer to transportation geotechnical infrastructure. In the next section, some successful geotechnical cases are presented.

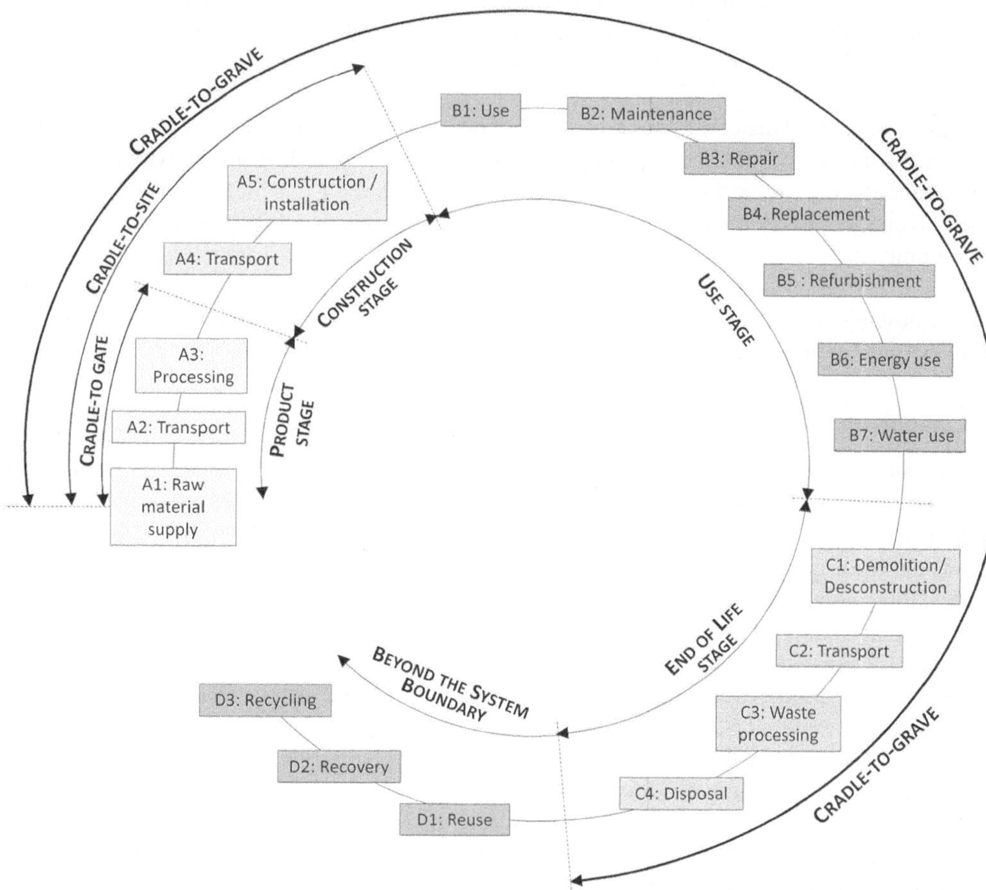

Figure 5.2 Links between the stages (including their activities) in a building's life cycle and LCA system boundaries. (Based on EN 15804:2012.)

5.2.2.1 *Incorporating LCA into geotechnical design*

Typical geotechnical works on transportation infrastructure include four phases: planning and design, construction, maintenance, service, and end of life. Figure 5.3 illustrates a possible methodology to associate the LCA-based analysis to the geotechnical design process, which can be subdivided into the following stages (Samuelsson et al. 2024):

- i Geotechnical design stage: This takes into consideration the client's requirements, the conditions/actions imposed by the superstructure (e.g., road, railway, building, bridge), the environmental/geotechnical/geological conditions, the de-sign codes, and other technical aspects. In general, this stage developed one or more design alternatives for a specific problem.
- ii LCA-based method: This allows the analysis and quantification of the environmental impact of each design alternative during its life cycle, which includes the steps of goal and scope definition, life-cycle inventory, life-cycle impact assessment, and interpretation (ISO 14040:2006).
- iii Identification and improvement stage: This allows the identification of materials and methodologies with higher impact and implementation of adjustments in order to mitigate the environmental impacts (e.g., change the conditions imposed by the superstructure, reduce the amount of materials, or replace the materials/methodology by other more environmentally friendly).
- iv Evaluation stage: This comprises the comparison of the design alternatives in order to define the best solution to implement in the final geotechnical design. In general, when other LCA assessments (e.g., social, economic) are associated with the LCA methodology, it is necessary to apply conversion and weight factors in order to convert all the assessments to a common unity.

In the following sections, some examples of the incorporation of the analysis of LCA, LCCA, and LCSA into geotechnical design are addressed.

5.2.2.2 *Life cycle assessment (LCA)*

To illustrate the application of LCA to geotechnical problems, two studies are summarized below: (i) the comparison of four techniques to repair a highway slope (Leal et al. 2020) and (ii) the use of different binder contents in soil stabilization (Rocha et al. 2016).

Considering a cradle-to-site approach, Leal et al. (2020) performed an LCA for an hypothetical slope, consisting in the comparison of four different geotechnical techniques: granular rock fill replacement, planting live willow poles, reinforced soil with polypropylene fibers, and electrokinetic geosynthetics (or electro-osmosis). This study analyses only the climate change impacts (i.e., the global warming potential for a horizon of 100 years), considering the effects of raw/recycled material acquisition and processing, the transport of the materials, and their application on site. The results clearly show that the granular rock fill replacement induces the greatest environmental impact (109–174 $kgCO_2e/m^2$) for a 2.5 m failure depth. This is mainly due to the use and transportation of a great volume of materials, as the transport of materials is responsible for more than 50% for the total impacts. Relatively to the granular replacement, the impact of the willow pole technique was about 7%–8% (the lowest impact), electrokinetic geosynthetics is in the range 8%–13%, and the reinforcement of soil with fibers is between 19% and 32% of impact induced by granular replacement.

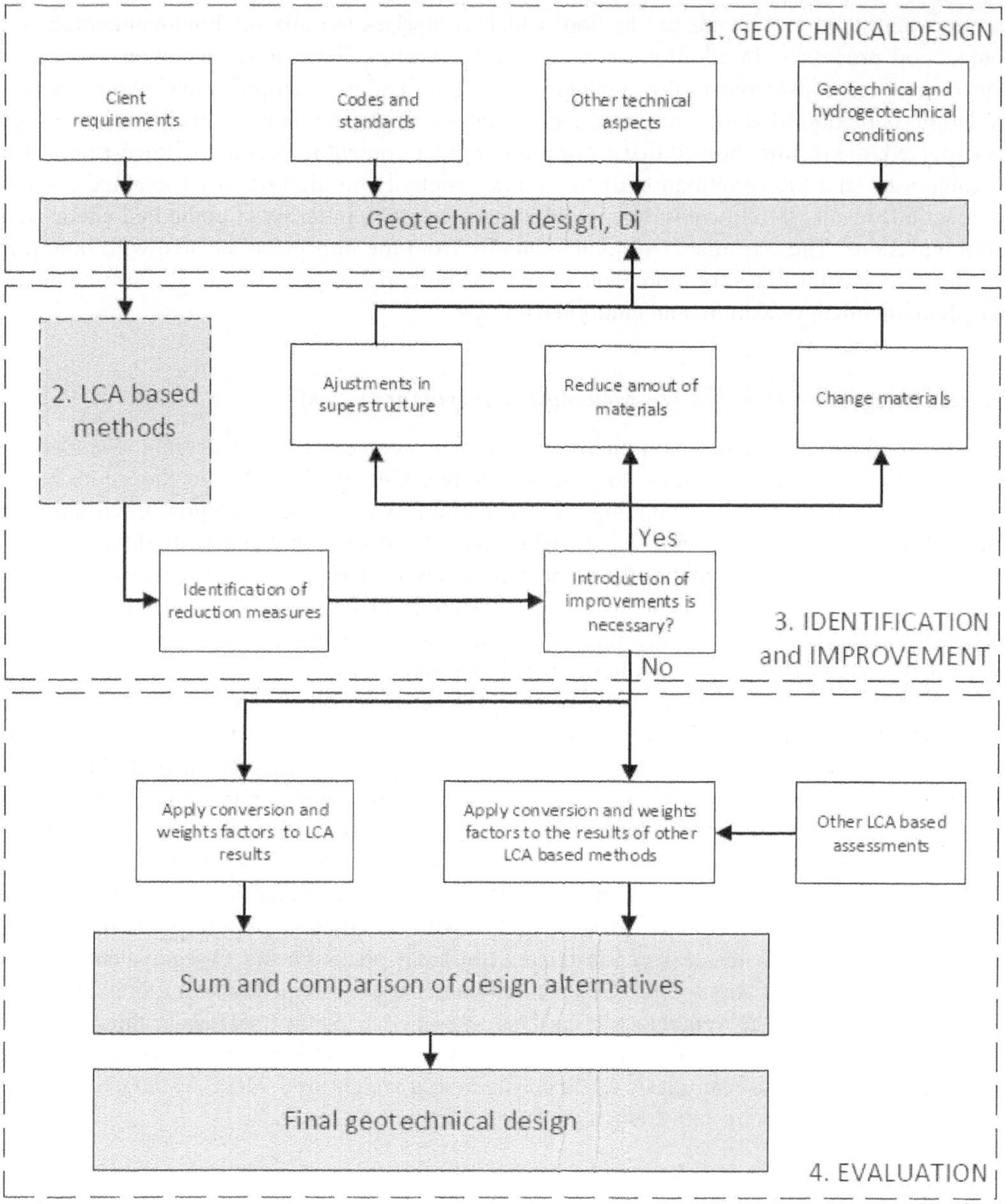

Figure 5.3 Scheme of incorporating LCA to geotechnical design. (Adapted from Samuelsson et al. 2024.)

Considering a cradle-to-gate approach (i.e., the product stage), Rocha et al. (2016) presented an LCA about the assessment of the environmental impact of clay-lime blends for the Paraguay data, comparing fifteen binder contents to obtain comparable performance in terms of mechanical properties, namely unconfined compressive strength, splitting tensile strength, and initial shear modulus. The study considers the processes of lime production and transportation, soil extraction and transportation, and blends production, which comprises the mixing, homogenization, and compaction processes. In addition to the global warming (GHG emission), other six impact categories were considered in this analysis, e.g., acidification, eutrophication, photochemical oxidation, depletion of abiotic resources, stratospheric ozone depletion, and embodied energy. As expected, the results showed that a minimum binder content results in the lowest impact in all categories, and the combination of low binder content and high density (obtained through compaction) results in a low environmental impact, namely, in terms of embodied energy and GHG emissions. These results arise fundamentally from the impacts associated with lime production, which contributes for more than 75% of the potential impacts of global warming potential, photochemical oxidation, and embodied energy.

5.2.2.3 Integrating LCA and life cycle cost assessment (LCCA)

In order to facilitate the decision-making process in geotechnical engineering, Samuelsson et al. (2024) presented a methodology to integrate LCA and LCCA into the geotechnical design. Succinctly, in this methodology, LCA method evaluates the potential environmental impact (as described in Section 5.2.2.2), while the LCCA evaluates the costs during the life cycle of the work. The compatibility of both methods implies the use of conversion factors (e.g., cost/1000 kgCO_2-equivalent) to express the environmental impacts and the cost to the same unit; additionally, weighted factors to LCA and LCCA results are applied in order to allow the comparison of different design alternatives in the same unit.

Considering a cradle-to-site approach (A1–A5: product and construction stages), this methodology was applied for evaluating the environmental impact and costs for three different embankment fill methods (crushed bedrock, cement-stabilized till, and foam glass) for construction a high-speed railway (Samuelsson et al. 2024). This study analyses only the climate impacts (CO_2 emissions), considering the effects of material production, the transport of the materials, and their installation on site. The results showed that the stages of material production and transport stages induce high CO_2 emissions and high costs. The exception is the cost of the cement for the cement-stabilized till method, which is about 22% of the total costs of the method. Samuelsson et al. (2024) also showed that the use of two different conversion factors (different prices for kgCO_2-equivalent) did not affect the qualitative seriation of the three methodologies used for embankment filling. However, the results indicate that the weight factors used has a great effect on the final results; thus, considering a weight of 0.5(LCA)-0.5(LCCA), the cement-stabilized till is the best methodology, followed by crushed bedrock and foam glass; while considering a weight of 0.97(LCA)-0.03(LCCA), all the three methodologies are equivalent in terms of climatic-costs impacts.

5.2.2.4 Life cycle sustainability assessment (LCSA)

The LCSA allows assessing and quantifying the sustainability of a product, system, or work integrating the environmental life-cycle assessment (LCA), LCCA, and social life-cycle assessment (S-LCA) during its life cycle. Raymond et al. (2021b), based on an approach cradle to site, presented a LCSA to compare the impacts of five ground improvement methods used to mitigate the liquefaction of a sandy soil: compaction grouting (CG), deep dynamic compaction (DDC), deep soil mixing (DSM), vibro compaction (VC), and vibro replacement (i.e., stone columns) (VR). This study analyses the environmental impacts (in terms of air emissions, water pollutants, and solid wastes) due to the effects of materials use (extraction, processing, and transportation), energy consumption (e.g., diesel fuel or electricity), use of vehicles/equipment, and construction activity.

Several impact factors were considered in the analysis of the five ground improvement methods, which comprise: (i) environmental indicators: resource depletion (i.e., primary energy and water consumption), climate change (global warming potential), acidification, eutrophication, ecotoxicity, human toxicity (e.g., from carcinogens and noncarcinogens), ozone depletion, particle pollution, and smog formation; (ii) soil quality indicators: erosion, infiltration reduction, groundwater regeneration reduction, and biotic production loss; (iii) cost indicator expressed by the project cost; and (iv) social indicator described by the social cost of GHG emissions. It should be highlighted that the analysis of the soil quality was only performed in terms of a qualitative assessment.

The results of the LCSA analysis showed that the most impactful method is the DSM followed by CG, since both are the most expensive and the less environmental friendly methods, mainly due to the highest primary energy consumption and global warming potential. These impacts come fundamentally from the use of Portland cement, which can be mitigated by replacing cement by other more sustainable binders (e.g., fly ash, blast-furnace slag, silica fume).

5.3 Case study: life cycle assessment of a geotechnical work in the United Kingdom

The general approach for design while considering LCA is shown in Figure 5.3. Three main approaches are identified, i.e., adjustments in the superstructure, reduction of the volume of materials used, or change in materials selected. Among these, perhaps the simplest approach is to consider materials with equivalent geotechnical performance (e.g., equivalent functional unit) but having lower carbon footprint. Table 5.2 shows typical technical geotechnical solutions and potential low-carbon alternatives. It can be observed that even with straightforward material substitutions, there are significant potential carbon savings that can be exploited.

For instance, the consideration of waste materials such as fly ash or steel furnace slag in lieu of more conventional binders can result in substantial carbon savings and also promote a greater circular economy practice in the construction industry. Furthermore, the consideration of natural fiber geotextiles in cases where separation, filtration, reinforcement, and drainage functions long term are less important can result in substantial carbon savings. In addition, the use of these geotextiles is more sympathetic to the local environment, as there is no residual contamination because the natural fiber geotextile degrades, which in turn has some advantages at the decommissioning stage of the assets.

In this case study, the two design scenarios considered exploit the carbon reduction and environmental impacts comparing different technical options for the remediation of a typical slip failure of a rail embankment in the United Kingdom (Figure 5.4). This type of failures is relatively common in England (hundreds annually) and is associated with deterioration of earthwork geostructures due to climate change (Walker et al. 2022).

Scenario 1 consists of a common remediation approach for slip failures incorporating a slope regrade (with benching), a geotextile as a separator, and rock fill as a backfill material (Figure 5.4a). For this scenario substitution of the rock fill material for a lower carbon alternative (e.g., type of rock fill and recycled ballast) will be considered as well as the use of natural fiber geotextiles. Scenario 2 considers a different design solution using bored piles (circular pile diameter of 0.25 m and installed at 1 m center-to-center spacing to a depth of 6.5 m) with reinstatement of the slope to 2:1 (H:V) gradient. For this scenario, substitution of concrete binders including

Table 5.2 Common technical solutions adopted in geotechnical works and potential low-carbon alternatives

Current solution	*Low-carbon alternative*	*Technical considerations*	*$kgCO_2e/kg$ reduction potential (reference)*
Ordinary Portland cement (OPC)	Geopolymer (metakaolin)	Similar performance but cast in situ technology not mature (Nawaz et al. 2020)	94% (McLellan et al. 2011)
Soil stabilization with OPC or hydrated lime[b]	Biopolymer lignosulfonate[a]	Similar performance for equivalent % addition by weight for mitigation soil erosion and swelling (Alazigha et al. 2018)	72% (Lime) 69% (OPC) (Laveglia et al. 2022; Modahl and Soldal 2021)
	Partial replacement with fly ash[a]	Similar performance for a reduction on the % addition by weight of lime (Heitor et al. 2021)	97% (Lime) 96% (OPC) (Setiawan et al. 2022)
	Partial replacement with steel furnace slag	Similar performance for a reduction on the % addition by weight of lime (Heitor et al. 2021)	57% (Lime) 52% (OPC) (0.4 $kgCO_2e/kg$ sequestration potential, Chen et al. 2022)
Synthetic geotextiles	Natural fiber geotextile	Similar performance short term but long-term performance limited (jute, Al-Rawabdeh et al. 2021)	98% (Flax) 98%(Hemp) 98% (Jute) (Beus et al. 2019; Wernet et al. 2016)

[a] Waste material and the LCA boundary do not include production of the waste as it results from another production loop.
[b] % addition by weight considered equivalent for all binders examined.

geopolymer and partial replacement will also be examined. The performance of the two technical solutions using the material properties listed in Table 5.3 was assessed using limit equilibrium methods using Rocscience Slide 2, and the factor of safety obtained was approximately the same for both solutions. As the merit of the technical solution considered is equivalent, comparisons are made in terms of their carbon (e.g., kg CO_2e) and environmental performance. It was assumed the ground water level to be at the interface between the embankment and formation level.

In this analysis, the LCA is based on a cradle-to-gate approach (phase (i) and (ii), see Section 5.2.2). However, it should be noted that depending on the location of the site (phase (iii), see Section 5.2.2), there can be important carbon and environmental impact implications associated with transport to site, maintenance, operational, and decommission (e.g., cradle to grave). The analysis of embodied carbon presented is based on the carbon calculator tool developed by the UK Rail Safety and Standards Board (RSSB), i.e., the rail carbon tool (RCT) (RSSB 2024). The RSSB RCT is a web-based tool that has a large database of carbon factors relevant to materials employed in rail infrastructure including geotechnical elements. It should be noted that most factors considered in the tool have been benchmarked for UK practice and are based on Hammond and Jones (2011) database. The user inputs the total material quantity for each element, identifies the material type from the in-built database, and the tool calculates the equivalent carbon for the project (cradle to gate), assuming a material density. The carbon indices for the scenarios examined including

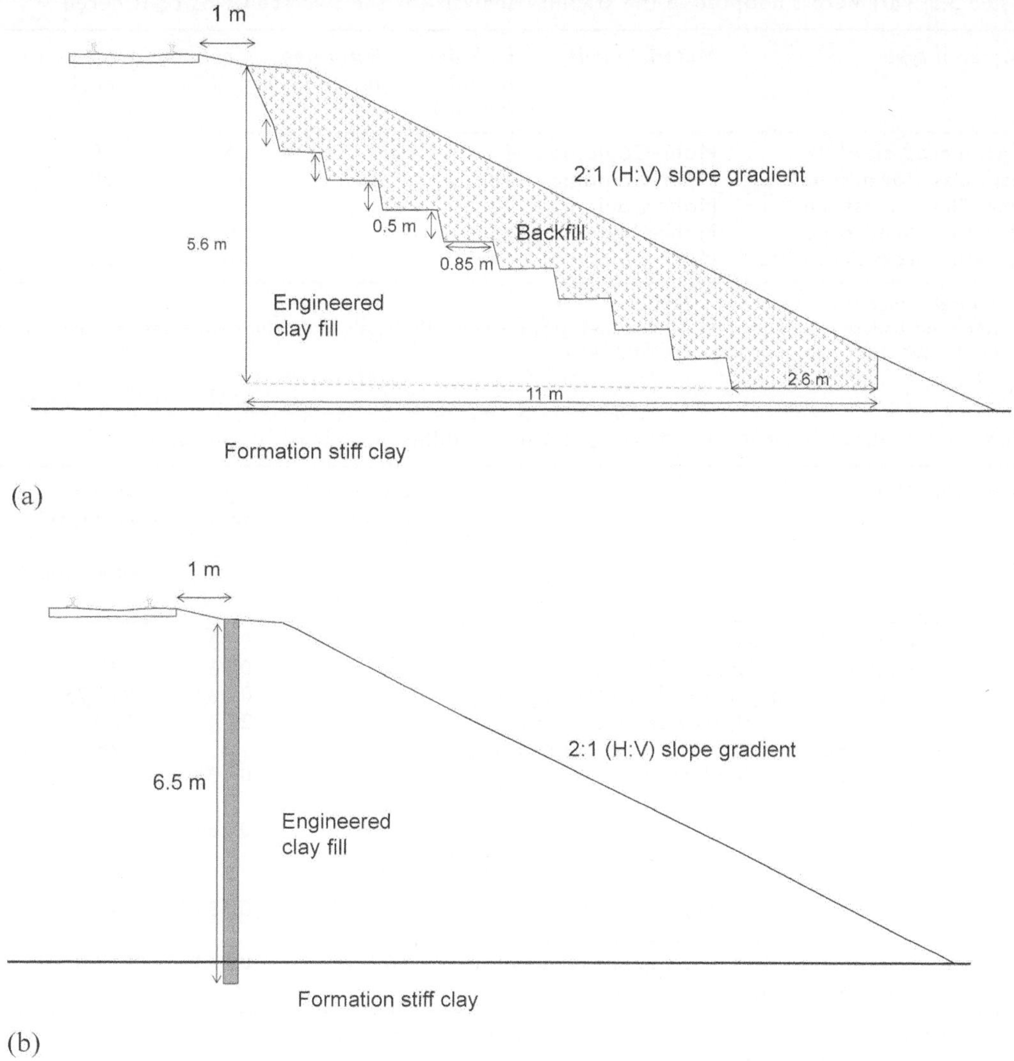

Figure 5.4 Design scenarios considered for the remediation of a typical slip failure on an embankment in the north of England. (a) Scenario 1: geotextile and rock fill; (b) Scenario 2: bored piles.

material substitutions and the associated kgCO_2e per linear meter of remediation are shown in Table 5.4.

For scenario 1, it can be observed that while the rock fill materials kgCO_2e/kg is much lower than the geotextile, the kgCO_2e per linear meter of remediation is much larger. This shows that direct comparisons of kgCO_2e/kg between materials can be often misleading, and the correct attribution of volume and density should be considered. In addition, as expected the recycled ballast material case provided the greatest carbon savings. Nonetheless, it should be noted no values for recycled ballast were available, and a value for recycled aggregates was considered instead. For scenario 2, while the differences in the binder can result in substantial differences

Table 5.3 Parameters adopted in the stability analysis for the two scenarios considered

Material type	*Material model*	*Bulk unit weight (kN/m³)*	*Saturated unit weight (kN/m³)*	*Cohesion, c′ (kPa)*	*Friction angle, ′ (°)*
Engineered clay fill[a]	Mohr-Coulomb	17	-	5	10
Stiff clay (formation)[a]	Mohr-Coulomb	18	18.5	15	20
Backfill – Limestone	Mohr-Coulomb	22.3	-	0	35
Backfill – Sandstone	Mohr-Coulomb	21.2	-	0	30
Backfill – recycled ballast[b]	Mohr-Coulomb	14.72	-	0	55

[a] Typical values taken from Look (2007).
[b] Recycled ballast considered consist of 63% granite and 21% basalt and values were taken from Abadi et al. (2023) for confining pressure of 30 kPa.

Table 5.4 Embodied carbon (cradle-to-gate) of the different scenarios considered

Scenario	*Material*	*Unit area/ volume per linear meter of remediation*	*$kgCO_2e$/ kg*	*$kgCO_2e$ per linear meter of remediation*[a]
1a	Rock fill – Limestone (2274 kg/m³)	16.5 m³	0.09	3382.5
1b	Rock fill – Sandstone (2163 kg/m³)	16.5 m³	0.06	2145
1c	Rock fill – Recycled ballast (1500 kg/m³)[b]	16.5 m³	0.037	915.75
1d	Geotextile – Polyethylene (separator)[c]	15 m²	2.54	360
1e	Geotextile – natural fiber flax (separator)[d]	15 m²	1.8	78
2a	Concrete pile (binder: OPC) including steel rebar[e]	0.1 m³	0.198	49.1
2b	Concrete pile (binder: OPC with 30% replacement blast furnace slag) including steel rebar[e]	0.1 m³	0.166	41.5
2c	Concrete pile (binder: Geopolymer metakaolin) including steel rebar[f]	0.1 m³	0.184	46
2e	Engineering fill reinstatement (values for rammed soil)	16.5 m³	0.024	712.8

[a] Hammond and Jones (2011) database available on the Rail Carbon Tool (RCT).
[b] Waste material, value taken for recycled aggregates (0.037 $kgCO_2e$/kg) from Aumônier et al. (2010).
[c] Polyethylene - HDPE -941 kg/m³ Depth: 0.01 m Length: 1 m Width: 1 m (2.54 $kgCO_2e$/kg).
[d] Wet Felted Fiber Board -290 kg/m³ Depth: 0.01 m Length: 1 m Width: 1 m (1.8 $kgCO_2e$/kg).
[e] RC 25/30 (25/30 MPa) with 110 kg rebar per m³ concrete, 2500 kg/m³.
[f] Metakaolin geopolymer (25/30 MPa) reinforced with 110 kg rebar per m³, 2500 kg/m³.

in carbon, e.g., 94% reduction with geopolymer in lieu of OPC (Table 5.2), the same is not true when reinforced concrete is considered (Table 5.4). This is associated with the composition of the concrete and the carbons associated with the steel rebar. In this case study, 13% of binder was considered in the concrete and 110 kg rebar per m³ concrete. Thus, the difference between the use of different binders is relatively small. These results demonstrate the importance of carefully considering all constituents of the concrete. The $kgCO_2e$ per linear meter of remediation data shown in Table 5.4 is also plotted in Figure 5.5. Surprisingly, scenario 2 having the concrete pile option showed the lowest $kgCO_2e$ per linear meter of remediation. While the consideration of different binders yielded different embodied carbon values, the concrete with

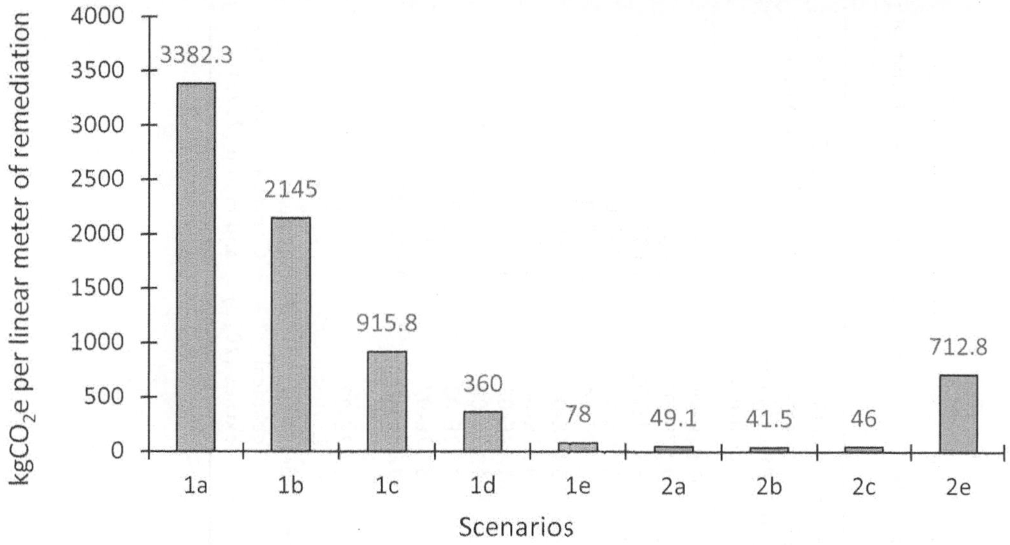

Figure 5.5 Comparison in terms of $kgCO_2e$ between the different scenarios considered.

30% replacement of blast furnace slag (2b) in combination of engineering fill reinstatement (2e) showed the lowest value for embodied carbon of all the options considered. While this may not seem intuitive at first sight, Heath et al. (2014) reported that concrete with partial replacement by fly ash and blast furnace slag release fewer CO_2 and global warming potential (GWP) than metakaolin geopolymer since metakaolin geopolymerization process uses more silicates. Nonetheless, it should also be noted that the level of complexity of both scenarios is quite different, and the use of rock fill could be preferred from a delivery and plant mobilization standpoint. While all the scenario 1 combinations yielded larger embodied carbon than scenario 2, the combination considering recycled ballast and natural fiber geotextile with 1275.75 $kgCO_2e$ per linear meter of remediation compared favorably with scenario 2 values. Finally, the carbon associated with transport and plant mobilization (cradle-to-grave approach) for both scenarios was not considered, and it is likely that important carbon emissions result from those processes that need to be included to allow for an accurate comparison. This is critical because the location of the site and local availability of materials could result in important carbon savings. Nonetheless, the results of the analysis of this simple case study showed that there are carbon savings to be exploited by considering material substitutions without that resulting in the lower performance.

The associated environmental impacts follow the categories suggested in EN 15804 (2021) and are based on the information reported in the Environmental Product Declaration (EPD) for the different materials adopted in both case studies from a cradle-to-gate perspective (A1–A3, Figure 5.2). For instance, global warming potential (climate change) in terms of fossil fuel (GWP-FF), biogenic (GWP-B), land use and associated change (GWP-LU), ozone depletion potential (ODP), acidification potential for soil and water (accumulated exceedance) (AP), eutrophication potential aquatic freshwater (EP-F), eutrophication potential aquatic marine (EP-M), eutrophication potential accumulated exceedance (EP-AC), photochemical ozone creation potential (POCP), abiotic depletion potential – nonfossil resource (ADP-R), abiotic depletion potential – fossil fuels (ADP-FF), and water deprivation potential (WDP) based on deprivation water consumption. Table 5.5 shows the environmental impacts for scenario 1. Similarly, to the observations of embodied

Table 5.5 Environmental impact indicators for different materials based on EPD suggested by EN 15804 (results per declared unit of 1 metric ton of rock fill material and 1 m^2 for geotextiles)

Impact	*Unit (per linear meter of remediation)*	*Scenario 1a Rock fill – Limestone (2274 kg/m^3)*	*Scenario 1b Rock fill – Sandstone (2163 kg/m^3)*[a]	*Scenario 1c Rock fill – Recycled ballast (1500 kg/m^3)*[b]	*Scenario 1d Geotextile – Polyethylene (separator)*[c]	*Scenario 1e Geotextile – natural fiber (separator)*[d]
GWP	kg CO_2 eq.	2585.1969	2.89E+02	4.653	2.81E+00	1.88E+00
GWP-FF	kg CO_2 eq.	2577.6927	3.57E+02	4.529	2.87E+00	0.00E+00
GWP-B	kg CO_2 eq.	0	−5.35E+01	0.110	−5.21E-02	1.70E-03
GWP-LU	kg CO_2 eq.	0.0525294	2.53E-01	0.022	1.50E-03	5.94E-03
ODP	kg CFC 11 eq.	2.5439E-06	6.78E-05	3.94E-07	2.07E-14	6.40E-09
AP	mol H+ eq.	0.4239873	3.57E+00	0.036	3.81E-03	2.38E-02
EP-F	kg P eq.	0.00086298	2.25E+01	0.001475	4.25E-06	3.17E-04
EP-M	kg N eq.	0.1238193	3.07E+00	0.02673	1.14E-03	1.30E-02
EP-AC	mol N eq.	1.8310248	1.68E+01	0.135135	1.19E-02	1.01E-01
POCP	kg NMVOC eq.	0.3827142	4.28E-02	0.011113	4.22E-03	1.44E-02
ADP-R	kg Sb eq.	0.00184228	2.36E-01	1.95E-05	5.27E-07	1.84E-05
ADP-FF	MJ, net calorific value	303.54489	5.00E+02	127.215	1.25E+02	2.61E+01
WDP	m^3 world eq. deprived	5.140377	3.00E+01	3.1185	2.07E-02	7.97E-01
Reference		Roadstone (2023)	Hutton Stone (2022)	D.A. Mattson (2022)	Beaulieu Technical Textiles (2022)	European Confederation of Flax and Hemp (2022)

[a] Values for walling blocks.
[b] Waste material, value taken for recycled aggregates.
[c] Nonwoven geotextile.
[d] Values for flax fiber scotched.

carbon, the scenario with the use of recycled aggregates (e.g., ballast) results on the lowest impact on all categories. Furthermore, the impact contribution from the geotextiles is substantially lower than the backfill materials. Surprisingly, while the natural fiber geotextile (herein considered flax) has lower values in GWP, in other categories, it does not compare favorably with polyethylene-based geotextiles. This may be associated with less established manufacturing processes for these fibers.

Table 5.6 shows the environmental impact for scenario 2. The consideration of different binders yielded different environmental impacts. While from the GWP perspective the binder incorporating the use of metakaolin-based concrete is the lowest, it compares unfavorably with OPC in the other impact categories. This is mainly associated with the high environmental impact of the activator solutions. Among the three binders, the 30% replacement of blast furnace slag (2b) shows the lowest values on all impact categories. From all the scenarios considered, the combination that is more favorable from an environmental impact standpoint is the use of recycled ballast aggregates with flax geotextile. This is not surprising as this combination was compared favorably to the lowest embodied carbon technical solution. It should be noted, nonetheless, that it is important to consider environmental impacts in the LCA assessment to fully capture the footprint associated. In this study, EPD data were used for comparison, but it should be noted that the important impact is also attributed to transport of materials, and it is likely that rock fill

Table 5.6 Environmental impact indicators for different materials based on EPD suggested by EN 15804 (results per declared unit of 1 metric ton of rock fill material and 1 m^2 for geotextiles)

Impact	*Unit (per linear meter of remediation)*	*Scenario 2a Concrete pile (binder: OPC) including steel rebar (110 kg rebar per m³ concrete)*	*Scenario 2b Concrete pile (binder: OPC with 30% replacement blast furnace slag) including steel rebar*	*Scenario 2c Concrete pile (binder: Geopolymer metakaolin) including steel rebar*	*Scenario 2d Engineering fill reinstatement (values for rammed earth)*
GWP	kg CO_2 eq.	55.99	50.00506	39.199132	154.2585
GWP-FF	kg CO_2 eq.	55.99	48.19876	ND	ND
GWP-B	kg CO_2 eq.	0.00016	−0.00160062	ND	ND
GWP-LU	kg CO_2 eq.	0.009782	0.00893289	ND	ND
ODP	kg CFC 11 eq.	1.5E-06	1.05005E-06	7.5002E-06	5.486E-14
AP	mol H+ eq.	0.163365	0.1599148	0.3267296	0.51612
EP-F	kg P eq.	0.010003	0.007903202	0.05501761	0.121341
EP-M	kg N eq.	0.040711	0.0327007	0.22390885	ND
EP-AC	mol N eq.	0.455491	0.3648907	0.68323605	ND
POCP	kg NMVOC eq.	0.115304	0.0940942	0.26519966	0.087714
ADP-R	kg Sb eq.	2.02E-05	1.47763E-05	4.4398E-05	8.026E-06
ADP-FF	MJ, net calorific value	425.5341	406.2387	ND	2037.75
WDP	m^3 world eq. deprived	5.814843	4.567143	ND	ND
Reference		Lafarge (2021) Tata Steel Limited (2022)	Lafarge (2021) JSW Cement Limited (2019)	Haberta and Ouellet-Plamondon (2016)	Ökobaudat (2018)

ND, not declared.

scenarios would accrue the largest values in A4 (Figure 5.2). From this standpoint, the selected option would also be preferred as the ballast could be recycled without the need for disposal during the maintenance operations.

The analysis of a simple rail embankment failure remediation showed that from an embodied carbon standpoint, there are potential carbon savings to be exploited without compromising geotechnical performance. In fact, the scenario that showed the lowest carbon and environmental impacts per linear meter of remediation was the one employing the use of concrete and having the highest level of technical complexity. This case study illustrated that the consideration of LCA can be introduced in geotechnical design in a relatively straightforward manner from an embodied carbon and environmental impacts, considering a cradle-to-gate (A1–A3, Figure 5.2) approach. While this practice does involve a shift in the design outcomes that have been traditionally focused mainly on cost, it is an important step to support the transition to lower carbon future to support society's decarbonization targets. Currently, the lack of information on carbon factors and environmental product declarations for different materials and/or processes on site is perhaps the greatest challenge that is hindering wide-scale implementation. Nonetheless, as more details about carbon factors and environmental product declarations become available and more carefully auditing is carried out by contractors and suppliers supported by standards and legislation, it is likely that full LCA analysis incorporating not only the construction of the asset but also its maintenance and end of life will become more standard in the next decades.

5.4 Final remarks

This chapter provides an in-depth literature review on eco-design and life cycle assessment (LCA) in transportation geotechnics, encompassing products, methodologies, and relevant studies.

The eco-design concept is a broad approach that aims to integrate environmental considerations into the design process not only of materials and products but also for systems and infrastructure. The eco-design in transportation geotechnics is a relatively new topic and should encompass all stages of a product's (system or infrastructure) life cycle, including manufacturing, distribution, usage, maintenance, end of life, and recycling. Existing regulations remain inadequate with insufficient focus on sustainability and circularity. New eco-design regulations should promote digital product passports, high-quality recycling, and increased use of recycled materials. In the context of transportation geotechnical infrastructure, which often have a service life exceeding 100 years, it is essential to integrate eco-design principles throughout their lifespan encompassing the product's manufacturing, construction, service/maintenance, refurbishment, and deconstruction phases. Furthermore, the implementation of eco-design in transportation infrastructure should embrace also the use of smart urban and territorial planning, prioritize the reuse and upgrading of existing infrastructure and materials, design adaptable, and flexible infrastructure that incorporates climate resilience and nature-based solutions, promote low-carbon construction techniques, and maximize the efficiency of infrastructure and materials and emission reductions. Given that, transportation infrastructure is capital intensive, sustainable financing should become a driving force in ensuring the long-term sustainability. Although eco-design initiatives may increase initial costs for manufacturers, governments and consumers, they can be offset by long-term financial savings.

Several examples of associated sustainable practices that incorporate eco-design concept in transportation geotechnical infrastructure were presented, including: (i) the use of recycled materials; (ii) soil stabilization using bio-based methods (such as plants, biopolymers, and microorganisms); (iii) the development and implementation of sustainable drainage systems, including permeable pavements; (iv) the application of geothermal energy in transportation infrastructure pavements; (v) the use of more sustainable geosynthetics; (vi) the use of sensors, artificial intelligence, and digital technologies to optimize infrastructure maintenance; (vii) the integration of renewable energy solutions, such as solar panels and wind turbines; (viii) the development of self-healing geotechnical materials; and (ix) the utilization of 3D printing technology to enable the use of local materials for manufacturing products and systems.

LCA is a comprehensive and systematic approach for evaluating the environmental impacts of products, systems, or infrastructures, over their entire life cycle (e.g., from cradle to grave). As far as infrastructure is concerned, the challenge is to reach a compromise between an adequate and comprehensive approach considering not only environmental aspects but also economic and social aspects. To this end, there are different LCA approaches, considering the LCCA and LCSA. In any case, LCA approaches are employed to compare alternative products, systems, or infrastructures by weighing their trade-offs while minimizing the risk of shifting burdens to other types of impacts, life-cycle stages, geographic regions, or generations. In summary, an integrated methodology combining LCA with economic and social aspects can be an effective and powerful way to successfully implement infrastructure projects.

A complete LCA analysis will normally require a great deal of data and as a result will be time consuming and expensive. In extreme cases, a study can take several years and cost millions of euros. In practice, a simplified approach of LCA is often used, tailored to the product, service, system, or infrastructure. The challenge is how to simplify the LCA model without causing a substantial reduction of its accuracy, i.e., the LCA model must be reliable to support its conclusions.

A proposed methodology for integrating LCA-based analysis into the geotechnical design process can be divided into the following four stages: (i) geotechnical design stage, which considers all requirement actions and conditioning factors; (ii) LCA-based method, which quantifies the environmental impact of each design alternative throughout its life cycle; (iii) identification and improvement stage, which identifies materials and techniques with higher impacts and promotes mitigation adjustments; and (iv) evaluation stage, which compares design alternatives to determine the best solution. Despite the benefits that can be derived from incorporating sustainability principles into the LCA of geotechnical assets in transportation infrastructure, this is still an unusual reality nowadays within the geotechnical community. Most published works using LCA-based approaches are related to ground improvement and/or soil stabilization works (≈27%), followed by earth-retaining walls (≈15%) and deep foundations and earthworks (≈10%).

To illustrate the application of LCA to geotechnical challenges, two scientific papers are briefly presented and discussed in the chapter. The first employs a cradle-to-site approach to compare four techniques for repairing a highway slope (Leal et al. 2020): granular rock fill replacement, planting live willow poles, reinforced soil with polypropylene fibers and electrokinetic geosynthetics (or electro-osmosis). The LCA results clearly indicate that granular rock fill replacement has the most significant environmental impact. The second study adopts a cradle-to-gate approach to assess the environmental impact of different binder contents in soil stabilization (Rocha et al. 2016). The findings reveal that a minimum binder content induces the lowest impact, while a combination of low binder content and high density results in a low environmental impact in terms of embodied energy and greenhouse gas emissions.

Finally, a cradle-to-gate LCA approach is used in a case study comparing two design scenarios (with different technical options) for the remediation of a typical slip failure of a rail embankment in the UK. Scenario 1 consists of a common remediation approach for slip failures, involving slope regarding, geotextiles, and rock fill. Scenario 2 considers alternative design solutions using bored piles. The analysis of the remedial measures for a simple rail embankment failure indicates that there are potential carbon savings that can be realized without compromising geotechnical performance. In fact, the scenario that showed the lowest carbon and environmental impacts per linear meter of remediation was the one employing the use of concrete piles. It was stated that, currently, the lack of information on carbon factors and environmental product declarations for different materials and or processes on site are perhaps the greatest challenge that is hindering wide-scale implementation. Nonetheless, as more details about carbon factors and environmental product declarations become available, and more carefully auditing is carried out by contractors and suppliers supported by regulation, it is likely that complete LCA analysis incorporating not only the construction of the infrastructure but also its maintenance, and end of life will become more standard in the next decades.

In summary, the chapter underscores the urgent need for sustainable geotechnical practices in transportation infrastructure. The eco-design and LCA concepts are closely interlinked and can be taken as intrinsic parts of a novel methodology based on the life cycle thinking approach that aims to provide a framework to support decision-makers toward a more sustainable, resilient, and equitable infrastructure design. Although cost constraints and data availability remain challenges, policy regulations, technological advancements, the implementation of digital product passports, and industry shifts are expected to accelerate the adoption of sustainable geotechnical solutions in the coming years.

Bibliography

Abadi, T., Madhusudhan, B.N., Li, H. and Pen, L.L. 2023. Reusing life-expired railway ballast: Laboratory testing, shape analysis, and petrographic evaluation. *Journal of Geotechnical and Geoenvironmental Engineering*, 149 (1): 04022123.

Alazigha, D.P., Indraratna, B., Vinod, J.S. and Heitor, A. 2018. Potential use of lignosulfonate for expansive soil stabilisation. *ICE Journal of Environmental Geotechnics*: 1–9. https://doi.org/10.1680/jenge.17.00051.

Alotaibi, E., Arab, M.G., Abdallah, M. and Nadia Nassifl, N. 2022. Life cycle assessment of biocemented sands using enzyme induced carbonate precipitation (EICP) for soil stabilization applications. *Scientific Reports*, 12: 6032. https://doi.org/10.1038/s41598-022-09723-7

Al-Rawabdeh, A., Vinod, J.S., Heitor, A. and Sarojiniamma, B.K. 2021. Numerical simulation of coir geotextile reinforced soil under cyclic loading. *Latest Developments in Geotechnical Earthquake Engineering and Soil Dynamics.* Springer Transactions in Civil and Environmental Engineering, Springer, Singapore. https://doi.org/10.1007/978-981-16-1468-2_7

Araújo, M.T., Ferrazzo, S.T., Consoli, N.C. and Araújo, M.T. 2024. Environmental, economic, and social impacts of sugar cane bagasse and eggshell wastes for soil stabilization. *Environmental Science and Pollution Research*, 31: 15973–15985. https://doi.org/10.1007/s11356-024-32299-w.

Architecture2030. 2024. *Why the Built Environment.* https://www.architecture2030.org/why-the-built-environment/, accessed April 21, 2024.

Arena, N. 2019. Life-cycle assessment applied to construction of Thames Tideway East tunnel, London, UK. *Proceedings of the Institute of Civil Engineers: Engineering Sustainability*, 17: 416–423.

Armistead, S.J., Rawlings, A.E., Smith, C.C. and Staniland, S.S. 2020. Biopolymer stabilization/solidification of soils: A rapid, micro-macro, cross-disciplinary approach. *Environmental Science & Technology*, 54 (21): 13963–13972.

Ashfaq, M. and Moghal, A.A.B. 2022. Cost and carbon footprint analysis of fly-ash utilization in earthworks. *The International Journal of Geosynthetics and Ground Engineering*, 8: 21. https://doi.org/10.1007/s40891-022-00364-4.

Ashfaq, M., Moghal, A.A.B., Basha, B.M. and Baig Moghal, A.A. 2021. Carbon footprint analysis on the expansive soil stabilization techniques, pp. 213–222. https://doi.org/10.1061/9780784483411.0.

Aumônier, S., Hartlin, B. and Peirce, A. 2010. *Carbon Footprint of Marine Aggregate Extraction.* Marine Estate Research Report. London, Environmental Resources Management Limited, The Crown Estate.

Bai, J., Ma, Y., Song, Q., Hu, Z., Li, Y. and Duan, H. 2024. Life cycle environmental assessment of three excavated soil and rock (ESR) treatment methods: A case study in Shenzhen city. *Sustainability*, 16: 2022. https://doi.org/10.3390/su16052022.

Balaguera, A., Carvajal, G.I., Albertí, J., and Fullana-i-Palmer, P. 2018. Life cycle assessment of road construction alternative materials: A literature review. *Resources, Conservation and Recycling*, 132: 37–48. https://doi.org/10.1016/j.resconrec.2018.01.003.

Balasbaneh, A.T. and Marsono, A.K.B. 2020. Applying multi-criteria decision-making on alternatives for earth-retaining walls: LCA, LCC, and S-LCA. *The International Journal of Life Cycle Assessment*, 25: 2140–2153.

Balasbaneh, A.T., Yeoh, D., Juki, M.I., Ibrahim, M.H.W. and Abidin, A.R.Z. 2021. Assessing the life cycle study of alternative earth-retaining walls from an environmental and economic viewpoint. *Environmental Science and Pollution Research*, 28: 37387–37399. https://doi.org/10.1007/s11356-021-13190-4.

Basu, D., Misra, A. and Puppala, A.J. 2014. Sustainability and geotechnical engineering: Perspectives and review. *Canadian Geotechnical Journal*, 52 (1): 96–113.

Basu, D., Misra, A. and Puppala, A.J. 2015. Sustainability and geotechnical engineering: Perspectives and review. *Canadian Geotechnical Journal*, 52 (1). https://doi.org/10.1139/cgj-2013-0120

Beaulieu Technical Textiles. 2022. *Woven geotextile EPD*. The International EPD® System, registration number S-P-05133, Comines-Warneton, Belgium.

Bednarski, Ł., Sieńko, R., Grygierek, M. and Howiacki, T. 2021. New distributed fibre optic 3DSensor with thermal self-compensation system: Design, research and field proof application inside geotechnical structure. *Sensors*, 21 (15): 5089.

Beus, N., Carus, M. and Barth, M. 2019. *Carbon Footprint and Sustainability of Different Natural Fibres for Biocomposites and Insulation Material – Study Providing Data for the Automotive and Insulation Industry*. Comprehensively revised second edition, nova-Institut GmbH, Hürth (Germany).

Blanck, G., Cuisinier, O. and Masrouri, F. 2016. Life cycle assessment of non-traditional treatments for the valorisation of dry soils in earthworks. *International Journal of Life Cycle Assessment*, 21: 1035–1048. https://doi.org/10.1007/s11367-016-1076-y.

Brundtland, G.H. 1987. *Our Common Future: Report of the World Commission on Environment and Development.* Geneva, UN-Document A/42/427.

Chang, I., Im, J. and Cho, G.C. 2016. Introduction of microbial biopolymers in soil treatment for future environ-mentally-friendly and sustainable geotechnical engineering. *Sustainability*, 8 (3), 251.

Chang, I., Lee, M., Tran, A.T.P., Lee, S., Kwon, Y.-M., Im, J. and Cho, G.-C. 2020. Review on biopolymer-based soil treatment (BPST) technology in geotechnical engineering practices. *Transportation Geotechnics*, 24: 100385.

Chau, C., Soga, K., Nicholson, D., O'Riordan, N. and Inui, T. 2008. Embodied energy as an environmental impact indicator for basement wall construction. *Proceedings of GeoCongress 2008: Geosustainability and Geohazard Mitigation*, Reston, VA. https://doi.org/10.1061/40971(310)10.

Chau, C., Soga, K., O'Riordan, N. and Nicholson, D. 2012. Embodied energy evaluation for sections of the UK channel tunnel rail link. *Proceedings of the Institution of Civil Engineers – Geotechnical Engineering*, 165 (2): 65–81. https://doi.org/10.1680/geng.9.00018.

Chen, J., Xing, X., Wang, Y., Zhang, W., Guo, Z. and Su, W. 2022. Application of iron and steel slags in mitigating greenhouse gas emissions: A review. *Journal of Science of the Total Environment*, (844): 157041. https://doi.org/10.1016/j.scitotenv.2022.15704.

Chica, L., Villada, J.P., Arcila, J.M. and Restrepo, J.W. 2023. Technical viability and life cycle assessment to excavated soil and CDW blends for hydraulic applications. *KSCE Journal of Civil Engineering*, 27: 1412–1422. https://doi.org/10.1007/s12205-023-1632-5.

COM/2022/142-final. 2022. Proposal for a regulation of the European parliament and of the council establishing a framework for setting ecodesign requirements for sustainable products and repealing Directive 2009/125/EC (European Commission).

Cordi, A.M., Gallagher, P.M. and Spatari, S. 2013. Environmental life cycle performance of recycled materials for sustainable slope engineering. *Proceedings of the Geo-Congress 2013: Stability and Performance of Slopes and Embankments III*, San Diego, CA, 3–7, pp. 1490–1501.

Correia, A.A.S., Caldeira, J.B., Branco, R. and Morais, P.V. 2023. Enhancing the strength of mine residue soil by bioremediation combined with biopolymers. *Applied Sciences*, 13 (18): 10550 https://doi.org/10.3390/app131810550

Correia, A.G., Winter, M.G. and Puppala, A.J. 2016. A review of sustainable approaches in transport infrastructure geotechnics. *Transportation Geotechnics*, 7: 21–28.

Craig, M.S., Mitchell, J.K. and Dove, J.E. 2015a. Energy and carbon assessment of ground improvement works. I: Definitions and background. *Journal of Geotechnical and Geoenvironmental Engineering*, 142 (3): 04015083. https://doi.org/10.1061/(ASCE)GT.1943-5606.0001410.

Craig, M.S., Mitchell, J.K. and Dove, J.E. 2015b. Energy and carbon assessment of ground improvement works. II: Working model and example. *Journal of Geotechnical and Geoenvironmental Engineering*, 142 (3): 04015084. https://doi.org/10.1061/(ASCE)GT.1943-5606.0001411.

Cristelo, N., Silva, C., Vieira, C. and Lopes, M.D. 2016. Geotechnical and geoenvironmental assessment of recycled construction and demolition waste for road embankments. *3rd International Conference on Transportation Geotechnics (ICTG) (Procedia Engineering, 143;* Guimaraes, PORTUGAL*)*, pp. 51–58.

Curpen, S., Teutsch, N., Kovler, K. and Spatari, S. 2023. Evaluating life cycle environmental impacts of coal fly ash utilization in embankment versus sand and landfilling. *Journal of Cleaner Production*, 385: 135402. https://doi.org/10.1016/j.jclepro.2022.135402

Rocha, C.G., Passuello, A., Consoli, N.C., Quinónez Samaniego, R.A. and Kanazawa, N.M. 2016. Life cycle assessment for soil stabilization dosages: A study for the Paraguayan Chaco. *Journal of Cleaner Production*, 139: 309–318.

Dąbrowska, J., Kiersnowska, A., Zięba, Z. and Trach, Y. 2023. Sustainability of geosynthetics-based solutions. *Environments*, 10 (4): 64.

Damians, I.P., Bathurst, R.J., Adroguer, E.G., Josa, A. and Lloret, A. 2017. Environmental assessment of earth retaining wall structures. *Environmental Geotechnics*, 4: 415–431.

Das, J.T., Puppala, A.J., Bheemasetti, T.V., Walshire, L.A. and Corcoran, M.K. 2018. Sustainability and resilience analyses in slope stabilization. *Proceedings of the Institute of Civil Engineers: Engineering Sustainability*, 171 (1): 25–36.

Das, J.T., Samuel, R.A., George, A.M., Chakraborty, S., Bheemasetti, T.V. and Puppala, A.J. 2018. Establishing a threshold sustainability index for a geotechnical construction. *International Foundation Congress and Equipment Expo 2018- Case Histories, Lessons Learned, and General Practice (ASCE)*, pp. 222–232.

De Belie, N. and Wang, J. 2016. Bacteria-based repair and self-healing of concrete. *Journal of Sustainable Cement-Based Materials*, 5 (1–2): 35–56.

de Melo, D.L., Alissa Kendall, A. and Jason T DeJong, J.T. 2024. Evaluation of life cycle assessment (LCA) use in geotechnical. *Environmental Research: Infrastructure and Sustainability*, 4, 012001. DOI 10.1088/2634-4505/ad2154

Deng, X., Li, Y., Liu, H., Zhao, Y., Yang, Y., Xu, X., Cheng, X. and Wit, B.D. 2021. Examining energy consumption and carbon emissions of microbial induced carbonate precipitation using the life cycle assessment method. *Sustainability*, 13 (9): 012001.

Ding, N., Meng, X., Zhang, Z., Zhang, Z., Ma, J., Shan, Y., Zhong, Z., Yu, H., Li, M. and Jiao, W. 2024. A review of life cycle assessment of soil remediation technology: Method applications and technological characteristics. *Reviews of Environmental Contamination & Toxicology*, 262 (1): 1–24. https://doi.org/10.1007/s44169-023-00051-z

Directive-125/EC. 2009. Establishing a framework for the setting of ecodesign requirements for energy-related products. *Official Journal of the European Union*, (285): 10–35.

Du, C., Dutta, S., Kurup, P., Yu, T. and Wang, X.S. 2020. A review of railway infrastructure monitoring using fiber optic sensors. *Sensors and Actuators A: Physical*, 303: 111728.

ECESP. 2021. *Circular Buildings and Infrastructure – State of Play Report*. European Circular Economy Stakeholder Platform, Brussels, Belgium.

Egan, D. and Slocombe, B.C. 2010. Demonstrating environmental benefits of ground improvement. *Proceedings of the Institution of Civil Engineers*, 163: 63–69.

Elsing, A., Fraser, I., Stucki, M., Büsser, S., Itten, R., Frischknecht, R. and Wallbaum, H. 2012. Comparative life cycle assessment of geosynthetics versus conventional construction materials, a study on behalf of the E.A.G.M., case 2, foundation stabilization. *5 European Geosynthetics Congress, Vol. 1, Topic Transport, Valencia 2012*.

European Commission. 2020. *Taxonomy: Final Report of the Technical Expert Group on Sustainable Finance*. European Commission, Brussels, Belgium.

European Commission, EUCO. 29/19 2019. *European Council Meeting (12 December 2019) – Conclusions*. European Commission, Brussels, Belgium.

European Committee for standardization 2013 EN 15804:2012+A1:2013 Sustainability of construction works – Environmental product declarations – Core rules for the product category of construction products Edition 1 (Brussels)

European Confederation of Flax and Hemp. 2022. *Detailed Life Cycle Assessment (LCA) of European Flax™ Scutched Fibre*, Alliance for European Flax-Linen & Hemp (A C L A).

Faruqi, A., Hall, C.A. and Kendall1 A. 2023. Sustainability of bio-mediated and bio-inspired ground improvement techniques for geologic hazard mitigation: A systematic literature review. *Frontiers in Earth Science*, 11: 1211574. https://doi.org/10.3389/feart.2023.

Ferrazzo, A.T., Araújo, M,T. Consoli, N.C , and Rocha,C.G., 2024a. Geotech social impacts: Development, application, and comparative analysis- *Environmental Impact Assessment Review*, 108, 107577. https://doi.org/10.1016/j.eiar.2024.107577.

Ferrazzo, A.T., Araújo, M,T. and Consoli, N.C 2024b. Which solution is more sustainable: Waste foundry sand stabilized with alkali-activated binder or Portland cement?, *Journal of Building Engineering*, 84, 108448. https://doi.org/10.1016/j.jobe.2024.108448.

Finnveden, G., Hauschild, M.Z., Ekvall, T., Guinée, J., Heijungs, R., Hellweg, S., Koehler, A., Pennington, D. and Suh, S. 2009. Recent developments in life cycle assessment. *Journal of Environmental Management*, 91 (1): 1–21. https://doi.org/10.1016/j.jenvman.2009.06.018.

Flammini, F., Naddei, R., Pragliola, C. and Smarra, G. 2016. Towards automated drone surveillance in railways: State-of-the-art and future directions. *Advanced Concepts for Intelligent Vision Systems*, J. edited by Blanc-Talon, Cosimo Distante, Wilfried Philips, Dan C. Popescu and Paul Scheunders Springer International Publishing, Cham, pp. 336–348.

Frischknecht, R., Büsser-Knöpfel, S., Itten, R., Stucki, M. and Wallbaum, H. 2013. Comparative life cycle assessment of geosynthetics versus conventional filter layer. *Proceedings of the 18th International Conference on Soil Mechanics and Geotechnical Engineering*, pp. 3203–3206. www.issmge.org/uploads/publications/1/2/3203-3206.pdf.

Ghadir, P., Zamanian, M., Mahbubi-Motlagh, N., Saberian, M., Li, J. and Ranjbar, N. 2021. Shear strength and life cycle assessment of volcanic ash-based geopolymer and cement stabilized soil: A comparative study. *Transportation Geotechnics*, 31: 100639.

GIH. 2021. *Advancing the Circular Economy through Infrastructure – Transition Pathways for Practitioners in Circular Infrastructure*. Global Infrastructure Hub, London, UK (a G20 initiative), p. 14.

Giri, R.K. and Reddy, K.R. 2014. LCA and sustainability assessment for selecting deep foundation system for high-rise buildings. *ICSI2014*, pp. 621–630.

Giri, R.K. and Reddy, K.R. 2015. Sustainability assessment of two alternate earth-retaining structures. *IFCEE*, pp. 2836–2845.

Goldenberg, M. and Reddy, K.R. 2017. Sustainability assessment of conventional and alternate landfill cover systems. *Geotechnical Frontiers*, 323–332.

Gu, X., Makasis, N., Motamedi, Y., Narsilio, G.A., Arulrajah, A. and Horpibulsuk, S. 2022. Geothermal pavements: Field observations, numerical modelling and long-term performance. *Géotechnique*, 72 (9): 832–846.

Gu, X., Makasis, N., Narsilio, G.A., Arulrajah, A. and Horpibulsuk, S. 2024. Impact of rainfall on the thermal performance of geothermal permeable pavement systems. *Transportation Geotechnics*, 45: 101194.

Haberta, G. and Ouellet-Plamondon, C. 2016. Recent update on the environmental impact of geopolymers. *RILEM Technical Letters*, 1: 17–23.

Hammond, G. and Jones, C. 2011. *Inventory of Carbon & Energy (ICE) Version 2.0 Summary Tables*. University of Bath, Bath, UK.

Harbottle, M., Al-Tabbaa, A. and Evans, C. 2007. A comparison of the technical sustainability of in situ stabilisation/solidification with disposal to landfill. *Journal of Hazardous Materials*, 141: 430–440.

Hausberger, L., Cordes, T. and Gschösser, F. 2023. Life cycle assessment of high-performance railway infrastructure, analysis of superstructures in tunnels and onOpen tracks. *Sustainability*, 15: 7064. https://doi.org/10.3390/su15097064.

Heath, A., Paine, K. and McManus, M. 2014. Minimising the global warming potential of clay based geopolymers. *Journal of Cleaner Production*, 78: 75–83. https://doi.org/10.1016/j.jclepro.2014.04.046.

Heitor, A., Parkinson, J. and Kotzur, T. 2021 The role of soil stabilisation in mitigating the impact of climate change in transport infrastructure. *MDPI Applied Sciences: Extreme Sciences and Engineering II*, Special issue, 11(3): 1080. https://doi.org/10.3390/app11031080.

Henrickson, C., Horvath, A., Joshi, S. and Lave, L. 1998. Economic input-output models for environmental life-cycle assessment. *Environmental Science and Technology*, 32(7): 184A–191A.

Hoerbinger, S. and Obriejetan, M. 2015. Environmental life cycle assessment model for soil bioengineering measures on infrastructure slopes. *Proceedings of the EGU General Assembly Conference Abstracts*, Vienna, Austria, p. 7924.

Huang, L., Bohne, R.A., Bruland, A., Jakobsen, P. and Lohne, J. 2014. Life cycle assessment of Norwegian road tunnel. *The International Journal of Life Cycle Assessment*, 20: 176–184.

Husain, S.F., Wiggins, R., Kim, Y., Alhadidi, Y. and Tutumluer, E. 2024. Resiliency against flooding in pavement geotechnics: Practice review of state transportation agencies. *Transportation Research Record*, 2678(11): 1719–1732.

Hutton Stone. 2022. *Darney Walling Stone EPD*. Looper EPD System. Hexham, Northumberland, Reino Unido

Indraratna, B., Arachchige, C.M.K., Rujikiatkamjorn, C., Heitor, A. and Qi, Y.J. 2024. Utilization of granular wastes in transportation infrastructure. *Geotechnical Testing Journal*, 47 (1): 409–424.

Inui, T., Chau, C., Soga, K., Nicolson, D. and O'Riordan, N. 2011. Embodied energy and gas emissions of retaining wall structures. *Journal of Geotechnical and Geoenvironmental Engineering*, 137: 958–967.

ISO 14040. 2006. *Environmental Management—Life Cycle Assessment—Principles and Framework*. International Organisation for Standardization. Geneva, Switzerland.

ISO 14044. 2006. *Environmental Management—Life Cycle Assessment—Requirements and Guidelines*. International Organisation for Standardization. Geneva, Switzerland.

Javeed, Y., Goh, Y., Mo, K.H., Yap, S.P. and Leo, B.F. 2024. Microbial self-healing in concrete: A comprehensive exploration of bacterial viability, implementation techniques, and mechanical properties. *Journal of Materials Research and Technology*, 29: 2376–2395.

JSW Cement Limited. 2019. *Average Ground Granulated Blast- Furnace Slag EPD*. The International EPD® System, registration number S-P-01415. Stockholm, Sweden.

Kendall, A., Raymond, A.J., Tipton, J. and DeJong, J.T. 2018. Review of life cycle-based environmental assessments of geotechnical systems. *Proceedings of the Institution of Civil Engineers: Engineering Sustainability*, 171 (2): 57–67. https://doi.org/10.1680/jensu.16.00073.

Kumar, A. and Parihar, A. 2023. Design and life cycle assessment of retaining wall with used foundry sand as backfill, pp. 55–63. In E. Rathje, B. M. Montoya, & M. H. Wayne (Eds.), *Geo-Congress 2023: Soil Improvement, Geoenvironmental, and Sustainability — Selected Papers from Sessions of Geo-Congress 2023* (Geotechnical Special Publication No. 339). American Society of Civil Engineers.

Lafarge. 2021. *Ready Mix Concrete EPD*. The International EPD® System, registration number S-P-04987. Stockholm, Sweden.

Laveglia, A., Sambataro, L., Ukrainczyk, N., de Belie, N. and Koenders, E. 2022. Hydrated lime life-cycle assessment: Current and future scenarios in four EU countries. *Journal of Cleaner Production*, 369: 133224. https://doi.org/10.1016/J.JCLEPRO.2022.133224.

Leal, D., Winter, M.G., Seddon, R. and Nettleton, I.M. 2018. *Innovative Geotechnical Repair Techniques—Comparative Life Cycle Assessment.* TRL Limited, p. 30. Crowthorne, Berkshire, UK.

Leal, D., Winter, M.G., Seddon, R. and Nettleton, I.M. 2020. A comparative life cycle assessment of innovative highway slope repair techniques. *Transporation Geotechnics*, 22: 100322.

Lee, M. and Basu, D. 2015. Sustainability assessment of mechanically stabilized earth walls. *15th Pan-American Conference on Soil Mechanics and Geotechnical Engineering*, Buenos Aires, Argentina, pp. 830–837.

Lee, M. and Basu, D. 2016. Impacts of the design methods of drilled shafts in sand on the environment, pp. 673–682. IFCEE 2018 (International Foundations Congress & Equipment Expo), Orlando, FL, EUA.

Lee, M. and Basu, D. 2018. Environmental impacts of drilled shafts and driven piles in sand. *IFCEE.* https://doi.org/10.1061/9780784481578.061.

Lee, M. and Basu, D. 2022. Environmental impacts of drilled shafts in sand. *Proceedings of the Institution of Civil Engineers—Engineering Sustainability*, 176 (1): 39–52. https://doi.org/10.1680/jensu.21.00091

Lee, M. and Basu, D. 2023. Environmental impacts of drilled shafts in sand. *Proceedings of the Institution of Civil Engineers – Engineering Sustainability*, 176 (1): 39–52.

Li, L. and Chen, K. 2017. Quantitative assessment of carbon dioxide emissions in construction projects: A case study in Shenzhen. *Journal of Cleaner Production*, 141: 394–408.

Li, X.J. and Zheng, Y.D. 2020. Using LCA to research carbon footprint for precast concrete piles during the building construction stage: A China study. *Journal of Cleaner Production*, 245: 118754.

Liu, C., Ahn, C.R., An, X. and Lee, S. 2013. Life-cycle assessment of concrete dam construction: Comparison of environmental impact of rock-filled and conventional concrete. *Journal of Construction Engineering and Management*, 139(12), A4013009-1–A4013009-11. https://doi.org/10.1061/(ASCE)CO.1943-7862.0000752

Look, B. 2014. *Handbook of Geotechnical Investigation and Design Tables.* Second edition. CRS Press, Boca Raton, Florida, EUA.

Look, B.G. 2007. *Handbook of Geotechnical Investigation and Design Tables.* Taylor and Francis, London, UK.

Lubrecht, M.D. 2012. Horizontal directional drilling: A green and sustainable technology for site remediation. *Environmental Science & Technology*, 46: 2484–2489.

Maghazei, O. and Steinmann, M. 2020. Drones in railways: Exploring current applications and future scenarios based on action research. *European Journal of Transport and Infrastructure Research*, 20 (3): 87–102.

Martin, K.K., Khodadadi, T.H., Chester, M. and Kavazanjian, E. 2020. Hotspot life cycle assessment for environ-mental impacts of EICP for ground improvement. *Geo-Congress 2020,* Minneapolis, Minnesota, EUA, pp. 321–329.

Mattson, D.A. 2022. *Recycled aggregates EPD.* The International EPD® System, registration number S-P-06094, Stockholm, Sweden.

McLellan, B.C., Williams, R.P., Lay, J., van Riessen, A. and Corder, G.D. 2011. Costs and carbon emissions for geopolymer pastes in comparison to ordinary Portland cement. *Journal of Cleaner Production*, 19 (9–10): 1080–1090.

Mendonça, A., Morais, P.V., Pires, A.C., Chung, A.P. and Venda Oliveira, P.J. 2021. Reducing soil permeability using bacteria-produced biopolymer. *Applied Sciences*, 11 (16): 7278.

Miguel, G.D., Saldanha, R.B., da Silva, A., Festugato, L., Chaves, H.M. and Mendes, C.C. 2023. Comparison of life-cycle assessment: Environmental and cost approach to soil improvement. *Proceedings of the Institution of Civil Engineers – Ground Improvement*, 176 (3): 182–198. https://doi.org/10.1680/jgrim.21.00006a.

Misra, A. 2010. *A Multicriteria Based Quantitative Framework for Assessing Sustainability of Pile Foundations.* University of Connecticut. https://opencommons.uconn.edu/cgi/viewcontent.cgi?article=1051&context=gs_theses

Misra, A. and Basu, D. 2011. *Sustainability in Geotechnical Engineering.* Internal Geotechnical Report 2011–2, p. 50. https://opencommons.uconn.edu/cgi/viewcontent.cgi?article=1000&context=cee_techreports.

Modahl, I.S. and Soldal, E. 2021. *The 2019 LCA of products from Borregaard, Sarpsborg.* Report No. OR.14.21 Open Version; 2021. https://norsus.no/wp-content/uploads/OR.14.21-Open-The-2019-LCA-of-products-from-Borregaard-Sarpsborg-.pdf?v=1, accessed September 12, 2022.

Nawaz, M., Heitor, A. and Sivakumar, M. 2020. Geopolymers in construction: Recent developments. *Construction and Building Materials*, 260: 120472.

Ng, C.W.W., Zhang, Q., Zhou, C. and Ni, J.J. 2022. Eco-geotechnics for human sustainability. *Science China-Technological Sciences*, 65 (12): 2809–2845.

Noh, S., Son, Y. and Park, J. 2018. Life cycle carbon dioxide emissions for fill dams. *Journal of Cleaner Production*, 201: 820–829. www.sciencedirect.com/science/article/pii/S0959652618324417?via%3Dihub.

Ökobaudat. 2018. *Rammed earth wall EPD*. https://www.oekobaudat.de/OEKOBAU.DAT/datasetdetail/process.xhtml?uuid=59622fc0-d719-43bf-a16b-ae924ae0d276&version=20.19.120, accessed September 25, 2024.

Olinic, T., Stanciu, A.M., Butcaru, A.C. and Luchian, V. 2023. Bio-reinforcement of slopes. *Scientific Papers-Series E-Land Reclamation Earth Observation & Surveying Environmental Engineering*, 12: 25–30.

Ongpeng, J.M.C. and Ginga, C.P. 2022. Life cycle assessment and carbon footprint analysis of recycled aggregates in the construction of earth-retaining walls during reconstruction. *Advances of Footprint Family for Sustainable Energy and Industrial Systems*, edited by J. Ren. Springer International Publishing, pp. 15–34, Cham, Switzerland.

Pacheco, V.L., Bragagnolo, L., Reginatto, C. and Thomé, A. 2022. Microbially induced calcite precipitation (MICP): Review from an engineering perspective. *Geotechnical and Geological Engineering*, 40 (5): 2379–2396.

Palmeira, E.M., Araújo, G.L.S. and Santos, E.C.G. 2021. Sustainable solutions with geosynthetics and alternative construction materials—A review. *Sustainability*, 13 (22): 12756.

Parente, M., Correia, A.G. and Cortez, P. 2016. Metaheuristics, data mining and geographic information systems for earthworks equipment allocation. *3rd International Conference on Transportation Geotechnics (ICTG) (Procedia Engineering, 143;* Guimaraes, PORTUGAL*)*, pp. 506–513.

PAS-2080. 2023. *Carbon Management in Buildings and Infrastructure*. British Standards Institution, London, UK.

Pehnt, M. 2006. Dynamic life cycle assessment (LCA) of renewable energy technologies. *Renewable Energy*, 31 (1): 55–71. https://doi.org/10.1016/j.renene.2005.03.002.

Perri, G., de Rose, M., Domitrovíc, J. and Vaiana, R. 2023. CO_2 impact analysis for road embankment construction: Comparison of lignin and lime soil stabilization treatments. *Sustainability*, 15: 1912. https://doi.org/10.3390/ su15031912.

Pettinaroli, A., Susani, S., Castellanza, R., Collina, E.M., Pierani, M., Paoli, R. and Romagnoli, F. 2023a. A sustainability-based approach for geotechnical infrastructure. *Environmental and Climate Technologies*, 27 (1): 738–752.

Pettinarolli, A., Susane, S., Castellanza, R., Collina, E.M., Pieranis, M., Paoli, R. and Romagnoli, F. 2023b. The impact of a permeation grouting technique quantitatively assessed through a process-focused life cycle assessment. *Environmental and Climate Technologies*, 27 (1): 797–812. https://doi.org/10.2478/rtuect-2023-0058.

Peuportier, B., Leurent, F. and Roger-Estrade, J. 2020. *Eco-Design of Buildings and Infrastructure: Developments in the Period 2016–2020*, Vol. 4. CRC Press, Boca Raton, Florida, EUA.

Phillips, E.K., Shillaber, C.M., Mitchell, J.K., Dove, J.E. and Filz, G.M. 2016. Sustainability comparison of a geo-synthetic-reinforced soil abutment and a traditionally-founded abutment: A case history, pp. 699–711, Phoenix, Arizona, EUA.

Pinske, M.A. 2011. *Life Cycle Assessment of Ground Improvement Methods*. MSc Thesis, University of California Davis.

Pons, J.J., Penadés-Pl`a V., Yepes, V. and Martí, J.V. 2018. Life cycle assessment of earth-retaining walls: An environmental comparison. *Journal of Cleaner Production*, 192: 411–420.

Pujadas-Gispert, E., Sanjuan-Delmás, D. and Josa, A. 2018. Environmental analysis of building shallow foundations: The influence of prefabrication, typology, and structural design codes. *Journal of Cleaner Production*, 186: 407–417.

Pujadas-Gispert, E., Vogtländer, J.G. and Moonen, S.P.G. 2021. Environmental and economic optimization of a conventional concrete building foundation: Selecting the best of 28 alternatives by applying the pareto front. *Sustainability*, 13: 1496.

Puppala, A.J. 2016. Advances in ground modification with chemical additives: From theory to practice. *Transportation Geotechnics*, 9: 123–138.

Purdy, C.M., Raymond, A.J., DeJong, J.T., Kendall, A., Krage, C. and Sharp, J. 2022. Life-cycle sustainability assessment of geotechnical site investigation. *Canadian Geotechnical. Journal*, 59: 863–877. https://doi.org/10.1139/cgj-2020-0523.

Rafalko, S.D., Sankey, J.E. and Freitag, N. 2010. Sustainability measures for MSE Walls and baseline environ-mental impact evaluations. *American Society of Civil Engineers*, 208: 486–493.

Rail Safety and Standards Board (RSSB). 2024. *Rail Carbon Tool*. https://www.rssb.co.uk/sustainability/rail-carbon-tool, accessed May 20, 2024.

Raymond, A. J., Kendall, A., DeJong, J. T., Kavazanjian, E., Woolley, M. A., & Martin, K. K. 2021a. Life cycle sustainability assessment of fugitive dust control methods. *Journal of Construction Engineering and Management*, 147(3): 04020181. https://doi.org/10.1061/(ASCE)CO.1943-7862.0001993

Raymond, A.J., DeJong, J.T., Kendall, A., Blackburn, J.T. and Deschamps, R. 2021b. Life cycle sustainability assessment of geotechnical ground improvement methods. *Journal of Geotechnical and Geoenvironmental Engineering*, 147 (12): 04021161.

Raymond, A.J., Pinskse, M.A., Kendall, A. and DeJong, J.T. 2017. Life-cycle assessment of ground improvement alternatives for the Treasure Island, California, redevelopment. *Geotechnical Frontiers*: 345–354.

Raymond, A.J., Tipton, J.R., Kendall, A. and DeJong, J.T. 2019. Review of impact categories and environmental indicators for life cycle assessment of geotechnical systems. *Journal of Industrial Ecology*, 24: 485–499. https://doi.org/10.1111/jiec.12946.

Raza, F., Alshameri, B. and Jamil, S.M. 2021a. Assessment of triple bottom line of sustainability for geotechnical projects. *Environment Development and Sustainability*, 23 (3): 4521–4558.

Raza, F., Alshameri, B. and Jamil, S.M. 2021b. Engineering aspect of sustainability assessment for geotechnical projects. *Environment Development and Sustainability*, 23 (4): 6359–6394.

Roadstone. 2023. *Aggregates & Granular Fill (Belgard Quarry) EPD*. The international EPD system, registration number S-P-05739. Stockholm, Sweden.

Rocha, C.C , Marin, E.J.B., Samaniego, R.A.Q., and Consoli, N,C, 2020. Decision-Making Model for Soil Stabilization: Minimizing Cost and Environmental Impacts. *Journal of Materials in Civil Engineering*, 33 (2). https://doi.org/10.1061/(ASCE)MT.1943-5533.0003551.

Rocha, C.C , Saldanha, R,B,, Araújo, M,T, and Consoli, N,C, 2022. Social and environmental assessments of Eco-friendly Pavement alternatives. *Construction and Building Materials*, 325, 126736. https://doi.org/10.1016/j.conbuildmat.2022.126736.

Roux, C. 2020. Eco-design of buildings and infrastructure – Developments in the period 2016–2020. *Chapter 1: Consequential Life Cycle Assessment Applied to Buildings*, edited by Bruno Peuportier, Fabien Leurent and Jean Rog-er-Estrade. CRC Press, London, p. 518. https://doi.org/10.1201/9781003095071.

Saieg, P., Sotelino, E.D., Nascimento, D. and Caiado, R.G.G. 2018. Interactions of building information modeling, lean and sustainability on the architectural, engineering and construction industry: A systematic review. *Journal of Cleaner Production*, 174: 788–806.

Saldanha, R.B., da Rocha, C.G., Caicedo, A.M.L. and Consoli, N.C. 2021. Technical and environmental performance of eggshell lime for soil stabilization. *Construction and Building Materials*, 298: 123648.

Saldanha, R.B., Caicedo, A.M.L., Araújo, M.T., Scheuermann Filho, H.C., Moncaleano,C.J., Silva, J.P.S.. and Consoli, N.C. 2023. Potential use of iron ore tailings for binder production: A life cycle assessment. *Construction and Building Materials*, 365: 130008, https://doi.org/10.1016/j.conbuildmat.2022.130008.

Salemans, J.W.M. and Blauw, M. 2010. *Life Cycle Analysis for Four Different Ground Improvement Techniques*. Ref. 1201072-003-OA-0001, Version 2, 11, Deltares.

Samuelsson, I., Larsson, S. and Spross, J. 2021. Life cycle assessment and life cycle cost analysis for geotechnical engineering: Review and research gaps. *IOP Conference Series: Earth and Environmental Science*, 710: 012031.

Samuelsson, I., Spross, J. and Larsson, S. 2022. Assessment of climate impact and costs comparing two railway embankment fill methods. *5th International Conference on New Developments in Soil Mechanics and Geotechnical Engineering*, vol. 305, pp. 181–189.

Samuelsson, I., Spross, J. and Larsson, S. 2024. Integrating life-cycle environmental impact and costs into geotechnical design. *Proceedings of the Institution of Civil Engineers – Engineering Sustainability*, 177 (1): 19–30.

Sánchez-Garrido, A.J., Navarro, I.J. and Yepes, V. 2022. Evaluating the sustainability of soil improvement techniques in foundation substructures. *Journal of Cleaner Production*, 351: 131463. https://doi.org/10.1016/j.jclepro.2022.131463.

Scoular, J., Ghail, R., Mason, P., Lawrence, J., Bellhouse, M., Holley, R. and Morgan, T. 2020. Retrospective InSAR analysis of east London during the construction of the Lee Tunnel. *Remote Sensing*, 12 (5): 849.

Seol, Y., Lee, S., Lee, J.-Y., Han, J.-G. and Hong, G. 2021. Excavation method determination of Earth-retaining wall for sustainable environment and economy: Life cycle assessment based on construction cases in Korea. *Sustainability*, 13: 2974.

Setiawan, A.A., Hardjasaputra, H. and Soegiarso, R. 2022. Embodied carbon dioxide of fly ash based geopolymer concrete. *The 8th International Conference of Euro Asia Civil Engineering Forum 2022, IOP Conf. Series: Earth and Environmental Science 1195 (2023) 012031*. IOP Publishing. https://doi.org/10.1088/1755-1315/1195/1/012031.

Shi, X., Kou, L., Liang, H., Wang, Y. and Li, W. 2024. Evaluating carbon emissions during slurry shield tunneling for sustainable management utilizing a hybrid life-cycle assessment approach. *Sustainability*, 16: 2702. https://doi.org/10.3390/su16072702.

Shillaber, C.M., Mitchell, J.K. and Dove, J.E. 2016. Energy and carbon assessment of ground improvement works. II: Working model and example. *Journal of Geotechnical and Geoenvironmental Engineering*, 142: 04015084.

Shillaber, C.M., Mitchell, J.K., Dove, J.E. and Ostrum, Z.A. 2017. Framework to account for uncertainty in energy and carbon assessment of ground improvement works. *Journal of Geotechnical & Geoenvironmental Engineering*, 143(5), 04017002. https://doi.org/10.1061/(ASCE)GT.1943-5606.0001649

Soga, K. 2010. Infrastructure embodied energy and gas emission of geotechnical infrastructure. *Kyoto Seminar 2010 on Geotechnics and Earthquake Geotechnics towards Global Sustainability (Geotechnical Geological and Earthquake Engineering, 15;* Kyoto Univ, Kyoto, Japan*)*, pp. 59–74.

Song, X., Carlsson, C., Kiilsgaard, R., Bendz, D. and Kennedy, H. 2020. Life cycle assessment of geotechnical works in building construction: A review and recommendations. *Sustainability*, 12: 8442. https://doi.org/10.3390/su12208442

Spaulding, C., Masse, F. and LaBrozzi, J. 2012. Ground improvement technologies for a sustainable world, pp. 891–898. In *Proceedings of GeoCongress 2008: Geosustainability and Geohazard Mitigation* (Geotechnical Special Publication No. 178). American Society of Civil Engineers, Oakland, Califórnia, EUA.

Spencer, C.A., van Paassen, L. and Sass, H. 2020. Effect of jute fibres on the process of MICP and properties of biocemented sand. *Materials*, 13(23): 5429. https://doi.org/10.3390/ma13235429

Speranza, G., Ferrari, A. and Lassalle, P.L. 2022. Life cycle environmental assessment of retaining walls in unsaturated soils Geomechan. *Energy & Environmental Science*, 30: 100241.

Steedman, R.S. 2010. Geotechnics and society carbon, a new focus for delivering sustainable geotechnical engineering. *Kyoto Seminar 2010 on Geotechnics and Earthquake Geotechnics towards Global Sustainability (Geotechnical Geological and Earthquake Engineering, 15;* Kyoto Univ, Kyoto, Japan*)*, pp. 75–88.

Storesund, R., Massey, J. and Kim, Y. 2012. Life cycle impacts for concrete retaining walls vs bioengineered slopes, In K. R. Reddy, M. V. Khire, & A. N. Alshawabkeh (Eds.), *GeoCongress 2008: Geosustainability & Geohazard Mitigation* (Proceedings; Geotechnical Special Publication No. 178, pp. 875–882). American Society of Civil Engineers. New Orleans, Louisiana, EUA.

Suer, P., Hallberg, N., Carlsson, C., Bendz, D. and Holm, G. 2009. Biogrouting compared to jet grouting: Environ-mental (LCA) and economical assessment. *Journal of Environmental Science and Health, Part A*, 44: 346–353.

Suhartomo, A., Simatupang, J.W., Widjaja, B. and Gapsari, F. 2019. Feasibility study on structural health monitoring systems using fiber-optic sensors (FOS) technology for transportation infrastructures in Indonesia. *IOP Conference Series: Materials Science and Engineering*, 494 (1): 012054.

Tata Steel Limited. 2022. *Steel Rebar EPD*. The International EPD® System, registration number S-P-05018, Stockholm, Sweden.

Tinoco, J., Correia, A.A.S., Venda Oliveira, P.J., Correia, A.G. and Lemos, L.J.L. 2016. A data-driven approach for qu prediction of laboratory soil-cement mixtures. *Procedia Engineering*, 143: 566–573.

UN – Environment-Programme 2023. *Building Materials and the Climate: Constructing a New Future*. Global Alliance for Buildings and Construction United Nations, Yale.

Vanícek, I., Jirásko, D. and Vanícek, M. 2016. Added value of transportation geotechnics to the sustainability (design approach). *Procedia Engineering*, 143: 1417–1424.

Venda Oliveira, P.J., Anunciação, G.R. and Correia, A.A.S. 2022. Effect of cyclic loading frequency on the behavior of a stabilized sand reinforced with polypropylene and sisal fibers. *Journal of Materials in Civil Engineering*, 34 (1): 06021008.

Venda Oliveira, P.J., Correia, A.A.S. and Garcia, M.R. 2013. Effect of stress level and binder composition on secondary compression of an artificially stabilized soil. *Journal of Geotechnical and Geoenvironmental Engineering*, 139 (5): 810–820.

Vieira, C.S. 2022. Sustainability in geotechnics through the use of environmentally friendly materials. *Sustainability*, 14 (3): 1155.

Vijayakumar, M., Cheng, J., Murat, K., Elgamal, A., Mackie, K. and Tatari, O. 2016. Carbon Footprint: *Liquefaction Effects on a Private Residence Geo-Chicago*. https://ascelibrary.org/doi/abs/10.1061/9780784480120.064.

von der Tann, L., Stordal, F. and Feizi, S. 2022. First steps in the development of standardised processes for life cycle assessments of geotechnical works. *IOPscience*, p. 1122.

Walker, C., Heitor, A. and Clarke, B. 2022. Influence of weather-driven processes on the performance of UK transport infrastructure with reference to historic geostructures. *MDPI Applied Sciences – Special Issue on Unsaturated Testing and Modelling*, 12 (15): 7461. https://doi.org/10.3390/app12157461.

WEC. 2024. *World Economic Forum – Engineering and Construction*. https://intelligence.weforum.org/topics/a1Gb0000000pTDgEAM?utm_source=4fb996aa-d3f3-4927-a5e7-ea9cc60a8e3a&utm_medium=InSIW&utm_campaign=https://www.weforum.org/&utm_content=iframe_implementation, accessed April 21, 2024.

Wernet, G., Bauer, C., Steubing, B., Reinhard, J., Moreno-Ruiz, E. and Weidema, B. 2016. The ecoinvent database version 3 (part I): Overview and methodology. *The International Journal of Life Cycle Assessment*, 21 (9): 1218–1230. www.link.springer. com/10.1007/s11367-016-1087-8.

Werth, K. and Höhny, S. 2012. Comparative life cycle assessment of geosynthetics versus conventional construction materials, a study on behalf of the E.A.G.M., case 3, landfill construction drainage layer. *EUROGEO 5*, Valencia, Spain.

Wilberforce, T., Olabi, A.G., Sayed, E.T., Elsaid, K., Maghrabie, H.M. and Abdelkareem, M.A. 2023. A review on zero energy buildings – Pros and cons. *Energy and Built Environment*, 4 (1): 25–38.

Wu, H., Aguilar, V. and Montgomery, J. 2022. Applications of machine learning to foundation design for transportation structures. *Transportation Research Record*, 2676(3): 03611981221122779. https://doi.org/10.1177/03611981221122779

Wu, W. and Skye, H.M. 2021. Residential net-zero energy buildings: Review and perspective. *Renewable and Sustainable Energy Reviews*, 142: 110859.

Yao, Y., Xu, P., Li, J., Hu, H. and Qi, Q. 2024. Advancements and applications of life cycle assessment in slope treatment: A comprehensive review. *Sustainability*, 16: 398. https://doi.org/10.3390/su16010398.

Yoo, C. 2023. Geosynthetic solutions for sustainable transportation infrastructure development. *Sustainability*, 15(22): 15772. https://doi.org/10.3390/su152215772

Yuguda, T.K., Li Y., Xiong, W. and Zhang, W. 2020. Life cycle assessment of options for retrofitting an existing dam to generate hydroelectricity. *The International Journal of Life Cycle Assessment*, 25: 57–72.

Zahmak, A., Abdallah, M., Jarah, B. and Arab, M.G. 2021. Environmental performance of alkali-activated binders for ground improvement. *Transportation Geotechnics*, 31: 100631. https://doi.org/10.1016/j.trgeo.2021.100631.

Zastrow, F., Molina-Moreno, F., García-Segura, T., Martí, J.V. and Víctor Yepes, V. 2017. Life cycle assessment of cost-optimized buttress earth-retaining walls: A parametric study. *Journal of Cleaner Production*, 140(3): 1037–1048. https://doi.org/10.1016/j.jclepro.2016.10.085.

Zhang, R., Long, M. and Zheng, J. 2019. Comparison of environmental impacts of two alternative stabilization techniques on expansive soil slopes. *Advances in Civil Engineering*: 2019: 9454929. https://doi.org/10.1155/2019/9454929

Part II

Intelligent and digital transformation

João Moutinho

This part provides an in-depth discussion of digital and intelligent transformation in the field of Transport Geotechnics, particularly with regard to how digitalization is changing monitoring, data analytics, and decision-making in practice. Digital technologies are continually changing how geotechnical professionals monitor, analyse, and respond to infrastructure conditions – offering unprecedented real-time analysis capabilities that signal a long-awaited advancement in the industry. This evolution promises new opportunities for enhancing the safety, efficiency, and sustainability of geotechnical projects as well as the use of exciting new opportunities such as artificial intelligence. Through an in-depth examination of various stages of digital transformation application – from the methods of data collection and interpretation to the implementation of intelligent tools for process optimization – this section highlights the powerful influence technology has on modern geotechnical practices.

The transformation process commenced with the replacement of traditional geotechnical monitoring methodologies by modern connected devices, such as IoT sensors, and the adoption of remote sensing technologies. Previously, monitored geotechnical variables were conventional geotechnical constructs, such as inclinometers and piezometers, which, although accurate, rarely provided an overview of the entire site. The incorporation of digital solutions, however, has created a new, ever-present layer of data that is constantly being accumulated, enabling engineers to respond to or manage the threats of risk or changes within the geotechnical environment in real time (Miranda et al. 2025). This transformation encourages a paradigm shift towards infrastructure monitoring with resilience, whereby changes can be made faster and factually sooner. The industry is now able to harness the power of edge, fog, and cloud computing to process massive data streams in real time, thanks to IoT sensors. Furthermore, remote sensing technologies and other technologies such as InSAR, LiDAR, and photogrammetry allow for monitoring to be taken to new scales and depths, making it feasible and highly efficient to understand soil behaviour and structural integrity. In this section, detailed case studies are provided to illustrate the practical application of these digital technologies, underscoring the tangible impact that digitalization has already achieved in real-world geotechnical engineering projects.

Beyond digital monitoring, data analysis in geotechnics is becoming increasingly precise and effective through the adoption of innovative methodologies and tools. Real-world case studies, examined in depth, serve to demonstrate solutions that facilitate decision-making, making daily tasks more manageable and seamlessly integrating technical expertise with data-driven insights (Pinheiro et al. 2025). The result is a refined and highly customized analysis of project-specific conditions that optimizes resource management while enhancing safety outcomes. Slope monitoring, numerical modelling, and stability analysis are among the methodologies explored,

DOI: 10.1201/9781003530220-7

along with their value to professionals and students who strive to stay informed about the latest trends and innovations in the sector.

Data standardization is identified as one of the most relevant components of digital transformation (Cuz et al. 2025). The systematic organization of data, including the collation of large volumes of data samples through surveys, facilitates quicker data transfer and better integration. With standards such as the one developed by the Association of Geotechnical and Geoenvironmental Specialists (AGS) on the creation and application of formats for data transfer and the transmission of geospatial information between systems and platforms, it is possible to create a common geotechnical information management system that is used throughout the entire life cycle of the project by all participants involved in its implementation. The introduction of such standards improved means of interoperability with Geographic Information Systems (GIS) as well as Building Information Modeling (BIM), thus ensuring integrated but organized and accessible data. This is also important for efficient data management to support better decision-making at every stage of a geotechnical project.

Furthermore, machine learning has emerged as a transformative force in data interpretation for transport geotechnics projects. Due to its impressive ability to process complex, nuanced datasets, machine learning technology enables quicker, more accurate solutions to a range of geotechnical challenges. In this context, various machine learning algorithms are being applied in areas such as slope stability classification and pavement condition forecasting (Tinoco et al. 2025). These advances enable decisions that are not only more informed and proactive but also critical to reducing risks and extending the operational lifespan of transport infrastructure. Several case studies are reviewed to illustrate the application of machine learning in these contexts, highlighting the diverse and increasingly crucial roles that different algorithms play in resolving complex geotechnical problems.

An additional focus of this exploration is the role of optimization systems and their application during the planning and construction phases of geotechnical projects. The adoption of such systems offers the ability to refine and optimize various project parameters, such as construction costs, project timelines, and carbon emissions, ultimately improving the project's environmental and economic outcomes (Parente et al. 2025). Traditionally, optimization and predictive systems have operated in offline modes, requiring manual updates to adapt to unexpected changes, such as equipment failures, variations in material types, or shifting weather conditions. However, as digital transformation progresses, the incorporation of emerging technologies, including artificial intelligence and digital twin systems, enables the development of real-time optimization frameworks that are adaptable to on-site changes and unforeseen events. Equipped with IoT sensors, predictive and optimization systems allow for the continuous monitoring of processes and equipment. A real-time optimization system, for instance, can monitor processes such as earthmoving, adjusting equipment operation to maintain peak efficiency in response to real-time data. This approach is supported by sensors and advanced algorithms that facilitate more intelligent, sustainable resource management that accounts for both economic and environmental factors that influence modern construction.

In summary, this part of the book provides a comprehensive, incremental, and structured view of how digital transformation is reshaping the field of Transport Geotechnics, outlining a pathway towards a more sustainable, secure, and innovative future for the sector. By offering practical case studies, in-depth technical analyses, and examples of emerging tools and methodologies, this section aims to inspire and guide both professionals and academics. The necessity of digitalized and intelligent infrastructure is recognized as essential for addressing the impending challenges within geotechnics. Each technological advancement, from IoT-enabled monitoring to real-time optimization frameworks, is positioned not only as an incremental improvement but as a critical step towards an integrated, forward-thinking approach to geotechnical engineering. The case studies and analyses presented in this part demonstrate the undeniable impact of these technologies, highlighting the transformative potential they hold for the future of the field.

References

Cruz, B., Cruz, J., Branco, J., Pinto, L., Pinho, N., Freitas, L., Chaminé, H. 2025. Geotechnical Data from Ground Investigations: Insights towards a Digital Geotechnics. *in* Gomes Correia, A., Pinto, A., Neves, J., Fortunato, E., (Eds.), *Transportation Geotechnics for Green, Digital, and Modern Infrastructures,* CRC Press (2026).

Miranda, T., Moutinho, J. and Dragos, I. 2025. Geotechnical Monitoring in the Digital Age: Integrating IoT, Remote Sensing, and Data Analysis. *in* Gomes Correia, A., Pinto, A., Neves, J., Fortunato, E., (Eds.), *Transportation Geotechnics for Green, Digital, and Modern Infrastructures,* CRC Press (2026).

Parente, M., Amândio, M. and Correia, A. G. 2025. Dynamic Resource Optimization in Earthworks. *in* Gomes Correia, A., Pinto, A., Neves, J., Fortunato, E., (Eds.), *Transportation Geotechnics for Green, Digital, and Modern Infrastructures,* CRC Press (2026).

Pinheiro, M., Silva, A. and Pereira, E. 2025. Exploring Cutting-Edge Approaches: Case Studies Unveiling New Trends and Tools in Real Geotechnical Data Analysis. *in* Gomes Correia, A., Pinto, A., Neves, J., Fortunato, E., (Eds.), *Transportation Geotechnics for Green, Digital, and Modern Infrastructures,* CRC Press (2026).

Tinoco, J., Cortez, P., Toll, D., Silva, R. 2025. Creating Knowledge by Transforming Geotechnical Data based on Machine Learning Algorithms. *in* Gomes Correia, A., Pinto, A., Neves, J., Fortunato, E., (Eds.), *Transportation Geotechnics for Green, Digital, and Modern Infrastructures,* CRC Press (2026).

Chapter 6

Geotechnical monitoring in the digital age

Integrating IoT, remote sensing, and data analysis

Tiago Miranda, Ionut Dragos Moldovan, and João Moutinho

6.1 Introduction

Globalization and demographic dynamics are, nowadays, two of the main forces shaping the development of contemporary societies. Both offer opportunities but also raise significant challenges, particularly in creating imbalances and pressures in an environment that is increasingly sensitive to anthropogenic effects and their impacts. In this context, understanding and anticipating global changes is crucial to ensuring the sustainability of the global terrestrial system. According to the 2030 Agenda presented by the United Nations, of the 17 Sustainable Development Goals, Goal 13 aims to "Strengthen resilience and adaptive capacity to climate-related hazards and natural disasters in all countries." As a consequence of the rapid economic growth, research on underground space development and use, ground improvement, geological engineering control, and environmental geotechnical engineering is increasing at a fast pace.

According to the latest projections IPCC 2024, in the next 50 years, Earth will experience an unprecedented increase in temperature, changes in precipitation patterns, rising sea levels, and more frequent extreme weather events. Therefore, geotechnical assets in the form of cuts, embankments, hillsides, or retaining structures are crucial in ensuring a safe and efficient communication system. The impact of climate change on geotechnical structures is, today, a real threat that needs to be addressed. Therefore, the European Union European Commission, 2013 has, in the last decade, adapted its strategy to address the need for infrastructure adaptation to climate change and to build more resilient and smart structures (Smethurst et al. 2017; IPCC 2024).

Taking advantage of computer technology advances, computational power, and increased availability of electronic components, it is now possible to address the previously mentioned geotechnical engineering problems. This can be achieved either by applying and developing numerical modeling or by using new monitoring instruments and data analysis techniques.

Geotechnical problems have different levels of complexity; therefore, monitoring can serve as a key tool in solving a significant part of them. Understanding the geotechnical variables, used during the design or exploitation stages, can help assess the extent of the problem and the risk of instability. Geotechnical monitoring and its digitalization are crucial to ensure a less risky infrastructure development *era* to come.

The aim of this chapter is to provide a workflow-based approach of topics related to digitization, digitalization, and digital transformation, such as: (i) geotechnical variables, (ii) monitoring conventional instruments, (iii) new monitoring instruments and sensors, (iv) data communication and software, (v) from raw data to real-world meaning data, and (vi) new trends on geotechnical monitoring techniques.

DOI: 10.1201/9781003530220-8

6.2 Geotechnical variables

6.2.1 *Introduction*

In engineering projects, geotechnical variables are critical for the stability and behavior of the infrastructure. Therefore, a thorough understanding and complete characterization of these variables are essential for safer and more efficient foundation design, construction, and long-term behavior of the infrastructure. Geotechnical variables include intrinsic soil parameters such as moisture content, shear strength, shear modulus, and permeability, as well as variables that quantify the soil's response to applied loads, such as pressure, strain, and displacement, among others. Accurate measurement, through laboratory and in situ tests, and continuous monitoring of these variables help engineers and technicians predict and mitigate potential geotechnical problems, ensuring the stability and safety of the infrastructure.

6.2.2 *Common geotechnical variables*

To fully and efficiently characterize the ground masses, a variety of geotechnical variables must be assessed. The most common of these include the friction angle, cohesion, elasticity modulus, uniaxial compressive strength, and water content. These are frequently measured, as they are also required for most numerical models. Although depending on the ground type and geotechnical project, additional variables can be obtained through laboratory tests. Table 6.1 presents a summary of the most common geotechnical variables and how they can be measured.

6.2.3 *Methods for measuring and monitoring geotechnical variables*

Geotechnical parameters can be obtained using laboratory tests and in situ tests, with examples illustrated in Figure 6.1.

In construction projects, geotechnical parameters may change over time due to changing environmental and operational conditions. For this reason, the most critical parameters must be reassessed and updated periodically, either through direct measurements or by using AI algorithms, as discussed in Chapter 8. Continuous or long-term measurement of geotechnical parameters is typically performed in situ and involves greater complexity compared to one-off laboratory tests. These setups often require arrays of instruments and techniques for measurement, data transmission, and subsequent interpretation of the results. The next section presents conventional geotechnical instruments, followed by a discussion of the challenges and trends associated with continuous measurements involving embedded sensors (Section 4) and remote sensors (Section 6.5).

6.3 Conventional instruments

6.3.1 *Overview of conventional geotechnical instruments*

In addition to laboratory and in situ testing, a variety of instruments are essential for monitoring and assessing geotechnical variables and site conditions. Over the years, a range of conventional instruments has been employed across various geotechnical applications. These instruments include inclinometers, piezometers, hydraulic load cells, topographic markers using total station readings, extensometers, temperature sensors, and others. Measuring displacements, loads, and water levels is vital to ensure the efficiency and safety of foundation and transportation works.

Table 6.1 Geotechnical variables summary description and test identification

Variable description	*Variable symbol*	*Method*	*Unit*	*Note*
Displacement	δ	Measurement monitoring inclinometer, accelerometer, total stations, etc.	mm, cm, m	Movement of soil or structure due to applied loads or environmental factors.
Shear strength	τ	Direct shear test, triaxial test, CPT, Vane shear test	kPa, MPa	Resistance of soil against shearing forces.
Water content	ω	Oven drying method	%	Ratio of the weight of water to the weight of dry soil.
Water pressure	u	Piezometer, transducers	Water column	Pressure exerted by water within the soil pores.
Soil moisture content	θ	TDR, gravimetric method	%	The amount of water contained in the soil, expressed as a percentage of the soil's dry weight. Influences the soil's compaction, shear strength, and permeability.
Uniaxial compressive strength	σ	Uniaxial compressive test	kPa, MPa	Maximum axial compressive stress a soil can withstand.
Permeability	k	Permeameter test	m/s	Measure of the ability of a material to allow fluids to pass through it.
Friction angle	ϕ	Direct shear test, triaxial test	degrees	Angle of shearing resistance of the soil.
Cohesion	C	Direct shear test, triaxial test	kPa, MPa	Measure of the intermolecular forces within a soil.
Relative density	D_r		%	Degree of compactness of granular soil.
Elasticity modulus	E	Uniaxial compressive test	kPa, MPa, GPa	Stiffness of soil, indicative of deformation under load.
Shear modulus under very small strains	G_0	Bender element test, resonant column test	kPa, MPa, GPa	Maximum stiffness of the soil to shearing loads. Shear modulus decreases at higher shear strains
Grain size distribution		Hydrometer analysis	% passing	Distribution of particle sizes within a soil sample.
Liquid limit	LL	Atterberg limit test	%	Water content at which soil changes from plastic to liquid state.
Plasticity index	PI	Plasticity limit test	%	Difference between liquid limit and plastic limit; measure of soil plasticity.
Pore content	n	Measurements before and after drying	%	Volume of voids over total volume; indicative of soil porosity.
Void ratio	e	Measurements before and after drying	-	Ratio of the volume of voids to the volume of solid particles in a soil sample.

TDR, Time-domain reflectometry.

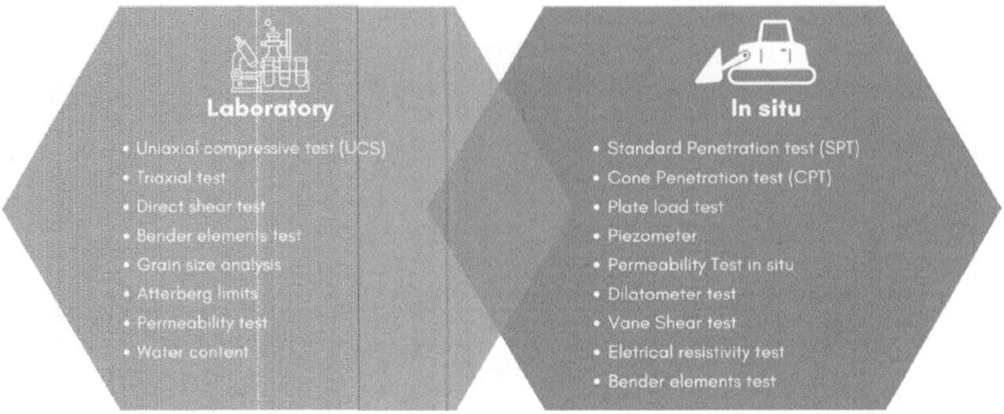

Figure 6.1 List of laboratory and *in situ* tests.

These instruments typically require manual operation and provide low-frequency readings due to the weight of the equipment and the complexities of site manipulation. As a result, geotechnical variables are obtained at discrete reading points, necessitating the use of interpolation methods to estimate values across the entire site (Dunnicliff 1993). The integration of these conventional instruments into geotechnical monitoring programs ensures a comprehensive understanding of site conditions, enabling informed decision-making and proactive management of potential risks. Despite advancements in technology, these instruments remain fundamental tools in the geotechnical engineer's arsenal, providing reliable and accurate data essential for the successful execution of geotechnical projects (Terzaghi et al. 1996).

6.3.2 *Types of conventional instruments*

The most common conventional instruments are based on displacement, pressure, and load measurement. That includes the use of inclinometers, which allow for the measurement of horizontal movement below the ground surface. Primarily used on slopes, embankments, or pile solutions, inclinometers consist of a thin-wall polyvinyl chloride, also called PVC pipe, installed vertically. Movements are measured after inserting a device known as "Torpedo" that passes through the entire length of the gauge well from bottom to top, taking readings at the fixed predetermined intervals. The Torpedo contains a pair of precision accelerometers that sense the inclination of the access tube, positioned 90° to each other. By comparing these values, it is possible to detect any lateral movement on a millimetric scale (Dunnicliff 1993). Other common instruments include piezometers, which measure pore water pressure and assess groundwater conditions, load and pressure cells for monitoring load variations in retaining systems, and extensometers, mostly used in tunnels to measure deformations. Additionally, almost any geotechnical project includes the information obtained from topography surveys, mainly due to the ease of installation and accuracy of the data. However, it has its limitations and is human dependent.

6.3.3 *Advantages and limitations*

The instruments mentioned above are the most commonly used for monitoring geotechnical projects, whether for foundations or slopes. These conventional instruments offer numerous advantages that explain their widespread use and popularity. These advantages include: (i) their established reliability and accuracy based on decades of track record with historical data providing higher reading confidence (Dunnicliff 1993); (ii) their wide availability and familiarity, making them accessible and understandable to all engineers and technicians; (iii) most of them are cost-effective, mainly because they are not fully reliant on technology and are dependent on the human factor; and (iv) manual control and flexibility, as technicians or engineers can adjust and adapt to different site conditions.

Regarding its disadvantages, the main concern is the labor-intensive nature of these instruments, as most are manually operated. There is an important trade-off between reading frequency and cost, which results in a lower frequency of data collection compared to automatic systems. This is particularly relevant because the purpose of these instruments is to monitor and, consequently, reduce the geotechnical risk. With a lower reading frequency, critical events may be missed (Terzaghi et al. 1996); Additionally, these instruments provide only point measurements, requiring interpolation and lower data accuracy. They are also sensitive to environmental factors such as temperature, humidity, and even electromagnetic interference. Finally, some instruments require complex installations that involve heavy equipment. In summary, while conventional geotechnical instruments provide reliable and well-understood measurements, their limitations, such as manual operation, lower data frequency, and environmental sensitivity, need to be considered. Balancing the use of these instruments with the adoption of more modern technologies, such as automated systems, can offer a more comprehensive monitoring solution for future geotechnical projects.

6.4 Continuous monitoring using IoT sensors

6.4.1 Introduction

Despite the technological advancements across nearly all fields of science and engineering over the past few decades, geotechnical monitoring technology remains in its infancy. Although this topic is currently widely disseminated, the technology commonly used today is virtually the same as it has been for several decades. Traditional monitoring instruments, often referred to as conventional or classic monitoring, such as inclinometers, piezometers, topographic markers, load cells, fissure meters, and accelerometers, are still predominantly in use, see Section 6.3. The lack of investment and economic capacity by companies explains the stagnation in the development and specialization in this area.

Additionally, the reading process, depending on the equipment, can involve a series of operations, logistics, and human resources, often requiring a certain level of technical training. This process can be classified as somewhat rudimentary, given the technological advancements and digitalization occurring worldwide. In recent years, some improvements have been made to existing technologies, particularly in the remote and continuous communication of measurements without human intervention. These advances, although recent, often require the complete replacement of existing equipment, representing a significant investment that cannot be adopted on a large scale. This poses a problem for infrastructure management companies, which, in recent years, have needed to monitor such elements in greater numbers, driven by the rapid expansion in road and rail infrastructures, for instance.

As previously mentioned, advancements in data acquisition and communication technologies have led to solutions that enable real-time monitoring with continuous and remote data acquisition. Most of these solutions are ready for use, while others are in the final stages of research and development (R and D). Conventional sensors have undergone significant evolution, making remote and continuous data acquisition automation a reality, as opposed to manual and sporadic

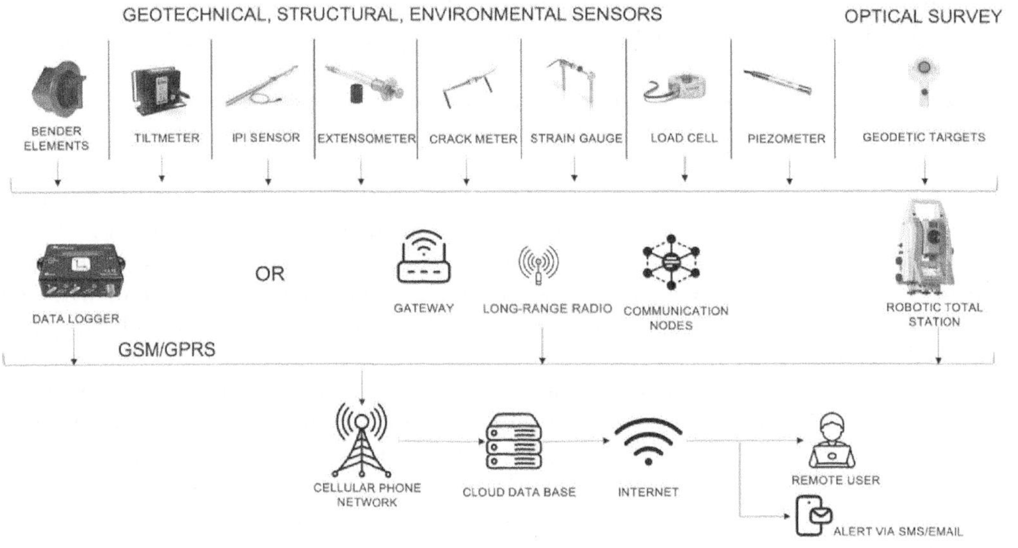

Figure 6.2 Example of a communication scheme for geotechnical automatic sensors.

data collection. Currently, there are two main architectures for automatic and continuous data acquisition from sensors (see Figure 6.2): (i) through a data logger, to which each sensor is connected by cable, or (ii) through a wireless system, with one or more concentrators, called gateways, receiving information from multiple devices, called nodes, located near the sensors and usually connected by cable.

Locally, data can either be stored 1 or be transmitted 2 in a raw or processed state. In the latter, processing can be something as simple as "de-noising" or "averaging" or as complex as "machine learning." This aim of processing is to avoid storing or transmitting unnecessary "garbage" data, redundant data, or oversized data while distributing the computational effort between the nodes, rather than centralizing it on a server in the cloud. Local processing may also be crucial for time-sensitive applications, where latency is a key factor, and the round-trip time for data transmission elsewhere would be insufficient.

Three different architectures for data processing are presented to ensure more efficient, reliable, and real-time data analysis: edge, fog, and cloud computing.

i Edge computing involves processing data very close to where it is generated, sometimes inside the device that encapsulates the sensor, minimizing latency and bandwidth use. In the context of geotechnical sensors, such as those used for monitoring soil moisture, pore water pressure, or ground movement, edge devices can immediately analyze sensor data on-site. This allows for a rapid detection of potential issues, such as landslides or foundation settlements, enabling prompt response actions. For instance, in a remote construction site, edge devices can collect and process data from various geotechnical sensors. This immediate processing helps filter out irrelevant data and highlights critical information, reducing the data volume that needs to be transmitted to central servers, lowering communication costs, and enhancing data privacy. If cameras are being used to monitor visual characteristics of geotechnical works, it is preferable, and sometimes indispensable, to conduct the computer vision analysis locally, extracting the relevant information, instead of transmitting large volumes of data, possibly with privacy-sensitive content;

ii Fog computing serves as an intermediary layer between edge device sensors and the cloud, bringing cloud capabilities closer to the data source, providing additional storage and computational power at the local network level. In Figure 6.2, this layer would be located at the data logger level included within the device or preceding it. For geotechnical sensor monitoring, fog nodes can aggregate data from multiple edge devices, perform preliminary analyses, and make localized decisions. This layer is particularly beneficial in large-scale monitoring systems, such as those used in dam or bridge monitoring, where real-time data processing and immediate feedback are crucial. By reducing the amount of data sent to the cloud, fog computing helps manage network congestion and ensures that time-sensitive information is processed quickly;

iii Cloud computing is the traditional centralized approach that offers extensive storage and powerful processing capabilities. The typical data center approach is ideal for handling large volumes of geotechnical data over extended periods. However, it may not be the most efficient architecture for the majority of situations, especially when there is a large network of sensors, and each sensor generates a substantial volume of data. Nevertheless, with the right capacity, cloud platforms can store large sets of historical data, perform complex analyses, and run advanced predictive models. As it usually stores raw data, no information is lost, as no processing is conducted before the data transmission to the cloud. For scientific studies, for instance, this is typically an advantage as it allows researchers to consider all data without any filtering or discarding. Cloud computing also facilitates collaboration among engineers and researchers by providing centralized access to data and computational resources, enabling them to develop and refine geotechnical models and strategies.

Depending on requirements and context, combining these computing paradigms creates a robust framework for geotechnical monitoring. Edge computing ensures immediate, localized data processing; fog computing enhances this by aggregating and analyzing data at a regional level, providing quicker insights; and cloud computing offers the scalability and advanced analytics required for comprehensive, long-term monitoring and predictive modeling.

6.4.2 Communication technologies for sensor networks in geotechnical monitoring

These new sensor communication methods, as mentioned earlier, can be implemented using two primary options: LoRaWAN networks or mobile data networks such as GPRS, 3G, 4G, and LTE. To minimize communication costs and energy consumption, especially in scenarios where many sensors are dispersed over several kilometers, LoRaWAN is becoming the preferred and more viable option. However, in situations where sensors are widely scattered, mobile data communication is still the most commonly used, mainly due to its simplicity and ease of implementation (Table 6.2).

Depending on the characteristics and specifications of each site, different communication architectures can be used and applied. For example, a client-server model, where the user interface is accessed via a web application, is presented in Figure 6.3. The architecture presented includes:

1 Sensors: Installed in geotechnical assets, sensors translate physical measurements into signals that capture critical data such as displacement, moisture, and pressure.

Table 6.2 Comparative analysis of different communication mechanisms

Features	*Traditional cellular*	*LoRaWAN*	*Cat-M1*	*Local area network*	*Narrow band IoT*	*Personal area network*
Range	Long	Long	Long	Short	Short stationary devices	Very short
Rates	High	Low	High	High	Low	Low
Battery life	Low	Long	Low	Low	Good	Good
Cost	High	Low	Medium	Medium	Low	Low

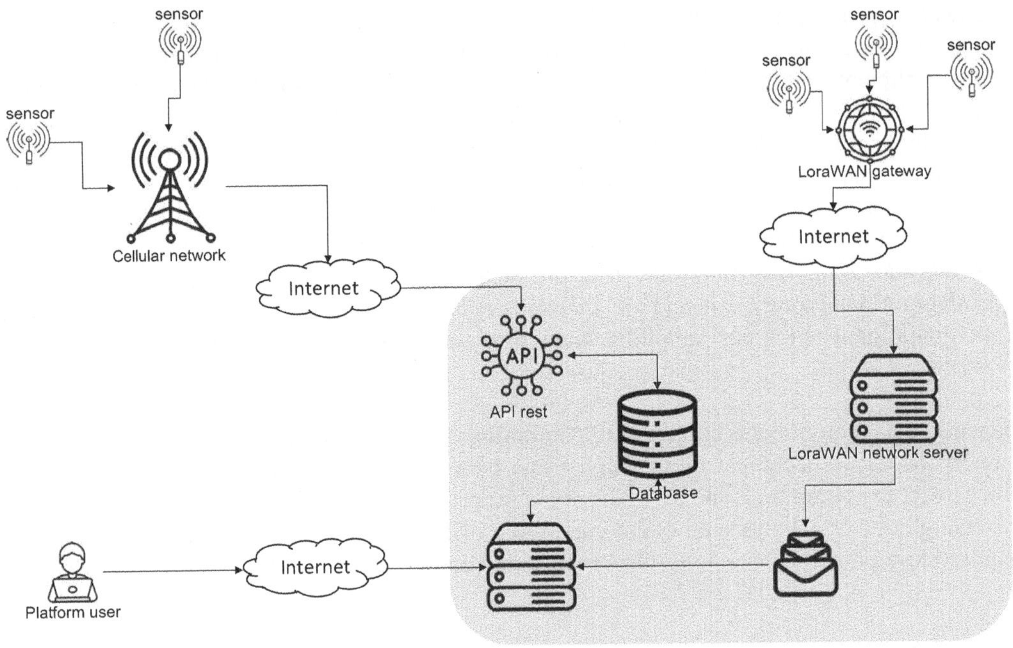

Figure 6.3 Example of a high-level architecture of a web platform.

2 Communication protocols:
   - LoRaWAN: Sensors transmit data to the platform using the MQTT protocol via LoRaWAN.
   - Mobile data: Sensors use HTTP/HTTPS protocols through a REST API for data transmission over mobile networks.

3 Message queue: Ensures reliable and ordered delivery of data packets to the processing platform.
4 Web application: Provides a user-friendly interface for data visualization and interaction.
5 Relational database: Stores the captured data, facilitating efficient data management and retrieval.

The data captured by sensors installed in the geotechnical asset are received by the platform through a message queue using the MQTT protocol if transmitted via LoRaWAN packets.

Alternatively, if the data is sent via mobile networks, it is received through a REST API using HTTP/HTTPS.

The integration of sensors and communication networks within the system follows a streamlined process:

1. Data capture: Sensors collect geotechnical data from the monitored assets.
2. Data transmission: Depending on the deployment scenario:
   - LoRaWAN: Data are sent in packets using the MQTT protocol to the message queue.
   - Mobile data: Data are transmitted via HTTP/HTTPS to the REST API.
3. Data reception: The platform receives data from the message queue or REST API, ensuring secure and reliable transmission.
4. Data storage: Captured data are stored in a relational database, maintaining its integrity and availability for analysis.
5. User interaction: Users access and analyze data through the web application, enabling informed decision-making.

Although there is a possibility to use different communication protocols, each option has its advantages and limitations. Using the LoRaWAN protocol results in lower operational costs due to minimal data usage and reduced energy consumption, making it suitable for dense sensor networks. Additionally, its high coverage and low power consumption are ideal for battery-operated sensors.

Regarding the mobile data protocol, the wide coverage and reliable connectivity are advantages where LoRaWAN might be insufficient. This protocol is ideal for handling large data volumes, making it beneficial for applications requiring high bandwidth.

Choosing the appropriate communication protocol for sensor network is crucial for optimizing performance, cost, and energy efficiency.

6.4.3 Type of IoT sensors and instruments

6.4.3.1 Shape array MEMS accelerometers

The shape array (SAA) is an advanced geotechnical instrument designed to measure deformation in structures such as tunnels, dams, embankments, and retaining walls. It consists of a series of rigid segments connected by flexible joints, allowing it to be installed vertically, horizontally, or in an arc to suit various monitoring needs. This remote instrument provides a more reliable alternative to the conventional inclinometer, which requires installation in a ground tube with a precision of 0.5 mm per 30 m. Regarding installation, the instrument must be installed on site and must be cut to the required length to monitor. All segments are sensorized with a spacing varying from 250 to 500 mm. Additionally, the instrument can be produced with "silent segments" that can be attached to the start or end of the segments. This is an important feature if the goal is to reuse the instrument with different geotechnical projects that have different lengths to monitor.

This instrument offers (Figure 6.4) several advantages when compared to the traditional inclinometer, namely the ease of installation, no need for on-site calibration, and automated, continuous, and frequent monitoring. It is more flexible and durable due to its flexible joints, it is highly resistant to deformation, and, unlike the in-place instruments, it rarely experiences breaks or data communication losses. One of its most relevant features is its installation versatility,

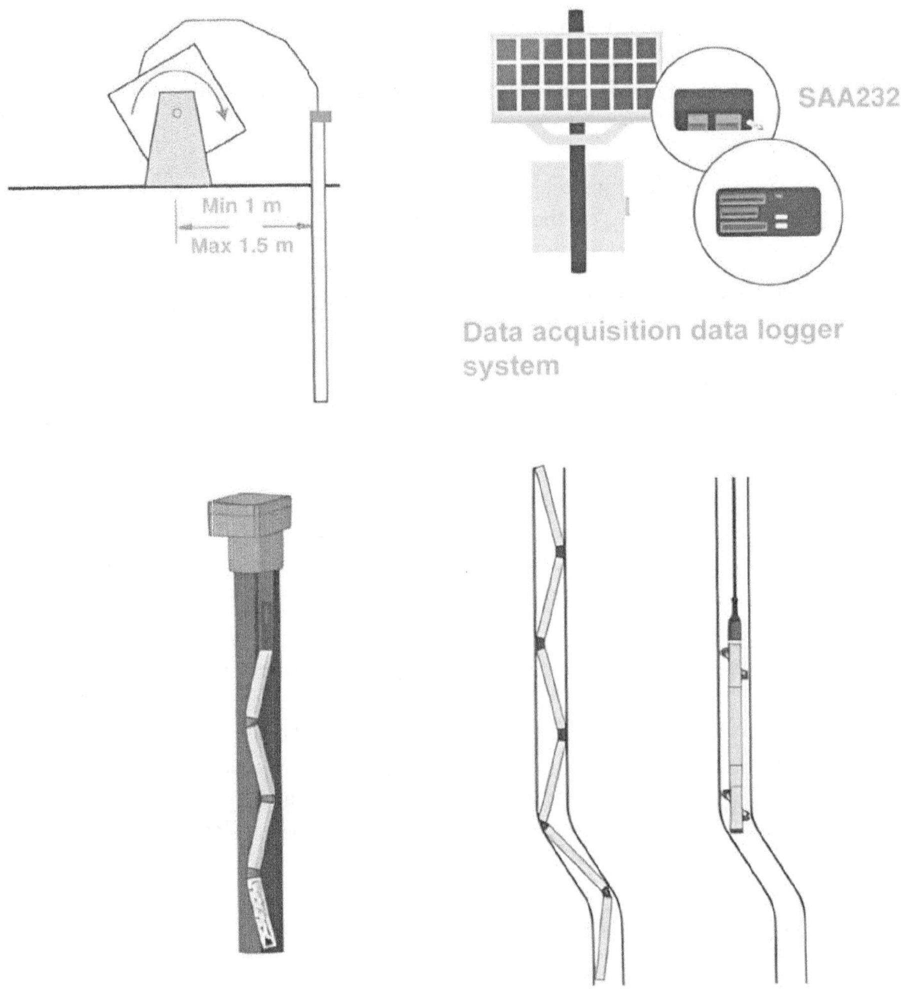

Figure 6.4 Example scheme of a shape array function and geometry.

since apart from vertical casing a tube of 27 mm or larger, it can also be installed horizontally to measure settlements or in arc to measure convergences. The integration with modern data transmission technologies further enhances its utility, providing real-time insights into structural health.

6.4.3.2 Bender elements

Bender elements are piezoelectric transducers used to measure the shear modulus of geomaterials under very small strains. A typical setup consists of two bender elements: the transmitter, which is subjected to a voltage input signal that triggers its lateral motion, and the receiver, which detects the shear wave that propagates through the material (Figure 6.5). The arrival of the wave at the receiver generates an electric signal, which, along with the input signal, can be used to calculate the propagation time and the shear wave velocity.

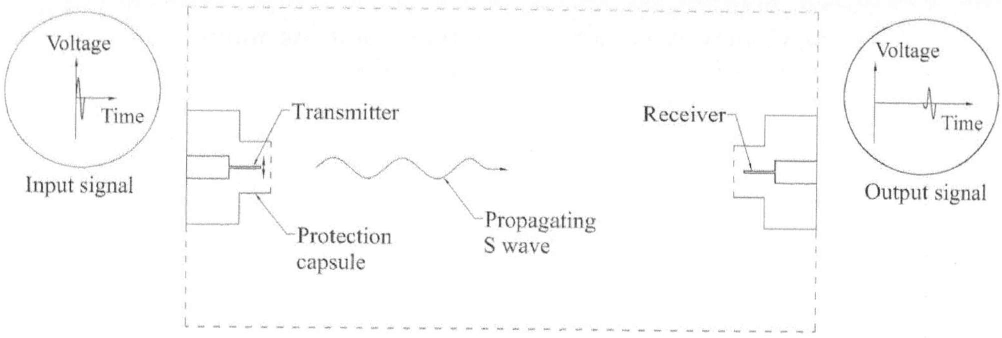

Figure 6.5 Working principle of bender elements.

Bender elements were proposed by Shirley 1978 as a laboratory device, with the experimental setup gradually improving (e.g., Moldovan and Gomes Correia 2021, Project INTENT 2023). Recently, they have been extended to field applications, such as installation into the granular layers of runway infrastructures (Kang et al. 2021). Following up on this new technology, the recent RandD Project INTENT 2023 aims to use bender elements, alongside conventional sensors and advanced numerical models, for the continuous health monitoring of the dynamic stiffness shear modulus of unbound granular layers of road pavements. The monitoring data is transmitted via a wired analog connection from the sensors to a nearby system capable of acquiring and processing the signal amplification, filtering, digitization, structuring the data, and transmitting it to a cloud database. On the cloud, the data is used to power machine learning algorithms for the continuous and automatic health monitoring of the pavement.

6.4.3.3 Tiltmeter

Similar to inclinometers, tilt monitors (see Figure 6.6) are affixed to structures and used to measure the tilt of walls or other structural members. The measurements can then be graphically plotted and compared with the baseline initial tilt. Recently, these types of sensors have been increasingly used either to replace some of the topographic marks or as a comprehensive method to measure the wall or structure inclination. There are several model types, ranging from 2D tilt sensors to 3D tilts with laser sensors. All the sensors measure tilt relative to an initial value, but they can also be converted to displacements if the installation height is used as part of the calculation, allowing for comparison with project-specific movement criteria.

6.4.3.4 Vibrating wire sensors

Another group of instruments widely used in geotechnical applications is based on the vibrating wire concept, see scheme in Figure 6.7. These instruments are crucial in geotechnics due to their high sensitivity and ability to transmit data over long distances (Xue et al. 2016). For example, vibrating wire load cells are used to measure the load or force in structural elements, such as anchors or piles. These cells consist of a high-strength steel wire that is tensioned between two fixed points. When a force is applied to the cell, the tension in the wire changes, altering its frequency. By establishing a relationship between the frequency and the applied load, accurate

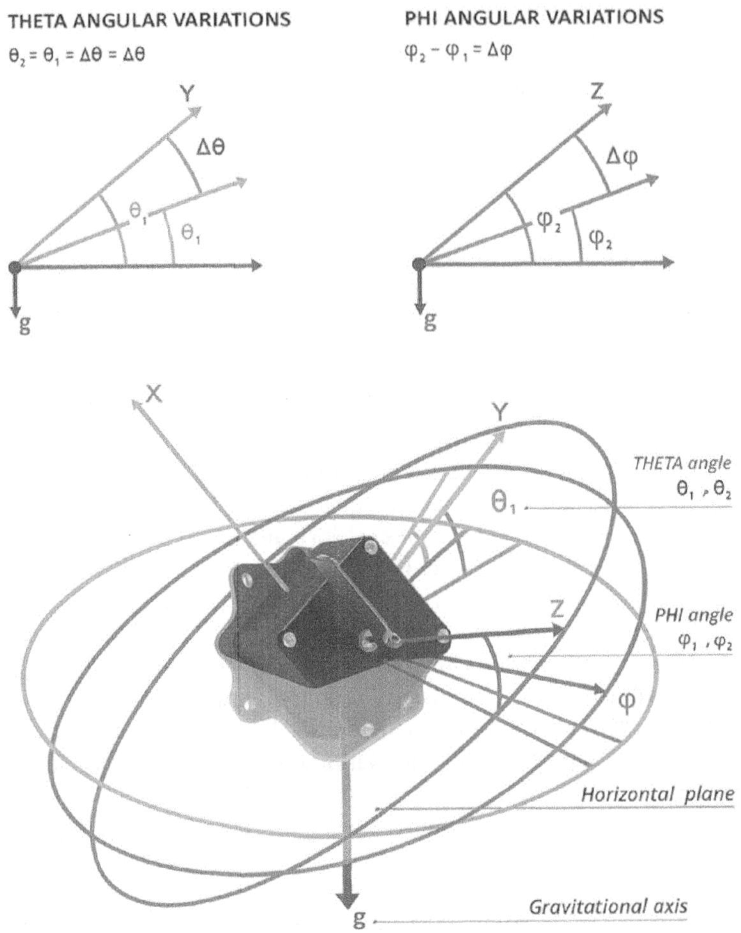

Figure 6.6 Example of a tiltmeter function method of measurement.

readings can be obtained. Similarly, vibrating wire piezometers operate on the same principle but measure water pressure instead.

6.4.3.5 Fiber optics

Fiber optics have been explored in the last decade as a method for measuring strain in geotechnical applications. This technique offers both spatial and temporal resolution in its measurement profile vertical or horizontal. The integration of fiber-optic FO sensors into soil or structures enables accurate and spatially resolved data acquisition. Advancements in technology enable strain measurements in the microstrain range with impressive spatial resolutions (Jeroen et al. 2018). The widespread use of fiber-optic cables along transport infrastructure is one of the main reasons for exploring the advantages of this technique for geotechnical monitoring. Recent studies have suggested the potential of the technique although it still requires further practical applications (Pei et al. 2012).

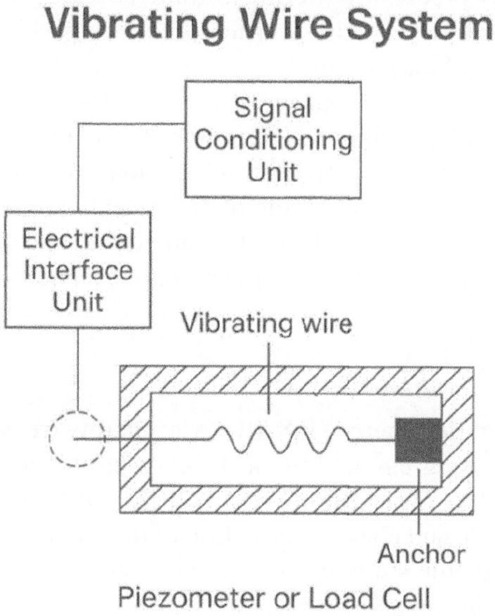

Figure 6.7 Vibrating wire system scheme.

6.5 Introducing remote sensing for geotechnical engineering

6.5.1 Overview

Traditional monitoring instruments are invasive and, in most applications, only provide point measurements in time and space, along with significant installation costs (ICCT 2012; Neuman et al. 2015). Considering the spatial extents of the areas to be monitored, limited in spatial coverage can, in many situations, lead to accidents ranging from small to large scale.

Remote sensing comprises a set of geospatial technologies that enable the collection of information about an object without physical contact distance information. This technology is based on the reflection and emission of electromagnetic radiation from all parts of the spectrum. In recent years, remote sensing has played an increasingly significant role in various fields, including commerce, science, and public policy. Remote sensors provide a global perspective on the state of terrestrial systems and support data-informed decision-making. From the interpretation of aerial images collected via unmanned aerial vehicles (UAVs) to the analysis of satellite-collected information, this technology can be applied on both local and global scales.

In the context of monitoring physical and environmental variables, remote sensing offers a wide range of solutions, including laser scanning both total station and mobile, light detection and ranging (LiDAR), and satellite data. However, this technology, especially satellite data, remains underexplored for infrastructure monitoring applications such as geotechnical engineering. Recent studies have started to use space technology to detect surface displacements with considerable accuracy, yielding positive results in global observations. Nevertheless, the challenge persists in applying this technology to more localized applications, such as geotechnical infrastructures.

It is worth noting that surface movement detection is primarily achieved through synthetic aperture radar (SAR) sensors and multitemporal interferometric synthetic aperture radar (MT-InSAR) techniques (Sousa 2013; Carlá et al. 2018). InSAR technology enables surface mapping via radar, generating high-resolution digital elevation maps and facilitating relatively rapid, large-scale, and highly precise monitoring millimeter scale. In addition to high resolution, these solutions can offer relatively low costs when used extensively compared to the currently existing in situ and point-based technologies.

This section will delve into the state-of-the-art in remote sensing technologies and their applications in geotechnical monitoring (Adel et al. 2021; Zhou 2023). It will explore the current capabilities, recent advancements, and the potential future integration into standard geotechnical practices. Through a comprehensive review of the literature and recent case studies, we aim to present a clear picture of how remote sensing can enhance the monitoring and management of geotechnical infrastructures, ultimately contributing to safer and more resilient engineering solutions.

6.5.2 *Interferometry*

Synthetic aperture radar interferometry (InSAR) is a microwave imaging system capable of penetrating clouds and operating day and night. It relies on the phase difference between two synthetic aperture radar (SAR) observations obtained from different sensor positions. A SAR signal contains both amplitude and phase information of the radar. The amplitude of the returned signal in each pixel indicates the strength of the reflections from the objects within that area. This strength depends on the material properties, surface roughness, and orientation of the objects. A stronger reflection suggests the presence of objects that are good reflectors of microwave signals, such as metallic structures or smooth surfaces facing the radar. The phase of the returned signal represents the distance the microwave signal has traveled to and from the objects within the pixel area. This distance provides precise information about the surface's topography. Since SAR systems emit coherent microwave signals, the phase information can be used for interferometric measurements, which are essential for generating high-resolution topographic maps and detecting surface deformations. The phases of SAR images are derived from the difference between the satellite's antenna and ground targets. This technology enables the creation of digital terrain models using variations in signal acquisition and satellite position.

According to a document published by the European Space Agency ESA on InSAR principles (Carlá et al. 2018), the first interferometry work originated in 1991 with the ERS-1 satellite and subsequently in 1995 with the ERS-2. In 2002, the largest Earth observation satellite, ENVISAT, was launched, containing optical and radar instruments to observe and monitor the Earth's surface, atmosphere, and oceans.

In detail, the SAR imaging system is obtained through an antenna directed toward the Earth's surface in a plane perpendicular to the orbit. The antenna's inclination is referred to as the off-nadir angle, usually varying between 20° and 50°. Due to the Earth's curvature, the incidence angle is equal to the off-nadir angle (see Figure 6.8).

A SAR image can be analyzed as a mosaic of small element pixels, forming a two-dimensional array composed of columns and rows. Each pixel corresponds to a small area of the Earth's surface known as a resolution cell. Each of these pixels' aggregates information on the amplitude and phase of the microwave field reflected by all dispersed elements rocks, vegetation, buildings, etc. The returned signal is a superposition of all the reflected waves from these objects, including both the amplitude and the phase of the waves.

The size of the resolution cell depends on the radar's resolution capability, determined by factors like the radar wavelength, bandwidth, and the synthetic aperture length. The information from all objects within a resolution cell is combined to form a single pixel. This means that each pixel value results from the coherent addition of the microwave signals reflected from all the scatters within that cell.

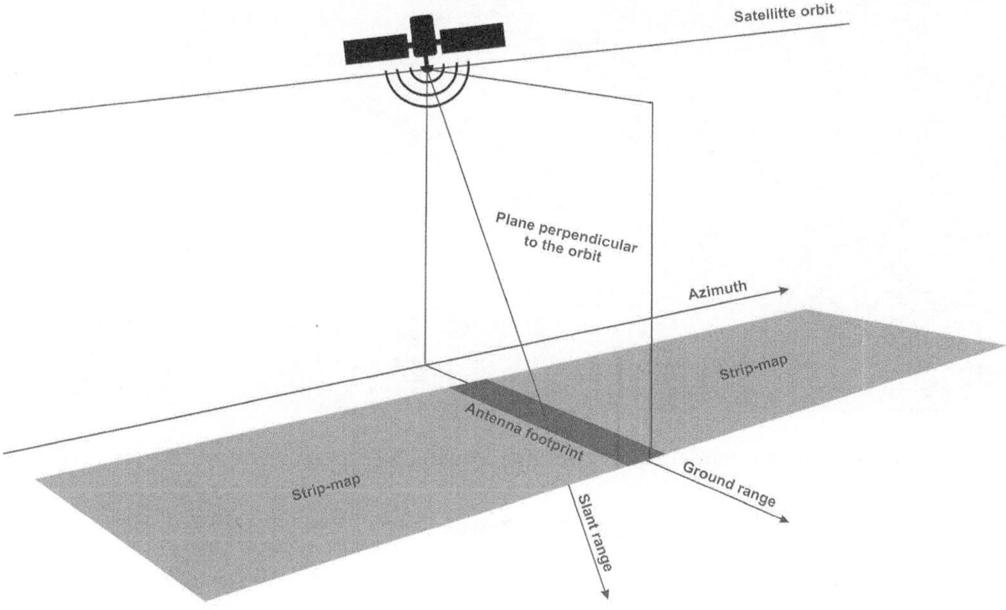

Figure 6.8 Example of a SAR system.

This data can be used to detect small movements in structures such as buildings, bridges, and dams, making it useful for structural health monitoring—identifying and monitoring potential structural issues before they become critical. It can also be applied to land surface deformation, helping to observe ground movements that might affect infrastructure stability.

6.5.3 *Laser and photogrammetry*

LiDAR is a three-dimensional mapping technology that uses a laser and a rotating mirror typically 360° horizontally and 270° vertically to quickly scan an object or a scenario. Similar to photogrammetry, there are two types of LiDAR: terrestrial, which uses a stationary fixed station on a tripod, and aerial, which collects information from an airborne platform. The reflection of laser beams results in a 3D point cloud of the analyzed surface or object. The emission and reflection of the laser beam record the distance from the equipment to the object with submillimeter precision. In just a few minutes, millions of points are obtained, each with xyz coordinates and corresponding RGB intensity values (Bauer et al. 2005; Biasion et al. 2005; Hughes et al. 2013). Subsequently, the point cloud must be processed to extract all the necessary information.

Modern lasers already include high-resolution cameras, enabling the creation of a digital terrain model DTM from the point cloud (see Figure 6.9) (Bailey et al. 2022; Mulas et al. 2023).

Photogrammetry is an optical method that derives the geometric properties of an object from photographic images. Similar to LiDAR, photogrammetry can be either terrestrial from a fixed point or aerial, using drones or UAVs (UAVs; Yan et al. 2023).

Compared to LiDAR, photogrammetry has the advantage of using colored images instead of the black and white characteristics of LiDAR point clouds, creating a more realistic 3D model (Figure 6.10). Additionally, in terms of cost, photogrammetry hardware and equipment are generally less expensive than those for LiDAR although data processing software for photogrammetry tends to be more expensive.

Figure 6.9 Digital terrain model (DTM) obtained through photogrammetry (Geocrit, 2023).

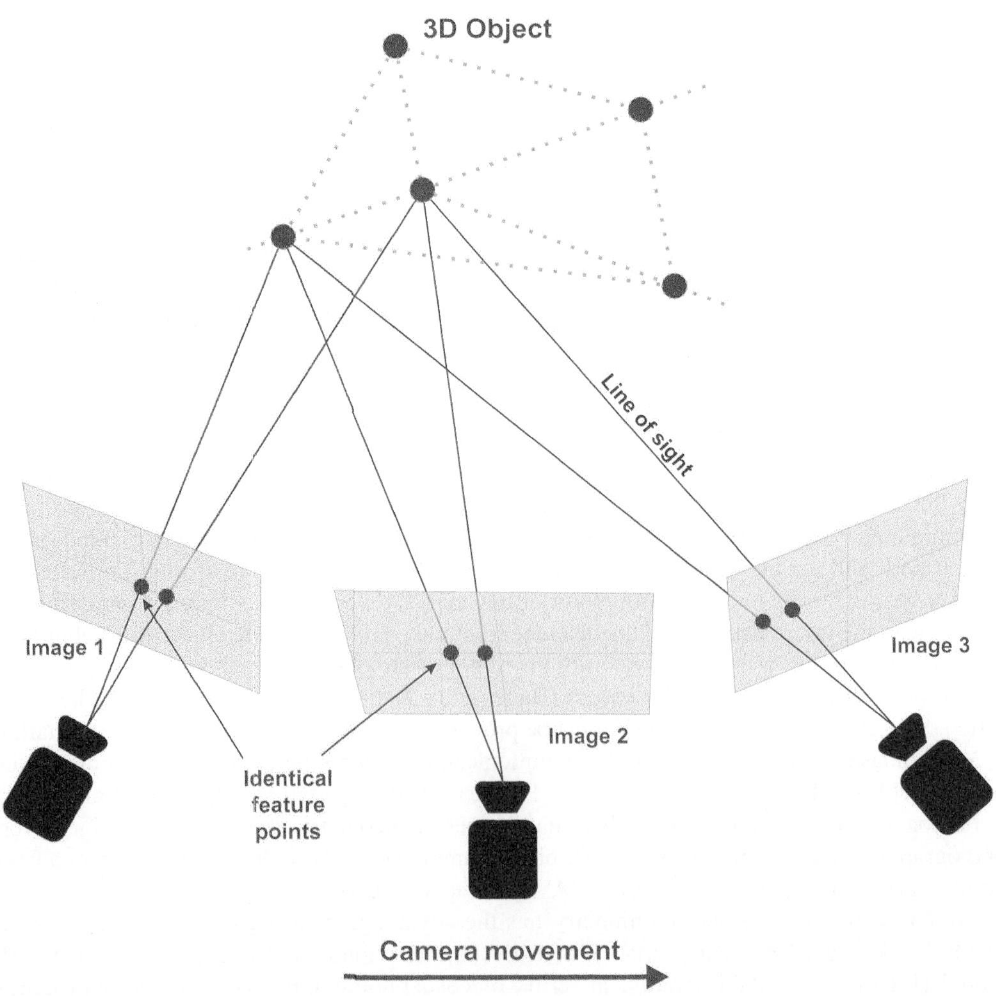

Figure 6.10 Basic principle of photogrammetry (Clemson Institute 2014).

To obtain 3D coordinates of a slope or element, at least two images from different angles with significant overlap are required. Ideally, multiple photographs should be used to achieve both transversal and longitudinal overlap.

This technique plays a crucial role in landslide risk assessment by providing 3D models for monitoring and early warning systems. By using either unmanned aerial systems (UAS) or structure-from-motion (SfM), it is possible to accurately measure large-scale deformations, usually higher than 5 cm (Yan et al. 2023).

6.5.4 *Examples of application of remote sensing for geotechnical engineering*

The use of remote sensing for geotechnical monitoring has become increasingly prevalent. The demand for high-precision geometric and geotechnical information through in situ or laboratory tests is one of the key drivers for technological development in this area. Moreover, when direct measurements are impossible due to inaccessible terrain, remote sensing becomes the only viable solution. The availability of different technologies enables the adoption and use of the most appropriate method for each specific situation. Technologies such as InSAR and LiDAR not only enable the documentation and characterization of infrastructure but also allow for continuous monitoring over time with minimal human and logistical efforts.

Given the new climate change scenarios, with increasingly erratic rainfall patterns and temperatures, the frequency of rockfalls and slope stability issues is rising, particularly in road networks where incidents can have severe consequences (Biasson et al. 2005).

6.5.4.1 InSAR-based concrete example

Currently, InSAR is used to detect ground movement such as subsidence over large areas or significant infrastructures like dams, mines, or tunnels. Previous rockfall events highlighted the need for more detailed monitoring, but due to the area extensions, remote methods were found to be the most useful ones. An empirical conditional vulnerability extraction model (ECVE), which is supported by the combination of the land use and land cover map a, the slope map c, the terrain orientation map d, and horizontal and vertical deformation maps e for the region mentioned, can be obtained. The identified zones by the ECVE model are shown in f–h, with detailed enlargements, identifying points of vertical deformation i and horizontal deformation j, which should be given closer attention. In b, the digital elevation model used to extract maps c and d is presented. Figure 6.11 shows the results of a DinSAR/PSI analysis obtained from Sentinel 1 images since January 2021 until March 2023. The analysis revealed that horizontal displacements are four to five times around 5 to 6 mm/year higher than vertical displacements.

6.5.4.2 LiDAR-based concrete example

Using a recent landslide event that occurred in the north of Portugal in 2022, the advantages of using LiDAR and photogrammetry to assess and manage the geotechnical risk were proven to be essential. By using a Faro Focus S350 laser scanner, several scans were conducted of the affected area after the landslide event. Those scans had allowed the identification of several morphological changes, identify erosion zones and looseness of material, as well as the monitoring of rock blocks that remained at risk (see Figure 6.12). By comparing two-point clouds

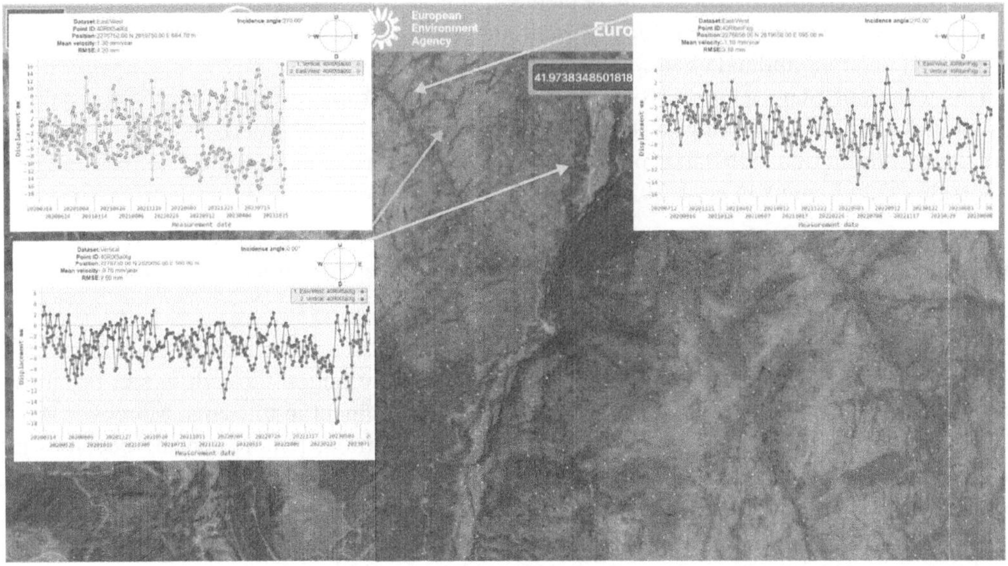

Figure 6.11 Time series of vertical a and horizontal b deformation for two points located in Arcos de Valdevez, covering the period from January 2021 to March 2023. (Source: Final Report Geocrit Project, GeoCrit 2023).

with one-year spacing, it was possible to obtain crucial information about stability changes on the remaining slope, highlighting critical areas that had been needing a continuous monitoring. Through the analysis of this models, it was possible to draw a risk mitigation plan, which included evacuating people from houses in the identified risk areas.

6.6 Conclusion

In conclusion, this chapter has presented a comprehensive workflow-based approach to the integration of digitization, digitalization, and digital transformation within the realm of geotechnical monitoring. The discussion encompassed a range of topics, starting from the physical variables to be monitored, passing through sensors, processing, communication, and ending with the accumulation of data in databases. It approached the traditional methods of monitoring using conventional instruments, outlining their limitations and the necessity for evolution in the field. The introduction of new monitoring instruments and sensors, namely those enabling remote sensing, marked a significant advancement, providing more precise and real-time data crucial for effective geotechnical analysis.

Furthermore, the chapter emphasized the importance of robust data communication systems and advanced software for efficient data management and interpretation. The transition from raw data to meaningful, real-world information was emphasized, showcasing the need for sophisticated analytical tools and techniques to convert large volumes of data into actionable insights, contemplated in further chapters. The narrative also explored emerging trends in geotechnical monitoring techniques, focusing on innovative solutions that promise to enhance accuracy and reliability.

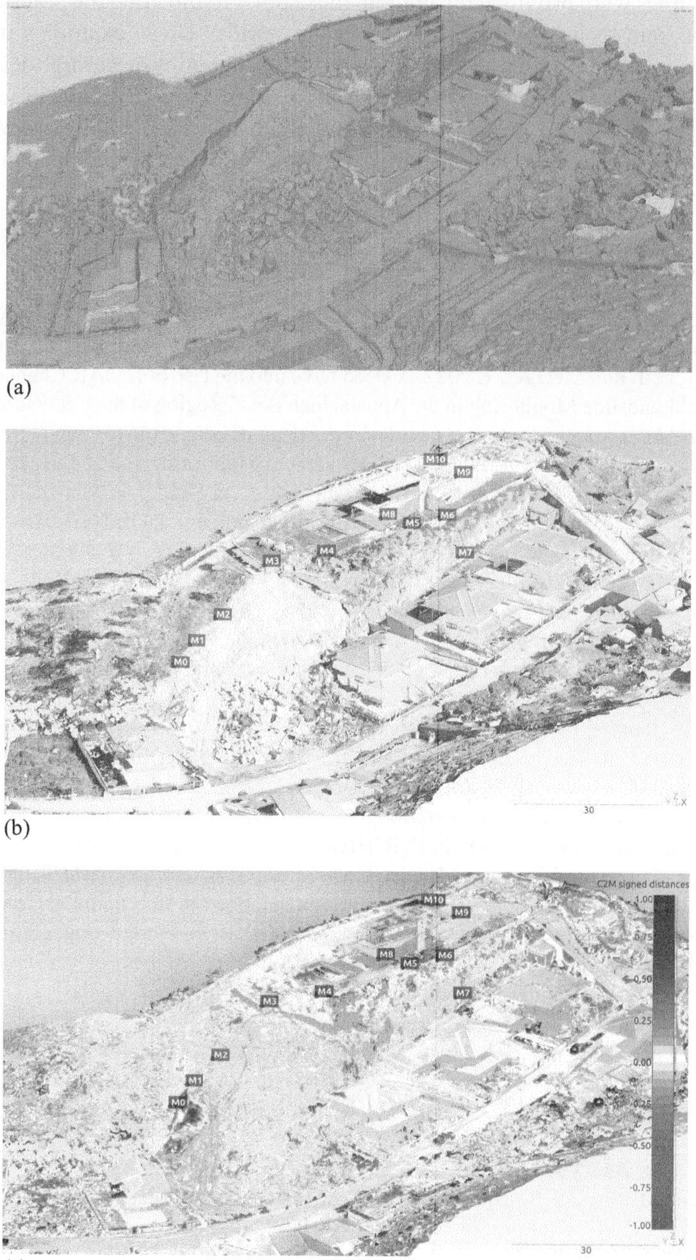

Figure 6.12 (a) Mesh model of the region of interest obtained during 2022; (b) dense point cloud from 2023; (c) absolute distances between 2023 and 2022, with the X^+ axis oriented to the East, Y^+ axis oriented to the North, and Z^+ axis oriented vertically upwards.

Concrete examples were provided throughout the chapter to illustrate the practical applications of these concepts and technologies in real-world scenarios. These examples underscored the transformative impact of digital advancements on geotechnical monitoring, demonstrating the potential for improved safety, efficiency, and decision-making in geotechnical engineering projects. The workflow-based approach presented serves as a valuable framework aiming to leverage digital transformation in geotechnical monitoring, ensuring a cohesive and integrated implementation of new technologies and methodologies.

Reference list

Adel, S., Aref, S. and Hamed, N. 2021. Application of Remote Sensing in Earth Sciences – A Review. *International Journal of Science and Engineering Applications*. https://doi.org/10.7753/IJSEA1005.1001

Bailey, T., Markley, C. T. and Metzger, T. 2022. A Deep Dive into the Use of LiDAR Change Detection Analysis for Regional Landslide Monitoring in the Appalachian Basin Region of the US. *Volume 3: Operations, Monitoring, and Maintenance; Materials and Joining*. https://doi.org/10.1115/ipc2022-86790

Bauer, A., Paar, G. and Kaltenbock, A. 2005. Mass Movement Monitoring Using Terrestrial Laser Scanner for Rock Fall Management. In: van Oosterom, P., Zlatanova, S., & Fendel, E. M. (Eds.), *Geo-information for Disaster Management* (pp. 393–406). Springer, Berlin/Heidelberg. DOI: 10.1007/3-540-27468-5_28.

Biasion, A., Bornaz, L. and Rinaudo, F. 2005. Laser Scanning Applications on Disaster Management. *Proceedings of the ISPRS Workshop on Laser Scanning 2005,* Enschede, The Netherlands, *September 12–14, 2005, International Society for Photogrammetry and Remote Sensing (ISPRS) XXXVI, Part 3/W19*: 50–55.

Pei, H.F., Yin, J.H., Zhu, H.H., Hong, C.Y., Jin, W. and Xu, D.S 2012. Monitoring of lateral displacements of a slope using a series of special fibre Bragg grating-based in-place inclinometers. *Measurement Science & Technology, volume 23, no. 2*, Article ID025007, 2012.

Carlà, T., Farina, P., Intrieri, E., Ketizmen, H. and Casagli, N. 2018. Integration of Ground-Based Radar and Satellite InSAR Data for the Analysis of an Unexpected Slope Failure in an Open-Pit Mine. *Engineering Geology, 235*, 39–52. https://doi.org/10.1016/j.enggeo.2018.01.021

Dunnicliff, J. 1993. *Geotechnical Instrumentation for Monitoring Field Performance*. Wiley-Interscience.

Hughes, D., McKinley, Donohue, S., Lynch, Bell, Harley, Graham, C. J. and Sonebi, M. 2013. Geotechnical Monitoring Using a Combination of Lidar, Real-Time DGPS, and Electrical Tomography to Assess Geotechnical Risk of a Coastal Landslide. Paper presented at the 6th International Conference on Structural Health Monitoring of Intelligent Infrastructure (SHMII 2013), Hong Kong, China.

INTENT. 2023–2026. *Intelligent Health Monitoring of Road Infrastructures Using Bender Elements Embedded in Pavements.* RandD Project funded by Fundação para a Ciência e a Tecnologia. https://intent.ulusofona.pt/

Intergovernmental Panel on Climate Change IPCC, 2024: Climate Change 2022: Impacts, Adaptation and Vulnerability. Contribution of Working Group II to the Sixth Assessment Report of the Intergovernmental Panel on Climate Change [H.-O. Pörtner, D.C. Roberts, M. Tignor, E.S. Poloczanska, K. Mintenbeck, A. Alegría, M. Craig, S. Langsdorf, S. Löschke, V. Möller, A. Okem, B. Rama (eds.)]. Cambridge University Press. Cambridge, UK and New York, NY, USA, 3056 pp., doi:10.1017/9781009325844.

Jeroen, V. R., Iten, M., Spruit, R. and de Boer, A. 2018. Geotechnical and Structural Applications of Fiber-Optic Sensing. https://doi.org/10.1007/978-3-319-59471-2_149

Kang, M., Qamhia, I., Tutumluer, E., Hong, W. T. and Tingle, J. 2021. Bender Element Field Sensor for the Measurement of Pavement Base and Subbase Stiffness Characteristics. *Transportation Research Record, 26758*, 394–407.

Moldovan, I. and Gomes Correia, A. 2021. Optimisation of Receiver's Location in Bender Element Experiments Using Computational Wave Filtration. *Soil Dynamics and Earthquake Engineering, 143*, 106591.

Mulas, M., Ciccarese, G., Marcato, G., Bossi, G., Tonidandel, D., Mair, V. and Corsini, A. 2023. Detecting Recent Dynamics in Large-Scale Landslides via the Digital Image Correlation of Airborne Optic and LiDAR Datasets: Test Sites in South Tyrol Italy. *Remote Sensing*. https://doi.org/10.3390/rs15122971

Neumann, J. E., Price, J. and Chinowsky, P. 2015. *Climatic Change, 131*, 97. https://doi.org/10.1007/s10584-013-1037-4

Shirley, D. J. 1978. An Improved Shear Wave Transducer. *Journal of the Acoustical Society of America, 635*, 1643–1645.

Smethurst, J. A., Smith, A., Uhlemann, S., Wooff, C., Chambers, J., Hughes, P. N., Lenart, S., Saroglou, H., Springman, S. M., Löfroth, H. and Hughes, D. 2017. Current and Future Role of Instrumentation and Monitoring in the Performance of Transport Infrastructure Slopes. *Quarterly Journal of Engineering Geology and Hydrogeology, 503*, 271–286.

Sousa, J. J. and Bastos, L. 2013. Multi-Temporal SAR Interferometry Reveals Acceleration of Bridge Sinking Before Collapse. *Natural Hazards and Earth System Sciences, 13*, 1–9. https://doi.org/10.5194/nhess-13-659-2013

Terzaghi, K., Peck, R. B. and Mesri, G. 1996. *Soil Mechanics in Engineering Practice*. John Wiley and Sons.

Xue, D. and Tao, W. 2016. Vibrating Wire Sensor Signal Denoising and Analysis Based on HHT. https://doi.org/10.2991/ICSMA-16.2016.65

Yan, Q. and Yue, D. 2023. A Method for Measuring Large-Scale Deformation of Landslide Bodies Based on Nap-of-the-Object Photogrammetry. *The International Archives of the Photogrammetry, Remote Sensing and Spatial Information Sciences*. https://doi.org/10.5194/isprs-archives-xlviii-5-w1-2023-29-2023

Zhou, J. 2023. *Application of Remote Sensing Technology in Urban Environment Monitoring*. In Advances in Civil Engineering and Environmental Engineering, Vol. 2. CRC Press.

GeoCrit. (2023). *GeoCrit: Next-Generation Systems for Monitoring and Management of Critical Geotechnical Infrastructures – Technical-Scientific Report No. 4 (August 2022 – January 2023)* [Internal technical report]. Portugal.

Chapter 7

Exploring cutting-edge approaches

Case studies unveiling new trends and tools in real geotechnical data analysis

Marisa Pinheiro, Eduardo Pereira, and André Silva

7.1 Introduction

7.1.1 *Monitoring overview*

As explained in Chapter 5, due to the recent and fast-paced growth of transportation infrastructure, the complexity of geotechnical projects has increased significantly and is becoming more demanding. Moreover, climate change will increasingly play an important role regarding safety and economic impacts of extreme climatic events that will occur more frequently and will have significant impacts on critical infrastructures. In this context, the monitoring of these assets is of paramount importance. Traditionally, monitoring of geotechnical projects has relied on a variety of conventional instruments for data collection such as piezometers, inclinometers, and load cells.

Despite the proven effectiveness of these instruments, they have limitations, such as the need for frequent maintenance, high costs, relation to advantageous environmental conditions, and often the inability to provide real-time data. These constraints have driven the development of new state-of-the-art sensors and instruments, which are revolutionizing the way engineers monitor and analyze geotechnical projects.

New sensors and modern instruments, such as fiber optic sensors, micro-electro-mechanical systems (MEMS), and remote sensing techniques, offer improved precision, greater resistance to adverse conditions, and the ability to transmit real-time data. These advances are transforming geotechnical practice, enabling more effective and efficient project monitoring, from building foundations to large infrastructure projects like tunnels and dams. Recent applications of these innovative technologies in geotechnical projects demonstrate their potential to enhance safety, reduce costs, and increase the reliability of construction.

7.1.2 *Physical techniques versus digital techniques*

Instrumentation and geotechnical monitoring are crucial for risk management in engineering projects, but decisions regarding the switch from manual to automatic systems have significant impacts on cost and efficiency. Although implementing automatic systems may require a higher initial investment, the benefits of having a substantial amount of information available 24/7 justify the expense, especially in scenarios where risk management is critical.

However, it is important to recognize that automation is not a solution in itself but rather a powerful tool for optimizing processes. While automation greatly enhances the efficiency of data collection and processing, human supervision remains essential. The primary focus of

DOI: 10.1201/9781003530220-9

automation is to streamline data collection, allowing for faster and more accurate processing. This shift significantly reduces the time spent on these tasks, enabling experts to focus on the crucial aspects of evaluation and interpretation.

Limited budgets and the lack of expertise and qualified labor are further driving the adoption of automation, which reduces reliance on specialized human resources and minimizes the margin for human error. By automating repetitive tasks, automation facilitates more informed and timely decision-making, as professionals can dedicate more of their attention to analyzing data and making critical judgments. Thus, automation should be viewed as a complement to human oversight, helping to mitigate risks and optimize resources, rather than as a replacement for human expertise in geotechnical monitoring.

7.1.3 Data acquisition and visualization

Performing readings and gathering data are important for geotechnical projects, but in a data era, the way that this data are processed and analyzed is extremely relevant. New IoT sensors and automatic instruments are able to communicate continuously and in real time, though this results in a massive amount of data that are virtually impossible to analyze manually without digital tools. As technology develops new data-driven algorithms, some web-based software platforms are emerging to cover some aspects in geotechnical monitoring. These kinds of platforms or software enable users to visualize, analyze, and interpret data collected from a wide range of sensors. As addressed in Chapter 6, the connectivity of the platform can be based on different communication protocols such as Wi-Fi, Bluetooth, Mobile, or LOraWan. Data integration using a standardized system is imperative to ensure consistency and coherence. Depending on the hardware supplier, the raw data need to be filtered and cleaned to remove noise and errors and also be converted and normalized in engineering units, since instruments and sensors often measure variables in frequency, pressure, or degrees. Data are then fully stored and displayed, usually using graphical elements so that patterns and historical data analysis can be performed.

Depending on the platform or software, customized dashboards display the data in a friendly format using graphs, charts, gauges, or maps. A section with the mapping of all instruments can also be useful from a user's point of view (check Figure 7.1 for an example of a sensor network). Other interactive elements, such as filters, zooming or drilling down, are also features that can facilitate the expert analysis.

A key additional feature can be the integration of alerts and notifications. Most platforms can be configured to send alerts and notification based on predefined thresholds, enabling users to respond promptly when facing risk scenarios.

In geotechnical instrumentation and monitoring, the distinction between real-time and near-real-time data is crucial for effective decision-making. Despite marketing efforts to sell geotechnical monitoring systems as real time, most of them actually operate in *near real time.* Real-time data provides immediate feedback, allowing for instantaneous responses to potential issues, which is vital in scenarios where rapid changes in conditions can lead to significant risks. Near-real-time data, though slightly delayed, still offers timely insights that are often sufficient for ongoing monitoring and less critical applications. Both approaches have their merits, but the choice between them depends on the specific needs and risk profile of the project. Although this delay is generally insignificant for many scenarios, it can be critical in situations where an immediate response is essential.

Figure 7.1 Example of a sensor/instrument network mapping with a color scale to serve as warning of the instruments on alert or alarm conditions.

Before data can be interpreted, it must first undergo a rigorous validation process.

Step 1: **Validation** involves checking the accuracy and reliability of the data to ensure that it is free from errors, anomalies, or inconsistencies. This step is essential to avoid making decisions based on faulty information.

Step 2: **Interpretation** then follows, where validated data are analyzed to derive meaningful insights and informed decision-making. By ensuring that the data are both accurate and timely, the subsequent interpretation can lead to more effective risk management and project outcomes.

7.1.4 Decision support indicators

Regarding decision support systems for infrastructure, specifically within road and rail transport networks, the existing literature provides various methodologies and tools for decision-making in relation to bridges, structures, and pavements. Some notable publications include (Garcia-Aristizabal et al. 2015, Ghuman and McCardle 2012).

The incorporation of information from sustainability indicators, divided into socio-economic, environmental, and structural areas, is relevant to support the decision-making process. Currently, decision support systems addressing climate change scenarios in geotechnical infrastructure are almost nonexistent. However, considering recent technological advancements, especially in the InternetofThings (IoT), data communication, and satellite data, the need for decision support systems is increasingly urgent and trending in the transportation industry. Combining information from different technologies (based on the temporal and spatial scale being analyzed) remains a challenge that needs to be overcome. Recently, some funded research projects have begun integrating multisource data, primarily using remote sensing. These projects have in common the goal to develop decision support systems to enhance the resilience of road infrastructures by combining ground and remote sensing. They conduct a multirisk assessment of all vulnerable points in a road, including slopes, bridges, tunnels, etc. Given these approaches, it is crucial to understand that any decision support system should consider all sustainability indicators related to social, economic, and environmental areas to avoid decisions based solely on execution or final costs. An example of a multisource decision support system is the Geocrit platform that presented a risk management method based on four groups of key performance indicators (KPI) obtained from different data sources. These indicators can be dynamic or static in time. KPIs included: climatic (precipitation, temperature, and extreme events), structural (condition index, risk, and site inspections), discrete *in situ* sensing (instruments in alert or alarm and deviation from the alarm threshold), and remote sensing (like satellite indicators, such as normalized difference vegetation index (NDVI), soil type, normalized difference water index (NDWI)). When it comes to climate indicators, those related to precipitation predictions need to be updated daily, as well as the warnings for extreme events. On the other hand, structural indicators, which are based on-site visits or scanning, exhibit more static behavior. The monitoring group takes into account the number of instruments in alert or alarm conditions, as well as any variation that was overpassed. Finally, the last group focuses on remote sensing information, with indicators like NDVI or soil type, which can be helpful to understand the ground conditions in case of risk management.

In the field of geotechnical instrumentation and monitoring, where the multidisciplinary nature and complex interaction between different areas of expertise are fundamental, there is a clear market trend toward more integrated solutions. Companies and suppliers are increasingly focused on offering "turnkey" packages and "one-stop shop" services that combine various technologies and expertise, simplifying the implementation and management of complex projects. This approach allows clients to obtain all necessary solutions from a single source, ensuring greater efficiency and cohesion across all stages of monitoring.

An effective example of this trend is the combination of large-scale monitoring techniques, such as interferometric synthetic aperture radar (InSAR), with in situ instrumentation. InSAR is particularly useful in transportation infrastructure, providing a macro view of large areas and detecting subtle ground movements over time. Additionally, InSAR allows for the verification of historical movements, making it the only tool capable of "looking into the past" to provide retroactive data on previous displacements (European Ground Motion Service-ESA). To achieve a complete and accurate analysis, this technology can be complemented by in situ instrumentation, which provides detailed and reliable data at specific points. Together, these approaches offer comprehensive and detailed coverage, essential for effective risk management in infrastructure projects.

7.2 In situ case studies

7.2.1 *Retaining wall system monitoring (Phoenix building case study, Lisbon, Portugal)*

7.2.1.1 *Introduction*

The "Phoenix Building" project comprises the construction of seven upper floors, one ground floor and three underground floors in a densely urbanized area of Lisbon, Portugal, with roads, railways, and other buildings and infrastructure in its surroundings. The ground floor is subdivided into three levels to connect with the surrounding streets. The adopted earth retaining solution consists of reinforced concrete diaphragm walls with a thickness of 50 cm.

Prior to the start of the construction, preliminary work was carried out, including removing soil and forming temporary slopes. The diaphragm walls were excavated mainly with a hydraulic grab, although in certain situations, vertical holes were drilled with a short auger and a Kelly rod.

During the provisional phase, the stability of the retaining wall was ensured through the application of prestressed anchors (2, 3, and 4 levels) and the installation of metal corner struts at different levels.

In the final phase of construction, the integrity of the retaining wall was maintained by the structure of the building in relation to the pressures exerted by the ground and road loads. The support systems, including anchors and struts, will be decommissioned after the main structure is completed.

Summarizing, the "Phoenix Building" project encompasses the construction of a building in a highly dense urban area with particular attention to retaining wall, excavation, and interlocking, with the deactivation of the temporary systems after the completion of the main structure.

7.2.1.2 *Instrumentation plan*

The instrumentation plan aims to ensure safety and economy during the excavation and retaining wall construction process allowing the analysis of their behavior during execution. The plan includes the measurement of horizontal displacements of the retaining wall, wall rotations, water level, load on the anchors, and vertical displacements of the surrounding roads, using devices such as automated inclinometers (ShapeArrays), tiltmeters, piezometers, load cells, and topographic marks.

The instruments were distributed across the four design levels, as illustrated in Figure 7.2, where the instrumentation locations and 3D perspective are shown (LEVEL-GAM 2024). The load cells were placed in reference alignments, and these sections also have an inclinometer. On the other hand, nine topographic marks were placed in total, three along the streets Sousa Lopes, three along avenue Álvaro Pais, and three along the IP (Portugal Infrastructures) railway lines (Table 7.1).

7.2.1.3 *Collection frequency*

Given the characteristics of the project, it was recommended that the installed devices be read daily (every 6 hours) during the excavation, retaining wall, and superstructure (with adjustments if necessary). The topographic marks were read monthly, as they were the only nonautomated instruments. The results were presented in tables and graphical plots on an online platform (LEVEL-GAM 2024-Level Sensing®) and interpreted by the geotechnical consultant in collaboration with the site and supervision technicians.

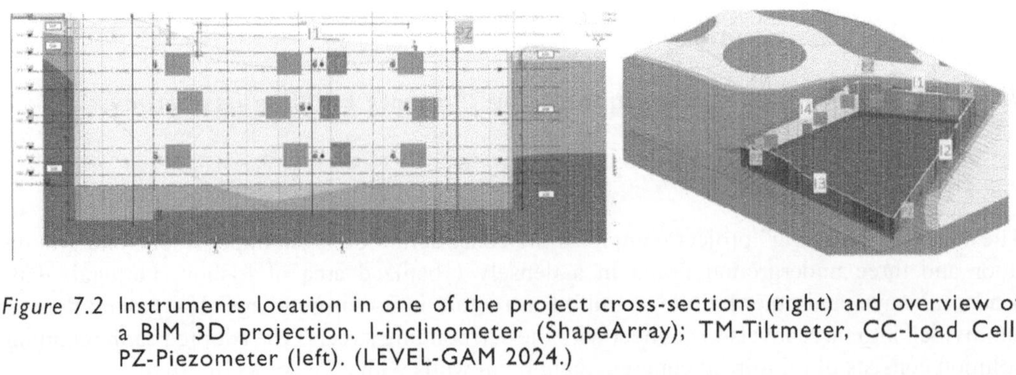

Figure 7.2 Instruments location in one of the project cross-sections (right) and overview of a BIM 3D projection. I-inclinometer (ShapeArray); TM-Tiltmeter, CC-Load Cell, PZ-Piezometer (left). (LEVEL-GAM 2024.)

Table 7.1 Summary table of the instruments and monitored variables

Quantities (un)	*Instrument*	*Measured variables*
4	Automated inclinometers (ShapeArrays)	Horizontal displacements of the retaining wall
30	Tiltmeters	Retaining wall rotations
4	Vibrating wire piezometers	Water level
16	Load cells	Load on anchors
9	Topographic prisms	Settlements of the surrounding structures

7.2.1.4 Alarm criteria

Based on the type of adopted geotechnical structures, risk factors, and the site's geology, the following alert and alarm criteria were established:

a Alert criterion: the maximum displacements of 25 mm in the horizontal direction and 10 mm in the vertical direction.
b Alarm criterion: the maximum displacements of 35 mm in the horizontal direction and 15 mm in the vertical direction.

7.2.1.5 Monitoring and visualization system automation

Automating geotechnical instrumentation using dataloggers and gateways has become increasingly used in the construction and maintenance projects of geotechnical structures. This technique allows monitoring data to be automatically collected by the dataloggers and transmitted to a gateway, which sends the data to an online visualization platform.

The used online platform (Level Sensing®) is a digital tool that enables real-time reading and interpretation of data collected by sensors installed in geotechnical structures. The platform can be remotely via web connection and stores the data in a cloud in a secure and organized manner (Figure 7.3).

7.2.1.6 Analysis and interpretation

The automated inclinometers played a pivotal role in the real-time monitoring of diaphragm wall stability and provided valuable correlations with other instrumentation systems such as load cells, tiltmeters, and piezometers.

During the design stage, finite element models of the geotechnical structures were developed using project-specific cross-sections as references. These models were instrumental in predicting the expected behavior, mainly in terms of displacements, within the retaining wall.

An integral facet of the analysis revolves around the pace at which excavation and anchor tensioning occurred. To further illustrate this aspect, Figure 7.4 presents the maximum cumulative displacements during the excavation phase, focusing on inclinometer I4, which exhibited the most substantial cumulative displacement.

Additionally, Figure 7.4 depicts the key milestones, including the testing dates of the closest anchors (N1, N2), the conclusion of the excavation phase in May 2022, and a transitional period spanning from the completion of excavation to the initiation of the structural phase. Moreover, a specific emphasis was placed on a notable period characterized by particularly intense rainfall between November and December 2022.

These temporal markers were crucial for a comprehensive understanding of the interplay between excavation and anchor tensioning and their influence on the behavior of the diaphragm wall, especially in the context of external environmental factors such as heavy rainfall.

7.2.1.7 Data discussion

As depicted in Table 7.2, the behavior of inclinometers I1 and I2 is closely aligned with the predictions from the numerical models. These inclinometers also exhibited noticeable consistency between the excavation phases and the anchor tensioning. Conversely, inclinometers I3 and I4 displayed deviations from the computed values by the numerical models, ultimately triggering alert and alarm criteria, respectively. It's worth noting that these inclinometers experienced a certain time lag between the excavation and tensioning phases, potentially attributable to their placement within a transitional zone spanning levels 2 and 3 of anchors. Particularly,

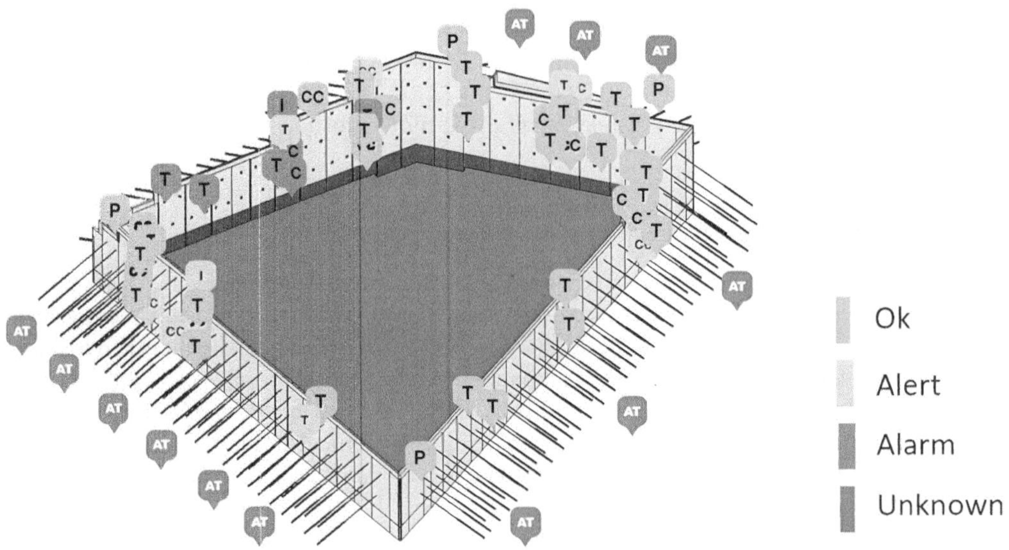

Figure 7.3 Level Sensing® online platform: Overview of the excavation, installed instruments and their status.

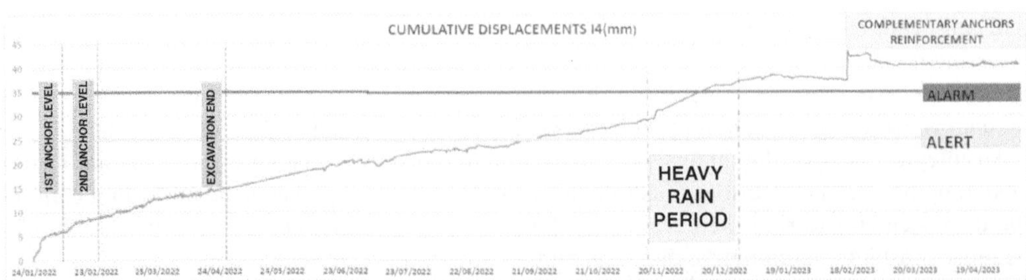

Figure 7.4 Cumulative displacement plot with emphasis on the installation's stages of the anchors (N1, N2), end of excavation (F.E), period of heavy rains (Nov/Dec 22) and last reading made available (May/23), after reinforcement of the panel with complementary anchor.

Table 7.2 Comparison between estimated and measured horizontal displacements

Shape Array	*Estimated displacement (mm)*	*Measured displacement in May/2022 (mm)*	*Measured displacement in Jan./2023 (mm)*	*Alert threshold (mm)*	*Alarm threshold (mm)*
I1	11.25	8.5	13.9	25.0	35.0
I2	11.25	5.4	8.5	25.0	35.0
I3	6.5	16.3	29.1	25.0	35.0
I4	6.5	15.0	38.5	25.0	35.0

inclinometer I3 exhibited the most pronounced deviation, possibly influenced by variations in the geological and geotechnical conditions.

Regarding the inclinometer behavior, it's essential to mention that excavation reached its maximum depth in May 2022 and has since been halted. During this period, an unexpected upward trend in displacements was observed, which deviated from the expected stabilization observed in I1 and I2. This trend notably intensified between November and December.

In response to these deviations, especially with the attainment of alert and alarm criteria, on-site investigations were conducted to identify the root causes of the inconsistent behavior in I3 and I4 compared to the initial expectations. It was determined that the malfunctioning of wedges within several anchors was the primary issue, preventing the proper locking of tensioned strands and, consequently, the stabilization of the walls. This issue was prevalent in various areas and was addressed at the end of December through corrective measures, including additional anchorages. Although the impact on inclinometers I3 and I4 was more pronounced, it resulted in the overall stabilization of the retaining wall behavior, as illustrated in Figure 7.3. Notably, sections I3 and I4 were equipped with only two anchors' levels, certainly limiting their capacity for redistribution.

Finally, the displacement acceleration between November and December was attributed to heavy rainfall during that period. It's important to highlight that the inclinometer instrument panels for I3 and I4 were installed on excavation slopes. The heavy rain, combined with the observed impermeability of the soil layers, likely led to increased loads from the back of the containment structure due to soil saturation on the slope. This, in turn, contributed to the observed displacements.

It's crucial to emphasize the observation method application benefits and also the remaining instrumentation, including tiltmeters and load cells, corroborated the observations made through the inclinometers. This underscores the significance of redundancy in instrumentation systems for validation and comprehensive structural behavior monitoring.

7.2.1.8 Main notes

The project's framework required the development of safe and cost-effective solutions that align with various constraints, including geological, geotechnical, and hydrogeological factors, as well as considerations for architectural preservation, and the integrity of nearby structures and infrastructures.

In this context, the Instrumentation and Observation Plan played a crucial role in managing the performance of the retaining wall structure and adjacent infrastructures. It highlighted its essential role in an urban setting, reflecting the characteristics of the current project.

Moreover, the importance of employing an automated instrumentation system was emphasized. This system not only proved its value in detecting deviations from expected behavior during periods of reduced oversight but also showcased its ongoing usefulness in monitoring and evaluating the effectiveness of implemented corrective measures.

7.2.2 *Tunnel construction in São Paulo, Brazil*

7.2.2.1 Introduction

In tunneling, as excavation progresses, it is crucial to understand the stress redistribution and the behavior of the surrounding soil to ensure that the level of deformations remains compatible with tunnel stability (Silva 2021; Silva 2022).

Traditionally, geotechnical instrumentation in tunnel engineering relies on data collected by field teams, typically two to three times a day. These data are then manually entered into spreadsheets and processed, becoming available to the client and consultants only after several hours. During this time, excavation and other adjacent activities continue, many of which may pose a direct risk to tunnel stability due to the ongoing work and potential interferences.

To minimize such risks, automated and remote geotechnical instrumentation offers an excellent alternative for tunnel monitoring. These systems can collect and provide data almost in real time, with the added benefit of preconfigured alerts that notify the team when any parameter exceeds preestablished limits.

The case under analysis involves a single tunnel constructed near the VSE Coxim ventilation shaft of Line 2 of the São Paulo Metro, Brazil. The tunnel has an elliptical shape with a larger horizontal diameter of approximately 6.2 m and was built using the New Austrian Tunneling Method (NATM). This tunnel is located adjacent to a larger-diameter main tunnel. The section discussed in this chapter was executed approximately 40 m from the tunnel entrance and 60 m from the excavation front (Figure 7.5).

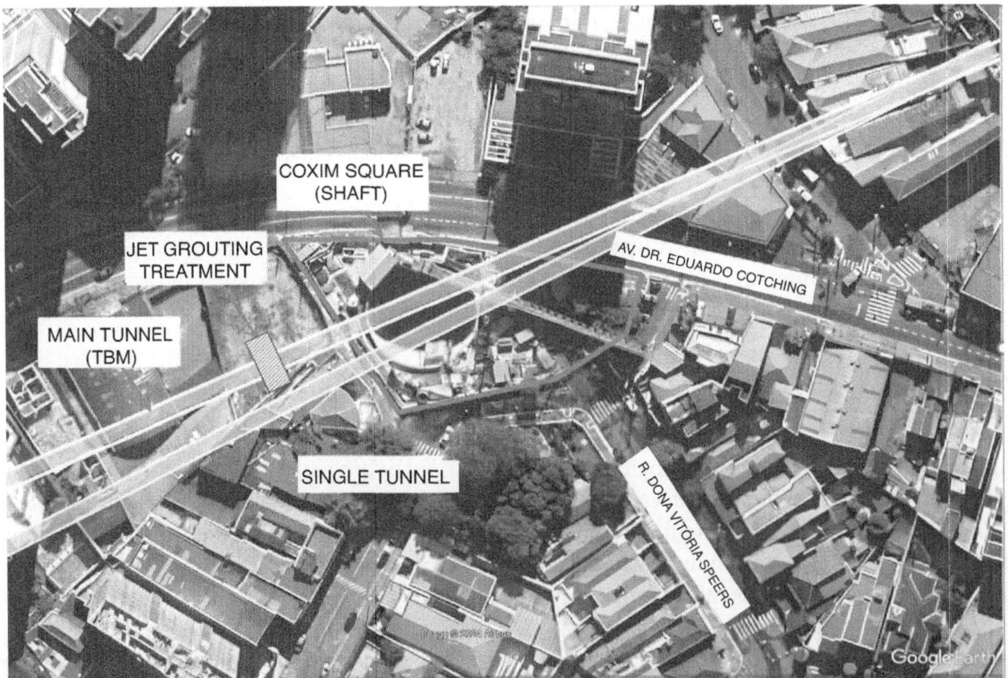

Figure 7.5 Tunnel site location.

The need for continuous and uninterrupted monitoring arose from the plan to install 113 JetGrouting (JG) columns in a section of the main tunnel, near the single tunnel. These columns were intended to reinforce the area ahead of the tunnel boring machine (TBM) arrival. JetGrouting is a well-known soil improvement technique that involves injecting cement slurry through high-pressure horizontal and vertical jets, mixing the existing soil with the slurry to form cement-soil columns or panels.

There was concern that during the installation of these JG columns, potential geological surprises or execution-related issues could cause excessive deformations in the single tunnel, potentially compromising its stability. Therefore, the deformations near the section closest to the JG work had to be closely monitored throughout the process.

The distance between the closest JG column and the tunnel wall was approximately 1.2 m, as illustrated in Figure 7.6, which shows a map of the work area. Figure 7.7 shows in detail the numbering of each JG columns for further analysis.

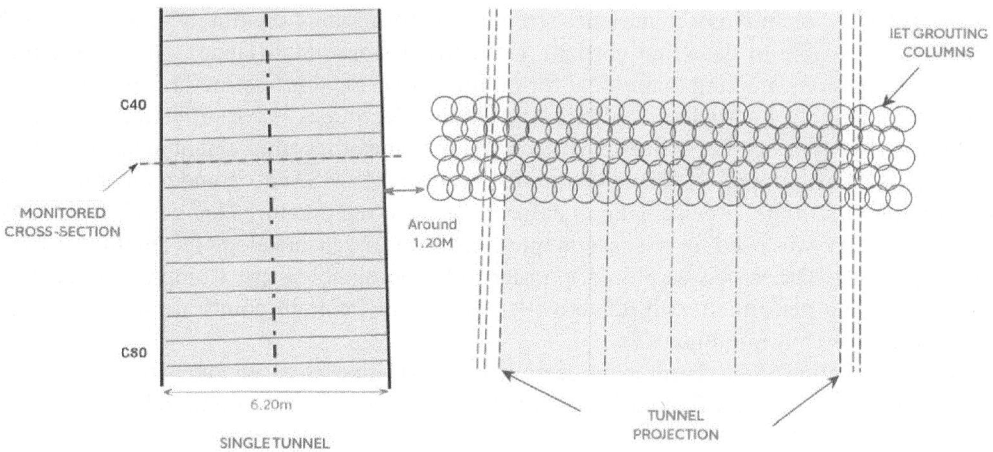

Figure 7.6 Location of the instrumented section and JG columns.

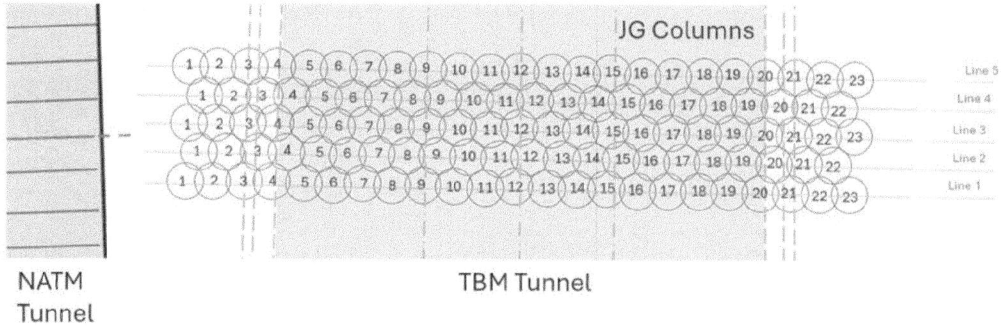

Figure 7.7 Location and numbering of JG columns.

7.2.2.2 *Instrumentation plan*

The conventional tunnel instrumentation, implemented from the start of the project using geodetic surveys and prisms, proved inadequate for providing a rapid response, when necessary, interventions were required, despite having a full-time team on site. This highlighted the need for continuous, uninterrupted monitoring, leading to the decision to adopt automated systems.

As excavation work in the single tunnel was ongoing, the new instrumentation could not interfere with these activities. It was essential to avoid installing equipment that might obstruct or be impacted by the movement of machinery within the tunnel.

Among the proposed solutions, the ShapeArray (SAA) linear chain sensors emerged as the most viable option. These sensors not only offered the required precision and met the need for a sufficient number of monitoring points, but they were also easy to install along the tunnel's length, ensuring minimal disruption to the excavation work.

7.2.2.2.1 SHAPE ACCELL ARRAY (SAA)

ShapeArrays (SAAs) are advanced electronic devices that utilize micro electromechanical systems (MEMS) technology to measure structural deformations accurately. They are highly versatile and capable of detecting vertical (settlement), horizontal (slope), and arc (convergence) deformations, making them ideal for monitoring critical infrastructure such as mines, dams, roads, tunnels, hydroelectric plants, and slopes. Each SAA is constructed from a series of 500-mm rigid pipe segments linked by flexible joints that allow movement in any direction. Every segment is equipped with its own microprocessor, MEMS sensor, and temperature sensor, providing real-time, precise data on deformation and temperature changes (Silva 2021).

An 11 m SAAV was used for the case in question, with 22 sensors along its length, with segment numbering. The SAA was placed in order to fit the tunnel shape. Compared to conventional monitoring performed with reflective targets, about 7x more points are monitored in a single section, as shown in Figure 7.8.

For the installation of the sensor in the section, a 1″ PVC pipe was used for coating and protection. This pipe was securely fixed using metal clamps directly onto the semicircle section. This

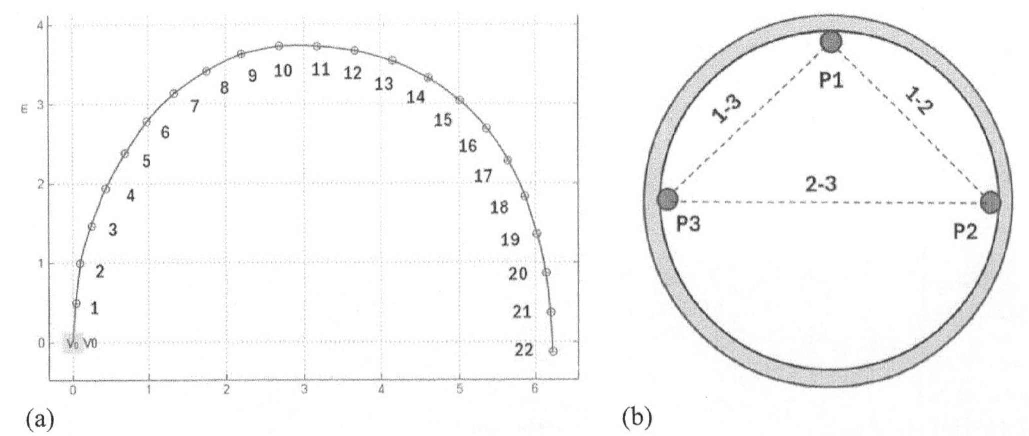

(a) (b)

Figure 7.8 Numbering of sensors along the monitored arc (right) and numbering of reflective targets and corresponding cords (left).

form of installation allowed the rapid release of equipment traffic inside the tunnel, requiring about 1.5 hours to perform this service.

The *datalogger* was installed about 30 m from the section and near the exit of the tunnel, in order to improve communication via *LoRa protocol* between the *datalogger* and the gateway, installed outside the access shaft. The connection between the *datalogger* and the SAA sensor was carried out via data cable.

Figure 7.9 is a photographic record of the equipment installation used in the automated monitoring.

The data collection interval was set to 5 minutes, with transmission to the data server occurring immediately after each reading. Once uploaded, the data were automatically processed and made available for near-real-time viewing through dedicated visualization software.

In collaboration with the client, alert criteria were established: a threshold of +/−3 mm between two consecutive readings (within the 5-minute interval). If this limit was exceeded, the responsible parties would be promptly notified via email (Figure 7.10).

Figure 7.9 Instrument and data collection system installation.

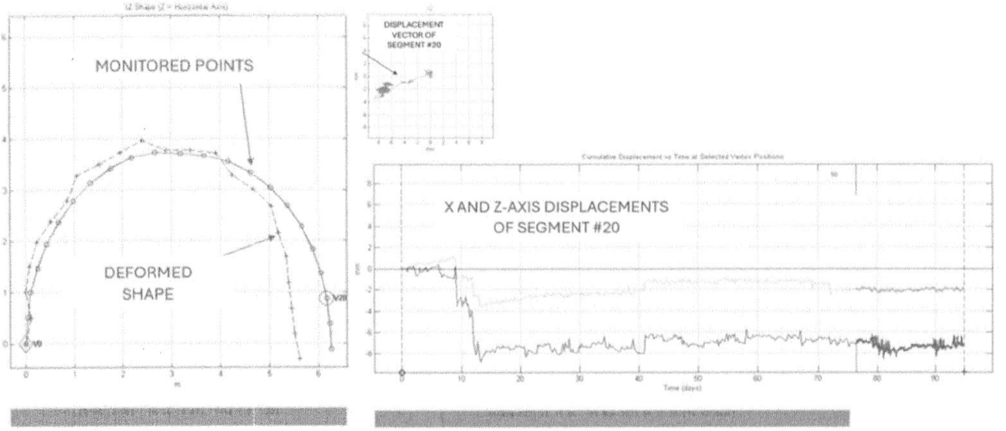

Figure 7.10 SAASuite—SAA data visualization software.

7.2.2.3 Monitoring and site events correlation

Automated deformation monitoring began on 5 August 2023, shortly after the system was installed. The installation of the jet-grouting (JG) columns had started on 12 July 2023, beginning with Column 11 of Line 1, approximately three weeks before the monitoring system was operational, as shown in Figure 7.10.

After the automated monitoring commenced, between 14 August 2023 and 19 August 2023, deformations exceeding the alert limits were detected during column installation, leading to several work stoppages. These variations were also confirmed through conventional methods using reflective targets. The columns installed during this period, which were linked to the stoppages, are shown in Figure 7.9.

The maximum convergence variation reached 8.8 mm in the 2–3 rope on 18 August 2023, prompting a halt in work from 23 August 2023 to 10 October 2023 to investigate the cause of the deformations. During this period, the contractor was replaced, and the JG project and procedures were reviewed, with key parameters adjusted to prevent further issues. Work resumed on 11 October 2023 with Executor 2 taking over.

With Executor 2, instances of cement grout leakage were observed both in the tunnel and through the deep horizontal pipes (DHPs), but deformations remained controlled and within the established alert thresholds. This highlighted the importance of contractor expertise and adjusted procedures in managing the execution process. The grout leaks were likely due to microfractures and more permeable materials redirecting the cement grout toward the DHPs and the tunnel. In these cases, work was temporarily stopped to assess the situation.

7.2.2.4 Data discussion

The comparison between an automated monitoring system and a conventional manual system for measuring tunnel displacements highlights significant differences in data frequency, precision, and the detection of critical displacements (Figure 7.11). The automated system collected 24,200 readings during the monitored period, with a frequency of one reading every five minutes, whereas the conventional system recorded only 113 readings at a daily

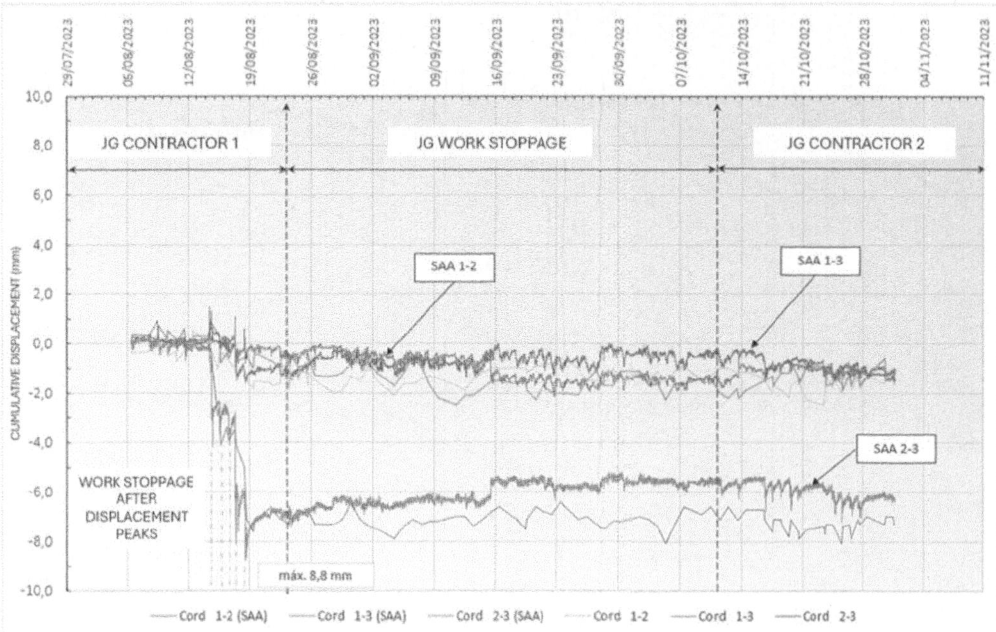

Figure 7.11 Comparison between automated and conventional convergences.

frequency. This stark contrast underscores the automated system's ability to deliver near real-time data, capturing transient events and trends that the less frequent manual readings could miss.

A key issue with the conventional system is its inability to detect displacements between scheduled readings, leading to data gaps. For example, substantial displacements were not recorded because they happened outside the manual system's daily intervals. By continuously monitoring the tunnel, the automated system could capture these events in real time, providing a more complete and accurate understanding of the tunnel's behavior (as shown in Figure 7.12).

The automated system consistently outperformed the conventional system in terms of precision and repeatability. The total station showed displacements nearing its 4 mm accuracy limit, while the automated system provided more precise and repeatable measurements throughout the monitoring period. This precision is crucial in geotechnical monitoring, where small, consistent measurements can reveal trends and help assess potential risks (Figure 7.12). Overall, the automated system's superior accuracy and continuous monitoring capabilities make it more effective for timely decision-making in tunnel management.

7.2.2.5 Main notes

Automated monitoring systems significantly improve upon conventional methods by offering higher accuracy, more frequent readings, and detailed insights into tunnel deformations. Though less precise, traditional manual readings were valuable for validating the automated data. A key result of this monitoring was the adjustment of JG column execution, using the observational method to make real-time, data-driven decisions. The study underscores the importance

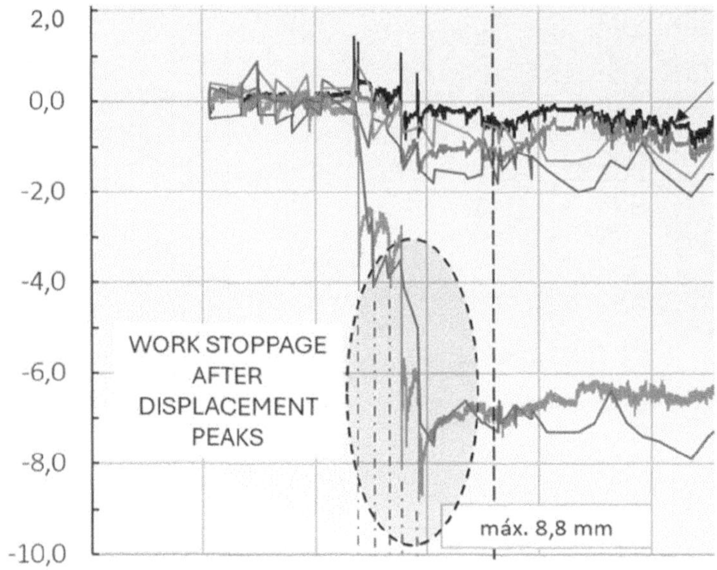

Figure 7.12 Important differences between the conventional and automated monitoring.

of skilled contractors and precise execution, alongside advanced monitoring technology, for successful outcomes.

A comparison of conventional and automated systems revealed significant differences in accuracy, precision, and sensor placement. The total station measurements struggled, particularly with displacements around +/−3 mm, so applying the same alert thresholds to both systems was impractical due to the automated system's ability to detect minor displacements more frequently.

Although manual readings were essential for validating the automated system's accuracy, they provided a familiar reference point that eased skepticism and helped build confidence in the new technology. The ShapeArray sensors offered highly reliable data, tracking tunnel deformations and displacement trends in real time, which was crucial for the project's success.

The case study also highlights the impact of construction quality on project outcomes. The expertise in installing geotechnical elements, such as JG columns, played a critical role in managing tunnel deformation, illustrating that both construction and monitoring technologies are key to project success.

7.2.3 *Aerial and remote slope monitoring and inventory in North of Portugal*

7.2.3.1 *Introduction*

In November 2022, a landslide occurred in Palmeira de Faro, Esposende, Portugal with devastating consequences, taking the lives of two people. The collapsed slope presented a maximum height of about 23 m and a length of 45 m and was excavated in a granitic rock mass strongly weathered, in some areas by mechanical means and, locally, with the support of blasting. The displaced material, with an estimated volume of 2000 m^3, described an oblique trajectory with respect to the slope with a maximum length of about 30 m (EGMS 2024, Geocrit 2023).

Afterward, multiple 3D photogrammetric surveys were conducted to monitor slope displacements over time to assess short- and medium-term geotechnical risks and potential further movements that could endanger other nearby residences. A total of four surveys—R1 (28 November 2022), R2 (13 February 2023), R3 (6 April 2023), and R4 (11 April 2023)—were carried out in order to detect any displacement of the slope, by comparing point clouds collected over time. All surveys were processed using the United States Geological Survey (USGS) standards.

7.2.3.2 Equipment and methodology

All photogrammetric surveys were conducted using the DJI MFT 15 mm, F1.7 camera, with a pixel size of 3.28 × 3.28 μm. Table 7.3 summarizes the characteristics of the three surveys analyzed: R1, R3, and R4. The surveys differ slightly in terms of flight altitude, number of images processed, and the ground sample distance (GSD) resolution. The R4 survey provided the highest resolution, at 0.91 cm/pixel, due to a lower flight altitude compared to R1 and R3 (Table 7.3).

For this analysis, the area of interest was reduced in R4 to approximately 2.97 ha, compared to the larger areas processed in R1 and R3. Point clouds generated from the images were processed using Agisoft Metashape® Professional, version 1.8.4. The resulting models were georeferenced using control points, with calibration parameters such as the focal length (*f*), principal point coordinates (cx, cy), and radial and tangential distortion coefficients optimized to reduce errors.

7.2.3.3 Calibration and error assessment

Once the photogrammetric models were generated, the root mean squared error (RMSE) for georeferencing was calculated. Based on control point locations, the planimetric error (*XY*) ranged from 2.54 to 2.89 cm, while the altimetric error (*Z*) varied from 1.58 to 4.70 cm. The overall RMSE for each survey is summarized in Table 7.4, highlighting the improvement in accuracy for R3 and R4 when compared to R1.

Control points played a critical role in optimizing the accuracy of each survey, with R4 showing the most precise results due to the updated flight parameters and processing methods. A global uncertainty of approximately ±3.49 cm was found when comparing all surveys relative to R1. The weighted average uncertainty was 3.49 cm for R1, 2.37 cm for R3, and 2.29 cm for R4.

7.2.3.4 Geomorphological change analysis

To assess geomorphological changes, digital elevation models (DEMs) were created for each survey, and the differences between R1, R3, and R4 were used to generate geomorphological

Table 7.3 Main features of all performed surveys

Survey	*Date*	*No. of images*	*Max. flight altitude*	*Processed area*	*GSD resolution*	*Valid project points*
R1	28/11/2022	344	54.8 m	3.71 ha	1.26 cm/pox	782 517
R3	06/04/2023	309	55.1 m	3.66 ha	1.28 cm/pix	849 296
R4	11/04/2024	335	39.8 m	2.97 ha	0.91 cm/pix	671 427

Table 7.4 Provides an overview of the RMSE for control and validation points across all surveys

Points	*Quantity*	*Survey*	*X (cm)*	*Y (cm)*	*Z (cm)*	*XY (cm)*	*Total (cm)*
Control	7	R1	1,55	2,93	1,18	3,31	3,52
	8	R3	1,02	1,65	0,48	1,94	2,00
	9	R4	0,86	1,68	0,91	1,89	2,10
Validation	1	R1	0,09	2,89	1,58	2,89	3,29
	1	R3	1,52	2,02	4,70	2,54	5,35
	2	R4	0,91	2,44	1,76	2,6	3,14

Using ETRS89/PT-TM06.

alteration maps (MAGs) using QGIS® (version 3.28). These maps highlighted regions of erosion (in red), accumulation (in blue), and stability (in white), with an uncertainty threshold of ±5.0 cm.

Comparing the periods between R1, R3, and R4 revealed significant erosion between monitoring points M0 and M1, with the volume of material lost increasing from 13.80 to 22.50 m³. The MAGs also indicated a widening of the detachment zone at the crest of the slope, particularly in the southwestern region (section d in Figure 7.13). The key geomorphological findings are:

- The western section of the slope continues to exhibit signs of material detachment and widening, with an extension to the southwest.
- The central and eastern sections showed minor displacement, mostly influenced by vegetation changes, and were largely stable in comparison to previous surveys.
- The rock mass in the central region exhibited some movement, estimated at 2 to 3 cm, with consistent findings across both photogrammetry and LIDAR monitoring.

Figure 7.13 displays the MAGs derived from R1–R4 comparisons, identifying areas of erosion and accumulation. Notably, the western section showed consistent material displacement over time, while other regions exhibited only minor changes.

7.2.3.5 *3D analysis and monitoring*

The CloudCompare® software was used to conduct a detailed 3D comparison of the point clouds. The "Cloud-to-Cloud distance" algorithm was employed to measure the absolute displacement of points, as well as displacements along the *X*, *Y*, and *Z* axes. The color-coded results indicate movements between 0 cm and 10 cm, with more significant movements represented in gray (10–50 cm) or no color for displacements beyond 50 cm.

The western section of the slope (Figure 7.14) experienced a displacement of approximately 4 cm in the northwest-southeast direction, consistent with the monitoring data collected via total station on multiple dates between February 2023 and April 2024. This movement also corresponds to a detachment of material at the crest.

In contrast, the central and eastern sections showed minor changes, with erosion measured between 2.5 and 4.0 cm. The rock mass, located near monitoring point M4, showed some instability, with possible movements of up to 10 cm recorded between May and November 2023, although more recent measurements indicate a stabilization trend.

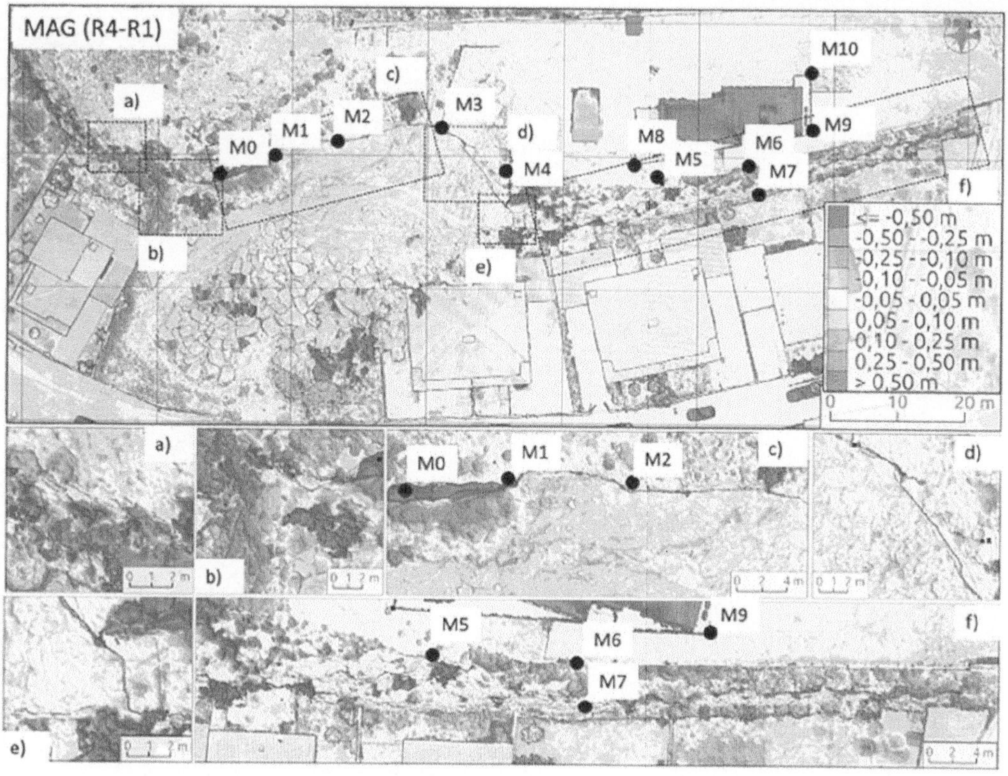

Figure 7.13 Geomorphological alteration map (MAG) of the terrain obtained by comparing the survey conducted on 26 November 2022 (R1) with the survey conducted on 11 April 2024 (R4), showing the locations of monitoring points M0–M10.

7.2.3.6 Main notes

The photogrammetric surveys, when compared with LIDAR and total station data, confirm localized geomorphological changes in the slope over the study period. The most significant displacements occurred in the western section, where detachment and erosion continue to affect the crest. The combination of photogrammetry, LIDAR, and total station monitoring has provided a comprehensive understanding of the slope's behavior, with consistent results across different methods.

The ongoing monitoring of critical areas, such as the rock mass in the central section and the western detachment zone, is essential to further assess risks and evaluate possible stabilization measures.

7.3 Future directions

The future of geotechnical instrumentation and monitoring is increasingly focused on the integration of advanced technologies, with an emphasis on performance and predictive models that enable more efficient and proactive risk management in infrastructure projects. These models have become essential tools in geotechnical engineering, as they not only analyze the past and present behavior of structures but also predict their future performance based on a vast array

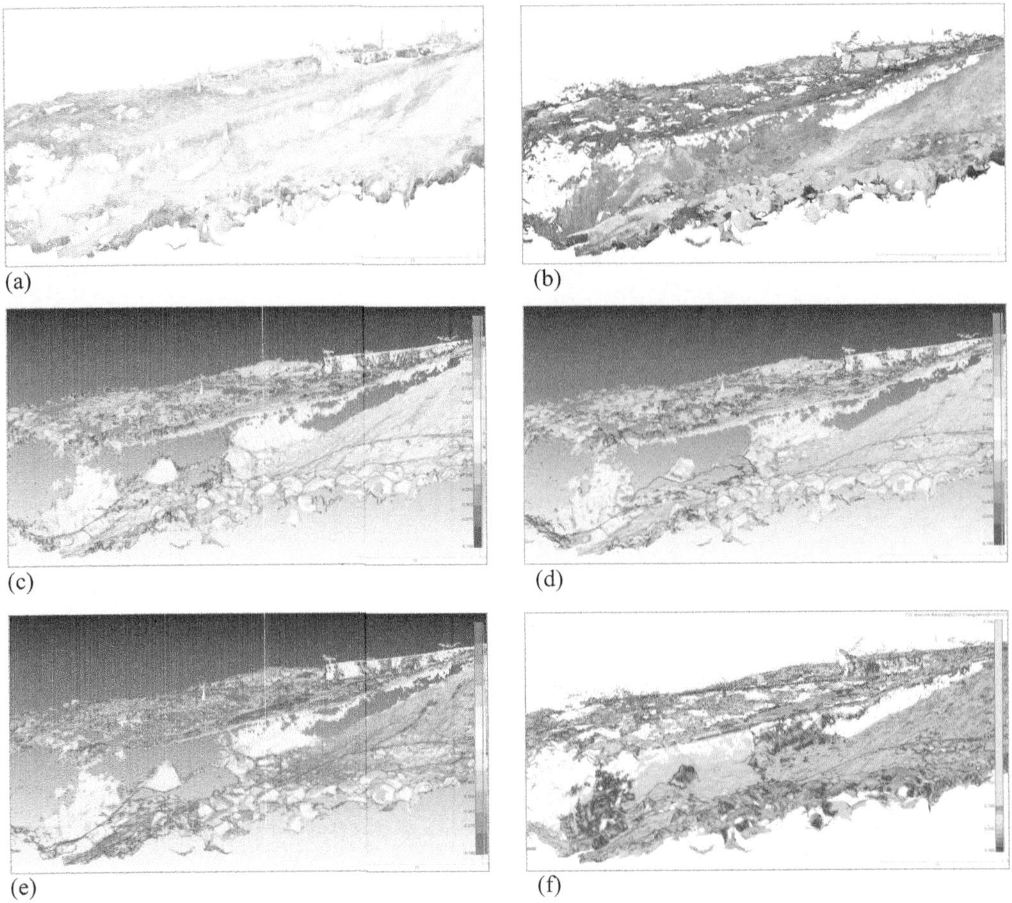

Figure 7.14 Western Section. (a) Dense point cloud from R1; (b) Dense point cloud from R4; geomorphological alterations obtained by comparing R1 and R4: (c) along the *X*-axis; (d) along the *Y*-axis; (e) along the *Z*-axis; (f) absolute value.

of monitored data. This data is obtained in the field through specialized instrumentation, in situ tests, and laboratory analyses, providing a robust foundation for creating precise and reliable models. With the growing capacity for processing and analyzing large volumes of data, we are entering an era where the accuracy and reliability of predictions are continuously improving.

One of the most promising trends in this context is the expanded application of the observational method, which relies on the continuous comparison between model predictions and actual field data. This approach allows for automatic adjustments to models based on *back analysis* and constant iteration of finite element models (FEM). The integration of these models with real-time monitoring data, obtained from geotechnical instrumentation, field tests, and laboratory analyses, creates a dynamic feedback loop. In this loop, predictions are continuously refined and updated, resulting in a deeper understanding of the behavior of structures and the ground over time.

However, for these innovations to be effectively implemented, it is crucial that legislation also evolves, adapted to the new technological demands. The way data and risk management of structures is conducted must be regulated in a manner that reflects these innovations, ensuring that best practices are followed, and that infrastructure safety is always prioritized. This legal adaptation will allow emerging technologies to be integrated safely and efficiently, promoting a more regulated and standardized environment for risk management.

Artificial intelligence (AI) plays a fundamental role in this scenario, facilitating the complex analysis and interpretation of the vast data sets derived from these various monitoring sources. AI algorithms can identify patterns and correlations that would not be easily detectable by traditional methods, enhancing engineers' ability to anticipate problems before they manifest. Additionally, the use of robust databases and the application of data science allow this information to be stored, processed, and accessed efficiently, making it a powerful tool for informed decision-making.

In the future, we can expect greater integration between geotechnical instrumentation, predictive models, and AI techniques, resulting in highly automated and intelligent monitoring systems. These systems will not only provide real-time data but also interpret that data, making autonomous adjustments to models and recommending corrective actions when necessary. This evolution promises to transform the way infrastructure projects are managed, increasing safety, reducing costs, and enabling a quicker and more effective response to the ever-changing conditions of the geotechnical environment.

Nevertheless, while automation brings significant advances to the field of geotechnical instrumentation and monitoring, it is crucial to highlight the indispensable role of human critical judgment in this process. No matter how advanced automated systems may be, they cannot replace the need for careful analysis and interpretation by experienced professionals. Incorrect or inaccurate data can easily contaminate analyses, leading to poor decisions that could compromise the safety and effectiveness of projects. Therefore, automation should be viewed as a powerful tool for data collection and processing, but the final decision, always humanized, must be based on a critical and thorough analysis of the results.

This critical judgment will require professionals in the field to have an increasingly broad and multidisciplinary education. In addition to traditional knowledge in geotechnics, engineers will need to develop competencies in areas such as computing, telecommunications, data transmission, electrical engineering, mechanics, robotics, and more. The integration of these disciplines will enable professionals not only to better understand the technologies involved but also to conduct more accurate and well-founded evaluations of the data generated, ensuring that decisions are the most appropriate for each specific situation. In sum, the future of geotechnical monitoring will combine the efficiency of automation with the indispensable human expertise, resulting in a more robust and reliable approach to risk management and infrastructure safety.

Acknowledgments

Tunnel construction in São Paulo

We are immensely grateful to all the professionals involved who made it possible to implement this technology for the first time in an infrastructure project, especially the entire Metro and Mendes Júnior teams. We also extend our heartfelt thanks to Ana Flávia Righini and Lucio Seiti Shibasaki, whose detailed records were invaluable in preparing this article. Additionally, we would like to acknowledge the entire field team of MMF Projectos, which is responsible for the system's installation, and Measurand for their technical support throughout the process.

Retaining wall system monitoring

The authors express their gratitude to the parties involved for granting permission to prepare and publish this article. It is important to emphasize that the implemented solutions were the result of a collaborative effort, with the following companies playing a crucial role: Rockbuilding-Soluções Imobiliárias S.A., as project manager, DST, as general contractor; Teixeira Trigo Lda., responsible for supervision; JetSJ as project consultant and designer; LEVEL for providing the online monitoring platform; and Geoint, for the instrumentation installation services.

References

European Ground Motion Service (EGMS). 2024. *Copernicus land monitoring service: European ground motion service*. Retrieved September 5, 2024, from https://land.copernicus.eu/pan-european/european-ground-motion-service

Garcia-Aristizabal, A., P. Gasparini and G. Uhinga. 2015. Multi-risk assessment as a tool for decision-making. *Future cities 4* (*Climate change and urban vulnerability in Africa*, Pauleit et al., Eds.): 229–258. doi: 10.1007/978-3-319-03982-4_7.

Geocrit. 2023. Sistemas de nova geração para a monitorização e gestão de infraestruturas geotécnicas críticas. Project final Report, June. POCI-01-0247-FEDER-047173.

Ghuman, P. and A. McCardle. 2012. *Bridge health monitoring using PSI*. European Space Agency Spec. Publ., SP-697.

Level-GAM. 2024. *Level sensing solutions*. Retrieved September 5, 2024, from https://www.level-gam.com

Silva, A. 2021. Monitoramento de Obras Subterrâneas com recursos a ShapeArrays. *5°Congresso Brasileiro de Túneis* (2021), São Paulo, Brasil.

Silva, A. 2022. Why we do (not) monitor – A Brazilian experience. *11th International Symposium on Field Monitoring in Geomechanics*, London, UK.

Chapter 8

Geotechnical data from ground investigations

Insights toward a digital geotechnics

Beatriz Cruz, Jorge Cruz, João Branco, Leandro Costa Pinto, Nicole Pinho, Liliana Freitas, and Helder I. Chaminé

8.1 Introduction

"Data-centric geotechnics should evolve from current knowledge and practice. As such, it is important to appreciate how data plays an increasingly central role in decision support as allowable (or working) stress design progressively changes to the reliability-based design." (Phoon et al. 2022).

The fast advancement of technology related to data acquisition and processing, such as data mining, big data analytics, and artificial intelligence tools, impacts our daily lives. Examples are virtual assistants such as Amazon Alexa, Apple Siri, or Google Assistant. Deaton (2018) identifies a significant concern that the prevailing standard for reporting geological and geotechnical data from ground investigations is limited to basic borehole log records.

The evolution of this record in the last 50 years has been dramatic, shifting from a sheet of paper to a digital fixed-layout flat file. The major disadvantage of this type of record is that it does not constitute accurate data, diminishing significant value for the receiver, typically the project owner, who relies on this method of transferring data and communication (e.g., Threadgold and Hutchison 1992, Deaton 2018, Phoon et al. 2022). This inefficiency in the current geotechnical practice is a pressing issue that needs to be addressed urgently. According to Deaton (2018), there remains a misconception in the industry about geotechnical data. For example, an archive of geotechnical elements in a digital fixed-layout flat document (typically PDF format) on a given repository server or maps should not be considered transfer data. This file encapsulates a complete description of a fixed-layout flat document containing a vast amount of valuable information in fixed-layout flat document format. Nevertheless, they are rendered unusable by this delivery format without re-entering the data into a database software or equivalent. Essentially, information in this format cannot be worked with and updated, whereas data provides the ability to handle, update, and create dynamic and exchangeable databases.

Therefore, transcribing data is a labor-intensive task that results in time and resource losses for the industry and can also introduce significant errors or typos. It is unrealistic to transcribe all data from a fixed-layout flat document for each project (e.g., Toll, 2003, 2008, 2009, Deaton 2018). As a result, a partial set of data is obtained from what is available to improve the construction process and design recommendations.

Deaton (2018) also points out several benefits for technicians or skilled consultants who incorporate digital data exchange based on commonly accepted standards, recommendations, and guidelines (e.g., Lefchik et al. 2007, Toll 2008, Threadgold 2015, Baynes et al. 2021, Turner et al. 2021, Baynes and Parry 2022, Chaminé and Fernandes 2023, and references therein). Firstly, they can rationalize and improve their internal processes to be more cost-effective,

DOI: 10.1201/9781003530220-10

improve data quality, and decrease potential lawful disputes. Having a single, commonly accepted data exchange standard allows for the integration of all aspects of the workflow in subsurface ground investigations, eliminating the need to "reinvent the wheel" for each different owner, as Threadgold and Hutchison (1992) highlighted. This is particularly relevant for specialized contractors carrying out projects or improving consultants' designs. Contractors have a complete and accurate understanding of the ground conditions and access to valuable engineering geology and geotechnical data to achieve substantial improvements during construction. This leads to better construction techniques, safer infrastructure, and reduced costs. This approach contributes to lowering costs and improving efficiency and should be a source of inspiration and motivation toward digital geotechnics.

The exponential evolution and advanced development of integrated platform software (such as CAD – Computer-Aided Design, GIS – Geographic Information System, BIM – Building Information Modeling, ML - Machine Learning, and AI – Artificial Intelligence), hardware, and cloud-based technologies are positively influencing the geotechnical engineering and geoscience fields (e.g., Threadgold and Hutchison 1992, Toll and Oliver 1995, Toll and Cubitt 2003, Brady et al. 2007, Toll 2008, 2009, Chaminé et al. 2013, 2016, Deaton 2018, Gondar et al. 2019, Gilder et al. 2021, Turner et al. 2021, Chaminé and Fernandes 2023, Sun et al. 2024). These digital advances in data processing provide opportunities for organizations that embrace technology and the ability to collect, transfer, manage, analyze, and geovisualize data. The geotechnical community should embrace the technological advancements essential in geotechnical practice.

Currently, valuable geotechnical data is stored in various forms and formats. According to Mokarram (2010), geotechnical experts encounter daily challenges in understanding and extracting the necessary information from geotechnical data, as well as in transferring and managing this data among team members and organizations. This approach prevents project optimization and makes it easier to reuse data already available from previous projects (Zimmermann et al. 2006). Ground investigations are generally expensive and time-consuming. Their data management could be improved, as data archiving, retrieval, and exchange can be complex and sometimes challenging. Therefore, there is an urgent need to create a well-structured and practical data format for standardized data transfer. This format should have significant sharing potential and eliminate technical ambiguities (e.g., Threadgold Hutchison 1992, Toll and Oliver 1995, Toll and Cubitt 2003, Zimmermann et al. 2006, Lefchik et al. 2007, Toll 2008, 2009, Mokarram 2010, Bardet et al. 2014, Bland et al. 2014, Cadden and Keelor 2017, Deaton 2018).

As a rule, the geotechnical community uses geotechnical data in different ways, i.e., some technicians only contribute to producing part of the data. For example, a driller reports the number of SPT (Standard Penetration Test) drives at each depth, and a field geotechnician compiles and archives the data collected in the field and georeferenced the location of the data on maps (e.g., Gonzalez de Vallejo and Ferrer 2011, Matos Fernandes 2020, Uprety and Lal 2021, Fernandes and Chaminé 2023). At the same time, the laboratory technician produces and organizes the data from the laboratory tests. The project engineer assembles the data collected from the team members and uses the dataset analysis for calculations, graphs, and models for the final report (Mokarram 2010). Figure 8.1 shows an example of a matrix of geotechnical users and entities engaged in a geotechnical project and their most likely data usage types.

Typically, team members use a network, constantly producing, modifying, and exchanging data (Mokarram 2010). For example, a CPT (Cone Penetration Testing) driller (driller A) uses a converter to transform data measured from CPT cone sensors. This data is tabulated in a specific format used only by the drilling company and is sent to the geotechnical engineer, as shown in Figure 8.2. The geotechnical engineer (engineer A) can then import this data into

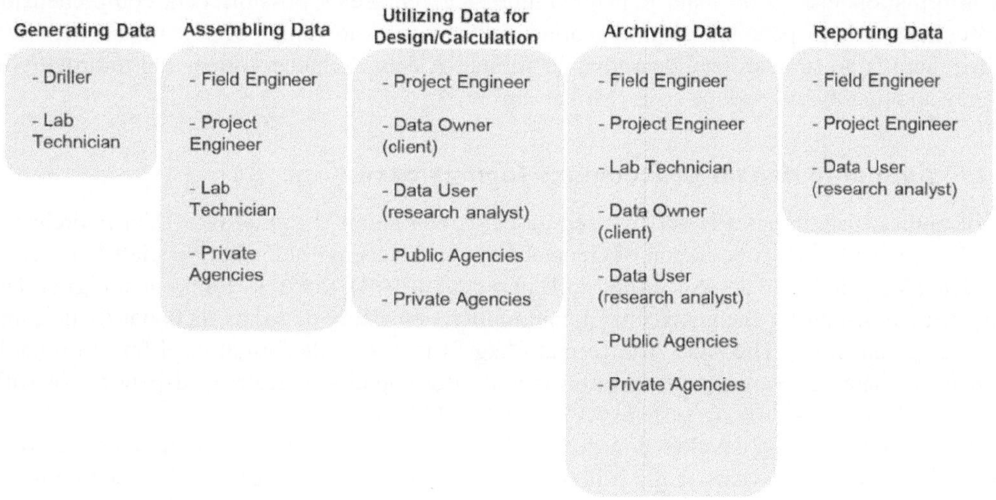

Figure 8.1 Data usage types of geotechnical entities and users involved in a site investigation project. (Adapted from Mokarram, 2010.)

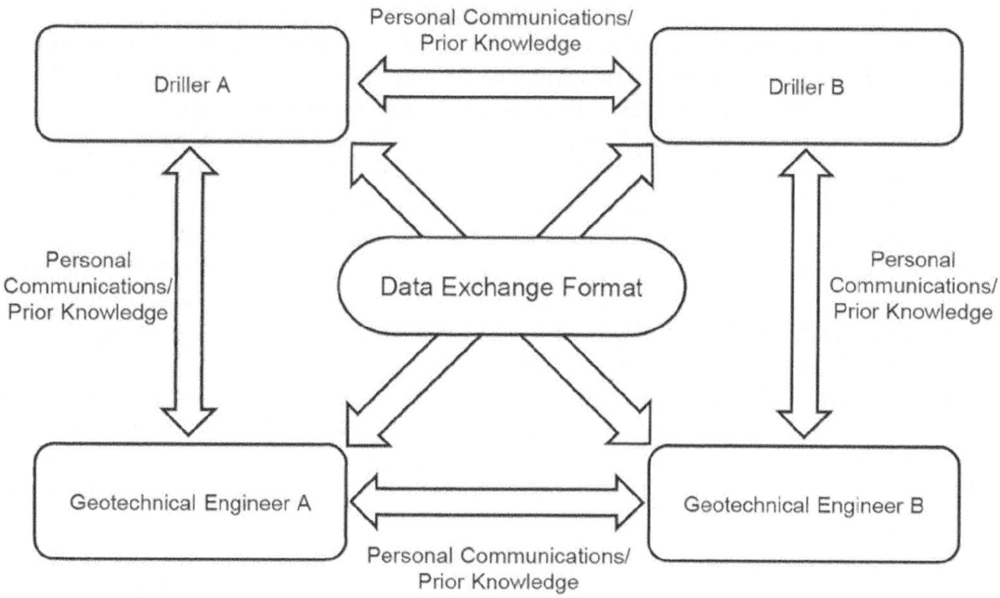

Figure 8.2 Possible data comprehension problems without personal communication and/or prior knowledge between team members when there is no standardized data transfer format. (Adapted from Mokarram, 2010.)

software for engineering computations. This creates a new data set with a format different from the CPT results. This new dataset can be exchanged with another practitioner (engineer B) for further processing. However, suppose the original CPT dataset is passed on to engineer B without personal communication with surveyor A to explain the dataset, there is a high probability of information loss in the data transfer. Figure 8.2 represents possible data comprehension difficulties without personal communication and/or prior knowledge between team members when there is no standardized data transfer format. A data exchange format can improve efficiency and quality and reduce costs (Mokarram 2010).

8.2 Available digital geotechnics format data

Geotechnical data transfer is widely used globally, beginning with the AGS Data Interchange format created by the Association of Geotechnical and Geoenvironmental Specialists based in the UK (AGS 2020, 2022). According to Bland et al. (2014), the AGS format is a digital data interchange format for the geotechnical community, consisting of a data dictionary, rules, and the file format itself. The AGS data format (".ags") is a text file format used for the reliable transfer of data between organizations in the site investigation industry, independent of software, hardware, or operating systems (AGS 2020, 2022).

This format was first developed in the UK in 1991 and has undergone significant improvements over the last 30 years (e.g., Toll and Oliver 1995, Toll and Cubitt 2003, Zimmermann et al. 2006, Toll 2008, 2009, Mokarram 2010, Bardet et al. 2014, Bland et al. 2014, Threadgold 2015, Keynetix 2016, Cadden and Keelor 2017, Deaton 2018, Kennedy 2018, AGS 2022), and references therein).

British Standards (BS) have codified the definition of geotechnical data in the key document (BS-8574:2014), a code of practice for managing geotechnical data for geotechnical engineering projects. Specifically, the BS defines geotechnical data as: "facts or figures obtained from all phases of a geotechnical project design, including derivations from other data" (BS-8574:2014). So, facts and figures may include text, numbers, and formulae (Deaton 2018).

Recognition of the lack of a standardized data transfer format was also identified by the US Department of Transportation (DOT). This format caused difficulties in receiving and sharing data internally across departments and externally with consultants and stakeholders, as each dataset was collected and recorded in various non-standardized formats. Thus, in 2006, the DIGGS project version 1.0 (Data Interchange for Geotechnical and Geoenvironmental Specialists) was created, with the cooperation of government agencies, universities, and industry partners under a joint Federal Highway Administration-State funding study led by the Ohio Department of Transportation (ODOT) (Cadden 2017).

According to Mokarram (2010), this version was built by combining existing data dictionaries developed by AGS, the Department of Civil Engineering at the University of Florida, and COSMOS (The Consortium of Organizations for Strong-Motion Observation Systems, a pilot XML-based exchange standard for Geotechnical Virtual Data Center). DIGGS provides a standardized data transfer structure (schema) using Extensible Markup Language (XML), which has been extended to include Geographic Markup Language (GML) to facilitate the exchange of data (e.g., readings and results) between hardware and software tools. DIGGS provides a standardized format for summarizing and sharing geotechnical data between individuals and organizations.

The ODOT (Ohio Department of Transportation) project was initiated to test the scheme and ensure it was compatible with current standards and testing protocols. DIGGS version 1.1 was released in April 2010 and corrected many technical issues identified in version 1.0. The version of DIGGS released in 2012 (DIGGS version 2) needed to be more easily understood by the industry and required greater support from software suppliers. It was recognized and recommended that industry-friendly tools be developed to facilitate the advancement of cartographic tools and demonstrate the ease of use of various geotechnical tests (e.g., Mokarram 2010, Cadden and Keelor 2017). The DIGGS initiative to establish an updated data format using XML resulted in a highly complex structure, with validation issues and inconsistencies that may hinder its acceptance within the community. As a point of confirmation, it should be mentioned that the AGS committee did not use DIGGS for AGS4 (Mokarram 2010).

The AGS format is the most advanced stage of development and the most widely accepted internationally, having been adopted in several countries such as New Zealand, Australia, Germany, Sweden, the Netherlands, Singapore, Italy, Romania, China, and Hong Kong, among others. The use of standards applied in engineering, the simplicity of the format, the structured schema, and a well-defined data dictionary make the AGS format one of the best codes for the standardized transfer of geotechnical data (AGS 2020, 2022). Figure 8.3 shows the different additions and updates to the AGS (UK) as well as its establishment in various parts of the world, such as Hong Kong (HK), Australia, and New Zealand. AGS (HK) was established in 1998. It currently promotes the format to meet current requirements, and suggests updates to the AGS data format (edition 4) for AGS-UK based on geological and geotechnical experience in Hong Kong (AGS-HK 2024).

In 2017, AGS3.2 NZ v1.1 was established in New Zealand, considering the original version of AGS3.0 (UK) (NZGS 2017). The AGS3.1RTA 1.1. created in Australia was based on AGS3.1 (UK) and the 3.2NZv1.0, with the inclusion of specifications for Australian soil and rock types. In 2012, the Geotechnical Society of New Zealand made a revision based on the AGS 4.0.3 edition, AGS4 NZ v1.0 edition, and 5 years later, they updated the edition to AGS4NZv1.0.1 edition (NZGS 2017). In 2021, Australia updated the format based on AGS4/AGS4.1 (UK) and AGS4NZv1.0.1. In June 2022, AGS4.1.0 AU 1.1 was replaced by AGS 4.1.1 AU 1.2 to rectify some errors and omissions in the previous version (AUGS 2022). Regarding AGS (2022), version 4.1 was a major update of the AGS-UK data format, reflecting new laboratory industry standards and British geotechnical and geoenvironmental practices. It included a substantial update of the data dictionary, introducing 24 new groups (some of which were existing ones). The AGS-UK 4.1.1 edition has seen very little updating compared to the 4.1 edition, with only

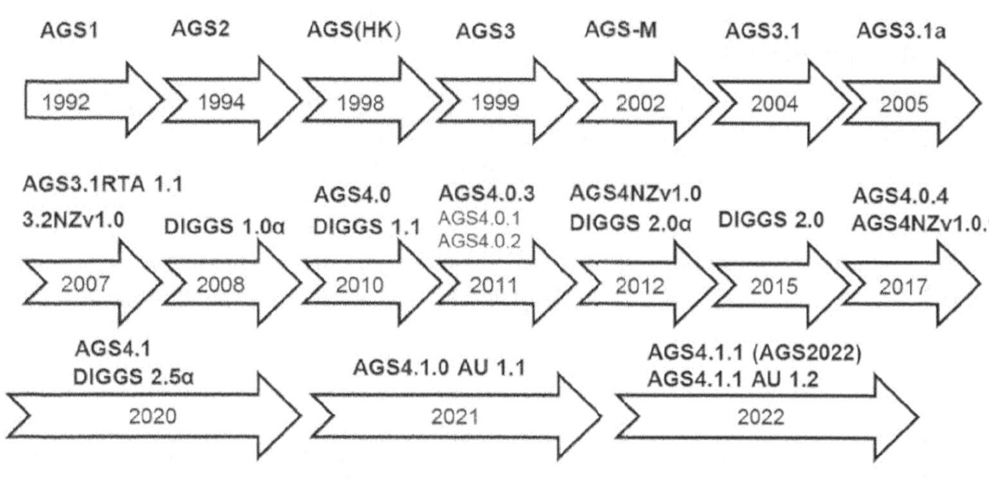

Figure 8.3 Schematic representation of AGS/DIGGS format updates. (Updated from Cruz et al., 2023.)

a few incorrect titles removed and inappropriate examples changed. It is worth mentioning that the AGS format follows British practice (details in AGS 2022).

Digital data transfer is essential for advancing the industry and keeping it aligned with technological progress. As organizations create vast repositories of geotechnical data, data mining, machine learning, artificial intelligence tools, and other analytical technologies will allow users to analyze, manipulate, and visualize geotechnical data in unimaginable ways. Organizations adopting this digital data transfer format can proactively address potential hazards or failures rather than react to them, whether it be predictions related to rock falls, slope geotechnics, pavement performance, or any other geological or geotechnical hazard that could have an impact on infrastructure (e.g.,, Toll 2008, 2009, Mokarram 2010, Deaton 2018, AUGS 2022).

8.3 AGS data format: The role in transportation geotechnics

8.3.1 *General remarks*

Geotechnical data is extensive. It changes, transforms, aggregates, expands, and is presented in many different types and formats. These data are usually compiled in various formats, such as spreadsheet files, geotechnical reports, and data repositories. The AGS format simplifies and speeds up the transformation of information into a data format that allows data to be standardized (e.g.,, Mokarram 2010, Bland et al. 2014, Bardet et al. 2014, Kennedy 2018, AGS 2020, 2022, AUGS 2022). For example, Highways England (UK) maintains a dynamic database containing hundreds of thousands of surveys on a map. It can be downloaded as AGS data format files so that any subscribed user can reuse existing historical data without re-entering it.

AGS (2020, 2022) allows for combining geological-geotechnical data with its spatial location. In this way, it enables a database with the locations of each geotechnical project and, within that project, in other words, from the AGS file, to obtain the location of the work carried out. That said, if there is a site next to an area with AGS data, it is feasible to use this data to understand the ground better and thus adjust the new geological-geotechnical campaign accordingly.

8.3.2 *AGS framework*

The AGS format focuses on the electronic transfer of data recorded during site investigations and activities related to the proper execution of the construction of any geotechnical project design. Typically, the data sent out is presented in forms such as exploratory borehole records, in situ/field test data, monitoring, and laboratory test results (details in AGS (2020, 2022)). This section follows and summarizes AGS's rules, suggestions, and guidelines (AGS 2020, 2022).

The AGS format follows rules and has a data dictionary, making it easier to use. The data dictionary is a structured list of variables recorded during mechanical prospecting and geotechnical testing. The AGS Data Management Working Group maintains the data dictionary. The application of BS-8574:2014, *i.e.*, the code of practice for managing geotechnical data for ground engineering projects, should be employed and updated to support consistent and transparent use and serve as the basis of the AGS Format in geotechnical engineering projects. Data items are the variables to be recorded and must be reported by the specifications used to carry out the work (for example, BS-5930:2015+A1:2020, BS-1377-2:2022, BS-EN-ISO-17892, among others). The data dictionary organizes these variables into groups, each intended for a specific part of the geotechnical investigation. The AGS4 standard file format consists of multiple data groups

```
"GROUP","CDIA"
"HEADING","LOCA_ID","CDIA_DPTH","CDIA_DIAM","CDIA_REM","FILE_FSET"
"UNIT","","m","mm","",""
"TYPE","ID","2DP","0DP","X","X"
"DATA","S1","12.00","200","","FS002"
"DATA","S2","14.00","200","","FS003"
```

Figure 8.4 Representation of the CDIA group in AGS code.

stored in CSV (Comma-Separated Values) format. Although it does not fully adhere to the CSV RFC 4180 specification (details in AGS (2020)), it is commonly referred to as CSV. When preparing files in this format, the rules presented in AGS4.1.1 must be followed (AGS 2022).

Figure 8.4 shows the first line identifies the CDIA group – Casing Diameter by Depth; the "HEADING" title line must be followed by a "UNIT" line, which contains the units used in the data fields, as well as the "TYPE" line that determines the type of unit; both lines are required for all data groups. The following "DATA" lines list the actual values. The "TYPE" and "UNIT" selected for a title are interconnected. The data types and units for each title are presented in the Data Dictionary. The types and units in the data dictionary must be used unless otherwise agreed between the data producer and receiver by the project specifications (AGS 2022).

8.3.2.1 Group hierarchy

The AGS4 groups are organized in a hierarchy (AGS 2022). At the top of the hierarchy is the PROJ group, with most of the other groups below it (see Figure 8.5). One of the groups immediately below PROJ is Location Details (LOCA). All in situ test data is directly below LOCA (for example, the SPT results in the ISPT Group). LOCA is called the "parent" Group of ISPT, and ISPT is called the "child" Group of LOCA. The main group for all laboratory tests is SAMP, which contains sample data (SAMP).

The LOCA group incorporates all the groups and families interested in managing data and technical information. In contrast, the LBSG Group only manages and controls laboratory tests, namely, the scheduling and custody of samples, and does not include the presentation of laboratory results themselves. These results are presented in simplified form in the SAMP family/group (which belongs to the LOCA family), demonstrating that the AGS code is not dedicated to managing laboratory activity (see Figure 8.5).

The LOCA group includes all the information and data resulting from prospecting. Here, we can find direct prospecting (boreholes, wells, drilling parameters, etc.), in situ tests (dynamic, static, load, etc.), sampling (type of samples and simplified laboratory test results), and time-related activity records.

Each group has only one parent in the hierarchy, but may have multiple child groups. Each child group is linked to its parent group by critical variables. The AGS format describes the data on two levels: summary and detail. For example, the groups DPRG and DPRB. The DPRG group describes the general information of the test, and the DPRB group has the data acquired in this test (AGS 2022).

Ten groups fall outside this hierarchy (technical data) and are related to the presentation and description of the data file. The PROJ, TRAN, ABBR, TYPE, DICT, FILE, UNIT, LBSG,

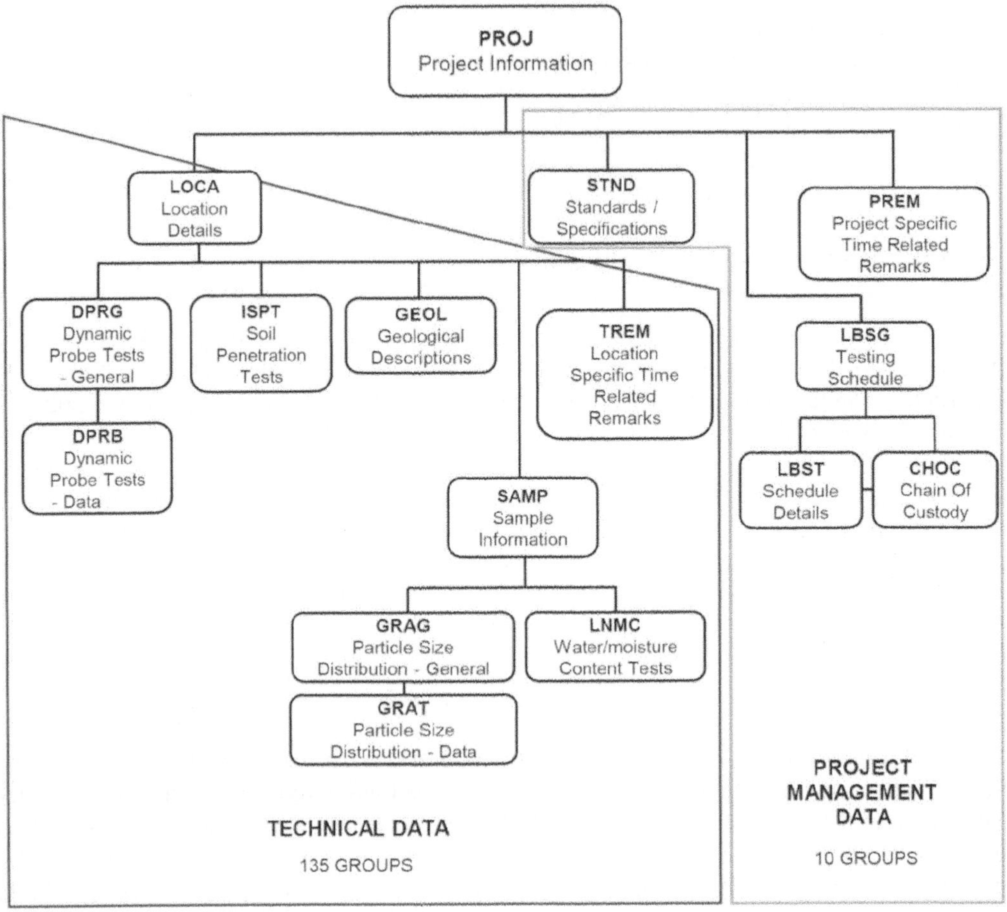

Figure 8.5 Flowchart illustrating the hierarchy of groups in an AGS project. (Adapted from AGS, 2022.)

PREM, and STND groups (AGS 2022) serve to describe the contents of the data file, as summarized in (Table 8.1):

- The PROJ, TRAN, ABBR, TYPE, and UNIT groups will always be included in an AGS file as they define the project, the submission details, and the abbreviations, data types, and units used in the data file;
- The DICT Group should only be included if user-defined groups or headings are present;
- The STND (Standards/Specifications) Group should be used to list the various standards and specifications that define the methods by which the data were collected;
- The LBSG (Testing Schedule Details) group should be used to enter the scheduling and custody of laboratory tests;
- The PREM Group (Project-Specific Time-Related Remarks) records time-related notes that apply to the entire project rather than a specific site or test.

Table 8.1 Mandatory groups in AGS file

Group name	*Content*	*Notes*	*Parent group*
PROJ	Project information	Required in all files	--
ABBR	Abbreviation definitions		--
UNIT	Definition of units		--
TYPE	Definition of data types		--
TRAN	Data file transmission information/data status		--
FILE	Associated files	Required in the file if FILE_FSET data is provided in any group	--
DICT	User-defined groups and headings	Required in all files that include user-defined groups and/or headings	--

Source: Adapted from AGS (2022).

8.3.2.2 Mandatory groups

In the AGS data format, seven mandatory groups define the characteristics of an AGS file, including the PROJ group (project information). Table 8.1 summarizes the seven mandatory groups.

Each group in the AGS has titles (headings) that must be included when entering the data. The titles have three possible states: key, required, and other. The key fields are mandatory for the exclusive definition of the data, identified by the asterisk (*) in the status column of each group. Required fields are essential to interpreting the data file and are marked by the letter "R" in the status column of each group. The other fields are included based on the objective and availability of the data file. This type of status has no sign or letter that defines it, being null in the group status column. Therefore, selecting the key and required titles for describing and interpreting data in each group is essential, as they are considered mandatory for constructing an AGS4 file. "Other" titles are optional and may or may not be selected by the user of the AGS4 file.

Therefore, the most appropriate code structure must be defined for a given project. Introducing groups, titles, and necessary variables should consider the desired data. This task must be coordinated with all those involved to include all the required geotechnical information for end consumers. Additions to the project's data dictionary must be included in the specification and agreed upon between data producers and recipients at the beginning of the project.

To illustrate, Table 8.2 represents PROJ – Project Information, a mandatory group for defining the AGS file. It presents only one key and mandatory title: PROJ_ID. The introduction of the remaining titles is optional. Selecting these titles will result in a better description of the project.

8.3.2.3 Applying the digital data format

The AGS consists of 146 groups and can be divided into four sets: LOCA, SAMP, the mandatory groups that define the AGS file, and other optional groups not included in the technical hierarchy. Groups in the hierarchy represent the technical data within the LOCA family. Figure 8.6 shows the different sets present in the AGS groups.

The LOCA (Location Details) group set includes 60 groups that characterize boreholes, in situ geoenvironmental tests, geological descriptions, geophysical studies, in situ geotechnical tests, installations, monitoring instruments, readings, and sampling information (SAMP). This SAMP set hierarchically depends on the LOCA group (Figure 8.6). It records all samples, for example, those collected in boreholes, field monitoring, and/or sub-samples created in the laboratory.

Table 8.2 PROJ – Project information group details

Status	*Heading*	*Unit*	*Type*	*Description*	*Example*
*R	PROJ_ID		X	Project identifier	121415
	PROJ_NAME		X	Project title	ACME Gas Works
	PROJ_LOC		X	Location of the site	High Street, Any town
	PROJ_CLINT		X	Client name	ACME Enterprises
	PROJ_CONT		X	Contractors name	ACME Drilling Ltd
	PROJ_ENG		X	Project Engineers	Consulting
	PROJ_MEMO		X	General project comments	
	FILE_FSET		X	Associated file reference (e.g., project specification, site location drawings)	

Source: AGS (2022).

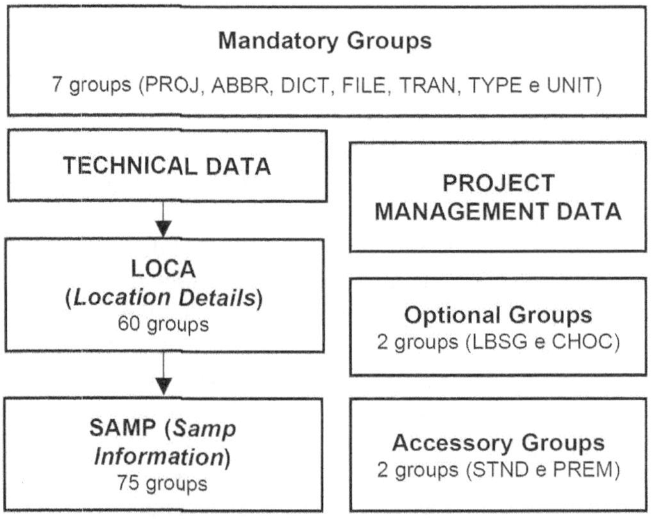

Figure 8.6 Sets of AGS groups. (Adapted from AGS, 2022.)

This group connects fieldwork and laboratory analysis by identifying samples. Within this set, 75 groups describe laboratory tests in geotechnics, primarily focusing on soils and rocks, while excluding bituminous materials and concrete aside (AGS 2022).

The accessory groups are PREM and STND. The PREM – Project-Specific Time-Related Remarks group can be applied to the daily recording of significant events, such as "torrential rain, flooded site" that have caused delays during the project (AGS 2022). The STND – Standards/Specifications group lists the standards, specifications, and other reference documents that describe the methods for collecting and reporting data in the AGS file.

At the start of the project, the laboratory's consideration of the items to be used in the SAMP group and the use of the LBSG (test schedule), LBST (details of the laboratory test schedule), and CHOC (chain of custody) groups, as to be made for better characterization of samples and efficiency of the relationship between fieldwork and laboratories.

The TREM – Location Specific Time-Related Remarks group is part of the LOCA set. It records the duration of activities, such as the drilling time for a particular borehole. Using this group would give an idea of the project's productivity. The DICT group is mandatory in an AGS file when the user creates new groups/titles. This group shows the flexibility of the format, as it allows users to incorporate additional groups and titles according to project-specific needs.

8.3.2.4 AGS data workflow

Figure 8.7 illustrates the AGS application phases. In the initial stage, fieldwork and project information outputs are essential. This valuable information is loaded in AGS data format, ensuring compliance with mandatory groups for this file type. The code must be checked to reach the final code without errors. The final code will contain geotechnical data and generate a data report. Thus, software can read AGS files, such as AutoCAD 2022/2024, which can incorporate geotechnical data into topographic data to create a geotechnical zoning. In addition, the AGS data format plays a vital role in managing geotechnical and geoenvironmental data as part of the Building Information Modeling (BIM) process. AGS considers that the combination of BS-8574:2014 compliance and the use of the AGS data format means that a ground investigation company is already BIM Level 2 compliant (Threadgold and Hutchison 1992).

The standardization of geotechnical data through AGS emphasizes the importance of defining groups, titles, units, and unit types at the project's outset to avoid ambiguity about the parameters to be used. The AGS format is user-friendly and straightforward to handle and understand. The ability to select groups for entering project data limits the number of groups to be used, depending on requirements. This makes it easier to use and also reduces file size for each project while maintaining the required mandatory groups to construct an AGS file correctly.

The use of standards, suggested methods and guidelines, such as ASTM (American Society for Testing and Materials: ASTM D2487-17, 1983), BS (British Standards), ISRM (International Society for Rock Mechanics: ISRM (1980, 2007, 2015), CFCFF (Committee on Fracture Characterization and Fluid Flow: CFCFF (1996), IAEG (International Association for Engineering Geology and the Environment: Baynes et al. (2021), and Euro codes (https://eurocodes.jrc.ec.europa.eu/), facilitates acceptance of the format by the technical and scientific community. The possibility of validation using free AGS file validators allows the structure and integrity of the format to be checked, minimizing the likelihood of errors being made. The fact

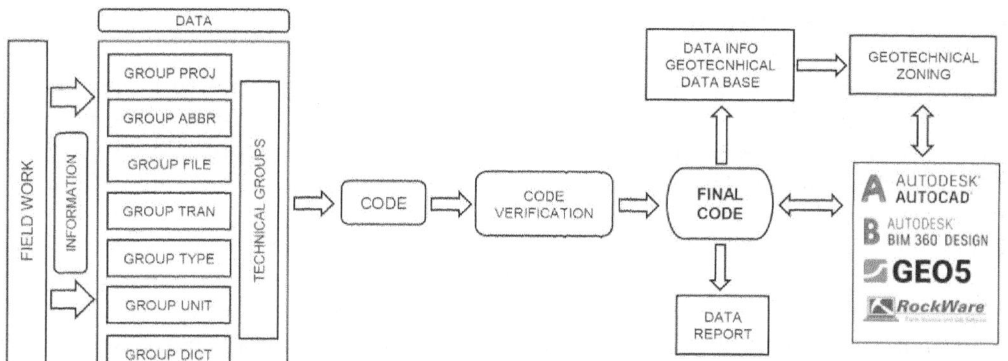

Figure 8.7 Conceptual diagram of AGS application.

that it is possible to transform an MS Excel file into an AGS file using the British Geological Survey's AGS file validator enables to automate data entry and reducing the risk of human error. Thus, encouraging greater use of this format. It also provides the ability to edit, add, and/or remove data from a project design, keeping the information up to date in a single file. It should be noted that there is still a need for more commercial tools to integrate and handle data in AGS format. However, tools can already process this format while still having obvious limitations, particularly when recognizing all the families/groups in the code.

For example, Autodesk, which owns AutoCAD and Revit (BIM software), already recognizes the AGS code in the geotechnical module ("Geotechnical modeler"). However, it can only import georeferenced points and geological and stratigraphic descriptions; it is not yet possible to import the other families/groups with associated geotechnical information (e.g., physical-chemical properties and mechanical characteristics of rock, soils, or geomaterials, among others). AutoCAD only allows importing the GEOL group by geological description or lithological codes (GEOL_LEG) to set up the survey points and the respective 3D modeling.

8.4 AGS in practice: the Portuguese reality

This work aims to highlight the importance of standardizing and digitizing geotechnical data. Internationally, the geotechnical community mostly accepts the AGS data format. This format was developed by AGS, which recognized the need for more data exchange between companies and institutions in the geotechnical and geoenvironmental fields over 30 years ago (details in AGS (2022)).

The Mota-Engil company (Geotechnical Department), based in the Porto area (Portugal), identified this problem well and, in recent years, has worked systematically closely with academia on this topic (particularly with the Department of Geotechnical Engineering and Laboratory of Cartography and Applied Geology of the School of Engineering (ISEP), Polytechnic of Porto and also with the Portuguese Committee of Geotechnics for Transportation of the Portuguese Geotechnical Society, SPG) to develop and implement in Portugal the best international practices of AGS format (Cruz 2022, Cruz et al 2023, Pinto 2023, Pinto et al. 2023, Pinho 2024). Nonetheless, it must be noted that there have been other attempts in Portugal to highlight the importance of using digitized standardized geotechnical terminology and symbols focused on the geology and geotechnics of the Lisbon region (e.g., Moitinho de Almeida et al. 2008, 2010, 2014).

The digitization and integration of geotechnical digital platforms represent a significant advance in how data is collected, processed, and used. A difficulty involved loading data in the AGS code into the Portuguese reality. For this purpose, a recently developed application, "ME-AGS GENERATOR" PT (version Beta), was designed and developed in MS. Excel using VBA – Visual Basic for Applications (details in Pinto (2023) and Pinto et al. (2023)). The MS. Excel spreadsheet was designed for the AGS format group working with it and then created in ".ags" format, but it follows the AGS recommendations (details in AGS (2020, 2022)). So, 'ME-AGS GENERATOR' PT (version Beta) gives users a friendlier experience from Portuguese reality when loading data. Since the names of the different headings in the AGS data format could be more intuitive, the development of this application has improved other users' interpretation of the headings. Hence, each of the different "sheets" has a "cover sheet" with the Portuguese translation of the various headings in the AGS data format to be more visually appealing to the user (Pinto et al. 2023).

Adopting the AGS format in Portugal would be necessary in the short to medium term for future management and development of engineering projects. However, this will require careful analysis of its application to our reality, including creating a broad working group incorporating regulatory bodies and representatives from academia and industry (Cruz et al. 2023). The AGS format is already being used in several countries. So, several European and American companies use it to standardize geotechnical data to increase their competitiveness in international project proposals and daily professional tasks.

The potential widespread use of this format in the Portuguese reality would allow for the creation of geotechnical databases with a powerful capacity for data exchange for various purposes. This approach would maximize the use and management of historical data simply and quickly. The archiving and electronic transfer of data in this format provides an efficient, speedy, and economical data management system, as well as unlimited management of analysis data.

8.5 Concluding remarks

The importance of digitization in geotechnics is evident at every stage of the process, from execution in the field to the digital infrastructure. Digitization promotes efficient and accurate data collection during geotechnical investigations (coupling field and laboratory data), facilitating its transfer and storage in standardized digital formats, such as AGS format. This improves the speed and accuracy of data analysis and allows it to be integrated into specialized software (e.g., GIS, BIM, MLand AI tools, etc.) that enhances the interpretation and manipulation of information. In addition, digitization ensures that all the data is archived in an organized and accessible way, providing a valuable resource for future reference.

Data standardization through digital systems guarantees the consistency and reliability of information, promoting more effective management of geotechnical projects and contributing significantly to the safety and success of engineering works.

Digitization in geotechnics represents a significant advance in handling geotechnical data, offering substantial efficiency, accuracy, and networking benefits. So, adopting digital geotechnical technologies is essential to meeting modern challenges and optimizing engineering design and construction processes. In fact, digitization improves data transfer efficiency and reduces the possibility of transcription errors. It facilitates access to data from anywhere, at any time. It speeds up data analysis and exchange processes and improves real-time database management. Also, it ensures data consistency through the use of standardized formats. Finally, it provides a solid basis for decision-making based on accurate and up-to-date data.

In short, this study highlights the need to continue developing digitization technologies and training professionals qualified to use these tools. Innovation and standardization can achieve more efficiency, accuracy, and collaboration in geotechnical practice.

Acknowledgments

Mota-Engil SA supported this study, and HIC was partially supported due to the projects LABCARGA|ISEP re-equipment program (IPP-ISEP|PAD'2007/08) and the Centre GeoBioTec|UA (UID/GEO/04035/2025). We thank the Association of Geotechnical and Geoenvironmental Specialists (AGS) for the website's shared and inspirational information. The authors are grateful to colleagues Isabel Fernandes (UL), Mário Quinta-Ferreira (UC), Sérgio Moreira (ISEP), José Teixeira (UP), Alexandre Pinto (UL), and António Gomes Correia (UM) for their support and input in some stages of the research.

Reference list

AGS – Association of Geotechnical and Geoenvironmental Specialists. 2020. Electronic transfer of geotechnical and geoenvironmental data Version AGS-UK version 4.0. https://www.ags.org.uk/data-format/ags4-data-format/ (accessed in July 2024)

AGS – Association of Geotechnical and Geoenvironmental Specialists. 2022. Electronic transfer of geotechnical and geoenvironmental data Version AGS version 4.1.1. https://www.ags.org.uk/content/uploads/2022/02/AGS4-v-4.1.1-2022.pdf (accessed in July 2024)

AGS-HK – Association of Geotechnical and Geoenvironmental Specialists (Honk Kong). 2024. Electronic transfer of geotechnical and geoenvironmental data Version AGS (Hong Kong) version 3.0. https://www.ags-hk.org/ags-data-format (accessed in July 2024)

ASTM – American Society for Testing and Materials. 1983. Standard practice for classification of soils for engineering purposes (Unified Soil Classification System). American Society for Testing and Materials, D2487-17.

AUGS – Australian Geomechanics Society. 2022. Electronic transfer of geotechnical and geoenvironmental data Version AGS version 4.1.1 AU 1.2. https://australiangeomechanics.org/ags-4-1-au/ (accessed in July 2024)

Bardet, J.-P., Liu, F., Mokarram, N. and Zand, A. 2014. An overview of geotechnical and geoenvironmental data exchange. In: Toll, D.G., Zhu, H., Osman, A., Coombs, W., Li, X. and Rouainia, M. (eds.), *Proceedings of the 2nd International Conference (ICITG), Durham, UK. Advances in Soil Mechanics and Geotechnical Engineering, Volume 3, Information Technology in Geo-Engineering.* IOS Press, Millpress, Amsterdam, pp. 3–14.

Baynes, F.J. and Parry, S. 2022. Guidelines for the development and application of engineering geological models on projects. International Association for Engineering Geology and the Environment (IAEG), Commission 25 Publication No. 1, 129 p.

Baynes, F.J., Parry, S. and Novotny, J. 2021. Engineering geological models, projects, and geotechnical risk. *Quarterly Journal of Engineering Geology and Hydrogeology*, 54. https://doi.org/10.1144/qjegh2020-080

Bland, J., Walthall, S.X. and Toll, D.G. 2014. The development and governance of the AGS format for geotechnical data. In: Toll, D.G., Zhu, H., Osman, A., Coombs, W., Li, X. and Rouainia, M. (eds.), *Proceedings of the 2nd International Conference (ICITG), Durham, UK. Advances in Soil Mechanics and Geotechnical Engineering, Volume 3, Information Technology in Geo-Engineering.* IOS Press, Millpress, Amsterdam, pp. 67–74.

Brady, S.R., Sinha, A.K. and Gundersen, L.C. (eds.). 2007. Geoinformatics 2007—Data to knowledge. In: *Proceedings: U.S. Geological Survey Scientific Investigations Report 2007–5199*, Reston, Virginia.

BS-1377-2. 2022. Methods of test for soils for civil engineering purposes – Classification tests and determination of geotechnical properties. British Standards, UK.

BS-5930#:2015+A1. 2020. Code of practice for ground investigations. British Standards, UK.

BS-8574. 2014. Code of practice for managing geotechnical data for ground engineering projects. British Standards, UK.

BS-EN-ISO-17892. 2014. Geotechnical investigation and testing. Laboratory testing of soil. British Standards, UK.

Cadden, A.P.E. and Keelor, B. 2017. Implementation and transition of data interchange for geotechnical and geoenvironmental specialists (DIGGS v2.0). The Ohio Department of Transportation, Office of Statewide Planning & Research, U.S. Department of Transportation, Federal Highway Administration, USA.

CFCFF – Committee on Fracture Characterization and Fluid Flow. 1996. *Rock Fractures and Fluid Flow: Contemporary Understanding and Applications*. National Research Council, National Academy Press.

Chaminé, H.I., Afonso, M.J., Teixeira, J., Ramos, L., Fonseca, L., Pinheiro, R. and Galiza, A.C. 2013. Using engineering geosciences mapping and GIS-based tools for georesources management: Lessons learned from rock quarrying. *European Geologist Journal*, 36:27–33.

Chaminé, H.I. and Fernandes, I. 2023. The role of engineering geology mapping and GIS-based tools in geotechnical practice. In: Chastre, C., Neves, J., Ribeiro, D., Neves, M.G. and Faria P. (eds.), *Advances on Testing and Experimentation in Civil Engineering*. Springer Tracts in Civil Engineering. Springer, Cham, pp. 3–27.

Chaminé, H.I., Teixeira, J., Freitas, L., Pires, A., Silva, R.S., Pinho, T., Monteiro, R., Costa, A.L., Abreu, T., Trigo, J.F., Afonso, M.J. and Carvalho, J.M. 2016. From engineering geosciences mapping towards sustainable urban planning. *European Geologist Journal*, 41:16–25.

Cruz, B. 2022. Uniformização de dados digitais — 'Digital Format Data': vantagens e aplicação na prática geotécnica. Instituto Superior de Engenharia do Porto, ISEP, Porto (MSc Dissertation). https://hdl.handle.net/10400.22/21336

Cruz, B., Cruz, J., Branco, J. and Chaminé, H.I. 2023. Digitalização, uniformização e transferência de dados geotécnicos: uma realidade incontornável. In: Duarte, I., Marcelino, J., Pinho, A., Faria, P., Lopes, L. and Pereira, P. (eds.), *Proceedings 18° Congresso Nacional de Geotecnia (18CNG)*. Universidade de Évora & Sociedade Portuguesa de Geotecnia, Évora.

Deaton, S.L. 2018. What are the benefits of geotechnical data interchange. In: *Proceedings of the 69th Highway Geology Symposium (HGS)*, Portland, ME, pp. 304–321.

Fernandes, I. and Chaminé, H.I. (2023) In situ geotechnical investigations. In: Chastre, C., Neves, J., Ribeiro, D., Neves, M.G. and Faria P. (eds.), *Advances on Testing and Experimentation in Civil Engineering. Springer Tracts in Civil Engineering.* Springer, Cham, pp. 29–54.

Gilder, C.E.L., Geach, M., Vardanega, P.J., Holcombe, E.A. and Nowak, P. 2021. Capturing the views of geoscientists on data sharing: A focus on the geotechnical community. *Quarterly Journal of Engineering Geology and Hydrogeology*, 54(2). https://doi.org/10.1144/qjegh2019-138

Gondar, J., Pinto, A. and Fartaria, C. 2019. The use of BIM technology in geotechnical engineering. In: Sigursteinsson, A., Erlingsson, S. and Bessason, B. (eds.), *Proceedings of the XVII ECSMGE-2019, Geotechnical Engineering Foundation of the Future*, ISSMGE and Icelandic Geotechnical Society, Reykjavík, Iceland, pp. 1–8.

Gonzalez de Vallejo, L.I. and Ferrer, M. 2011. *Geological Engineering.* CRC Press, Taylor-Francis Group.

ISRM – International Society for Rock Mechanics. 1980. Rock characterization, testing and monitoring: ISRM suggested methods. In: Brown, E.T. (ed.), *Suggested Methods Prepared by the Commission on Testing Methods, ISRM.* Pergamon Press, London, p. 211.

ISRM – International Society for Rock Mechanics. 2007. The complete ISRM suggested methods for characterization, testing and monitoring: 1974–2006. In: Ulusay, R. and Hudson, J.A. (eds.), *Suggested methods Prepared by the Commission on Testing Methods, ISRM.* ISRM - International Society for Rock Mechanics, Ankara, p. 628.

ISRM – International Society for Rock Mechanics. 2015. The ISRM suggested methods for rock characterization, testing and monitoring: 2007–2014. In: Ulusay, R. (ed.), *Suggested Methods Prepared by the Commission on Testing Methods, ISRM.* Springer, Berlin, p.280.

Kennedy, K. 2018. The importance of AGS data format. AGS – Association of Geotechnical and Geoenvironmental Specialists. https://www.ags.org.uk/2018/11/the-importance-of-ags-data-format/

Keynetix – Geotechnical Data Software Transformed. 2016. AGS data creation and checking made easy key AGS user guide, key AGS professional. Bentley, USA

Lefchik, T.E., Bobba, C. and Beach, K. 2007. DIGGS: Setting the standard for geotechnical and geoenvironmental data management. In: *Proceedings of the 58th Highway Geology Symposium (HGS),* Pocono Manor, PA, pp. 1–11.

Matos Fernandes, M. 2020. *Analysis and Design of Geotechnical Structures.* CRC Press.

Moitinho de Almeida, I., Almeida, G., Lopes, I. and Pinto, C. 2008. Subsurface 3D engineering geological mapping using a geotechnical database. In: Huang, A.-B. and Mayne, P.W. (eds.), *Proceedings 3rd International Conference on Site Characterization,* Taipei, Taiwan. CRC Press, pp. 789–792.

Moitinho de Almeida, I., Lopes, I., Marques, F., Almeida, G. and Pinto, C. 2014. Normalização da terminologia e simbologia na prospecção geotécnica. In: Cavaleiro, V., et al. (eds.), *Proceedings 14° Congresso Nacional de Geotecnia.* Sociedade Portuguesa de Geotecnia & UBI, Covilhã, pp. 1–10.

Moitinho de Almeida, I., Pinto C., Rita, M., Taborda R. and Jeremias, F.T. 2010. Utilização do sistema de informação geocientífico GEOSIS_LX na gestão dos dados geotécnicos de Lisboa. In: *Proceedings 12° Congresso Nacional de Geotecnia.* Sociedade Portuguesa de Geotecnia & UM, Guimarães, pp. 47–56.

Mokarram, N. 2010. Data exchange in geotechnical engineering. University of Southern California, USA (PhD Thesis). https://cee.usc.edu/wp-content/uploads/2017/02/Dissertation_Mokarram2010.pdf

NZGS – New Zealand Geotechnical Society. 2017. Electronic transfer of geotechnical and geoenvironmental data Version 1.0.1 (AGS Edition 4 NZ). https://www.nzgs.org/library/ags4-nz-electronic-transfer-of-data/

Phoon, K.-K., Ching, J. and Cao, Z. 2022. Unpacking data-centric geotechnics. *Underground Space*, 7:967–989.

Pinho, N. 2024. Digitalização em Geotecnia: da execução em campo à infraestrutura digital de dados (AGS+SIG). Instituto Superior de Engenharia do Porto, ISEP, Porto (MSc Dissertation). https://hdl.handle.net/10400.22/26321

Pinto, L.C. 2023. Uma abordagem de programação eficiente para a geração automatizada de ficheiros AGS4 em casos práticos geotécnicos. Instituto Superior de Engenharia do Porto, ISEP, Porto (MSc Dissertation). https://hdl.handle.net/10400.22/23825

Pinto, L.C., Moreira, S., Cruz, J., Branco, J., Cruz, B. and Chaminé, H.I. 2023. ME-AGS GENERATOR: aplicativo para a geração automatizada de ficheiros AGS4 em geotecnia. In: Mota, D., Silva, E., Santos, J.F., Marinho, J., Cardoso, M., Morgado, N., Carvalho, P. and Moreira S. (eds.), *Simpósio de Engenharia Informática 2023: Proceedings Book*. Instituto Superior de Engenharia do Porto, P. Porto, Porto, pp. 76–85.

Sun, Z., Brink, T., Carande, W. and Duzgun, S. 2024. Towards practical artificial intelligence in Earth sciences. *Computer Geosciences*. https://doi.org/10.1007/s10596-024-10317-7

Threadgold, L. 2015. AGS data format means many ground investigation companies are already BIM Level 2 compliant. *Ground Engineering*. https://www.geplus.co.uk/opinion/ags-data-format-means-many-ground-investigation-companies-are-already-bim-level-2-compliant-len-threadgold-11-09-2015/

Threadgold, L. and Hutchison, R. 1992. The electronic transfer of geotechnical data from ground investigations. In: *Colloque International Géotechnique et Informatique, Presse de l'Ecole National des Ponts et Chaussées*, Paris, pp. 749–756.

Toll, D.G. 2008. International data exchange: The future for geo-engineering. In: Singh, D.N. (ed.), *Proceedings of the 12th International Conference of International Association for Computer Methods and Advances in Geomechanics (IACMAG)*, Goa, India, pp. 1958–1965.

Toll, D.G. 2009. International data standards for geotechnical engineering. In: Hamza, M., Amza, M., Shahien, M., El-Mossallamy, Y. (eds.), *Proceedings of the 17th International Conference on Soil Mechanics and Geotechnical Engineering*. IOS Press, pp. 2690–2693. https://doi.org/10.3233/978-1-60750-031-5-2690

Toll, D.G. and Cubitt, A.C. 2003. Representing geotechnical entities on the World Wide Web. *Advances in Engineering Software*, 34(11–12):729–736.

Toll, D.G. and Oliver, A.J. 1995. Structuring soil and rock descriptions for storage in geotechnical databases: Geological data management. In: Giles J.R.A. (ed.), *Geological Society Special Publications, No. 97*. Geological Society Publishing House, pp. 65–71.

Turner, A.K., Kessler, H. and van der Meulen, M.J. (eds.). 2021. *Applied Multidimensional Geological Modeling: Informing Sustainable Human Interactions with the Shallow Subsurface*. Wiley-Blackwell, Chichester.

Uprety, B. and Lal, M. 2021. Standard penetration test in geotechnical engineering site investigations. *Journal of Emerging Technologies and Innovative Research*, 8(4):119–127.

Zimmermann, R., Ku, W.S., Wang, H., Zand, A. and Bardet, J.P. 2006. A distributed geotechnical information management and exchange architecture. *IEEE Internet Computing*, 10(5):26–33.

Chapter 9

Creating knowledge by transforming geotechnical data based on machine learning algorithms

Joaquim Tinoco, Paulo Cortez, David Toll, and Raquel Silva

9.1 Introduction

Transportation infrastructures play a crucial role in facilitating people's convenience when traveling, fostering economic activities, and promoting cultural exchanges. Consequently, well-maintained conditions are imperative for transportation infrastructures, which include roads, railways, and airports (Roshan et al. 2022a). Among these, road networks are particularly significant, serving as the primary mode for transporting passengers and goods (Roshan et al. 2022b). To ensure convenient passenger trips and safe cargo transport, designing and continuously monitoring the health conditions of transportation infrastructure is essential (Roshan et al. 2023, 2024). In light of this, there is a growing interest in employing machine learning (ML) algorithms for the analysis and evaluation of complex geotechnical data, due to their promising and efficient performance. This chapter focuses on two transportation geotechnics applications where ML algorithms have made significant contributions to their understanding, namely on slope stability condition assessment and surface condition of road pavements.

Slope stability is vital for both railways and road networks, as failures can result in economic losses and endanger human safety. However, network managers face several challenges in applying various slope failure detection methods, such as spatial complexity, uncertainties, transferability, and a lack of data relating to slope failures. To overcome these challenges, recent advancements have leveraged artificial intelligence (AI) techniques, including artificial neural networks (ANNs), support vector machines (SVMs), cubist algorithms, and extreme learning machines. In the context of earthwork hazard category (EHC) identification, a case study in Section 9.3 highlights the combination of two AI techniques: Genetic algorithms (GAs) for finding the optimal set of input variables and ANNs for learning complex patterns. This combination, called an evolutionary neural network (ENN), takes advantage of both AI techniques.

The importance of road networks in economic and social development is undeniable. Roads contribute to poverty alleviation by providing access to essential services. Recognized as a crucial public asset, road infrastructure significantly impacts economic progress. Automation has replaced manual visual inspection, with various technologies such as vibration-based, image-based, and acoustic-based methods employed for automatic data acquisition. AI and ML have emerged as promising tools for data analytics, particularly in pavement condition assessment and deterioration prediction. In Section 9.4, a case study addressing road surface condition assessment explores the effectiveness of three classification algorithms in differentiating road types using vibration-based data acquisition and ML algorithms.

DOI: 10.1201/9781003530220-11

9.2 Machine learning concepts

9.2.1 *ML algorithms*

Identifying and selecting the optimal ML algorithm for addressing specific real-world datasets is a dynamic and evolving area of research (Thomas et al. 2011). This endeavor is inherently intricate due to its dependence on the characteristics of the input data and the algorithm's capacity to discern intricate relationships within the dataset (Rodriguez-Galiano et al. 2015). The models generated by these algorithms are often regarded as black-box entities, wherein the algorithm independently derives the output without providing a transparent and interpretable explanation to human observers.

In the realm of case studies, diverse algorithms have been systematically evaluated, incorporating a spectrum of methodologies. Advanced and highly nonlinear techniques, including ANNs, GAs, and SVMs, have been explored. The application of these algorithms involves complex computational processes wherein the algorithms autonomously learn and adapt to intricate patterns and structures within the dataset, ultimately facilitating effective prediction. This investigation aims to contribute to the ongoing discourse on determining the most suitable ML algorithms by systematically examining the performance and adaptability of the diverse algorithmic frameworks across various real-world datasets.

9.2.1.1 Support vector machines

SVMs are an effective class of supervised learning algorithms that perform well in problems involving regression and classification. Fundamentally, SVMs work by projecting input features into a higher-dimensional space and using a hyperplane to try and distinguish data points into different groups. This hyperplane, which is positioned to optimize the margin of separation between the various classes, functions as the ideal decision boundary (Rodriguez-Galiano et al. 2015, Nunes and Mota 2019, Simplilearn 2022).

Crucially, within the SVMs framework, there exists a subset of points lying atop the margins or on the boundary itself, aptly named support vectors. These support vectors play a pivotal role in defining the hyperplane, as they encapsulate all pertinent information required for its reconstruction (see Figure 9.1). Remarkably, even in the absence of the remaining training points, the support vectors alone are sufficient to identify the optimal hyperplane. This intrinsic property highlights the robustness of SVMs, indicating that non-support vector points can be omitted from the training set without sacrificing the integrity of the classification model.

However, the efficiency of SVMs is contingent upon the careful selection of their parameters, including the choice of kernel function and regularization parameters (Meyer and Wien 2001, Karatzoglou et al. 2006, Mammone et al. 2009). Indeed, SVMs can be highly sensitive to these parameters, necessitating meticulous tuning to achieve optimal performance. This sensitivity underscores the importance of parameter optimization in harnessing the full potential of SVMs across diverse datasets and problem domains.

9.2.1.2 Artificial neural networks

ANNs, also known as neural networks (NNs) in short, are advanced computer models that are modeled after the complex mechanisms of the human nervous system. ANNs mimic the functionality of interconnected neurons to simplify ML tasks, based on the biological model of NNs. At the core of an ANN are neurons, also known as units or nodes, which serve as the fundamental processing elements responsible for information manipulation and propagation.

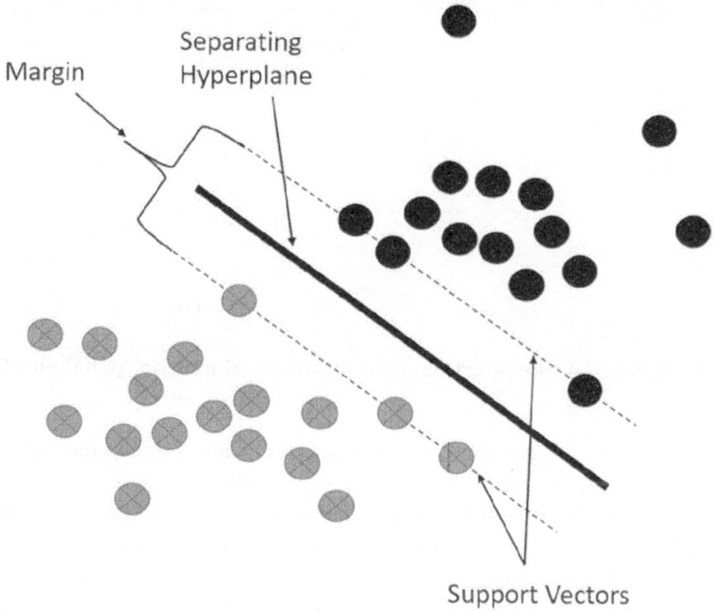

Figure 9.1 SVM schematization, from Meyer and Wien (2001).

Similar to synapses in the human brain, connections within an ANN enable the transmission of information from input nodes through hidden layers to the output layer. Each connection in the network is endowed with a weight, indicative of its significance in shaping the output. These weights collectively influence the network's ability to generate accurate predictions or outputs corresponding to a given input (Rodriguez-Galiano et al. 2015, Choi et al. 2020).

An ANN is composed of multiple interconnected processing modules that work in parallel to solve complicated problem domains using computational resources as a whole. The fundamental operation of an ANN is the complex interaction between neurons and connections, which are coordinated to generate desired outputs in response to various inputs.

Mathematically, an artificial neuron embodies several key components, as depicted in Figure 9.2. These include multiple incoming connections, akin to synapses on dendrites, through which the neuron receives input signals. Conversely, outgoing connections, analogous to synapses on axons, propagate processed information to downstream neurons. Furthermore, every neuron has an activation value that determines how responsive it is to external stimuli. This activation value is comparable to the membrane potential of a biological neuron.

Through the integration of these components inside a structured network design, ANNs are able to recognize intricate patterns and relationships that are present in data, which opens up a vast range of applications, from natural language processing to picture identification. One of the most significant advantages of ANNs is that they can extract patterns and detect trends too complex to be noticed by humans or other computer techniques.

9.2.1.3 Genetic algorithms

GAs represent randomized search algorithms designed to emulate natural selection and genetics. Modeled after the mechanics of biological evolution, GAs operate in string structures evolving over time, following the principle of survival of the fittest through a randomized yet structured

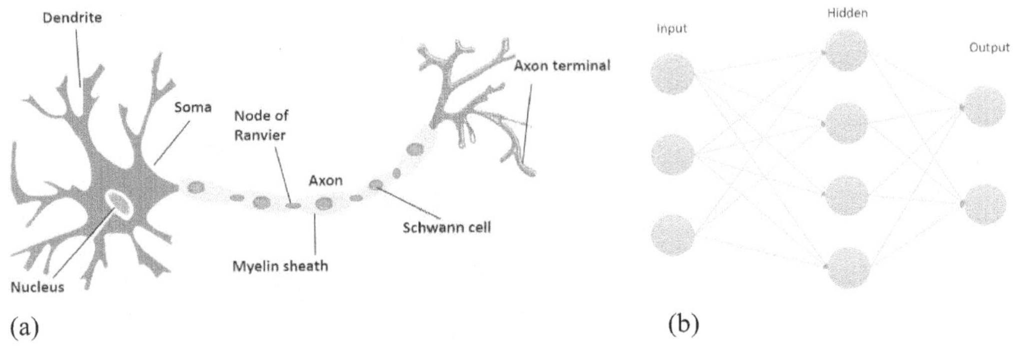

Figure 9.2 ANN analogy: (a) biological neuron, (b) artificial neuron. (Ghallou 2024.)

information exchange. In each generation, a new set of strings is generated, incorporating parts of the fittest members from the previous set (Golberg 1989).

In the context of data-driven applications, heuristic search methods, including GAs, are frequently employed for feature selection due to their capability to yield meaningful optimization outcomes with moderate computational resources. The complexity associated with models presenting numerous input features requires the identification of an optimal feature subset to enhance prediction accuracy. GAs, deeply rooted in AI and natural selection principles, initiate the process with random solutions and iteratively refine them toward optimality (Sun et al. 2006).

The GA operates as a stochastic algorithm grounded in probability theory. When applied to a stage-wise superstructure model, its search process is determined by a stochastic strategy. The global optimal solution for synthesizing heat exchanger networks can be probabilistically obtained. The search process begins with an initial set of stochastic solutions, referred to as the "population." Each solution, called a "chromosome," comprises "genes" representing optimal variables, such as mass flow rates of cold and hot streams (Alexandridis et al. 2005).

There are two primary operations in the genetic algorithm: genetic operations and evolution operations. Genetic operations involve the transfer principle of probability, selecting good chromosomes to propagate with a certain probability while allowing inferior chromosomes to perish. This guides the search direction toward the most promising region. With their stochastic search technique, GAs can simultaneously explore different regions of the search space, making them less prone to terminating in local minima (Alexandridis et al. 2005, Sun et al. 2005).

9.2.2 *Model generalization assessment*

Evaluating the generalization capabilities of ML models is fundamental for assessing their performance on new, unseen data. Numerous techniques are applied for this purpose, including *k*-fold cross-validation, regularization techniques, holdout, evaluation and tuning, cross-target generalization assessment, adaptation to unseen data, and cross-validation technique. Subsequent case studies have generally adopted the *k*-fold cross-validation method.

A *k*-fold cross-validation method uses all the available data and is, therefore, a more robust model that requires higher computation effort. In this case, the dataset is divided into k subsets of the same size. As demonstrated in Figure 9.3, the testing process is repeated k times, with $k-1$ subsets for training and the remaining subset for validating the model. In this way, each example serves a dual purpose as training and test data. The outcomes are averaged over the rounds, producing a more accurate estimate (Bonissone et al. 2009, Cortez 2022).

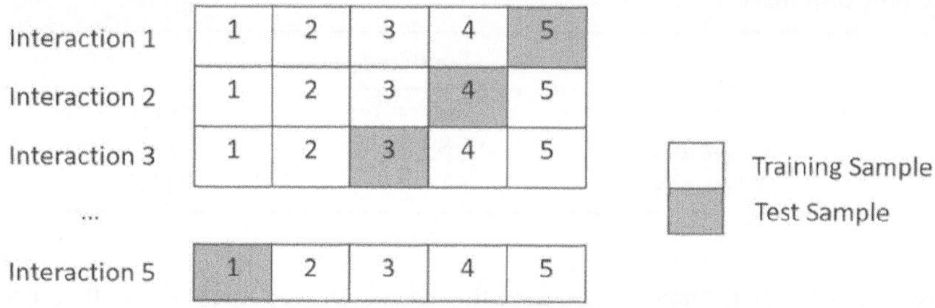

Figure 9.3 Cross-validation method scheme. (Adapted from Stuart and Peter 2016.)

Model evaluation is a vital process that assesses the model's performance and indicates how well it will generalize in future predictions. For this purpose, several metrics can facilitate its assessment and help identify its weaknesses and strengths. The choice of the best ML model for a given task is made by comparing several model candidates in the calculated evaluation metrics.

9.2.3 *Performance measures*

Accuracy is one of the most common metrics for classifier assessment by evaluating the overall algorithm effectiveness by calculating the ratio of correct predictions among the total number of predictions made (Transfer 2024):

$$\text{Accuracy} = \text{number of correctly classified examples/total number of cases}$$

By definition, false predictions do not affect the measure of accuracy. However, accuracy is not a suitable metric for evaluating classifiers in imbalanced datasets, where certain classes are underrepresented. In addition to accuracy, other metrics, namely, the confusion matrix, precision, recall (or sensitivity), specificity, F1-score, and receiver operator characteristic area under the curve (ROC-AUC), are used to measure the performance.

9.2.3.1 Confusion matrix

A confusion matrix (see Table 9.1) offers a clearer presentation of classifier performance, revealing weaknesses and providing guidelines for further research and model improvement (Transfer 2024). Each column in the confusion matrix represents a predicted outcome, and each row represents the actual instance. The diagonal of the matrix represents the number of correctly classified examples, while other elements indicate the number of incorrectly classified examples as other classes (Boutaba et al. 2018).

In its simplest form, a confusion matrix for a binary classifier shows four possible outcomes: True positive (TP), false positive (FP), true negative (TN), and false negative (FN). TP occurs when the model correctly predicts the result, FP when the model predicts a positive result but the actual result is negative, TN when the model predicts a negative instance correctly, and FN when the model predicts a negative result though the actual result is positive (Ghosh 2024).

Confusion matrices provide insights into the relationships between different data features and reveal the data structure. However, they may not be suitable for multi-label classification, where an object from a dataset can belong to multiple classes simultaneously.

Table 9.1 Confusion matrix

		Predicted class	
		Class = Yes	*Class = No*
Actual class	Class = Yes	TP	FN
	Class = No	FP	TN

Regarding other performance metrics, the recall, also known as the true-positive rate or sensitivity, gauges the model's ability to correctly identify instances of a specific class. Precision, alternatively known as positive predictive value, measures the model's accuracy when predicting a particular class. The formulas for recall and precision are defined as True Positives / (True Positives + False Negatives) and True Positives / (True Positives + False Positives), respectively.

The third metric, the *F1-score*, represents the harmonic mean of precision and recall and is calculated by 2 * (precision * recall)/(precision + recall). All three metrics range from 0 to 100%, with higher values indicating a more accurate predictor.

9.2.3.2 Receiver operating characteristic curve

The ROC is a curve plotting the true-positive rate (TPR) on the y-axis versus the false positive rate (FPR) on the x-axis (Novaković et al. 2017). TPR, or sensitivity is the probability that an actual positive will be correctly identified as positive, while FPR is the probability of falsely rejecting the null hypothesis.

The ROC curve is formed by connecting points representing TPR-FPR pairs for different test threshold values. The closer the curve is to the coordinate (0,1), the better its predictive capacity. An ROC curve close to the line of equality (Figure 9.4) indicates performance no better than random guessing, emphasizing the importance of distance from this line. The area under the receiver operating characteristic curve (AUC) provides a confidence level of the model's ability to predict positive outcomes accurately (Novaković et al. 2017, Krstinić et al. 2020).

9.2.3.3 Performance metrics

Regression tasks are often used in geotechnical engineering problems that require a supervised learning algorithm to match independent variables (inputs) with a given output target. As useful assessment instruments for examining the accuracy of models in data science, model performance measurement criteria in regression predictions and evaluation of variable prediction models (metrics as exemplified below) are examined (Carter et al. 2016).

- Mean absolute deviation (MAD): The average absolute difference between the values that were predicted and the actual values is measured by MAD. This parameter displays the model's accuracy as a measure that is independent of the magnitude of the values.

$$\text{MAD} = 1/N \times \sum_{i=1}^{N} \left| y_i - \hat{y}_i \right| \tag{9.1}$$

- Relative absolute error (RAE) measures the average absolute difference between the actual values of the variables and the MAD value. In relation to the size of the data, this metric displays how accurate the forecasts are.

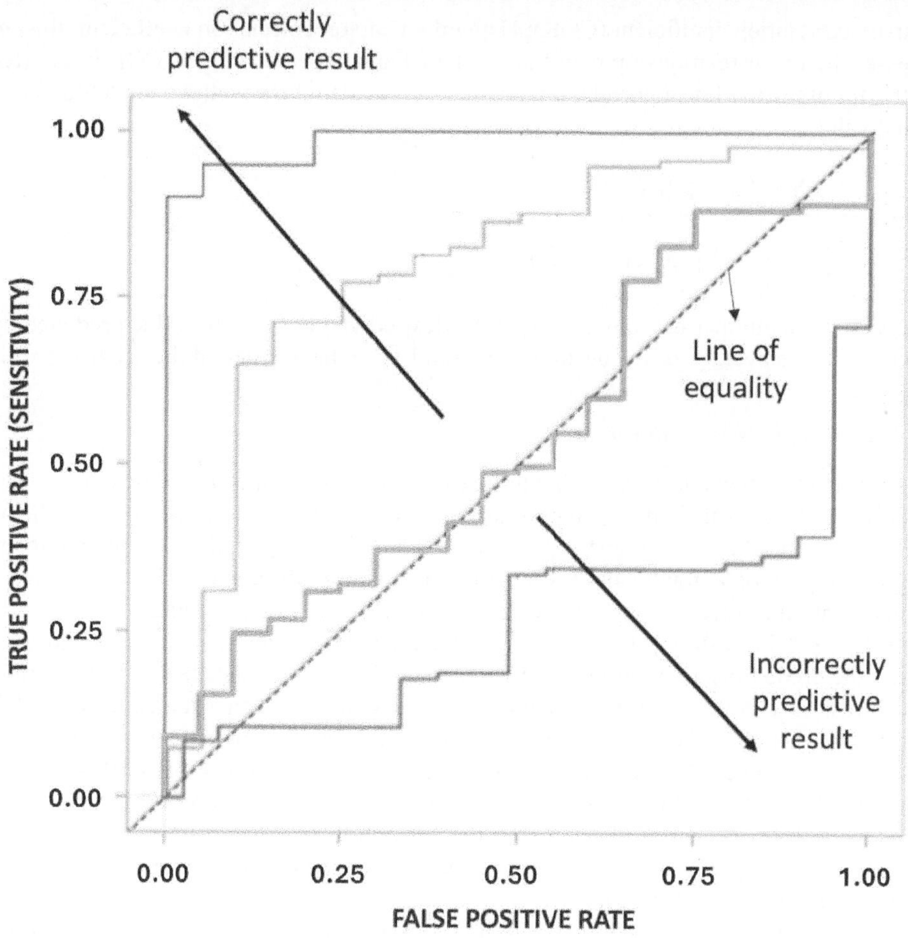

Figure 9.4 Example of ROC curve. (Adapted from Carter et al. 2015.)

$$\text{RAE} = \text{MAD} / \sum_{i=1}^{N} \left| y_i - \bar{y}_i \right| / N \times 100\ (\%) \tag{9.2}$$

- Root mean squared error (RMSE): This metric determines the root mean square of the variation between the values that were expected and those that were observed. This parameter serves as a benchmark for assessing a model's accuracy.

$$\text{RMSE} = \sqrt{\sum_{i=1}^{N} \left(y_i - \hat{y}_i \right)^2 / N} \tag{9.3}$$

- Relative root mean squared error (RRMSE): RRMSE plots the average absolute difference between the true values and the RMSE value. This metric assesses how accurate forecasts are in relation to the size of the data.

$$\text{RRMSE} = \text{RMSE} / \sqrt{\sum_{i=1}^{N} \left(y_i - \bar{y}_i \right)^2 / N} \times 100\ (\%) \tag{9.4}$$

- Pearson correlation coefficient (COR): Using the Pearson correlation coefficient, this metric assesses the linear relationship between the actual and expected values. COR is an effective metric for assessing linear correlation that makes it evident how well the model forecasts the actual values.

$$\mathrm{COR} = \frac{\sum_{i=1}^{N}\left(y_i - \bar{y}_i\right)\left(\hat{y}_i - \bar{\hat{y}}_i\right)}{\sqrt{\sum_{i=1}^{N}\left(y_i - \bar{y}_i\right)^2 \sum_{i=1}^{N}\left(\hat{y}_i - \bar{\hat{y}}_i\right)^2}} \tag{9.5}$$

where N is the number of examples, y_i is the desired value, $\hat{y}_i$ is the value predicted by the model, $\bar{y}_i$ is the average of the desired values and $\bar{\hat{y}}_i$ is the average of the predicted values.

9.2.3.4 *Regression error characteristic*

Regression analysis benefits greatly from the use of regression error characteristic (REC) curves, which offer a detailed insight into regression models' performance. REC curves, which plot error tolerance on the x-axis against the percentage of observations falling within that tolerance on the y-axis, provide a graphical representation of model performance. They originated from the generalization of receiver operating characteristic (ROC) curves to regression tasks (Bi and Bennett 2003). With different levels of error tolerance, researchers and analysts can evaluate how well the predictions of a regression model match the actual values thanks to this graphic.

In contrast to conventional assessment measures such as mean squared error (MSE) or R-squared, REC curves offer valuable perspectives on the compromises between forecast precision and error tolerance. Based on the particular needs of their application, analysts can select and tune models with confidence by looking at the REC curves. Moreover, REC curves are particularly useful in scenarios where the cost of prediction errors varies across different levels of tolerance, enabling stakeholders to optimize models based on their unique cost-benefit considerations (Bi and Bennett 2003).

REC curves serve as a powerful visualization tool in regression analysis, offering a comprehensive understanding of model performance and aiding in the decision-making process for model selection and tuning. Their ability to capture the trade-offs between prediction accuracy and error tolerance makes REC curves an essential component of the toolkit for researchers, analysts, and practitioners in various fields.

9.2.4 *Sensitivity analysis*

Sensitivity analysis (SA) emerges as an important method for improving the interpretability of models, particularly those driven by data and based on ANN algorithms. The need for interpretability is emphasized, especially in engineering applications where a full grasp of what models have learned is critical. In many instances, data-driven models, including ANNs, are perceived as black-box models, requiring the application of explainable artificial intelligence (EAI) methods. SA serves as a methodological approach to unravel the intricacies of model behavior and shed light on the relative importance of each attribute influencing the model's predictions (R Development Core Team 2023).

Inspired by experimental design, SA allows the systematic measurement of a model's responses when specific inputs are modified. This permits the quantification of the influence and significance of each attribute, contributing to a transparent understanding of the model's decision-making process. Unlike other approaches, such as rule extraction from trained ANNs, SA avoids the pitfalls of discretization or oversimplification of predictive model responses. It is adept at accurately measuring the trained ANN's responses, even when dealing with complex class decision boundaries, employing a computationally efficient mapping approach (Holzinger et al. 2022, R Development Core Team 2023).

The rationale behind SA is rooted in the observation that the significance of an input is directly proportional to the changes observed in the model's output. This metric, endorsed based on its theoretical soundness and superior experimental evidence, forms the basis for calculating the relative importance (Ra) of each input variable. In essence, SA serves as a bridge between the inherent complexity of ML models and the imperative for transparency and interpretability, offering a systematic means to understand the impact of input variables on model predictions (Cortez and Embrechts 2011, Confalonieri et al. 2021).

9.2.5 *Resampling*

When dealing with imbalanced classification tasks in which at least one target class label has fewer training samples than other target class labels, the straightforward application of a data mining (DM) training algorithm frequently results in data-driven models with improved prediction accuracy for the majority classes and decreased classification accuracy for the minority classes. Thus, methods such as oversampling and the synthetic minority oversampling technique (SMOTE), which modify the data to be used for training purposes with the aim of balancing the target class, are frequently adopted to handle datasets where the classes are not equally distributed (imbalanced). In particular, oversampling is a straightforward strategy that balances the final training set by adding random samples (with repetition) of the classes with less training data. SMOTE is a more advanced method that generates "new data" by establishing a neighborhood by examining the nearest neighbors and then sampling from that neighborhood. It assumes that the proximity of the original data makes it comparable. By creating synthetic instances, SMOTE helps to increase the representation of the minority class, making the dataset more balanced. This can lead to improved model performance, especially when the class imbalance is significant. It is important to note that SMOTE should be applied only to the training set and not to the entire dataset. Also, it is recommended to combine SMOTE with other techniques, such as undersampling the majority class or using more advanced algorithms, such as the SMOTE-ENN (SMOTE with edited nearest neighbors) method, to further enhance the effectiveness of the resampling process (Brook and Arnold 1985, Cortez and Embrechts 2013).

9.3 Slope stability

This ML case study application combines ANNs and GA with the goal of creating novel models to identify their stability, from this point referred to as EHC (Power et al. 2016), of rock and soil cuttings as well as soil embankments.

9.3.1 *Data characterization*

The EHC system is composed of four levels, "A," "B," "C," and "D," where the probability of failures increases from class "A" to "D." Three distinct databases were compiled for training and testing purposes, each containing information gathered during routine inspections and supplemented with geometric, geological, and geographic data for each slope. All

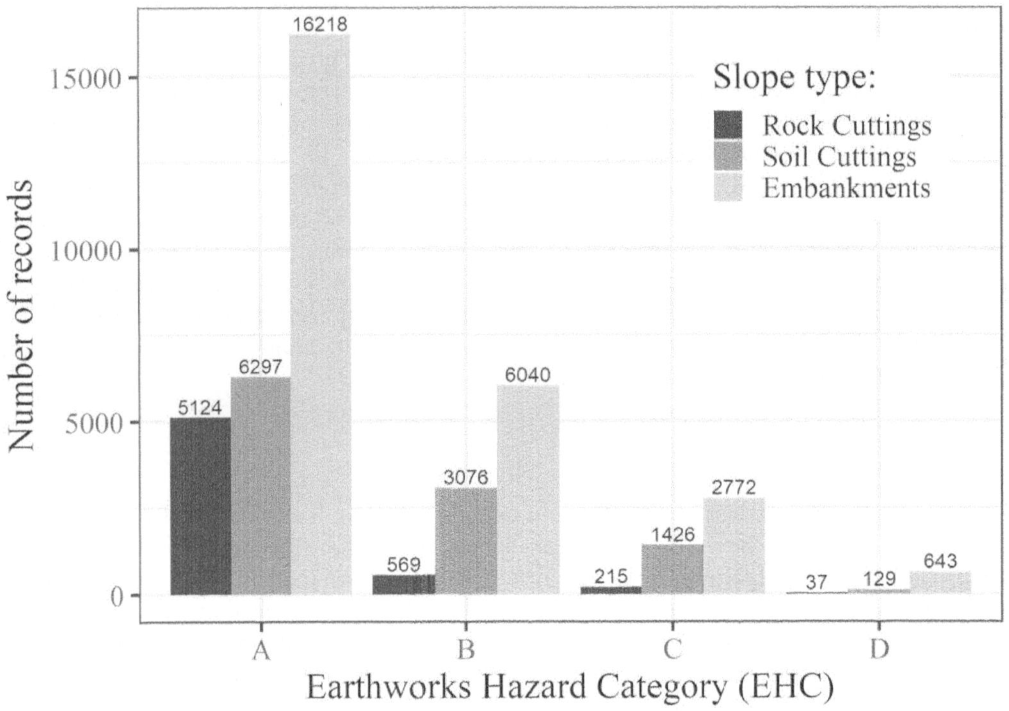

Figure 9.5 The data distribution of rock and soil cuttings from slopes and embankments by EHC class. (Tinoco et al. 2023.)

three databases were compiled by Network Rail employees and are related to the UK railway network. On the basis of their experience, Network Rail engineers attributed a class of the EHC system to each slope, which was taken as a stand-in for the slope's actual stability conditions for the year 2015.

Figure 9.5 represents the distribution of EHC levels. Its analysis reveals a significant number of available records for all three types of slopes, namely 5,945 and 10,928 records for rock and soil cuttings, respectively, and 25,673 records for embankment slopes. A highly asymmetric distribution (imbalanced data) of the records for each EHC class was also noted. For example, taking as reference the database for embankments, more than 63% of the data are classified as class "A," and only 2.5% belong to class "D."

9.3.2 *Methodology*

The methodology employed in the study is a comprehensive approach aimed at addressing the challenges in slope stabilization identification by EHC prediction. The study begins by adopting a nominal classification approach based on the success observed in previous attempts. Specifically, ANN algorithms are chosen due to their superior performance in tackling the EHC problem. Resampling techniques, including SMOTE, oversampling, and no resampling, are introduced during the modeling step to combat imbalanced data issues. The selection of these diverse resampling approaches is driven by the need to explore and determine the most effective strategy for handling imbalances in the dataset.

A pivotal aspect of the methodology involves the application of GA for optimization. The GA is employed to identify the optimal set of variables that minimizes the objective function, with a focus on maximizing the recall metric for all four EHC classes. The GA operates by defining an initial population, where each individual represents a potential solution using a binary chromosome representation. The objective function is associated with the recall metric, and the fitness function utilizes an ANN algorithm-based predictive model. This process results in the optimization of the set of attributes for EHC identification.

Following optimization, the study delves into ANN training, where five ANNs are fitted to the database for each Pareto optimal solution obtained from the GA. Notably, three sampling methods, namely no resampling, SMOTE, and oversampling, are applied to the training data, amounting to a total of 75 ANNs for each Pareto optimal solution. SA is then conducted on each model to discern the relative relevance of each attribute for EHC prediction, providing insights into the contribution of individual features to the model's predictive performance.

In parallel, the feature selection process involves considering over 50 variables derived from routine inspections and various sources related to rock, soil cuttings, and embankment slope scenarios (Tinoco et al. 2018a, 2018b). The ANN model itself adopts a multilayer perceptron with feedforward connections, incorporating one hidden layer with a variable number of processing units. A grid search is conducted to optimize the number of processing units, and the logistic function is applied to the neural function of the hidden nodes. The backpropagation algorithm, coupled with the Broyden–Fletcher–Goldfarb–Shanno (BFGS) optimization technique, is utilized for training the ANN, showcasing a thoughtful selection of parameters to enhance model performance.

The study further integrates evolutionary computational methods, employing a GA to iteratively refine the input feature set and enhance the predictive capabilities of the model. Additionally, the non-dominated sorting genetic algorithm-II (NSGA-II) is leveraged as an evolutionary multi-objective optimization (EMO) method to address the multi-criteria optimization problem associated with slope stabilization identification. Standard NSGA-II parameterization includes a population size of 100, a stop criterion after 100 generations, a crossover probability of 0.7, and a mutation probability of 0.2 (see Figure 9.6).

The *R* statistical environment was adopted to implement all experiments conducted in this research. Moreover, to facilitate the implementation of ANNs and NSGA-II algorithms, as well as validation techniques such as cross-validation, the *rminer* (Kaur et al. 2019) and *mco* (Mersmann et al. 2014) packages were used.

9.3.3 Results and discussion

The preliminary studies (Tinoco et al. 2018a, 2018b) provided some useful insights that helped to design the approach proposed in this work. Firstly, the best predictive results in both works were produced by the ANN model when performing a nominal classification. Secondly, while interesting predictive results were achieved by the ANN model, the performance was far from ideal. In particular, a high error was obtained for the infrequent EHC classes (which are related to more hazardous conditions). Thirdly, although the implementation of resampling techniques to handle the imbalanced data issue (which arises as the majority of slopes are in perfect condition) has not always been effective, in some instances, it aided the algorithms in better learning the problem. Such an asymmetric distribution of the data represents a key challenge of the work since it affects the model's response over all four EHC classes. Fourthly, it was observed that the identification

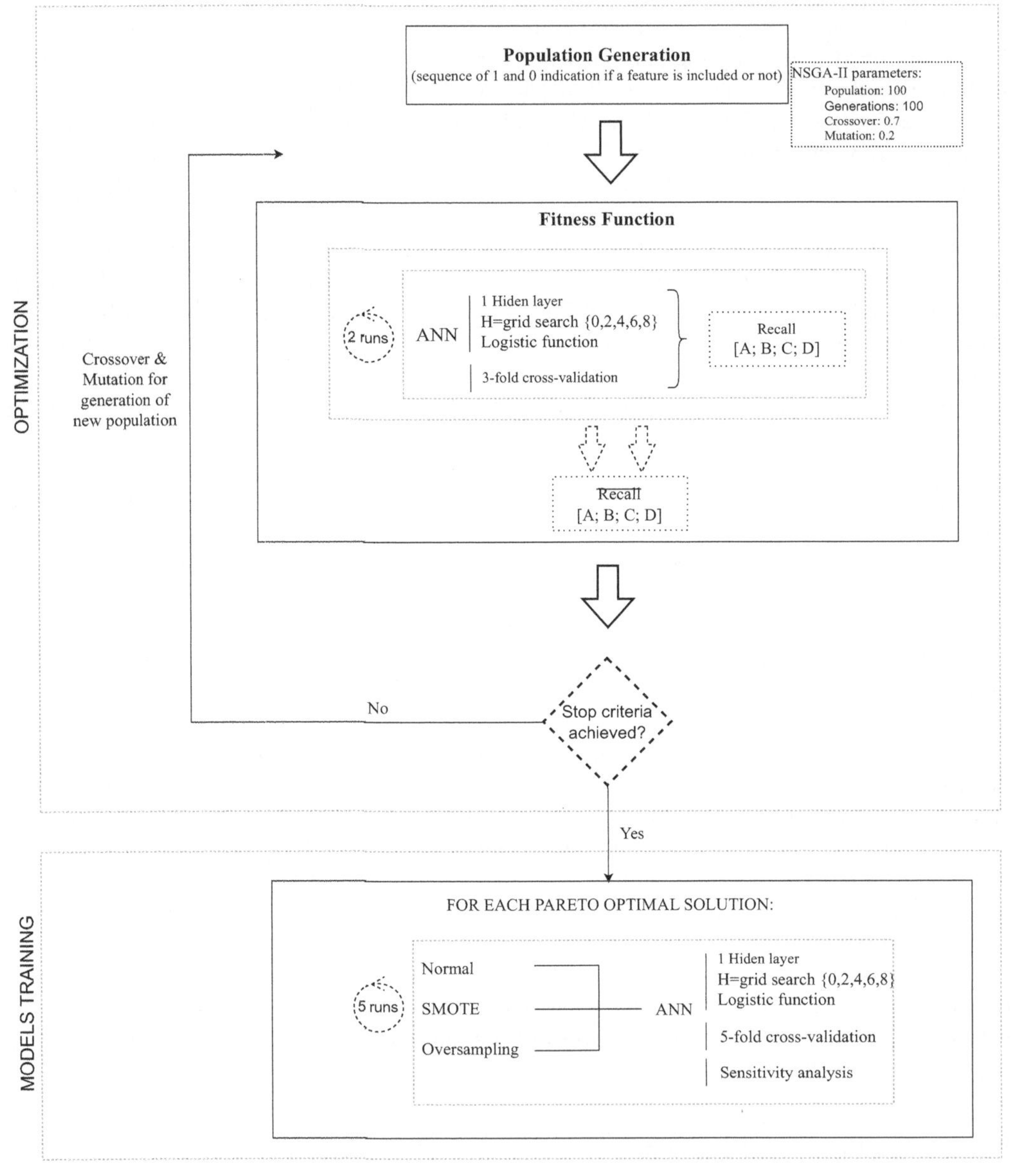

Figure 9.6 Flowchart of the applied methodology. (Tinoco et al. 2023.)

of slope stability is governed by a specific group of characteristics. Figure 9.7 compares the performance of ANN models when identifying the slope stability of rock and soil cuttings as well as soil embankments based on recall, precision, and F1-score metrics.

Figure 9.8 depicts the trade-off between recall values for each one of the four EHC classes for all Pareto optimal solutions covering the three types of slopes, that is, rock and soil cuttings and embankments. From its analysis, and as expected, the best solution is always a performance commitment between all EHC classes. This commitment is particularly evident between

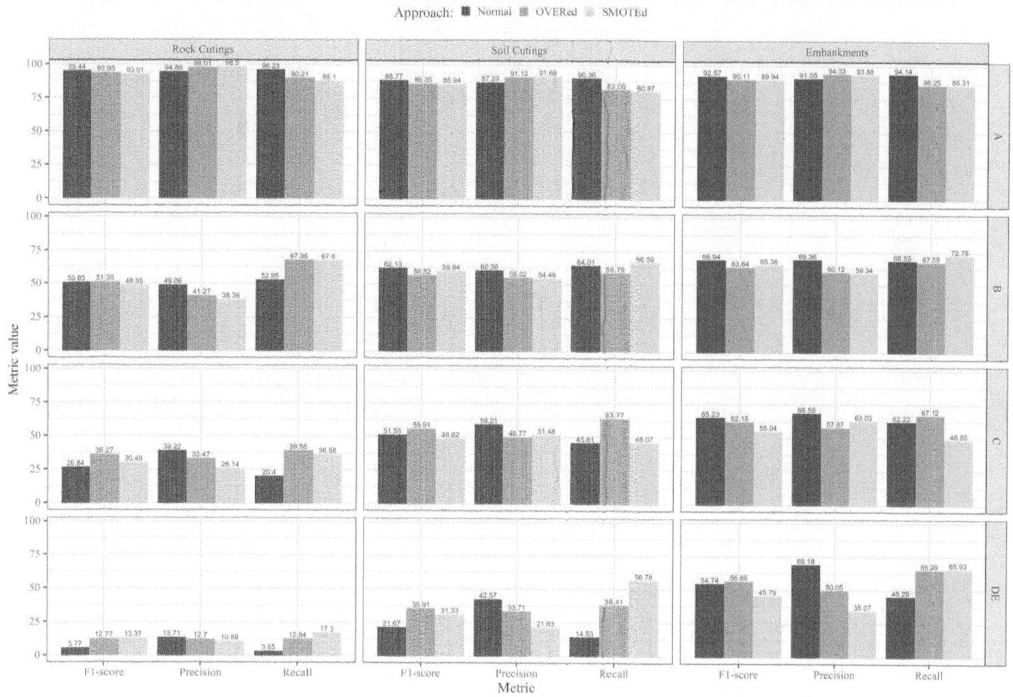

Figure 9.7 Comparison of ANN models based on recall, precision, and F1-score according to a nominal classification strategy for predicting slope stability of rock and soil cuttings and soil embankments. (Adapted from Tinoco et al. 2018a, b.)

classes “A” and classes “B,” “C,” and “D.” For example, to obtain a high performance for class “A,” the response for the other classes must be compromised in some way. The same behavior is also observed for the other possible EHC class combinations, although in some instances, specifically between the “B” and “C” classes, commitment is not as evident. In addition to the performance commitment between EHC classes, Figure 9.8 also demonstrates the inferior performance of all Pareto optimal solutions for classes “C” and “D,” namely for rock cuttings (below 0.4) when compared to class “A” (above 0.9).

GAs were applied in this study for feature selection purposes. They provided information regarding the optimal number of inputs for the accurate determination of slope stability, as well as the variables that should be used as model attributes.

Figure 9.9 demonstrates that, with the exception of embankments, all Pareto optimal solutions evaluated fewer than ten features and no more than forty inputs. There are some solutions for embankments that consider fewer than ten inputs, with a minimum of five attributes. Still, regarding embankments, it is interesting to observe that most of the solutions have considered, on average, 15 or 34 inputs. Moreover, for the rock and soil cuttings’ slopes, although this was not so evident, the number of variables considered by a significant part of the solutions ranged between 10 and 20 or between 30 and 40. These observations indicate that only about half of the more than 50 features available for model training were simultaneously considered as input to a model. In fact, the majority of Pareto optimal solutions in the study of rock cuttings considered between sixteen and seventeen features, while some were fed by as many as thirty-one

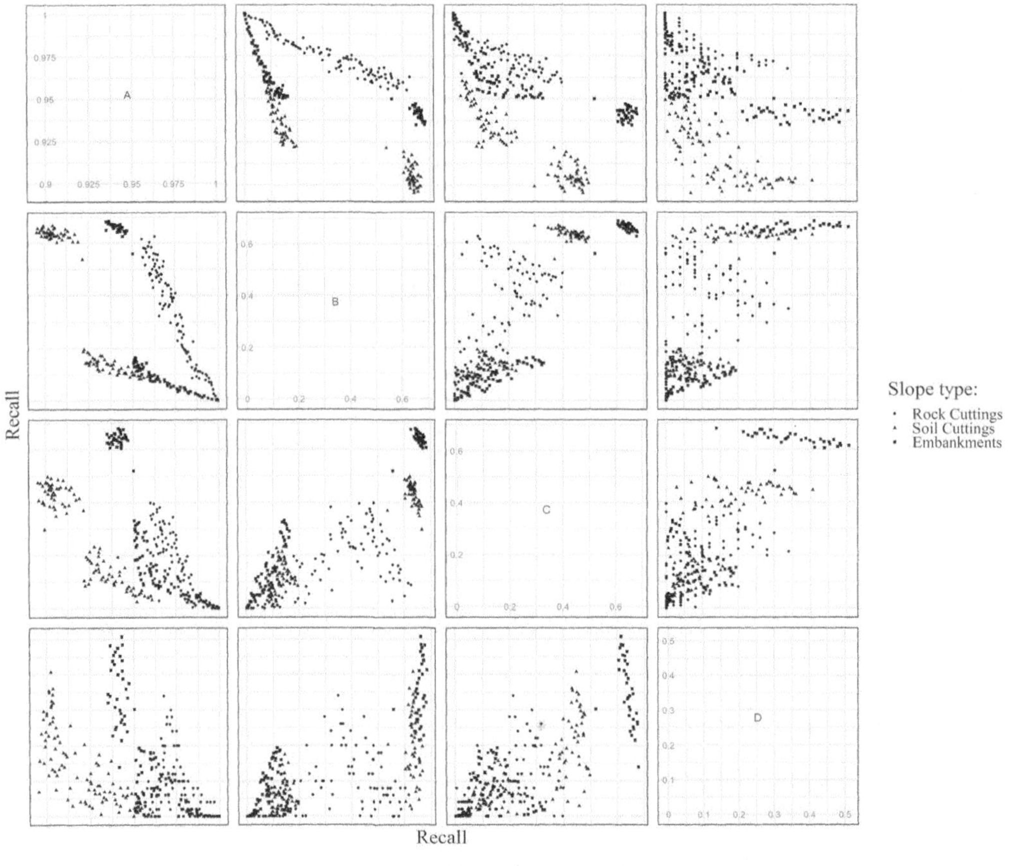

Figure 9.8 Recall the correlation matrix for each Pareto-optimal solution. (Tinoco et al. 2023.)

features. Regarding soil cutting, the most frequent number of variables taken by the Pareto optimal solutions was 12, and for embankments that number was 15.

A single model was chosen from all Pareto optimal solutions for each slope type based on the recall, precision, and F1-score performance metrics across all EHC classes. All Pareto optimal solutions were retrained using the SMOTE and oversampling techniques, as well as without resampling (normal).

Using the recall, precision, and F1-score metrics as benchmarks, Figure 9.10 compares ANNs models from the first ANN models (from now on referred to as the "old" models) (Tinoco et al. 2018a, 2018b) and the best ANN models selected among all Pareto optimal solutions (referred to from now on as the "updated" models). This comparison was made by calculating the difference between the metrics of the "updated" and "old" models. Hence, positive values correspond to the superior performance of the "updated" models. Even though the overall difference between the "updated" and "old" models is not statistically significant, the "updated" models have several key advantages over the "old" ones. On the one hand, using the recall metric as a benchmark, the "updated" models, particularly those following a SMOTE approach, demonstrate a significant improvement in the prediction of classes "C" and "D." On the other hand, this "updated" model utilizes a significantly reduced number of features. On average, fewer

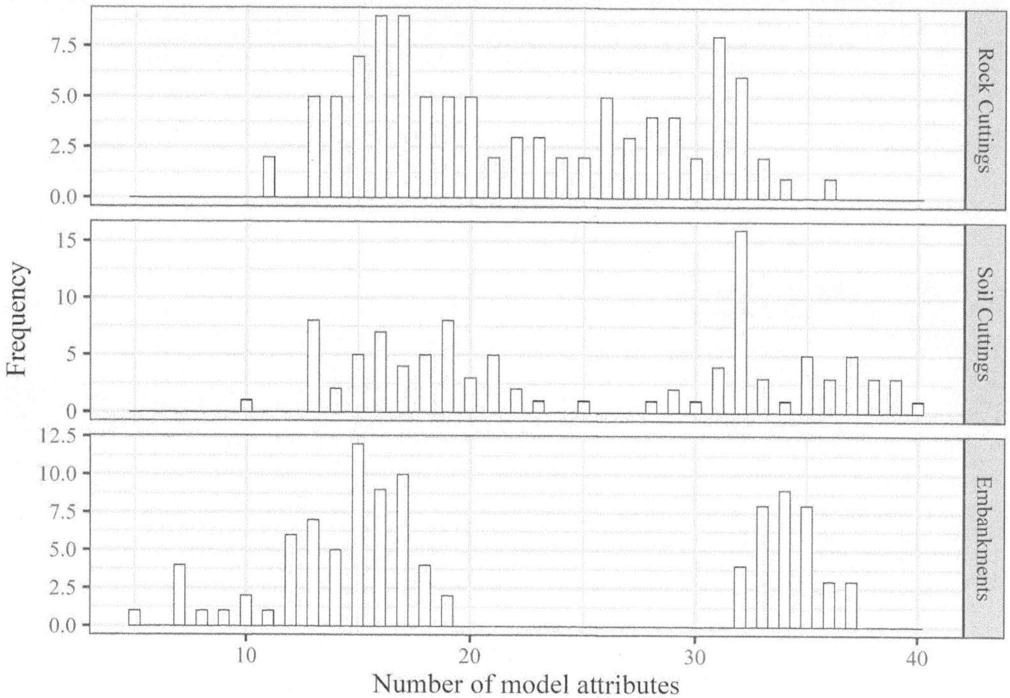

Figure 9.9 Number of features used as model attributes over all Pareto optimal solutions. (Tinoco et al. 2023.)

than 30 variables were considered instead of the more than 50 considered in the "old" models. These two aspects represent a significant accomplishment from a practical standpoint. First, less information is required to accomplish the same overall performance, which represents significant time and cost savings. Second, a better performance on classes "C" and "D" prediction was achieved with almost no performance penalization for the remaining classes.

This trade-off between the models' responses for the major and minority classes is a consequence of the highly asymmetric distribution of the database. However, it should be noted that the "updated" models achieved a better balance by increasing their performance for the minority classes, for which the probability of failure is higher, without significantly compromising their responses for the major classes.

When analyzing the effect of the training sampling approaches (oversampling and SMOTE), some effectiveness can be observed for classes "C" and "D," for which the probability of failure is higher (as these are minority classes), particularly for rock cuttings. Concerning classes "A" and "B," the implementation of a sampling approach does not produce the same effectiveness. When comparing SMOTE and oversampling, the effectiveness of the first method seems to be slightly superior. These results are in line with those observed during the first iteration, as reported in (Tinoco et al. 2018a, 2018b).

Figure 9.11 depicts the relationship between the observed and predicted EHC classes based on the "updated" SMOTE-based ANN model. Its superior performance in classes "A" and "B" is evident from its analysis. Particularly for soil cuttings and embankments, it can also be observed a very interesting response for class "D" (more than 60% of the slopes belonging to

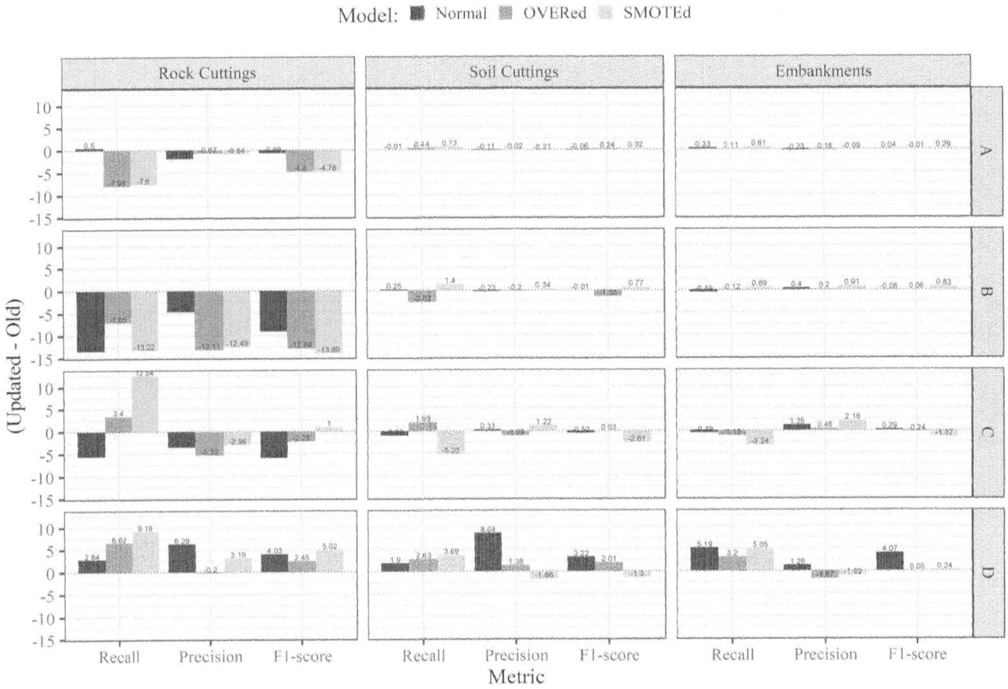

Figure 9.10 Comparison of the best ANN models ("updated" vs. "old") based on recall, precision, and F1-score according to a nominal classification strategy in slope stability prediction of rock and soil cuttings, slopes, and embankments. (Tinoco et al. 2023.)

class "D" were correctly identified). Furthermore, one can observe that when a slope's class is not correctly identified, it is categorized as belonging to the closest class. For instance, a slope of class "D" that is not predicted as "D" is typically classified as "C." These results are encouraging and drive new experiences toward improvements in the performance of models, particularly for minority classes.

Figure 9.12 shows the relative importance of the twenty most relevant variables according to the best "updated" models, as described above. Comparing this ranking with the one of the "old" models reported in (Cortez 2010, Ali et al. 2023), it can be observed that both rankings share around half of the more relevant features. The exceptions are the rock cuttings, where only 6 of the 20 most relevant variables were also identified as such in the "updated" models.

Focusing only on the 20 more relevant variables according to the "updated" models for each of the slope types, it can be observed that any variable has an individual influence higher than 10.5%. Concerning the rock cuttings, the variables "RS Length" and "RS Actual Angle" present a relative importance of 10.5%. In the case of soil cuttings, the variable "Max Height" stands out from the others, with an influence of 8.9%. Relating to embankments, the variable "Validate Track Movement" presents a relative influence of 10.4%, which is closely followed by "End Height," with 8.1%. Moreover, it can also be observed that at least one variable related to the slope angle or slope height appears at the top of the ranking.

Overall, these rankings follow what is known regarding the main factors relating to slope stability. Several empirical studies have shown that a high slope and its angle are among the main factors that contribute most to the stability of a given slope.

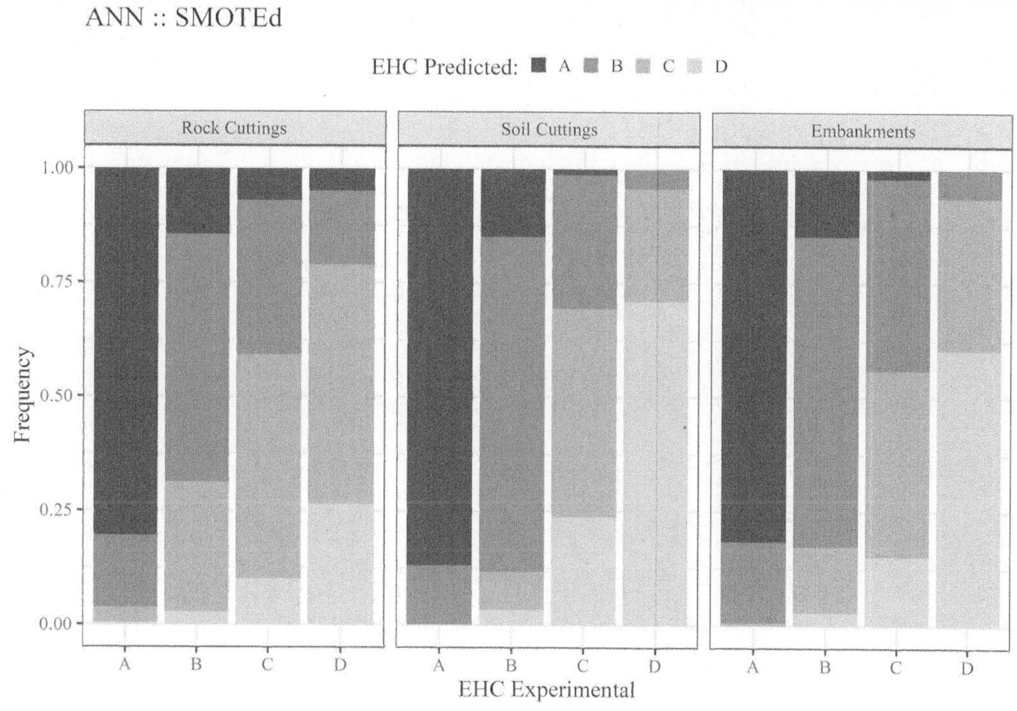

Figure 9.11 Best performance of ANN models ("updated" models) in EHC prediction of rock and soil cuttings, slopes, and embankments according to a nominal classification strategy and following a SMOTE approach. (Tinoco et al. 2023.)

9.3.4 Conclusions

Combining ANN and GA, an association also known as ENN was adopted for the purposes of feature selection and slope stability classification. This innovative approach allows ANNs and GAs to work together to assess the stability of a slope. Although the overall performance did not improve significantly, the new models presented some important advantages. First, to achieve the same overall performance, the number of input variables decreased to less than 30. This is a key point from a practical point of view since it endows the model with more flexibility when less information is available. Furthermore, this can allow the transport infrastructure management companies to collect less information to determine the stability level of their slope network, which, in turn, will allow them to reduce costs related to routine inspections. Second, better performance is achieved for the identification of classes "C" and "D," which correspond to a higher probability of failure, without significantly compromising the model's performance on the remaining classes. From a safety point of view, this represents a significant improvement, allowing for the strategic investment of the available budget into critical slope remediation. In addition, such a strategic maintenance plan can reduce traffic constraints due to slope failures and, consequently, the probability of severe incidents. In summary, these results are encouraging and constitute a promising research direction for more efficient identification of slope stability, particularly for minority classes. In addition, considering the high number of data available for each of the three slope types, a deep learning approach may provide an important contribution to the prediction models' efficiency.

Even though new efforts are still necessary, this study demonstrates once more the extreme difficulty of slope stability evaluation, namely by considering only visual features. Nonetheless,

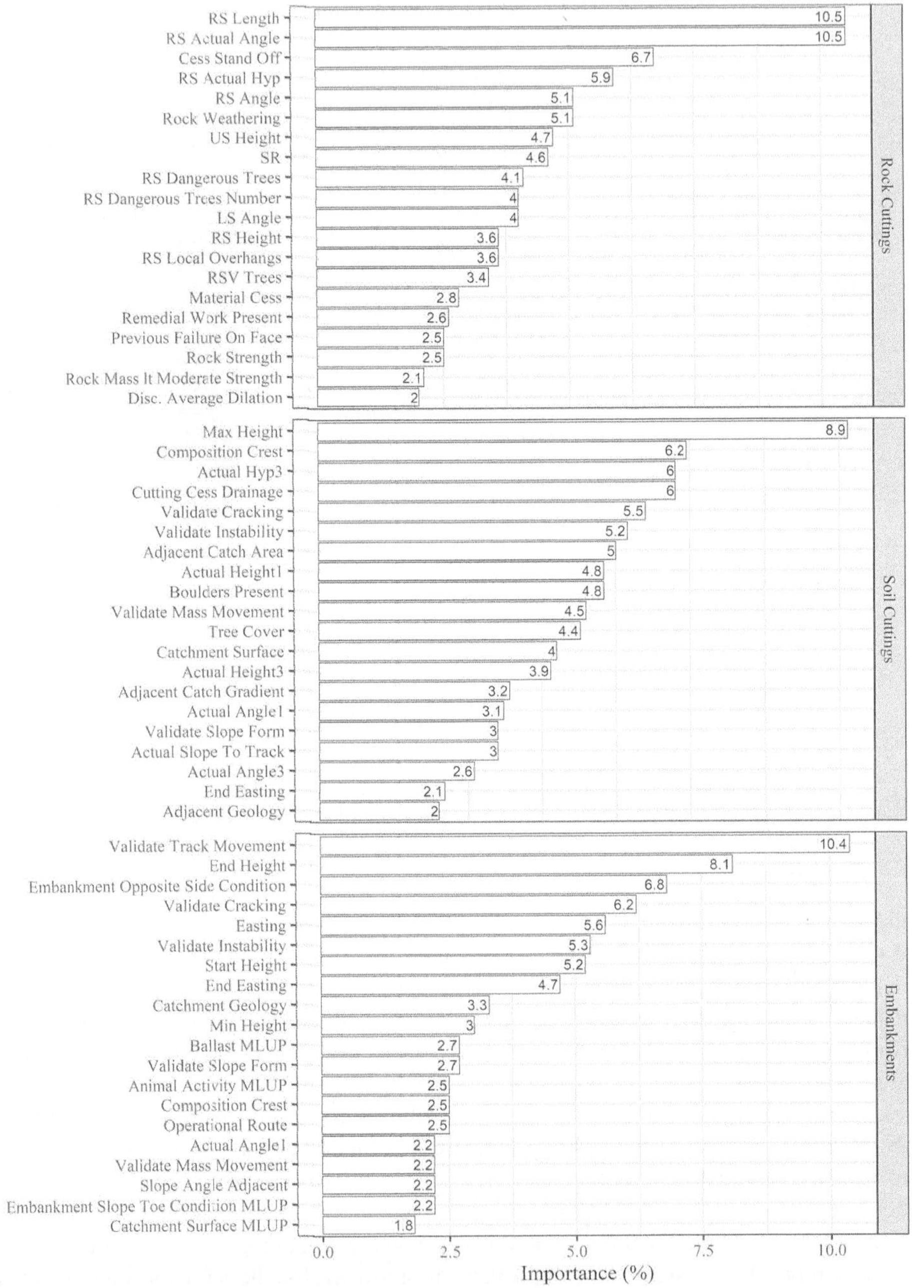

Figure 9.12 Bar plot of relative importance of variables according to the best ANN model ("updated" models) following a nominal classification strategy in predicting the slope stability of rock and soil cuttings and embankments. (Tinoco et al. 2023.)

this difficult task is of utmost importance, especially for developed nations that must maintain large-scale transportation networks despite limited budgets for maintenance and network operations.

9.4 Surface condition

This ML case study utilizes vibration data from sensors installed in construction trucks to predict the heavy vehicle fuel consumption and classify the road category based on pavement surface quality through which it circulated.

9.4.1 *Data and methodology*

9.4.1.1 Data acquisition

The data collection was based on a previous collaboration between industry and academia (Sampaio 2021). Two primary sensors were used to retrieve information such as vehicle location, load, speed, date, and time: A global navigation satellite system (GNSS) receiver and an inertial measurement unit (IMU). The GNSS receiver provided accurate location data, while the IMU measured speed, direction, and gravitational forces. The truckload was recorded manually by the driver before each trip or transport cycle. The prototype was developed and coupled together to reduce oscillations. Additionally, it was encapsulated to secure the entire system. The prototype was placed near the driver, on top of adhesive rubber material, while the GNSS receiver's antenna was mounted outside the truck. This positioning enhances signal quality and the accuracy of the received data. Figure 9.13a, (i) displays the encapsulated prototype, (ii) highlights the GNSS receiver antenna circled in yellow, and (iii) indicates the prototype's location.

Each truck driver was asked to fill out a form upon leaving the quarry, detailing information such as the trip's start time, the load carried by the truck, and the fuel consumption. Each trip from point (*A*) to point (*B*) was considered one journey and counted as a single trip. Notably, the trucks transported various materials, primarily bitumen mixtures, across different road types, including unpaved roads, national roads, and highways. The trip distances ranged from 20 to 60 km. Additionally, data such as vehicle time, GPS location, altitude, speed, and three-axis inclination were recorded throughout the trips.

To streamline data transfer, a microcomputer connected to the quarry's Wi-Fi enabled seamless file uploads to the server once the truck returned from its journey.

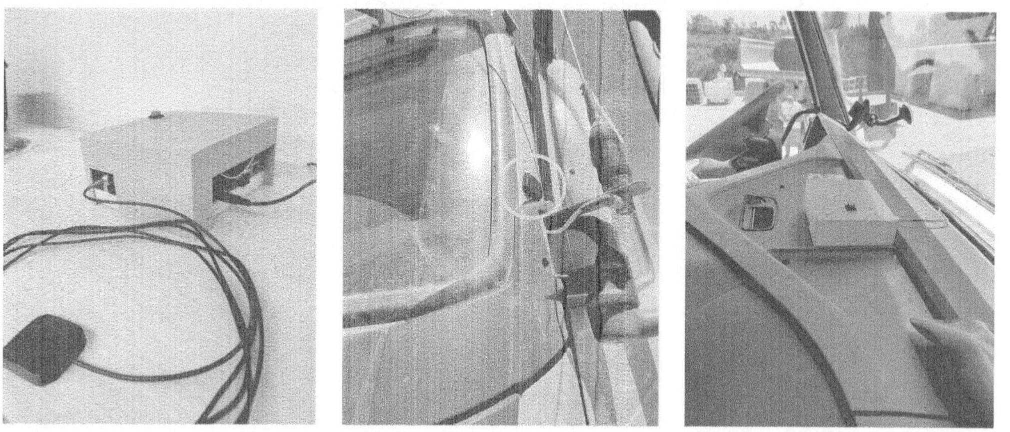

(a) (b) (c)

Figure 9.13 Data acquisition apparatus: (a) encapsulated prototype, (b) GNSS receiver antenna, (c) location of the prototype. (Sampaio 2021.)

9.4.1.2 *Data preparation*

9.4.1.2.1 FUEL CONSUMPTION

The data for the 21 routes was analyzed using a regression model. Altimetric profiles were generated by converting time and velocity into distance and validated using slope data. To analyze the elevation profiles, a 10-m sliding window was applied. The inclination was determined for each section. This methodology allowed each trip to be divided based on the total distance traveled on each type of slope (Table 9.2).

The processed database includes the following parameters:

1 Code names: Percentage of travel distance in each slope type;
2 Distance: Total distance of the journey in meters;
3 Av_Speed: Average speed values in meters per second;
4 Cargo: Mean of the cargo values in tons;
5 Fuel_Consump: Fuel consumption in liters;

9.4.1.2.2 ROAD TYPE CLASSIFICATION

A classification model was applied to 27 valid files to determine the road category. From these files, twelve had variable load and fifteen had zero load. To enable the classifier model to make accurate predictions, it was necessary to manually assign the road type label after analyzing each individual truck's position on Google Earth Pro. The positions were examined and classified based on the following criteria: (i) 0 if stationary (speed below 1 m/s), (ii) 1 if a paved highway, (iii) 2 if a paved road, and (iv) 3 if an unpaved road. Examples of criteria (ii), (iii), and (iv) can be seen in Figure 9.14.

Table 9.2 Types of slopes

Code name	*Slope description*	*Value range*
S1	Flat surface	−1% < Slope ≤ + 1%
S2	Slight rise	+1% < Slope ≤ + 5%
S3	Moderate rise	+5% < Slope ≤ + 10%
S4	High rise	Slope ≤ + 10%
S5	Slight descent	−5% < Slope ≤ −1%
S6	Moderate descent	Slope ≤ −5%

(a) (b) (c)

Figure 9.14 Examples of: (a) Criteria ii: paved road—highway, (b) criteria iii: paved road, (c) criteria iv: unpaved road.

The database has 5495 examples of road category 1, 10876 examples of road category 2, and 362 examples of road category 3. The initial dataset was unbalanced, with far fewer examples of unpaved roads (category 3) compared to paved roads (categories 1 and 2: highways and other paved roads, respectively).

It is important to emphasize that a one-hundred-meter sliding window methodology was employed after several iterations to determine the optimal distance for converting time and velocity into distance. This approach involved analyxing the acceleration and speed inputs for every hundred meters of the road, which resulted in identifying a standardized distance in which these values were averaged.

The processed database includes the following parameters:

1 Av_accel: Average vertical acceleration (z-axis);
2 Max_accel, Min_accel, and Maxmin_accel: Maximum, minimum, and the difference between the maximum and minimum, respectively, of the vertical acceleration;
3 Sd_accel: Standard deviation of the vertical acceleration;
4 Cargo: Mean of the cargo values in tons;
5 Speed: Mean of the speed values in meters per second;
6 Road type.

9.4.2 Model training and evaluation

Three models, namely random forest (RF), SVM, and NN, were employed to determine which one yielded better results. For the fuel consumption analysis, a three-fold cross-validation was applied with ten runs, while for the road type classification, a k-fold of ten was applied with thirty runs.

9.4.2.1 Fuel consumption

Figure 9.15 shows a comparison between the REC curves of the three models. The RF model outperforms the other two models in predicting fuel consumption in heavy vehicles, exhibiting a lower error than the others. It has a mean absolute error (MAE) of 1.723915, corresponding to a percentage error of approximately 7.6%. The RMSE is 2.127431, and the data's R-squared (R^2) value for the data is 61.4%, indicating a better fit. Additionally, Pearson's correlation coefficient (R) for the RF model is 78.4%, suggesting a strong correlation between the independent variables and the target variable. This robust correlation supports the hypothesis.

Based on the data presented in Figure 9.16, the RF model demonstrates a high level of accuracy in replicating the behavior of the target variable, particularly for fuel consumption values in the mid-range (20–27 L). However, there is a slight underestimation of values closer to the lower end (approximately 17 L) and an overestimation of values closer to the upper end (around 30 L). This discrepancy can be attributed to the limited number of registers within these ranges in the database. This limitation will be resolved as the database gradually expands.

After examining the significance graph (Figure 9.17), it is important to note that the RF model identifies the weight of the load carried by the truck (cargo) as the most important factor. This is then followed by the occurrence of slight and moderate inclines, as well as moderate declines. Higher weight and inclines lead to greater fuel consumption, while declines result in lower fuel usage. The validity of these factors is easily inferred.

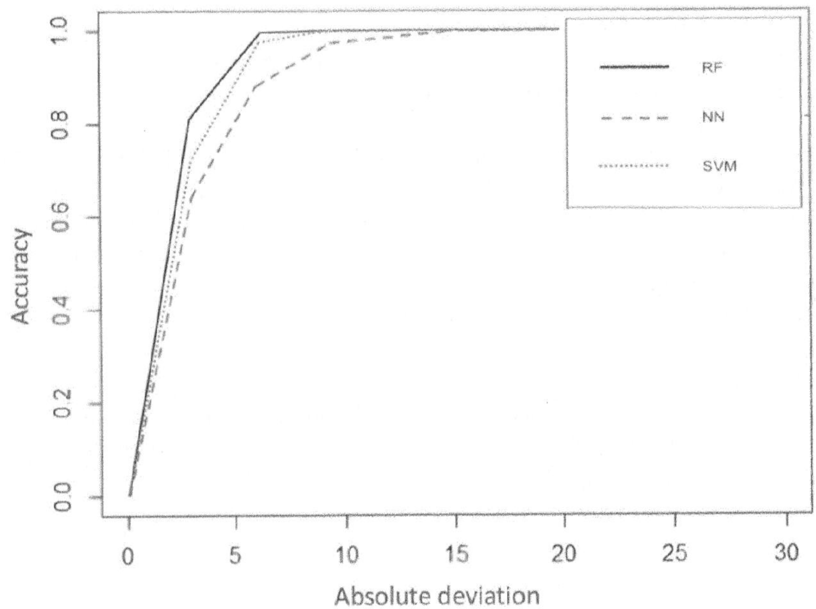

Figure 9.15 REC curves for fuel consumption prediction.

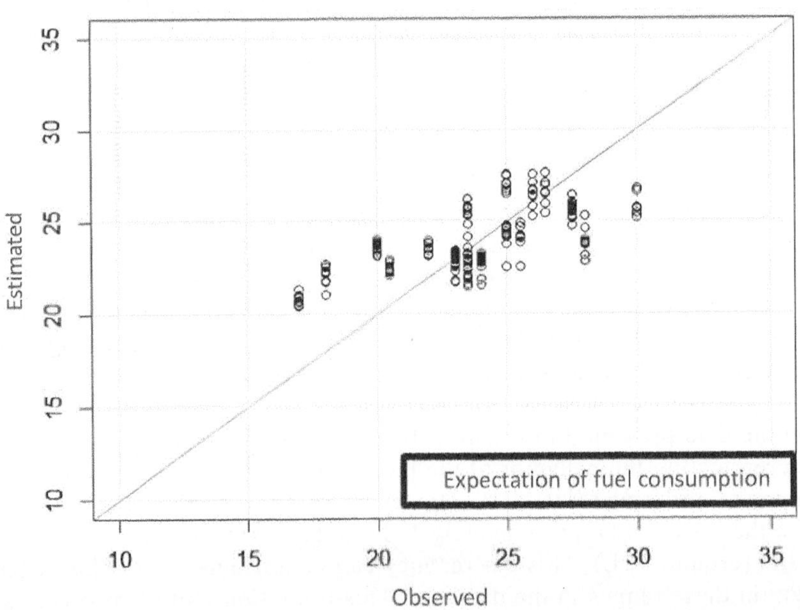

Figure 9.16 Estimated and observed data for fuel consumption, according to the RF model.

The model considers the variable related to steep climbs as less significant because it rarely occurs along the vehicle's path. Although it has low importance, excluding this variable from the training dataset slightly decreases the predictive ability of the final model. Therefore, it was retained in the training dataset to ensure optimal performance.

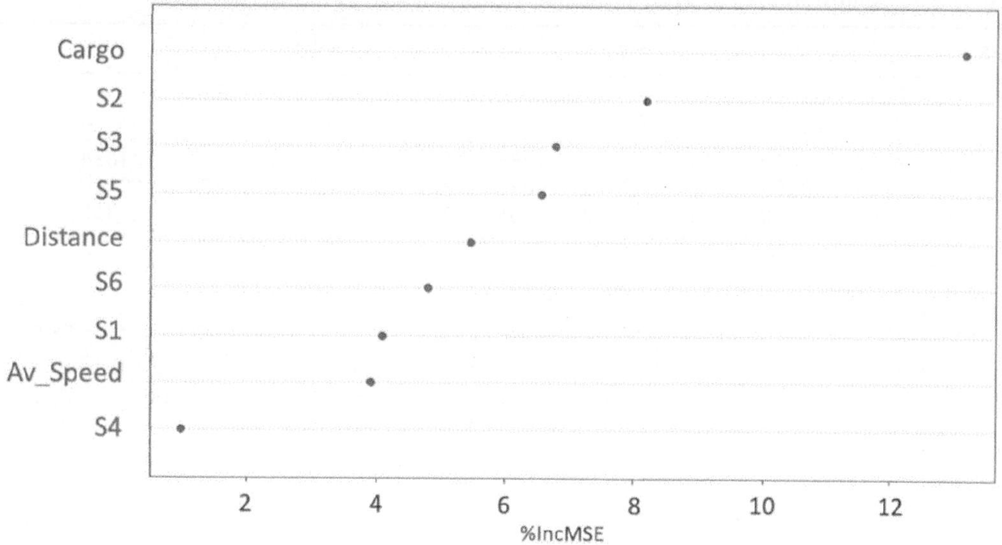

Figure 9.17 Importance graph for the RF model: fuel consumption.

9.4.2.2 *Road type classification*

The ML model achieved an accuracy of 91.6% for the RF, 69.83% for the NN, and 67.26% for the SVM model. While the RF model performed exceptionally well, the same cannot be said for the other models. The subsequent decrease in accuracy was anticipated, and further analysis will be presented next.

Of all the models evaluated, only the RF model attempted to predict road type 3 in the confusion matrix (Table 9.3), achieving a 64% success rate. It is worth mentioning that the model had very few misclassifications between class 1 and class 3 (almost zero), indicating its effective learning of the classification boundary between these classes. On the other hand, the other models faced challenges in predicting road type 3, which was expected due to the noticeable data imbalance in the dataset.

As expected, upon analyzing the importance graph (Figure 9.18), it is evident that speed carries significant weight in the results, along with cargo and the standard deviation of vertical acceleration. The effect of the speed variable and its importance in the models indicates that the classifier has identified and utilized the legally defined maximum speed for each road type to accurately determine the road category.

The ROC curve was plotted using the values from the confusion matrix (Figure 9.19). The latter graph is particularly noticeable in the RF model, closely resembling a perfect learner curve. Furthermore, the area under the ROC curve (AUC) is significantly larger, indicating better predictive performance. In contrast, the ROC curve for the NN and SVM models closely aligns with the forty-five-degree diagonal (line of equality), suggesting lower accuracy, as if their results were almost obtained by chance.

The authors tested other versions of the processed database to assess whether the ML models could accurately identify the road surface based solely on the vibration pattern and the truck's load, without including speed information. Although the model performed better when provided with speed information, it showcased the ML algorithm's capability to distinguish between various road types even in the absence of this information. In order to gain a more comprehensive

Table 9.3 Confusion matrix: road type classification

RF					*ANN*					*SVM*				
		Pred					*Pred*					*Pred*		
		1	*2*	*3*			*1*	*2*	*3*			*1*	*2*	*3*
Target	1	4885	600	10	Target	1	1858	3637	0	Target	1	861	24634	0
	2	629	10206	41		2	1097	9779	0		2	496	10380	0
	3	38	90	234		3	65	297	0		3	7	355	0

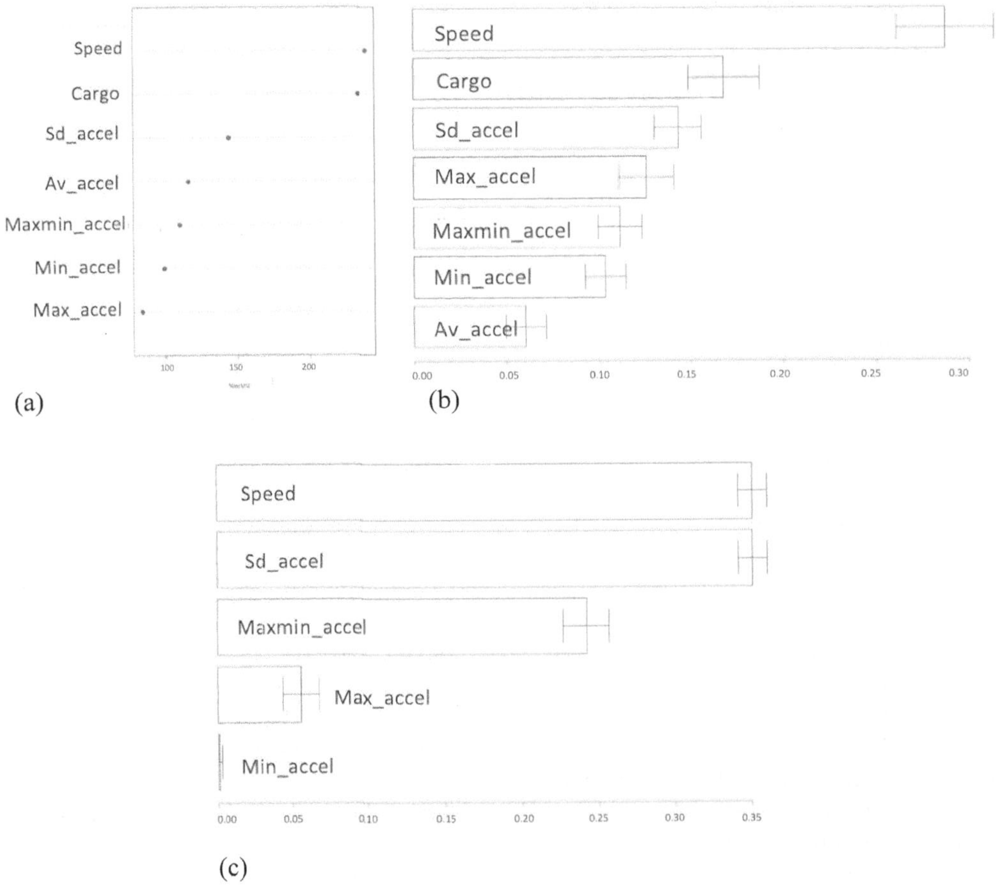

Figure 9.18 Importance graph for road type classification: (a) Random forest, (b) neural network, (c) support vector machine.

understanding of the models' capabilities, a new dataset was generated to evaluate their ability to handle a balanced set of data. To address the limited number of road type 3 examples (only 362), a script was created to randomly select an equal number of examples from the other classes. Although this approach resulted in discarding a significant amount of data, duplicating the road type 3 examples repeatedly was not a feasible option as it could potentially lead

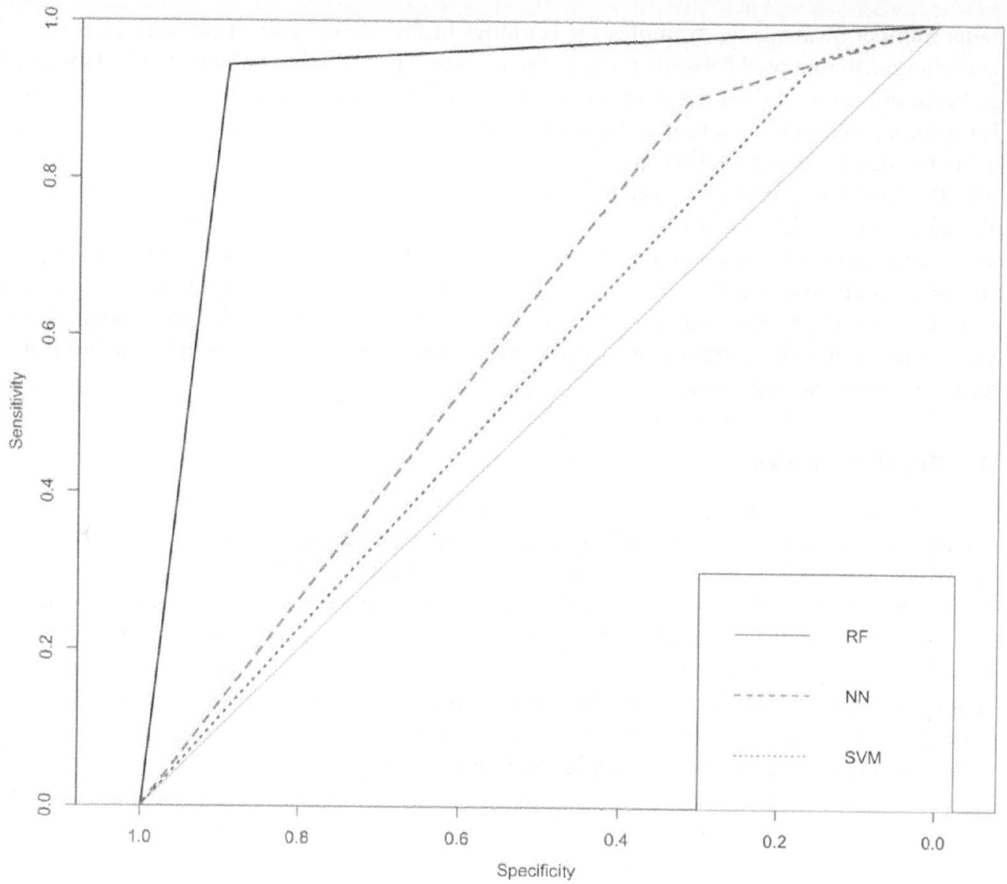

Figure 9.19 ROC curve: road type classification.

to overfitting of the corresponding ML models. In this case, the model achieved the highest number of correctly predicted samples in road type 3. This demonstrates that a more balanced initial dataset can yield promising results in detecting unpaved roads. Furthermore, it suggests that future works should consider using a larger dataset.

9.4.3 Conclusions

AI- and ML-based models were used in this study to analyze the vibration patterns of a construction truck during its transportation cycles. The goal was to accurately predict the truck's fuel consumption and the type of road surface/pavement it traveled on.

The results clearly indicate that the RF model outperformed in both cases. When predicting fuel consumption, the target variable showed strong correlations with the load carried, the longitudinal slope of the routes, and the total distance. This demonstrates that these input parameters are crucial for achieving more accurate and reliable fuel consumption predictions. The model accurately replicated the behavior of the target variable in mid-range fuel consumption. However, its accuracy was worse at the lower and upper ends of the scale, possibly due to a smaller number of available examples for learning. In the second part of the study, the model successfully differentiated between three different road types. The parameters that had the most significant impact on the results were speed, cargo, and standard deviation. For future research, it is recommended to use a larger and more balanced dataset, considering the promising results obtained in detecting unpaved roads.

Finally, the study can be expanded to include the detection of road quality by analyzing road pathologies and quality indexes, such as the International Roughness Index (IRI). Furthermore, in the future, additional sensors could be applied to various types of vehicles, especially in public transportation, to gather data for analysis. The analyzed results contain valuable information for drivers, municipalities, and companies in the construction sector. This information allows them to understand the current state, needs, and priorities of the transportation network and allocate resources accordingly.

9.6 Final remarks

This chapter presented two distinct case studies demonstrating the versatility and potential of ML techniques in addressing complex geotechnical problems in transportation infrastructure.

The first case highlighted the integration of ANNs with GAs as ENNs to classify slope stability levels. Although there was no significant improvement in overall performance compared to previous findings, the method's ability to reduce input variables and enhance the identification of critical slopes underscores its practical value, particularly in cost-sensitive applications, offering potential cost savings in routine inspections and strategic investment planning. This approach offers a promising direction for further research in slope stability assessment, especially in optimizing maintenance strategies with limited resources.

The second case study demonstrated the application of RF models to predict fuel consumption and road surface conditions based on sensor data from construction vehicles. The model's success in accurately predicting mid-range fuel consumption and differentiating between road types, emphasizing the significance of speed and cargo, illustrates the potential for ML in optimizing operational efficiencies in the construction sector. Expanding this research to include more comprehensive datasets and additional sensors could significantly enhance the model's utility, making it a valuable tool for infrastructure management.

Assessing slope stability and surface conditions is paramount in the construction and maintenance of transportation infrastructures. In this regard, AI algorithms, namely ANNs and GAs for slope stability evaluation, RF, and ANNs and SVMs for surface condition assessment. Based on the case studies presented in this chapter, the efficiency of each AI method varies, influenced by factors such as the specifics of available databases, the method employed for feature selection, the number of input variables, and, in certain AI methods, the number of hidden layers. The adoption of combined AI methods has been widely advocated to enhance reliability and reduce costs. ENNs, formed through the integration of ANNs and GAs, have shown promising outcomes. The developed innovative ENNs have successfully minimized the number of input variables, a crucial aspect from a cost analysis perspective (i.e., reducing the data collection burden for transportation authorities). Additionally, they effectively identified critical slope instability, contributing to planning actions for enhanced safety and hazard risk reduction. It has been observed that the performance of AI algorithms is closely tied to dataset variations. In a broader context, RF, ANNs, and SVMs have demonstrated practical applicability for categorizing all types of roads, including paved and unpaved, based on surface condition.

In summary, the application of ML in transportation geotechnics offers substantial benefits in terms of predictive accuracy, operational efficiency, and cost-effectiveness. However, the complexity and specificity of each case highlight the importance of selecting the appropriate ML technique for the problem at hand. Future research should continue to explore and refine these methods, expanding their applicability and effectiveness in real-world scenarios.

Acknowledgments

This research was funded by national funds through the Foundation for Science and Technology (FCT), under the grant agreement CEECINST/00018/2021 (https://doi.org/10.54499/CEECINST/00018/2021/CP2806/CT0006) attributed to Joaquim Tinoco, and was partly financed by FCT/MCTES through national funds (PIDDAC) under the R&D Unit Institute for Sustainability and Innovation in Structural Engineering (ISISE), under reference UIDB/04029/2020 (doi.org/10.54499/UIDB/04029/2020), and under the Associate Laboratory Advanced Production and Intelligent Systems ARISE under reference LA/P/0112/2020.

A special acknowledgment goes to Professor António Gomes Correia for his initiative, support, and supervision since 2008 in harnessing AI to tackle complex geotechnical problems.

References

Alexandridis, A., P. Patrinos, H. Sarimveis and G. Tsekouras. 2005. A two-stage evolutionary algorithm for variable selection in the development of RBF neural network models. *Chemometrics and Intelligent Laboratory Systems* 75: 149–162. https://doi.org/10.1016/j.chemolab.2004.06.004

Ali, Y.A., E.M. Awwad, M. Al-Razgan and A. Maarouf. 2023. Hyperparameter search for machine learning algorithms for optimizing the computational complexity. *Processes* 11, no. 2: 349. https://doi.org/10.3390/pr11020349

Bi, J. and K.P. Bennett. 2003. Regression error characteristic curves. In *Proceedings of the 20th International Conference on Machine Learning (ICML-03)*, Washington, DC.

Bonissone, P.P., R. Subbu and J. Lizzi. 2009. Multicriteria decision making (MCDM): A framework for research and applications. *IEEE Computational Intelligence Magazine* 4, no. 4: 48–61. https://doi.org/10.1109/MCI.2009.933093

Boutaba, R., M.A. Salahuddin, N. Limam, S. Ayoubi, N. Shahriar, F. Estrada-Solano and O.M. Caicedo. 2018. A comprehensive survey on machine learning for networking: Evolution, applications and research opportunities. *Journal of Internet Services and Applications* 9, no. 1: 1–99. https://doi.org/10.1186/s13174-018-0087-2

Brook, R.J. and G.C. Arnold. 1985. *Applied Regression Analysis and Experimental Design* (1st ed.). CRC Press, Boca Raton. https://doi.org/10.1201/9781315137674

Carter, J.V., J. Pan, S.N. Rai and S. Galandiuk. 2016. ROC-ing along: Evaluation and interpretation of receiver operating characteristic curves. *Surgery* 159, no. 6: 1638–1645. https://doi.org/10.1016/j.surg.2015.12.029

Choi, R.Y., A.S. Coyner, J. Kalpathy-Cramer, M.F. Chiang and J.P. Campbell. 2020. Introduction to machine learning, neural networks, and deep learning. *Translational Vision Science & Technology* 9, no. 2: 14. https://doi.org/10.1167/tvst.9.2.14

Confalonieri, R., L. Coba, B. Wagner and T.R. Besold. 2021. A historical perspective of explainable Artificial Intelligence. *WIREs Data Mining and Knowledge Discovery* 11, no. 1: e1391.

Cortez, P. 2010. Data mining with neural networks and SVM using the R/rminer tool. In *Advances in Data Mining: Applications and Theoretical Aspects*, P. Perner, Ed., Springer: 572–583. https://doi.org/10.1007/978-3-642-14400-4_44

Cortez, P. 2022. A tutorial on using the rminer r package for data mining tasks. Teaching Report, Version 2. Department of Information Systems, ALGORITMI Research Centre, Engineering School, University of Minho.

Cortez, P. and M. Embrechts. 2011. Opening black box data mining models using sensitivity analysis. In *Proceedings of the 2011 IEEE Symposium on Computational Intelligence and Data Mining*, Paris, France: 341–348.

Cortez, P. and M. Embrechts. 2013. Using sensitivity analysis and visualization techniques to open black box data mining models. *Information Sciences* 225: 1–17. https://doi.org/10.1016/j.ins.2012.10.039

Ghallou, I. 2024. Build your perceptron neural net from scratch. *Medium.* https://medium.com/@ismail-ghallou/build-your-perceptron-neural-net-from-scratch-e12b7be9d1ef (accessed September 2024).

Ghosh, S. 2024. The ultimate guide to evaluation and selection of models in machine learning. https://neptune.ai/blog/ml-model-evaluation-and-selection (accessed September 2024).

Golberg, D.E. 1989. *Genetic Algorithms in Search, Optimization, and Machine Learning.* Addison Wesley: 36.

Holzinger, A., A. Saranti, C. Molnar, P. Biecek and W. Samek. 2022. Explainable AI methods – a brief overview. In *xxAI – Beyond Explainable AI*, Springer: 13–38. https://doi.org/10.1007/978-3-031-04083-2_2

Karatzoglou, A., D. Meyer and K. Hornik. 2006. Support vector machines in R. *Journal of Statistical Software* 15, no. 9: 1–28. https://doi.org/ 10.18637/jss.v015.i09

Kaur, H., H.S. Pannu and A.K. Malhi. 2019. A systematic review on imbalanced data challenges in machine learning: Applications and solutions. *ACM Computing Surveys (CSUR)* 52, no. 4: 1–36. https://doi.org/10.1145/3343440

Krstinić, D., M. Braović, L. Šerić and D. Božić-Štulić. 2020. Multi-label classifier performance evaluation with confusion matrix. *Computer Science & Information Technology* 1: 1–14. https://doi.org/10.5121/csit.2020.100801

Mammone, A., M. Turchi and N. Cristianini. 2009. Support vector machines. *Wiley Interdisciplinary Reviews: WIREs Computational Statistics* 1, no. 3: 283–289. https://doi.org/10.1002/wics.49

Mersmann, O., H. Trautmann, D. Steuer, B. Bischl and K. Deb. 2014. Package "mco": Multiple criteria optimization algorithms and related functions. https://cran.r-project.org/package=mco (accessed September 2024).

Meyer, D. and F. Wien. 2001. Support vector machines. *R News* 1, no. 3: 23–26.

Novaković, J.D., A. Veljović, S.S. Ilić, Ž. Papić and M. Tomović. 2017. Evaluation of classification models in machine learning. *Theory and Applications of Mathematics & Computer Science* 7, no. 1: 39–46.

Nunes, D.E. and V.F.S. Mota. 2019. A participatory sensing framework to classify road surface quality. *Journal of Internet Services and Applications* 10, no. 13. https://doi.org/10.1186/s13174-019-0111-1

Power, C., J. Mian, T. Spink, S. Abbott and M. Edwards. 2016. Development of an evidence-based geotechnical asset management policy for network rail, Great Britain. *Procedia Engineering* 143: 726–733. https://doi.org/10.1016/j.proeng.2016.06.112

R Development Core Team. 2023. R: A language and environment for statistical computing. In *R Foundation for Statistical Computing*, Vienna, Austria. https://www.R-project.org

Rodriguez-Galiano, V., M. Sanchez-Castillo, M. Chica-Olmo and M. Chica-Rivas. 2015. Machine learning predictive models for mineral prospectivity: An evaluation of neural networks, random forest, regression trees, and SVMs. *Ore Geology Reviews* 71: 804–818. https://doi.org/10.1016/j.oregeorev.2015.01.001

Roshan, M.J., M. Abedi, R. Fangueiro, A.G. Correia and M.M. Silva. 2024. Evaluation procedure for damage detection by a self-sensing cement composite. *Measurement* 226: 114149. https://doi.org/10.1016/j.measurement.2024.114149

Roshan, M.J., M. Abedi, A. Gomes Correia and R. Fangueiro. 2023. Application of self-sensing cement-stabilized sand for damage detection. *Construction and Building Materials* 403: 133080. https://doi.org/10.1016/j.conbuildmat.2023.133080

Roshan, M.J., A.S.A. Rashid, N. Abdul Wahab, S. Tamassoki, S.N. Jusoh, M.A. Hezmi, N.N. Nik Daud, N. Mohd Apandi and M. Azmi. 2022a. Improved methods to prevent railway embankment failure and subgrade degradation: A review. *Transportation Geotechnics* 37: 100834. https://doi.org/10.1016/j.trgeo.2022.100834.

Roshan, M.J., A.S.B.A. Rashid, M.A. Bin Hezmi, M.N. Nejabi, S.N. Siti, S. Tamassoki and R. Razali. 2022b. Evaluation of cement stabilised residual soil on macro- and micro-scale for road construction. *Journal of Engineering and Applied Sciences* 69: 109. https://doi.org/10.1186/s44147-022-00165-6

Sampaio, M.O.V.P. 2021. Sensorização e machine learning para a previsão de consumo de combustível. Master's thesis. Instituto Superior de Engenharia do Porto (ISEP).

Simplilearn. 2022. Machine learning: Learn the basics of machine learning—free machine learning beginners course. https://www.simplilearn.com/learn-machine-learning-basics-skillup?term=machine+learning (accessed July 2022).

Stuart, R. and P.N. Peter. 2016. *Artificial Intelligence – A Modern Approach* (3rd ed.). Prentice Hall.

Sun, Y., C. Babbs and E. Delp. 2006. A comparison of feature selection methods for the detection of breast cancers in mammograms: Adaptive sequential floating search vs. genetic algorithm. In *Proceedings of the 2005 IEEE Engineering in Medicine and Biology*, 27th Annual Conference: 6532–6535.

Thomas, J., M. Lambert, B. Snyder, M. Janning, J. Haning, Y. Hu, M. Ahmad and S. Visa. 2011. Learning morphological data of tomato fruits. In *Proceedings of the 22nd Midwest Artificial Intelligence and Cognitive Science Conference*, Cincinnati, OH: 133–137.

Tinoco, J., A. Gomes Correia, P. Cortez and D. Toll. 2018a. Data-driven model for stability condition prediction of soil embankments based on visual data features. *Journal of Computing in Civil Engineering* 32, no. 2: 04018027. https://doi.org/10.1061/(ASCE)CP.1943-5487.0000770

Tinoco, J., A. Gomes Correia, P. Cortez and D. Toll. 2018b. Stability condition identification of rock and soil cutting slopes based on soft computing. *Journal of Computing in Civil Engineering* 32, no. 4: 04017088. https://doi.org/10.1061/(ASCE)CP.1943-5487.0000739

Tinoco, J., A. Gomes Correia, P. Cortez and D. Toll. 2023. An evolutionary neural network approach for slopes stability assessment. *Applied Sciences* 13, no. 14: 8084. https://doi.org/10.3390/app13148084

Transfer, K. 2024. Pytorch k-fold cross-validation using dataloader and sklearn. https://androidkt.com/pytorch-k-fold-cross-validation-using-dataloader-and-sklearn/ (accessed September 2024).

Chapter 10

Dynamic resource optimization in earthworks

Manuel Parente, Margarida Amândio, and António Gomes Correia

10.1 Introduction

During the last decade, the potential of artificial intelligence approaches has been evidenced, namely, within the areas of machine learning and modern optimization, as tools that can mitigate the high uncertainty and variability inherent to the construction sector. Even though it is increasingly common for construction companies and designers to resort to new technologies that promote efficiency and reduce costs, many still only resort to their own intuition and experience to estimate work rates of processes and equipment.

Machine learning tools have already demonstrated their potential to contribute to estimating construction equipment productivity and cost (Gomes Correia and Magnan 2012, Parente et al. 2014, Pereira et al. 2021, Silva et al. 2023) more accurately than those typically proposed in equipment suppliers' handbook (Figure 10.1) (Gomes Correia et al. 2013, Jassim et al. 2017, Kassem et al. 2021). Additionally, these tools have become instrumental in decision support systems, as the case of the application of modern optimization to construction site resource allocation problems (Montaser and Moselhi 2014, Parente et al. 2015, 2016, Gomes Correia et al. 2023).

The adoption of this type of system, stemming from previous work in the planning and design phases, opens several research and development directions, among which is the ability to find construction resource allocation solutions capable of optimizing several aspects simultaneously, such as construction costs, execution duration, and sustainability key indices (i.e., carbon emissions or waste reduction). However, existing solutions so far solely comprise offline forecasting and/or optimization systems, whose ability to adapt to unforeseen occurrences during construction (e.g., equipment breakdowns, presence of geomaterials different from those expected, climate change, and others) is limited and depends on the ability of the site manager to update the systems manually and seek new solutions (Parente et al. 2015).

Recently, the development of Industry 4.0 (I4.0) concepts has promoted the exploration of technologies such as Internet of Things (IoT), sensorization, and digital twins in the construction sector (Parente et al. 2020). In this context, the combination of the prediction and optimization approaches mentioned above with the new technologies associated with I4.0 allows the integration of systems (previously offline) with the ability to monitor equipment and process data in real-time. In fact, the integration of sensors in construction equipment, along with the implementation of IoT environments, provides knowledge not only of the position and activity of equipment such as excavators, geomaterial transport dumpers, spreaders, and compactors, but also of a range of parameters associated with their productivity in real-time (Lee et al. 2018, You et al. 2021, Gomes Correia et al. 2021, 2024). A very representative example of

DOI: 10.1201/9781003530220-12

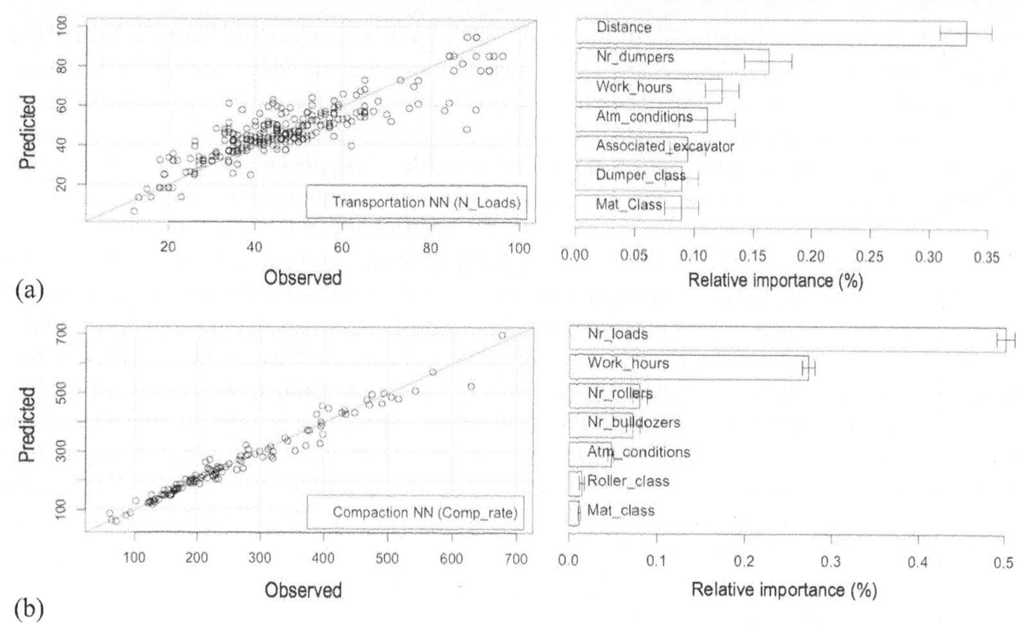

Figure 10.1 Machine learning-based prediction of earthwork equipment productivity for: (a) excavation and transportation teams; and (b) spreading and compaction teams. (Parente et al. 2014.)

these capabilities can be found in intelligent compactors, whose ability to adjust the applied compaction energy depending on the response of the geomaterial is combined with the creation of real-time compaction maps (Parente and Gomes Correia 2013), thus producing the necessary information to measure their productivity at each moment. Indeed, this not only forms the basis for scheduling and optimizing compaction tasks while concurrently analyzing the geomaterial's response, but the collected data can also be processed using machine learning and data mining techniques to enable precise predictions of optimal compaction settings, facilitating continuous control and classification of materials based on their characteristics. As a result, this process enhances the overall efficiency and provides quality control metrics that can be directly compared to design specifications, including potentially keeping track of the water content/degree of saturation in soils, which is essential for the compaction process (Tatsuoka and Gomes Correia 2018). Moreover, the entire system is supported by seamless connectivity between compactors, devices, and personnel, in line with the Internet of Things (IoT) framework. This connectivity enhances transparency, streamlines communication, and enables real-time operational management from anywhere with internet access, including via smartphones. In essence, intelligent compaction stands out as a transformative technology in the construction sector, especially during the execution phase, as it automates key processes such as data gathering, analysis, decision-making, and task execution. Thus, by resorting to these I4.0 concepts, this technology becomes crucial for modern construction projects, providing accurate assessments of compaction quality by addressing the dynamic interaction between compacting equipment and the underlying ground material (Gomes Correia and Gastine 2024).

In this context, it becomes possible to address the volatility and uncertainty inherent in geotechnical processes by adopting a reactive optimization approach, in which forecasting and optimization systems are updated in real-time with data from remote monitoring systems (Parente et al. 2018, Salem and Moselhi 2020, Liu et al. 2022, Gomes Correia and Gastine 2024). The resulting systems include alarm systems (i.e., watchdog systems), which provide information to the user about situations that may be potentially detrimental to the optimal progress of work. Examples of these are equipment operating below the estimated productivity or at risk of failure. Additionally, there are re-optimization algorithms capable of proposing minimum adjustments to the current solution of equipment allocation. These adjustments improve the workflow and productivity in construction processes, based on the data received from the installed sensors. This chapter presents a general framework for a digital twin optimization system that can be applied to any geotechnical process heavily dependent on heavy mechanical resources allocated to tasks over time. Its application to an earthwork case study is further demonstrated toward the end of this chapter. In this line, Section 10.2 introduces the architecture of the system, describing its underlying modules and technologies, as well as its process map that represents the information workflow. Section 10.3 discusses the application of the proposed system to a simulation based on an earthwork case study, with an analysis of optimization and re-optimization as a result of a simulated unforeseen behavior, referring to the unexpected reduction in the productivity of a resource during its activity. Finally, in Section 10.4, some conclusions are pointed out, as well as future directions of research.

10.2 Digital twin optimization system

10.2.1 Proposed system framework

One of the central aspects that enables the analysis of the optimal resource allocation configurations in any environment is the compilation and structuring in databases of the information necessary for the process. This information can be divided into static data, dynamic data, and historical data. On the one hand, static data includes information relating to the characteristics of the geotechnical process and the work to be carried out, as well as the availability and specifications of existing equipment. On the other hand, dynamic data refers to that obtained from the sensors installed in existing resources. Depending on the nature and conditions of the site, this data can be sent in real-time, reaching considerable volumes of instantaneous information at each instant of time that encodes the progress of operations and the state of active equipment. Finally, the historical data corresponds to the evolution of the work progress over time in order to record the processes carried out and choices adopted, including information such as the allocation solutions implemented in each phase of the work, or notifications regarding the completion of tasks by the production teams.

The optimization process consists of a multi-objective search for resource allocation solutions. The objectives of this optimization may address the minimization (or maximization) of any combination of mathematically quantifiable criteria, relevant examples being the total duration of execution of the geotechnical project, the total costs of execution, or other relevant aspects, such as the accumulated carbon emissions of a given possible allocation. Typically, in a multi-objective analysis, the result is represented by a Pareto front, characterized by a series of optimal solutions, each of which corresponds to an optimal compromise between the objectives considered (e.g., time vs. cost). This type of solution representation allows for greater flexibility of the model presented, since decision-makers have at their disposal several optimal solutions, being able to choose the one that best suits the constraints and goals of the project (e.g., budget, deadlines). It is important to note that the solutions resulting from the optimization process are not static over time, i.e., when a team finishes its work on one work front, it is immediately reassigned to another front. Since the productivity of teams in sequential tasks is synchronized, productivity bottlenecks are also avoided, ensuring that maximum productivity of resources is achieved.

Finally, for this data to be more easily deciphered and interpreted by the system users, there needs to be a visualization layer and a user interface. In this interface, the information presented to the user represents the processed data from active entities (i.e., active productive resources), which is based on the data from geolocation sensors and the outcomes from quality control monitoring sensors. In addition to enabling the control and monitoring of active resources through a digital twin of the site supported by sensors and a geographic information system (GIS), this data also feeds the optimization platform with the information needed to determine the optimal allocation of resources throughout the site and over time (Figure 10.2).

10.2.2 System management and optimization workflow

Geotechnical resource-dependent processes can be compared to a production line composed of a set of sequential activities that require different types of resources. Considering that these processes can also be characterized by high levels of uncertainty, it is of the utmost importance to plan them with sufficient flexibility to allow the adaptation of the plans to unforeseen situations that occur during the execution of the process. One way to achieve this goal is to adopt a reactive approach during both planning and execution of the process, assuming that the need to adjust the initial planning will be constant throughout the development of these processes in order to maintain productivity and costs at optimal values. Remote monitoring and control technologies for active resources can also support this reactive planning approach by providing real-time information on the possibility of unforeseen situations occurring, as well as their nature and origin.

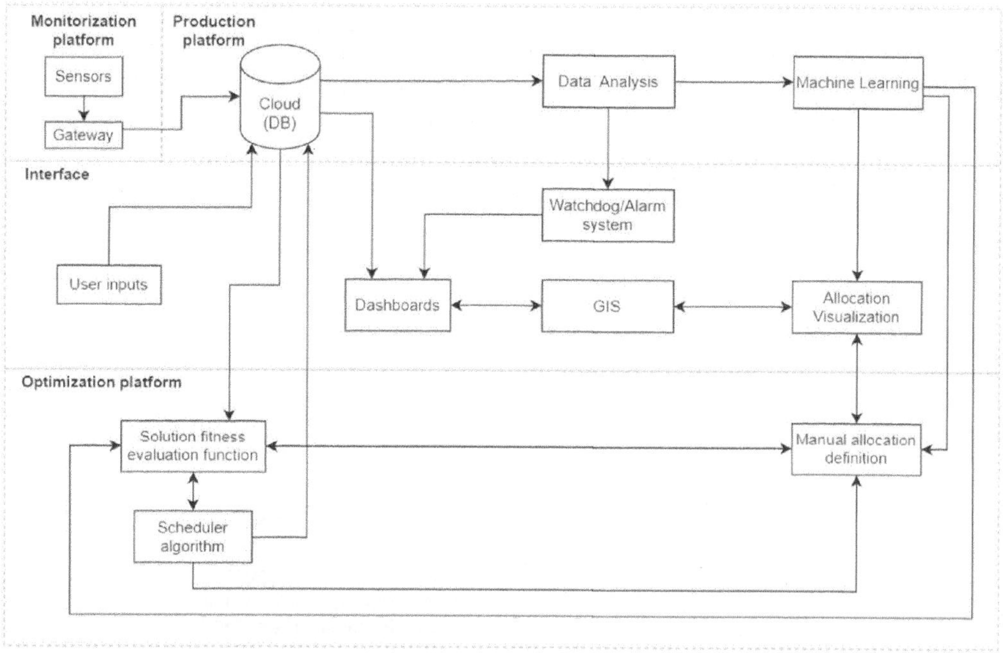

Figure 10.2 System architecture.

A possible process map and information flow aligned with the proposal of this chapter is described in Figure 10.3. The process begins with the implementation of an optimal resource allocation solution calculated offline. The initial offline solution is based on the productivity and cost estimates entered by the user in the interface for each type of equipment available before

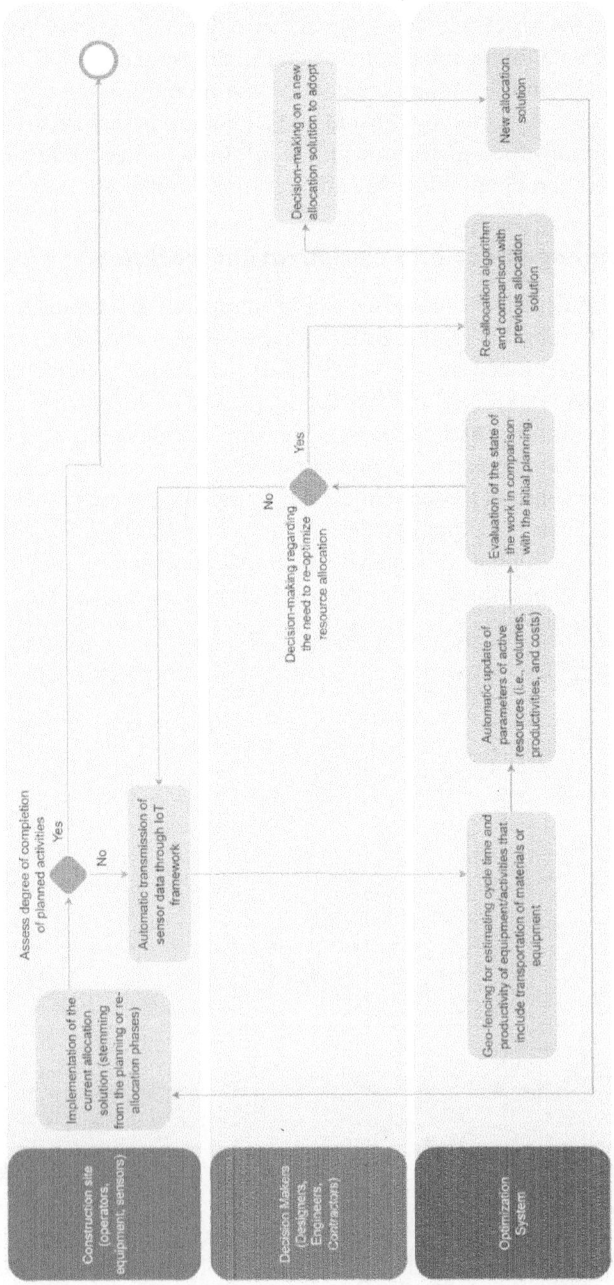

Figure 10.3 Process map and information flow of the proposed system.

the start of the process, as well as the quantification of work associated with each work front and the process constraints. Taking the earthworks process as an example, which will be further discussed in the case study Section (Section 10.3), this would translate into the identification of each embankment and excavation fronts, volumes of material to be excavated/embanked, types of geomaterials on the excavation fronts, and distances between work fronts as the process constraints. The methodology followed for this initial optimization is based on an implementation of the multi-objective optimization algorithm proposed in (Parente et al. 2015), providing the user with a Pareto front of viable solutions, with variable and optimal costs and execution durations (i.e., corresponding to optimal trade-offs between cost and duration), of which one of the solutions is chosen by the user and implemented on site. Each solution corresponds to the allocation of available resources over time to obtain the associated costs and durations, as demonstrated in Section 10.3.

From the moment the construction phase begins, the aforementioned monitoring system constantly records information in near real-time. This data is then processed and included in a dynamic database, which is complemented by user input information used to correct the productivity estimations based on the monitored data. Moreover, when this system is applied to a process involving sequential and interdependent tasks, monitoring a single team of equipment dedicated to one specific task is typically sufficient to infer the productivity of the equipment used in subsequent tasks. This simplification not only streamlines the system's implementation, making it less complex and reducing the likelihood of errors, but also enhances cost-effectiveness by limiting sensorization to just one type of equipment. Consequently, the extent of sensorization becomes a key consideration in the system's deployment. Monitoring more equipment and parameters simplifies productivity estimation and minimizes the need for human intervention. However, it also increases the effort required for data management due to the volume of diverse data levels that must be maintained. Conversely, a lower level of sensorization, while potentially yielding valuable insights, requires greater user involvement to validate the inferred data. For instance, in earthworks, due to the interdependence of productivity between sequential tasks, the monitoring of the transportation equipment between excavation fronts and embankments by means of GPS allows to infer the approximate productivity of all active teams in the respective production lines (Lee et al. 2018). In addition, since the investment in monitoring equipment is limited to the installation of GPS in trucks, this alternative is especially *low-cost*, both in terms of monitoring equipment costs and in terms of on-site data transmission requirements. This also enhances the system by enabling the generation of warnings and alarms in situations where equipment is not in its designated area, is stationary and inactive, or when there is a failure in GPS sensor data transmission. Concurrently, the generated information can be adapted to cater to different types of users, e.g., equipment operators can access specific information such as their average productivity, the productivity of downstream teams, the volume of tasks assigned to them, and estimated arrival times of the next truck in processes involving material transportation, among other details.

Whenever the productivity of active teams varies enough to cause negative discrepancies compared to the initial planning, the system generates warnings and/or alarms, identifying the team (or the individual equipment), the location, and the nature of the discrepancy. At any time, there is the possibility, on the main user's side, to update the estimates of both productivity, process constraints, and work progress on each work front manually (if necessary) to meet any additional information they are currently collecting. In addition, at any time the user can ask the system to suggest new equipment allocation solutions (i.e., re-optimization) whenever it is found that the discrepancies in productivity are too high or that there are new optimization conditions (e.g., different types of geomaterials, changes to the estimates of the volumes of geomaterials, faulty equipment that is no longer available, or new equipment that can be rented that becomes available, among others). In this way, the user can find new solutions that mitigate potential delays or even be more efficient than the current allocation solution for the new working/optimization conditions. It should be noted that, as will be described in more detail in Section 10.3, the re-optimization algorithm takes into account the disruption in the site as a consequence of the change in equipment allocation, calculating the effect of changing the position of several pieces of equipment that may have negative consequences on process efficiency, even if only temporarily.

10.3 Case study application

As established in previous work, the earthmoving process fits perfectly as a geotechnical process that can be compared to a production line (Parente et al. 2015) composed of different activities from the excavation and transportation of geomaterials (on the excavation fronts) to their spreading and compaction (on the embankment fronts). Therefore, earthworks are ideal geotechnical processes to demonstrate the applicability of the proposed system. The case study used to exemplify this application corresponds a real highway construction project carried out in Portugal (Parente et al. 2023). The application to the case study was divided into two main phases: the planning phase and the construction phase. While the first is related to the planning of the earthwork project before the start of the construction, the second pertains to the management of the earthwork construction and the decision-making process.

10.3.1 Case study description and system inputs

The specifications of the earthwork construction site are presented in Table 10.1. The resources available to carry out the work are summarized in Table 10.2 for the embankment and excavation equipment. Figure 10.4 shows the linear construction site in aerial view, with the respective excavation and embankment work fronts. In this case study, there was no need to optimize the excavation and transport volumes (i.e., Bruckner's diagram), as this decision had been previously made by the design team, and the transport distances/volumes had already been established.

Regarding excavation and transport volumes, it is relevant to take into account the specificities of an earthmoving project, which is strongly influenced by the geotechnical characteristics of the materials present on site, as well as by the compaction process in the construction of embankments. In fact, one of the constraints that generally characterizes earthworks is related to the fact that the materials used to build the embankments are the same as those resulting from the excavation processes on the corresponding fronts (with some exceptions, in particular, loan fronts). This seemingly simple restriction has several implications for the planning and execution of these types of projects.

Indeed, in this context, it is important to consider that the types of geomaterials excavated are the same that will be compacted in the embankment fronts. This implies that the allocation

Table 10.1 General information on earthwork operations

Global data	
Total length (m)	20, 368
Total processed volume (m^3)	15,23,206
Number of excavation fronts	15
Number of embankment fronts	24

Table 10.2 Equipment information[a]

Embankment equipment

Type	*Model*	*Operating weight (ton)*
Rollers (6)	CATERPILLAR CS 78 B (3)	18.70
	CATERPILLAR CS 683 E (3)	18.80
Spreaders (14)	CATERPILLAR D6 (9)	23.01
	CATERPILLAR D8 (4)	39.75
	CATERPILLAR D9 (1)	49.99

Excavation equipment

Type	*Model*	*Capacity (m^3)*
Excavators (28)	CATERPILLAR 330D LN (1)	2.07
	CATERPILLAR 330GC (4)	1.54
	CATERPILLAR 336GC (5)	1.88
	CATERPILLAR 352F (1)	3.10
	HITACHI ZX170W-6 (4)	0.60
	VOLVO EC 250 ENL (7)	0.95
	VOLVO EC 300 ENL (1)	1.32
	VOLVO EC 380 ENL (2)	1.42
	VOLVO EC 480 EL (2)	2.10
	VOLVO EW 180 E (1)	0.87
Trucks and *dumpers* (39)	MERCEDES ACTROS 3236K/42 (2)	19.20
	MAN TGS 35.360 8 × 4 (4)	13.00
	MERCEDES AROCS 3243K (15)	13.40
	MERCEDES AROCS 3243K AOE (2)	14.00
	VOLVO A-40 (11)	22.50
	VOLVO A40G (3)	24.00
	VOLVO A-30 D 6x6 (2)	18.00

[a]Total number of available equipment inside parenthesis.

Figure 10.4 Representation in plan of the work fronts and their geographical distribution along the work line. (Parente et al. 2023.)

of a particular compactor roller to an embankment front implies that its productivity will be limited not only by the productivities of the upstream teams, namely, excavation, transport, and spreading (i.e., as a result of the sequentiality and interdependence of the production lines' productivities), but also from the type of geomaterials present at the excavation fronts and transported to that embankment front, as planned in Bruckner's diagram.

Subsequently, determining the productivity of rollers becomes a complex process, as it is found that the productivity of a particular roller varies when allocated to different fronts. At the same time, a single embankment front can receive geomaterials transported from several excavation fronts with completely different geomaterials, which adds additional complexity to the estimation of the compaction teams' productivity. In these cases, the compaction productivity can be estimated by resorting to the weighted average of the quantity of each type of material received at the corresponding embankment front, which, in turn, is also a function of the productivities of the upstream teams (i.e., productivities of the excavation, transport, and spreading teams).

Thus, it can be easily inferred that compaction processes are of utmost importance regarding the progress of earthworks, not only as a consequence of the complexity of their execution in technical terms (and subsequent complexity in estimating their productivity), but also because, from the perspective of earthwork production lines, they correspond to the process generally identified as the bottleneck of the production line's productivity. As a result, compaction usually coincides with the production line process whose productivity and efficiency should be sought to be maximized, as this inherently implies maximizing the productivity and efficiencies of the upstream processes, thus being one of the main objectives of the proposed methodology and optimization system.

In this context, it should be noted that, despite the fact that an updated edition of the *Guide des Terrassements Routiers* (GTR) has recently become available (IDRRIM 2023), the estimated productivity of the compaction rollers at the time of development of this work was determined using the classification of rollers and geomaterials proposed by earlier editions of the 2000 edition of GTR (SETRA-LCPC 2000), while the productivity of the remaining equipment was estimated from the manuals and information provided by the respective manufacturer. The GTR is a compaction guide resulting from the compilation of empirical knowledge in the field of embankment construction originating in France. In addition to being the basic reference for the construction of embankments and pavement bed layers in France, the knowledge of this guide can be generalized to any type of geomaterial and compaction equipment in the world, as has already been demonstrated in several studies (Marques et al. 2008).

According to the guide, the estimation of the productivity of a compaction roller is a function of the geomaterial, the equipment specifications, and the site conditions (i.e., related to moisture status of the geomaterial). On the one hand, the classification of geomaterials is a function of characteristics such as the maximum size of the aggregates, the value of the methylene blue test, and the Atterberg limits, all of which are tests already traditionally adopted in the characterization of soils in the laboratory before and during earthworks. On the other hand, the classification of compaction rollers according to the guide is directly related to its specifications, namely, the type of roller (i.e., vibrator, static, tire), its operating weight, drum width, and vibration parameters, when applicable. This assessment of geomaterials and equipment seeks to be sufficiently detailed to allow the determination of the compaction energy required to obtain good compaction results, depending on the type of material and the in-situ conditions. Thus, it is possible to obtain from the GTR a value of the estimated theoretical productivity of a roller under certain working conditions quite accurately (i.e., assuming that the productivity of the roller is not limited by the amount of geomaterial available) even before starting the work, including for variable layer thicknesses, speed, and number of passes.

In this case study, both the available compaction rollers and the geomaterials at the excavation fronts had similar characteristics in terms of GTR classification, with the rollers being class V4 and the geomaterials class B2. The proposed system is configured to request the GTR classification as user input, both for the rollers and for the geomaterials at the excavation fronts. By using the GTR, the user can estimate the productivity of each roller allocated to a specific

Table 10.3 Required inputs for the available equipment (rollers)

Parameter	*Description*
Model	Equipment model
Registration	ID of the equipment
Costs	Costs associated with renting and operating the equipment (e.g., fuel consumption, labor) per operating hour
Efficiency Factor (*k*)	Efficiency factor, which takes into account the fact that the actual working time of the equipment in each hour is generally less than 60 minutes, depending on the operator's capacity, maintenance, site conditions, among others
Emissions	Carbon gas emissions per operating hour (optional)

embankment front based on the geomaterials that will be transported there and the working conditions at the site.

Given the complexity and multiplicity of possible solutions for the allocation of equipment to the work fronts, as well as the dynamic nature of the problem, the adopted solution must also be of a dynamic nature. The dynamic component of the problem is related to the fact that, during the execution of the earthworks, there are several excavation and compaction teams completing their work on respective fronts at different times. This leads to either a reassignment of these teams to other work fronts or their redistribution into other active teams, ensuring that the work continues optimally. For this reason, the proposed optimization algorithm provides optimal solutions in terms of minimizing cost and time for the entire planning and construction of the earthworks, divided into several construction phases (i.e., with each "construction phase" lasting until a work front reaches its conclusion), as opposed to a local optimization, which would only address each work phase individually.

To enable the calculation of optimal solutions in terms of both cost and time, the algorithm requires information on the specifications of available equipment resources, including compactor rollers, spreaders (e.g., bulldozers, graders), excavators, and trucks, as exemplified in Table 10.3. Additionally, it is necessary to provide information about the work fronts regarding their location, the volume of geomaterial to be processed, the maximum number of active equipment simultaneously for each work front (i.e., which allows the system to account for space limitations), and the distances between the excavation and embankment work fronts. It should be noted that the optimization algorithm makes its own selection of the optimal fleet according to the available equipment, i.e., it does not necessarily use all the available equipment if this implies less efficient solutions.

10.3.2 Planning phase implementation and results

Having access to the information related to the work to be carried out and the resources available as described in the previous subsection, the system is then capable of providing various possible optimal solutions corresponding to a compromise between objectives (i.e., cost vs. duration). Figure 10.5 shows the Pareto front of solutions obtained for the planning phase in this case study. The optimization algorithm used to obtain the solutions presented here was a non-dominated sorting genetic algorithm-II (NSGA-II) (Deb et al. 2002), implemented in C#, with the following optimization parameters: population size and maximum number of generations of 100, crossover probability of 0.8, mutation probability of 0.2, and number of minimization objectives of 2 (time and cost). The computational effort corresponded to about 3 hours of calculation using a personal laptop with common characteristics, namely, an Intel i7-10750H

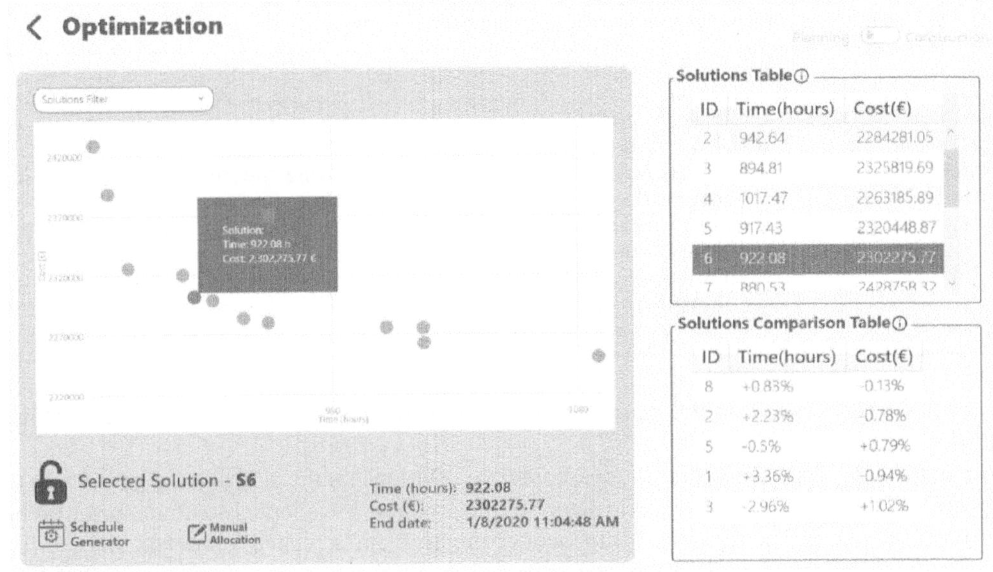

Figure 10.5 Screenshot of the system showing the Pareto front with the selected solution, including its characteristics (Solution 6 – red point). (Parente et al. 2023.)

2.60 GHz processor, with 16GB of RAM and Windows 11 as the operating system. Note that this calculation time can be improved by increasing computational resources or using current technologies, such as dedicated servers or cloud computing technologies. However, the current version is consistent with the intended use for the system even in construction execution phases, namely, performing calculations overnight, allowing for the reorganization of the equipment fleet the following morning, if necessary.

For each solution generated and represented in Figure 10.5, it is possible to consult the detailed planning throughout all the earthwork phases. Figure 10.6 shows various types of visualization of the same solution (Solution 6) in three different representations, namely, in table format (Figure 10.6a), in a lines-of-balance format (Figure 10.6b), or in a map layout (Figure 10.6c), in which the allocation of equipment throughout work fronts over time can be checked.

While the table format and the map format represent intuitive visualization modes that can be exported to spreadsheet or image formats, respectively, it is important to mention that the format shown in Figure 10.6b was adapted from conventional lines-of-balance diagrams. Indeed, it is an interactive graphical representation of the evolution of all the excavation work fronts throughout the progress of the earthwork project in the form of a matrix, where each row represents an excavation work front and each column represents a phase with the respective work front completed (which can be more than one). Along each row of the matrix, it is possible to follow the evolution of the volume of material to be excavated for each corresponding work front and throughout all phases. Each cell, composed of a local line chart representing the embankment work front receiving the excavated material, represents the current state of that excavation work front in a particular phase. The initial and final points of this line chart correspond to the initial and final volumes of material of that excavation work front in that phase, while its slope is a function of the productivity of the corresponding active team (i.e., the steeper the slope, the higher the productivity). For each excavation, the first and last cell containing a local chart

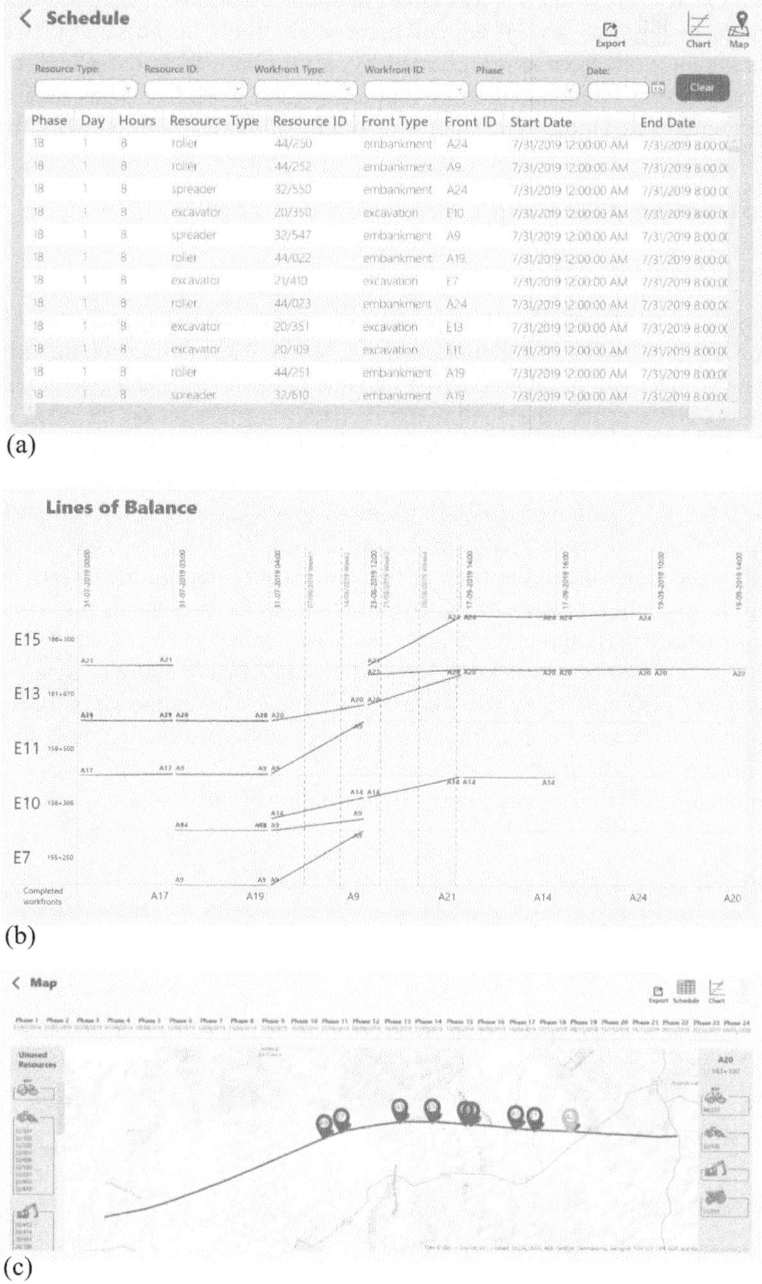

Figure 10.6 Screenshot of the system showing the allocation corresponding to Solution 6 for construction phase 18 in the format of: (a) table; (b) line-of-balance-based diagram; (c) map. (Parente et al. 2023.)

starts at zero (all the material volume is available) and ends when the line chart reaches the top of the cell (no volume is available, as the remaining volume of geomaterial at the excavation front is null). Thus, it is possible to trace the evolution of the excavation volume throughout the phases until no excavation volume is available. In some situations, the same excavation work front may supply material to several embankment work fronts in the same phase. Therefore, as each embankment work front is represented by a line chart, that cell (which represents the embankment work fronts being supplied by that excavation work front) has several lines, each line being representative of the evolution of the volume of that excavation work front.

10.3.3 Construction phase implementation and results

10.3.3.1 Resource monitoring

The monitoring approach adopted for this application was the previously mentioned low-cost version based on GPS monitoring of the transportation teams. This simple solution allows to infer the approximate productivity of all production line due to the interdependence of tasks. Once the construction process starts, and the conditions for data transmission on the site (IoT platform) are ensured, equipment data starts to be collected from the installed GPS sensors and transmitted to the user through the dashboard. This dashboard is customizable, and Figure 10.7 demonstrates the sum of the total vs. remaining volume of all work fronts, as well as the comparison between the predicted vs. measured productivity (processed from GPS data) for each compaction front. It is important to mention that, at the time of this work's preparation, there were no data available regarding the execution process of the construction, so the present section corresponds to a simulated construction phase through dynamic databases supported by a script for transmitting synthetic data (i.e., the data was previously generated synthetically and was sent to the system's dynamic databases over time to simulate the entry of sensor data by the second). Real-time data acquisition from the equipment is essential to use the system's construction module. This module relies on monitoring and quantifying the number of cycles each truck completes over time, thereby enabling an estimation of productivity based on the volume of material transported. To determine

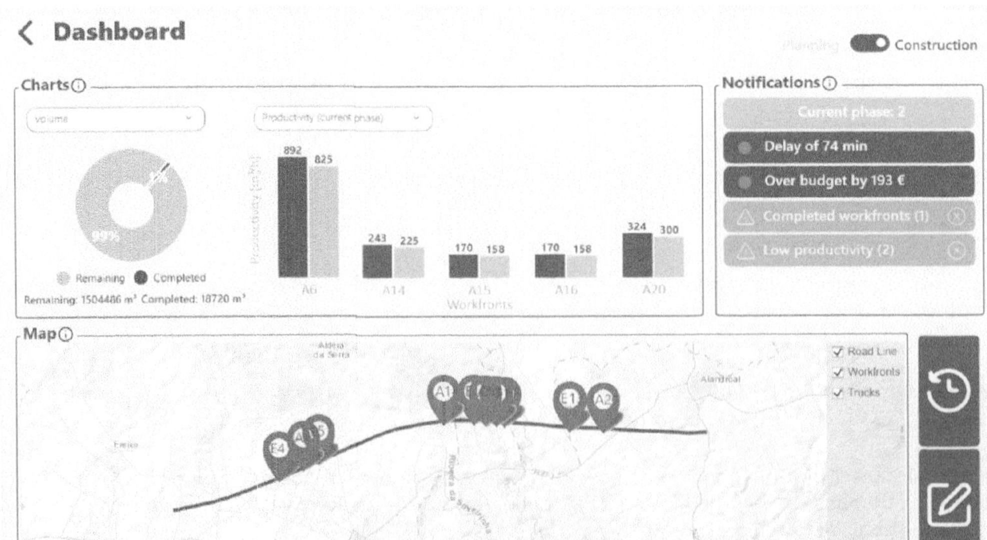

Figure 10.7 Screenshot of the system's dashboard corresponding to the project construction phase (simulated). (Parente et al. 2023.)

the number of load-unload cycles of each transportation equipment, geo-fencing techniques were adopted. These allow for the definition of an area around the excavation and or embankment fronts, delimited by lines defined by coordinates, so that the number of times one of the transport equipment crosses these lines can be counted. This data also allows for the calculation of the productivity of the excavation teams to which the transport equipment is assigned during the loading cycle, as well as the compaction teams that receive the excavated material.

The dashboard also allows access to the historical module and the manual editing function of the project's characteristics. This last option allows for revisions or adjustments to the current planning, such as equipment availability (since some equipment may at any time become unavailable due to breakdown), changes related to the work fronts (including changes to the types and/or volumes of geomaterials available at the work fronts, as they may be different from those predicted in the geotechnical studies), or updates to the productivity of the equipment following the monthly topographic surveys carried out on site. Depending on the type of edits to the project characteristics made, the system assesses their impact on the current planning and recalculates the costs and duration in each subsequent phase for the new working conditions, transmitting this information to the user in report format (Figure 10.8). At this point, the user has the option to continue with the current planning, accepting the new cost and duration forecasts, or request the system to re-optimize/reallocate resources for the remaining phases of the project, as will be detailed more thoroughly in Section 10.3.3.2.

As previously mentioned, one type of editing that is expected to be used periodically (e.g., monthly) by an earthwork construction company is related to the rectification of the current volumes of excavation and embankment following periodic topographic control. These measurements and corrections are a common procedure in earthworks for quality control, to assess progress and try to mitigate any potential disturbances from the outset. Since it is possible to directly measure, through topographic surveys, the actual amount of excavated volume, these values can be compared with those estimated by the system, which are calculated based on the load capacity of the allocated trucks and their number of cycles.

Thus, to rectify the difference between the average amount of material excavated and transported by the trucks and the maximum assumed capacity of the transport equipment, an efficiency factor, k, is assigned to the productivity of the latter (similar to what is done in the conventional calculation of the productivity of excavation, spreading, and compaction equipment). This factor corresponds to the ratio between the actual volume of excavated material, resulting from topographic surveys, and the value estimated by the system through the number of cycles and the capacity of each truck. In this case, an update of this nature was carried out, in which the volumes of excavation and embankment were slightly adjusted, for which the system initially calculates the new costs and durations for the new volumes of material while maintaining the current allocation solution. Following these updates, it is up to the user to decide, based on a report presented by the system corresponding to the effect of the update on the cost and duration of the work (i.e., assuming the currently active allocation), whether to proceed with the current allocation or to re-optimize and obtain a new set of optimal solutions for the new working conditions. For the purpose of this case study, at this stage of the simulation, it was decided to maintain the current allocation and continue the work.

10.3.3.2 Re-optimization of available resources

In any construction project, despite all planning and quality control measures, unforeseen events can occur that significantly alter the working conditions of the active teams, calling for a new assessment of resource allocation to ensure that work continues with maximum efficiency.

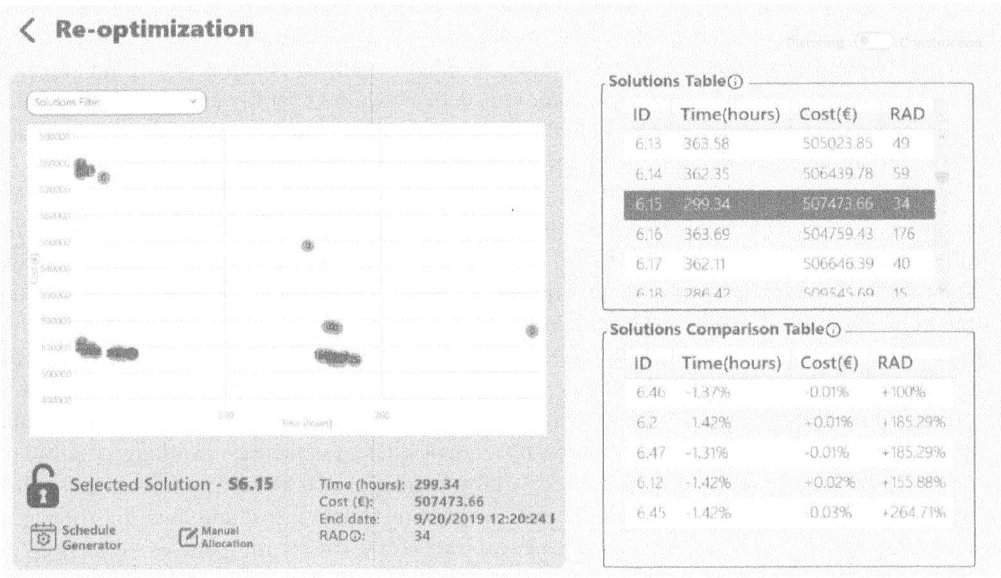

(a)

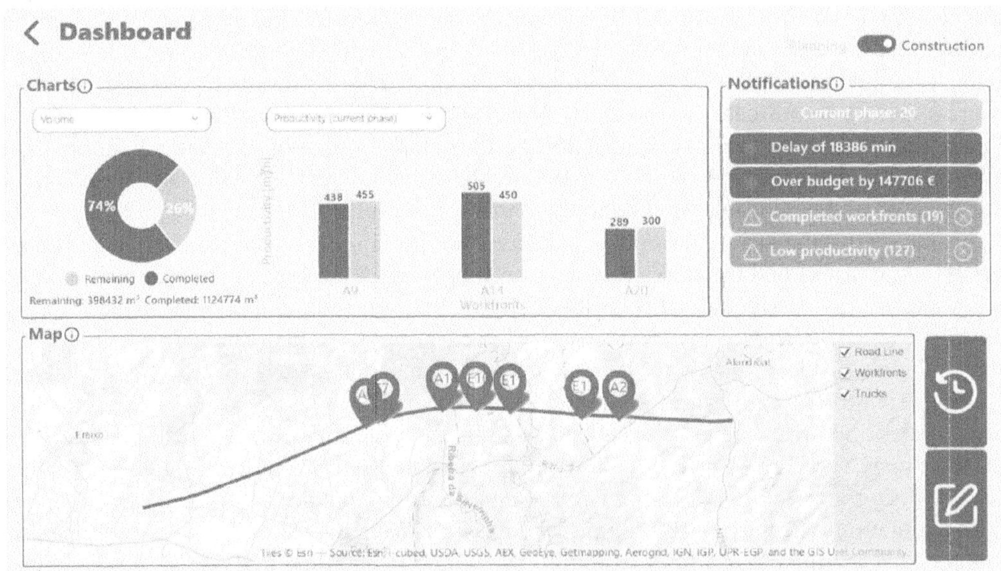

(b)

Figure 10.8 Screenshot of the system showing: (a) the new Pareto front as a result of resource re-optimization, with the selection of Solution 6.15; and (b) the current state of work progress in construction phase 20. (Parente et al. 2023.)

One of the most common unpredictable situations in earthworks is equipment malfunction, which leads to its unavailability to be allocated. Other reasonably common changes include the addition of new work fronts (e.g., borrow sites) to support or complete an ongoing embankment. As described in the previous section, the user can make these types of changes or adjustments at any time, subsequently facing the decision to re-optimize the available resources. It is noteworthy that, in certain cases, the edits made to the working conditions are sufficient to render the original resource allocation solution unfeasible. An intuitive example of such an edit can be seen in cases where the failure of a piece of equipment leaves a supposedly active work front without any resources (resulting in an infinite execution time for that work front). The case presented in Figure 10.8 is consistent with this type of situation, where a breakdown in truck 75 left a production line without the capacity to transport geomaterials between the corresponding excavation and compaction fronts, thus requiring the re-optimization of the fleet allocation.

The system's re-optimization algorithm differs significantly from the one used in the planning stages. Indeed, although the optimization parameters are the same as originally used and the same minimization objectives (in this case, cost and duration) are taken into account, the re-optimization algorithm always includes an additional minimization objective, denominated RAD (re-allocation disruption). The purpose of this third optimization objective is to minimize the disruption introduced in the work resulting from the re-allocation of equipment from their current front to a new work front. The RAD is always calculated in comparison to a previous active solution, with its value obtained through the normalization of the equipment's travel distance between the current front and the front suggested by the new solution (relative to a uniform minimum distance, which in this case was considered to be 100 m). Thus, a high RAD value can mean that many resources are being reallocated, or alternatively that some of the resources being reallocated are moving long distances, which can mean a longer period in which the equipment is not only effectively inactive but may also be causing obstacles to the normal functioning of the remaining equipment. By using this additional minimization objective, the system is forced to seek solutions that adjust the current allocation into an optimal allocation, while introducing the least possible disruption in the normal development of the work. In other words, the system seeks the minimum adjustments that allow converting the current solution into an optimal solution.

After the re-optimization process, the obtained re-optimization solutions, represented in Figure 10.8a, are evaluated in terms of the original minimization objectives (in this case, cost and duration), with the addition of a third value corresponding to the RAD, which would represent an additional axis of values in a 3D graph. However, during the development of this module, it was found that a 2D representation is more effective for the analysis and interpretation of the results by the user, allowing an informed choice of a new potential allocation solution to proceed with the work. In the case study described here, a new re-optimization solution corresponding to solution 15 was chosen. Note that, for nomenclature and historical consultation purposes, the new solutions obtained are designated as "S6.X," with "S6" being the previously selected solution and "X" corresponding to the new solution chosen from the re-optimization process, making its reading simpler and clearer, as presented in Figure 10.8a. In this case, after selecting solution S6.15, with a RAD value of 34, the earthwork processes continued according to the planning of this solution, as represented in Figure 10.8b.

10.3.3.3 Historical data

Whether during an ongoing construction or when accessing data from past earthwork projects, the historical data module displays relevant information from all previous projects. This includes solutions originally adopted during the planning phase, significant notifications made by the monitoring and alarm system, and solution changes resulting from re-optimizations. As shown in Figure 10.9, a vertical timeline is presented with key milestones and edits, along with a list of significant notifications that have arisen throughout the project. For each re-optimization request where the user selected a new solution, it is possible to access detailed planning of

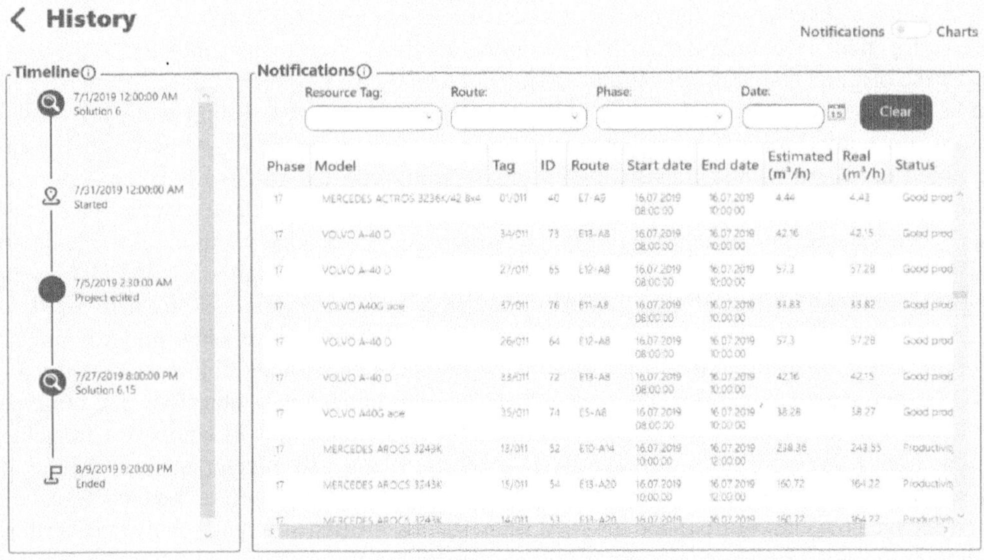

Figure 10.9 Historical data in a table format. (Parente et al. 2023.)

both previous and subsequent/current solutions. Regarding the list of notifications, it provides detailed information about each truck and productivity conditions. This information can be used in other types of analyses, such as audits, statistical analyses, or artificial intelligence and machine learning techniques, capable of deriving new insights and/or interesting patterns from these data for future studies.

10.4 Concluding remarks

This chapter introduced a versatile digital twin optimization system, designed to be applicable across a range of geotechnical processes that are resource dependent, with its efficacy and applicability demonstrated through an application in earthworks. The integration of IoT and optimization tools within this system enabled near real-time, continuous monitoring, enhancing management and decision-making capabilities in geotechnical operations. On the one hand, optimization techniques can be used during the planning phase to try to optimize the allocation of resources throughout the construction project, minimizing aspects such as cost and duration, based on productivity data and estimated costs. On the other hand, following the start of the construction phase, the implementation of an IoT platform makes it possible to monitor and adapt the allocation solution adopted in the design phase to new unpredictable conditions during the construction phase in real-time.

The application of this system in earthworks served as a proof of concept, illustrating its potential in optimizing resource allocation during the planning phase and minimizing key project constraints such as cost and duration. Moreover, the construction phase of the project highlighted the system's ability to dynamically adapt to unforeseen events, a critical aspect in the fluid and often unpredictable realm of this kind of processes. While the case study is based on real data from an earthwork project, the construction phase (Section 10.3.3.2) was simulated due to data availability constraints. Still, it effectively demonstrated the system's robust potential in operational management within the earthwork domain. The employment of GPS sensors and geo-fencing techniques for precise equipment monitoring, along with strategic real-time adjustments to allocation plans, emphasized the system's efficacy in mitigating project delays and ensuring optimal progress. A comprehensive project plan was successfully generated and re-optimized for all phases of the earthwork project, showcasing the system's responsiveness to simulated scenarios such as equipment failure.

One of the key innovations of this approach is the re-optimization methodology, which not only strives to maintain near-optimal equipment activity but also accounts for the disruptions caused by changes in equipment allocation, costs, and duration. This is achieved by incorporating the RAD as an additional optimization objective. Overall, this highlights the significant potential of the digital twin optimization system in geotechnical engineering, offering a novel, efficient approach to project execution that is adaptable to various geotechnical processes beyond earthworks.

Reference list

Deb, K., A. Pratap, S. Agarwal, and T. Meyarivan. 2002. A fast and elitist multiobjective genetic algorithm: NSGA-II. *IEEE Trans. Evol. Comput.* 6: 182–97. https://doi.org/10.1109/4235.996017.

Gomes Correia, A., P. Cortez, J. Tinoco, and R. Marques. 2013. Artificial intelligence applications in transportation geotechnics. *Geotech. Geol. Eng.* 31, no. 3: 861–79. https://doi.org/10.1007/s10706-012-9585-3.

Gomes Correia, A., and E. Gastine. 2024. How AI and IoT are transforming geotechnical site monitoring and construction. In *Proc. XVIII ECSMGE 2024*, Lisbon, Portugal. https://doi.org/10.1201/9781003431749-4.

Gomes Correia, A., and J.-P. Magnan. 2012. Trends and challenges in earthworks for transportation infrastructures. In *Adv. Transp. Geotechs 2*, eds. S. Miura, T. Ishikawa, N. Yoshida, Y. Hisari, and N. Abe, 1–12. London, UK: CRC Press.

Gomes Correia, A., J. Neves, E. Fortunato, and M. Parente. 2021. Pavement and rail track geotechnics and prospects for the digital age. *Geotecnia* 152: 113–41. https://doi.org/10.14195/2184-8394_152_4.

Gomes Correia, A., J. Neves, J. Moutinho, A.C. Freire, A. Paixão, B. Monteiro, M. Parente, and N. Cristelo. 2023. Geotechnics in the eco-digital transition of transport infrastructures. *Geotecnia extra* 2024: 41–78. https://doi.org/10.14195/2184-8394_extra2024_1.

IDRRIM. 2023. *Guide des Terrassements des Remblais et des Couches de Forme*. France: Cerema.

Jassim, H.S.H., W. Lu, and T. Olofsson. 2017. Predicting energy consumption and CO_2 emissions of excavators in earthwork operations: An artificial neural network model. *Sustainability* 9: 1257. https://doi.org/10.3390/su9071257.

Kassem, M., E. Mahamedi, K. Rogage, K. Duffy, and J. Huntingdon. 2021. Measuring and benchmarking the productivity of excavators in infrastructure projects: A deep neural network approach. *Autom. Constr.* 124: 103532. https://doi.org/10.1016/j.autcon.2020.103532.

Lee, S., S. Park, and J. Seo. 2018. Utilization analysis methodology for fleet telematics of heavy earthwork equipment. *Autom. Constr.* 92: 59–67. https://doi.org/10.1016/j.autcon.2018.02.035.

Liu, Y., K. You, Y. Jiang, Z. Wu, Z. Liu, G. Peng, and C. Zhou. 2022. Multi-objective optimal scheduling of automated construction equipment using non-dominated sorting genetic algorithm (NSGA-III). *Autom. Constr.* 143: 104587. https://doi.org/10.1016/j.autcon.2022.104587.

Marques, R., A. Gomes Correia, and P. Cortez. 2008. Data mining applied to compaction of geomaterials. In *Proc. 8th Int. Conf. Bear. Capacit. Roads, Railw. Airfields*, eds. E. Tutumluer and I.L. Al-Qad, 1: 597–605. London: Taylor & Francis Group.

Montaser, A., and O. Moselhi. 2014. Truck+for earthmoving operations. *J. Inf. Technol. Constr.* 19: 412–33.

Parente, M., M. Amândio, V. Moreira, and A. Gomes Correia. 2023. Earthworks management, monitoring and optimization system based on digital twin and simulation of a case study. *Geotecnia* 159: 43–63.

Parente, M., P. Cortez, and A. Gomes Correia. 2015. An evolutionary multi-objective optimization system for earthworks. *Expert Syst. Appl.* 42, no. 19: 6674–85. https://doi.org/10.1016/j.eswa.2015.04.051.

Parente, M., G. Figueira, P. Amorim, and A. Marques. 2020. Production scheduling in the context of Industry 4.0: Review and trends. *Int. J. Prod. Res.* 58, no. 17: 1–31. https://doi.org/10.1080/00207543.2020.1718794.

Parente, M., and A. Gomes Correia. 2013. Compaction management: Results of a demonstration project. *Adv. Mater. Res.* 779: 1697–700. https://doi.org/10.4028/www.scientific.net/AMR.779-780.1697.

Parente, M., A. Gomes Correia, and P. Cortez. 2014. Artificial neural networks applied to an earthwork construction database. In *Proc. 2nd Int. Conf. Inf. Technol. Geo-Eng. (ICITG 2014)*, Durham, UK.

Parente, M., A. Gomes Correia, and P. Cortez. 2016. Metaheuristics, data mining and geographic information systems for earthworks equipment allocation. *Adv. Transp. Geotech. III* 143: 506–13. https://doi.org/10.1016/j.proeng.2016.06.064.

Parente, M., A. Gomes Correia, G. Figueira, and A. Mehrsai. 2018. Towards improving earthworks production from an Industry 4.0 perspective: The role of remote information technologies and dynamic optimization techniques. In *Proc. 7th Transport Res. Arena (TRA 2018)*, Vienna, Austria. https://doi.org/10.5281/zenodo.1491368.

Pereira, G., M. Parente, J. Moutinho, and M. Sampaio. 2021. Fuel consumption prediction for construction trucks: A noninvasive approach using dedicated sensors and machine learning. *Infrastructures* 6, no. 11: 157. https://doi.org/10.3390/infrastructures6110157.

Salem, A., and O. Moselhi. 2020. AI-based cloud computing application for smart earthmoving operations. *Can. J. Civ. Eng.* 48, no. 3: 312–27. https://doi.org/10.1139/cjce-2019-0681.

SETRA-LCPC. 2000. *Guide des Terrassements Routiers – Réalisation des Remblais et des Couches de Forme*. Ministère de l'Équipement, du Logement et des Transports, France.

Silva, R., M. Parente, and J. Neves. 2023. Machine learning for the analysis of equipment sensor data in road construction projects. Accepted at *4th Int. Soc. Intell. Constr.* (ISIC 2024), Orlando, Florida, USA.

Tatsuoka, F., and A. Gomes Correia. 2018. Importance of controlling the degree of saturation in soil compaction linked to soil structure design. *Transp. Geotech.* 17: 3–23. https://doi.org/10.1016/j.trgeo.2018.06.004.

You, K., L. Ding, C. Zhou, Q. Dou, X. Wang, and B. Hu. 2021. 5G-based earthwork monitoring system for an unmanned bulldozer. *Autom. Constr.* 131: 103891. https://doi.org/10.1016/j.autcon.2021.103891.

Part III

Harbour facilities and modern transportation infrastructures

Alexandre Pinto

This part of the book highlights the importance of harbour facilities and modern transportation infrastructures in addressing the challenges faced by new societies, for instance: (i) the importance of ground improvement solutions to overcome very demanding geological and geotechnical restraints; (ii) the widespread use of geotechnical underpinning and strengthening solutions to improve seismic resistance associated with the use of bigger vessels with deeper drafts; (iii) the remarkable development and upgrading of centenary port infrastructures, and (iv) the application of the observational approach for geotechnical risk management. Four cases are presented highlighting the fundamental influence of geotechnical engineering to achieve those transversal goals.

The first case, presented in Chapter 11 (Watabe 2025), is related to the case studies of ground improvement of large-scale manmade islands for airport facilities in Japan. Where advanced and empirical geotechnical engineering methods that are crucial for coastal development, such as the construction of port and airport facilities through land reclamation on soft seabeds, are presented. This chapter is divided into two parts: (i) a predictive approach for consolidation settlement, covering primary to secondary consolidation, including a technique forestimating long-term consolidation settlement beyond conventional consolidation theory; and (ii) a comprehensive overview of ground improvement techniques for soft ground, including underlying principles and practical applications.

The second case, presented in Chapter 12 (Ramos 2025), describes an unconventional example of underpinning and strengthening solutions within a complex geotechnical environment at a port infrastructure located in a high seismicity region. The adopted solutions enabled substantial dredging beneath the foundation level of the existing infrastructure, as well as significant resistance enhancement, mainly seismic, in compliance with current regulations to allow the installation of a high-capacity vessel synchrolift. The project involved substantial load increases and seabed dredging of approximately 8m below the foundation level of the existing gravity-type block quay walls. In addition to extensive dredging, underpinning and strengthening of the quay walls were required, along with the construction of high-capacity complementary structures designed to interact with both the soil and the existing infrastructure. These improvements ensured their ability to support the increased loads from vessel lifting, onshore transfer, and dry-docking operations of up to 9000 tonne vessels.

The third case, presented in Chapter 13 (Lopes and Silva 2025), includes material previously published in Lopes and Dias (2021). The port of Leixões (Portugal): over 120 years of knowledge. Geotechnical Research 8(2): 42–53, https://doi.org/10.1680/jgere.18.00015. That work was published under the CC BY 4.0 license, which allows reuse with attribution. The current version includes new content and revisions, and the authors have clearly indicated all reused

DOI: 10.1201/9781003530220-13

material in accordance with the license terms. This case describes the remarkable development of the Port of Leixões, where geotechnical engineering has played a crucial role. It describes the construction of Portugal's second most important port, whose docks were entirely excavated at the mouth of the Leça River. It also explores the developments over more than 120 years, focusing on the main geotechnical problems faced and their mitigation. The port of Leixões was constructed in the 19th century as an alternative to the Douro port in northern Portugal. Despite the historical importance of Douro port in the 15th and 16th centuries (during the Age of Discovery), and later in the 18th century with the *Port Wine* trade, it had several restrictions, namely in its geomorphic and hydrodynamic characteristics. The first studies for the new port of Leixões date back to the 15th century, but it was in the 18th century that significant efforts were made to provide a viable alternative to Douro port. Due to the magnitude of the work, it was considered one of the largest engineering works of the 19th century. Several structures have been built, including vertical rock/concrete quay walls, piled quays, and dolphins, being the main restraints, as well as the implemented solutions described.

Finally, Chapter 14 (Carreto et al. 2025) demonstrates the use of the observational approach to manage the geotechnical risk associated with a soil excavation slope along the A1 motorway near Coimbra, the most important in Portugal, operated by Brisa Concessão Rodoviária (BCR). The study follows a landslide that occurred during the road construction in 1980. The landslide involved substantial mass movement, extending over 150 m in length and reaching a depth of approximately 20 m. Any intervention intended to cease the ongoing movements observed after post-initial stabilisation measures would incur exorbitant costs. Since these movements remained minimal and progressed at a relatively constant rate, it was considered that the risk was compatible with the use of the observational approach. Forty years later, intervention measures have been restricted to short-term, inexpensive actions that primarily focus on improving the slope's internal drainage.

In summary, this part of the book, through the presented cases, also provides an inspired view about the importance of harbour facilities and modern transportation infrastructures for the new society's challenges, including the geotechnical engineering remarkable contribution.

References

Carreto J., Gonzalez I., Martins, J.F., Cruz, M.Q., Fernandes E.M. and Caldeira, L. 2025. The use of the Observational Approach in the Risk Management of an Excavation Slope on the A1 Motorway Near Condeixa, During the Operation Phase. *in* Gomes Correia, A., Pinto, A., Neves, J., Fortunato, E., (Eds.), *Transportation Geotechnics for Green, Digital, and Modern Infrastructures,* CRC Press (2026).

Lopes, H. G. and Silva, J. 2025. The Port of Leixões (Portugal): Over 120 Years of Knowledge. *in* Gomes Correia, A., Pinto, A., Neves, J., Fortunato, E., (Eds.), *Transportation Geotechnics for Green, Digital, and Modern Infrastructures,* CRC Press (2026)

Ramos, P. M. 2025. Underpinning and Strengthening of an Existing Port Infrastructure to Allow the Installation of a High-capacity Vessel Synchrolift. *in* Gomes Correia, A., Pinto, A., Neves, J., Fortunato, E., (Eds.), *Transportation Geotechnics for Green, Digital, and Modern Infrastructures,* CRC Press (2026).

Watabe, Y. 2025. Learnings from Case Studies of Large-scale Manmade Island. *in* Gomes Correia, A., Pinto, A., Neves, J., Fortunato, E., (Eds.), *Transportation Geotechnics for Green, Digital, and Modern Infrastructures,* CRC Press (2026).

Chapter 11

Learnings from case studies of large-scale manmade island

Yoichi Watabe

11.1 Introduction

In coastal development, land reclamation is carried out by constructing a seawall and filling the soil behind it, and then, port facilities, industrial areas, commercial areas, leisure facilities, and even residential areas are constructed on the reclaimed land. Airport facilities are also constructed in East Asia and Southeast Asia, typically in Japan, China and Singapore. As most of these are constructed on soft clay deposits, ensuring the stability of the seawall and predicting consolidation settlement are key to the success of the construction. The design considers the applicability of ground improvement methods, adopts the most appropriate method to ensure the stability of the seawall, predicts consolidation settlement and applies drainage methods to promote consolidation to reduce construction time. The first part of this chapter outlines a method for predicting the consolidation settlement based on a classical simple theory with empirical values, which is outlined as common engineering sense, and predicting differential settlement using construction data is introduced as advanced construction management. The second part of the chapter outlines the principles and applicability of ground improvement methods used mainly for soft clay deposits and also introduces cement-treated soils used as landfill material made of dredged clay. The final part of the chapter provides an example of a major project where a variety of ground improvement methods were used in an integrated way, with the right techniques applied in the right places.

11.2 Fundamental and advanced prediction of consolidation settlement

11.2.1 Settlement prediction based on water content

11.2.1.1 Physical properties of a clay with a water content of 100%

In geotechnical engineering, it is essential for a geotechnical engineer to have an understanding of the approximate order of magnitude of soil parameters. The following are empirical common-sense values of physical properties for marine clays with a water content of 100%.

Let us consider the void ratio, e. The following equation relates the void ratio to the degree of saturation, S_r:

$$e = \frac{w}{S_r} \cdot \frac{\rho_s}{\rho_w} = \frac{w \cdot G_s}{S_r} \tag{11.1}$$

DOI: 10.1201/9781003530220-14

In this context, the degree of saturation, S_r, is assumed to be 100% if the clay is collected from a seabed or below the groundwater level. The density of water, ρ_w, is approximately 1.0 Mg/m^3, and the density of soil particles, ρ_s, in most clays composed of silicate minerals is approximately 2.7 Mg/m^3 (with a specific gravity, G_s, of 2.7). By substituting these values, the void ratio, e, is calculated as follows:

$$e = \frac{w}{S_r} \cdot \frac{\rho_s}{\rho_w} = \frac{w \cdot G_s}{S_r} = \frac{100 \times 2.7}{100} = 2.7 \tag{11.2}$$

This means that the void ratio of saturated soil with a water content of 100% is approximately $e = 2.7$. This value is practical for estimation purposes. For instance, if the water content is half ($w = 50\%$), the void ratio is also halved ($e = 1.35$). Similarly, the water content is 1/3 ($w = 33\%$), and the void ratio is also 1/3 ($e = 0.9$).

In the shallower regions of seabed clays (typically up to 20 m depth) the natural water content, w_n, and the liquid limit, w_L, are often almost identical. From this empirical rule, it can be deduced that, for instance, in the case of a marine clay with natural water content $w_n = 100\%$, the liquid limit is about $w_L = 100\%$, indicating a high plastic clay.

The effective unit weight γ' for the clay with $w_n = 100\%$ can be expressed as:

$$\gamma' = \frac{(\rho_s - \rho_w) g}{1 + e} \tag{11.3}$$

Substituting the values described above into equation (11.3), the effective unit weight, γ', is estimated as below:

$$\gamma' = \frac{(2.7 - 1.0) \cdot 9.8}{1 + 2.7} = 4.5 \left(\text{kN/m}^3\right) \tag{11.4}$$

It is both important and practical to possess fundamental values, such as the unit weight of marine clay deposits and the effective overburden stress. The effective overburden stress, σ'_{v0}, at a depth of, for instance, 10 m may be estimated as follows:

$$\sigma'_{v0} = \gamma' \cdot z = 4.5 \times 10 = 45 \left(\text{kN/m}^2\right) \tag{11.5}$$

11.2.1.2 Consolidation properties of a clay with a water content of 100%

When evaluating mechanical properties, particularly the consolidation properties of soft clay deposits, it is beneficial to remember fundamental reference values. In the following analysis, a 10-m thick clay deposit with a natural water content of 100% is considered as shown in Figure 11.1. At the site, the water depth is 2 m, and a 4-m thick sand layer has been placed for land reclamation. Consolidation settlement can be approximately predicted based on empirical values.

Based on the natural water content $w_n = 100\%$, the void ratio is estimated to be approximately $e = 2.7$, and the effective unit weight is approximately $\gamma' = 4.5$ kN/m^3, as previously described. Therefore, the effective overburden stress at the midpoint of the clay layer (i.e. $z = 5$ m) is approximately $\sigma'_{v0} = \gamma' \times z = 4.5 \times 5 = 22.5$ kN/m^2. In the shallower layer of the seabed soil, the liquid limit, w_L, is approximately equal to the natural water content, w_n. The following well-established empirical equation expresses the relationship between the compression index, C_c, and the liquid limit, w_L:

$$C_c = 0.009 \cdot (w_L - 10) \tag{11.6}$$

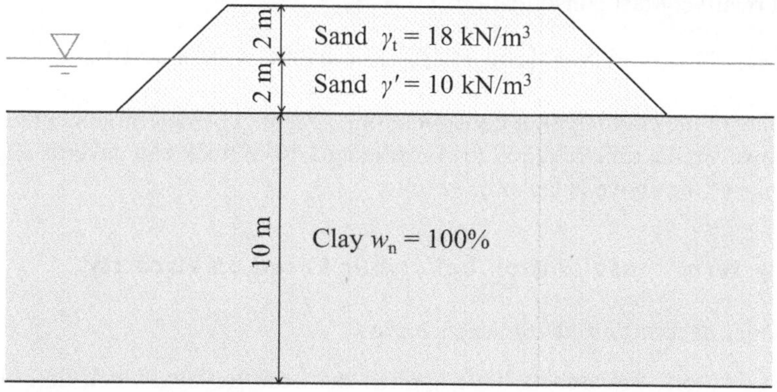

Figure 11.1 Illustration of a simplified case of reclamation on clay deposits.

The constants 0.009 and 10 are useful for European clays, although these values can vary depending on the type of clay. For Japanese marine clays, the C_c value tends to be larger, and with a typical $C_c = 1.0$ for Osaka Bay clay when $w_L = 100\%$.

Sand fill often has a bulk unit weight of approximately 18.0 kN/m³ above the groundwater level and an effective unit weight of approximately $\gamma' = 10.0$ kN/m³ below the groundwater level. The increase of consolidation pressure due to sand fill is calculated at $\Delta p = 2\ \text{m} \times 10.0\ \text{kN/m}^3 + 2\ \text{m} \times 18.0\ \text{kN/m}^3 = 56.0\ \text{kN/m}^2$. To simplify the calculations, the effect of the relative rise in the groundwater level due to consolidation settlement (i.e. the increment in buoyancy acting on the newly submerged part of the sand fill over time) is not considered. The following calculation is based on the initial condition with the largest consolidation pressure increment, which is also a safety assumption to calculate a larger settlement.

Using the above values, the amount of settlement can be calculated as follows:

$$S = \frac{C_c}{1+e} \cdot H \cdot \log \frac{\sigma'_{v0} + \Delta p}{\sigma'_{v0}} = \frac{1.0}{1+2.7} \cdot 10 \cdot \log \frac{22.5+56.0}{22.5} = 1.47\ (\text{m}) \tag{11.7}$$

If the effect of buoyancy on the newly submerged part of the sand fill due to settlement is to be considered, the effective consolidation stress increment should be recalculated multiple times, taking into account the rising groundwater level caused by the settlement. After this, the settlement should be recalculated; it will be slightly smaller than the previously obtained value.

Although the coefficient of consolidation, c_v, cannot be determined without a consolidation test, let us consider the time required for consolidation to reach a degree of 90% consolidation for both slow and fast consolidation cases, assuming $c_v = 10 \times 10^{-9}\text{m}^2/\text{s}$ and $c_v = 100 \times 10^{-9}\text{m}^2/\text{s}$, respectively. These values are also useful to remember as empirical values. For a clay layer with a thickness of 10 m, with a drainage condition on both the top and bottom boundaries, the maximum drainage distance H^* is 5 m. Using the time factor corresponding to a degree of consolidation of 90%, $T_{v90} = 0.848$, the time required up to a degree of consolidation of 90% is predicted by the following calculations:

In the case of relatively slow consolidation:

$$t_{90} = H^{*2} / c_v \times T_{v90} = 5 \times 5 / \left(10 \times 10^{-9}\right) \times 0.848 = 2.12 \times 10^9\text{s} = 67\ \text{year} \tag{11.8}$$

In the case of relatively fast consolidation:

$$t_{90} = H^{*2} / c_v \times T_{v90} = 5 \times 5/\left(100 \times 10^{-9}\right) \times 0.848 = 2.12 \times 10^8 \text{s} = 6.7 \text{ year} \quad (11.9)$$

The actual value of the coefficient of consolidation, c_v, cannot be determined without a consolidation test; however, the time required to consolidate a 10-m thick clay layer is expected to fall within the range of the above calculation results.

11.2.2 Long-term consolidation behaviour based on viscosity

11.2.2.1 Model description of the isotache law

Over the past 70 years, researchers have applied the isotache law, which describes a unique relationship between stress, strain and strain rate, including the strain rate dependency of the compression curve. This law focuses on the effect of viscosity in the long-term consolidation of clay. Studies by Watabe et al. (2008, 2012) and Watabe and Leroueil (2015) conducted long-term consolidation tests and multi-stepwise CRS consolidation tests on various types of worldwide clays, including Kansai International Airport marine clays. They modelled the strain rate dependency in long-term consolidation behaviours. The strain rate dependency has been rheologically modelled using the 'isotache law', which is essentially the same as the 'power creep law in conjunction with overstress concept' in classical mechanics, despite the two being expressed by using different equations. An outline of these models is described as follows.

The isotache law in the compression curve of consolidation behaviour was first identified by Šuklje (1957). The 'isostache law' is rheologically modelled by determining the strain rate dependency of the consolidation yield stress. This is done by fitting the 'shape' of the compression curve, which is normalised by the consolidation yield stress, σ'_p, based on the method proposed by Leroueil et al. (1985). The equations are expressed in terms of viscoplastic strain, ε_{vp}, and described by equations (11.10–11.12):

$$\varepsilon_{vp} = \varepsilon - \varepsilon_e \quad (11.10)$$

$$\frac{\sigma'}{\sigma'_p} = f\left(\varepsilon_{vp}\right) \quad (11.11)$$

$$\sigma'_p = g\left(\dot{\varepsilon}_{vp}\right) \quad (11.12)$$

Watabe et al. (2008) showed that the consolidation yield stress converges to the lower limit, σ'_{pL}, as the viscoplastic strain rate, $\dot{\varepsilon}_{vp}$, decreases, and is proposed in equation (11.13) for the strain rate dependency of the consolidation yield stress, σ'_p:

$$\ln\frac{\sigma'_p - \sigma'_{pL}}{\sigma'_{pL}} = c_1 + c_2 \ln\dot{\varepsilon}_{vp} \quad (11.13)$$

where c_1 and c_2 are constants. Together with the lower limit of consolidation yield stress, σ'_{pL}, these constants are referred to as isotache parameters. Watabe et al. (2012) conducted experiments on a variety of clays and found that the values of the isotache parameter are almost common, regardless of the clay type or condition. The parameters commonly applicable to many clays are $\sigma'_{pL} / \sigma'_{p0} = 0.70$ and $c_1 = 0.935$, where σ'_{p0} is the consolidation yield stress corresponding to $\dot{\varepsilon}_{vp} = 1.0 \times 10^{-7} s^{-1}$, which is roughly equivalent to the strain rate at an elapsed time of 24 hours in the incremental loading oedometer test. Using these values, the parameter c_2 is automatically determined as in equation (11.14):

$$c_2 = \frac{\ln \dfrac{\sigma'_p - \sigma'_{pL}}{\sigma'_{pL}} - c_1}{\ln \dot{\varepsilon}_{vp}} = \frac{\ln \dfrac{1-0.7}{0.7} - 0.935}{\ln\left(1.0 \times 10^{-7}\right)} \tag{11.14}$$

Equation (11.13) models the fact that the consolidation yield stress, σ'_p, converges to the lower limit, σ'_{pL}, as the strain rate becomes infinitesimally small. It can be easily transformed into an equation similar to the power creep law in conjunction with the overstress concept, as expressed in equation (11.15) (Watabe and Leroueil 2015):

$$\dot{\varepsilon}_{vp} = c_3 \left(\frac{\sigma'_p - \sigma'_{pL}}{\sigma'_{pL}} \right)^{c_4} \tag{11.15}$$

where $c_3 = \exp(-c_1/c_2)$ and $c_4 = 1/c_2$. Hereafter, calculations using this method will be referred to as 'calculation with isotache law evaluated by the rheological approach'.

On the other hand, the power creep law in conjunction with the overstress concept in classical mechanics is described as follows: Visco-plastic strain occurs when the current effective stress, σ', is larger than the current yield stress, σ'_y. This yield stress, as described later, corresponds to the lower limit of consolidation yield stress, σ'_{pL}, in the isotache law in the initial state. The model is based on the concept that creep deformation during consolidation will eventually terminate as the structure gradually becomes denser (harder). The strain rate is expressed by equation (11.16):

$$\dot{\varepsilon}_{vp} = a \left(\frac{\sigma' - \sigma'_y}{\sigma'_y} \right)^{b} \tag{11.16}$$

where a and b are constants.

Hereafter, calculations using this method will be referred to as 'power creep law' based on the classical mechanical model. Although formulations of the isotache law and the power creep law differ, as described above, these two laws are essentially the same.

In the above model equations, the compression curves obtained from the tests can be used directly. However, as with most soil constitutive models, the compression curve, expressed in strain, is assumed as a bi-linear relationship expressed by the compression index, λ_ε $(= \lambda/(1+e_0) = C_c/(1+e_0)/\ln 10)$, and the swelling index, κ_ε $(= \kappa/(1+e_0) = C_s/(1+e_0)/\ln 10)$. The parameters for the Pleistocene clay Ma9, collected at the construction site of Kansai International Airport in Osaka Bay in Japan, are shown in Table 11.1. These are representative values obtained in the previous studies and used in the following calculations.

The simulation results of the multi-stepwise CRS consolidation, where the strain rate is decreased / increased sequentially by one order of magnitude for each 1% increment of strain, is shown in Figure 11.2. The strain rate ranges from $3.3 \times 10^{-12} s^{-1}$ to $3.3 \times 10^{-5} s^{-1}$ after the strain reaches 5%. The results, calculated using both the isotache law and the power creep law, are superimposed on the same figure for comparison. However, the results coincide and cannot be distinguished from one another. For the power creep law calculation, it was assumed that $\sigma'_y = \sigma'_{pL}$ in the initial state. That means that σ'_y corresponds to the consolidation yield stress of the

Table 11.1 Input parameters for calculation (c_3 and c_4 are calculated from c_1 and c_2, respectively)

	Kansai International Airport
$\sigma'_{pL}/\sigma'_{p0}$	0.70
σ'_{p0} (kPa)	1408 (for equations 11.13 and 11.15)
σ'_{y} (kPa)	980 (for equation 11.16)
c_3, a	2.127×10^{-4}
c_4, b	9.043
σ'_{v0} (kPa)	1016
e_0	1.420
C_c (λ_ε)	1.000 (0.1790)
C_s (κ_ε)	0.100 (0.0179)

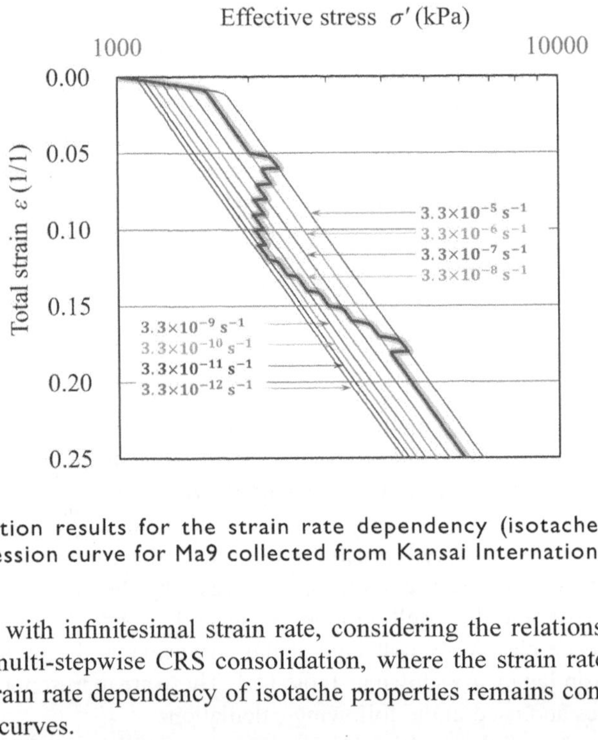

Figure 11.2 Calculation results for the strain rate dependency (isotache properties) of the compression curve for Ma9 collected from Kansai International Airport.

compression curve with infinitesimal strain rate, considering the relationship of $\sigma'_{pL} / \sigma'_{p0}$ (= 0.7). Even in the multi-stepwise CRS consolidation, where the strain rate is stepwise varied sequentially, the strain rate dependency of isotache properties remains consistent with the isotache compression curves.

According to the previous studies, the isotache parameters for marine clays at Kansai International Airport are commonly determined with $\sigma'_{pL} / \sigma'_{p0} = 0.7$, $c_1 = 0.935$, and $c_2 = 0.111$. Figure 11.3 shows an example of calculated results, illustrating the long-term consolidation behaviour. The data is plotted for the creep stage starting at a total strain of 15% in the CRS consolidation test with a strain rate of 1.0×10^{-6}s^{-1}. It can be seen that the secondary consolidation settlement of 6% follows the initial settlement of 15% at the constant strain rate of 1.0×10^{-6}s^{-1}. The time history can be calculated by applying the isotache law.

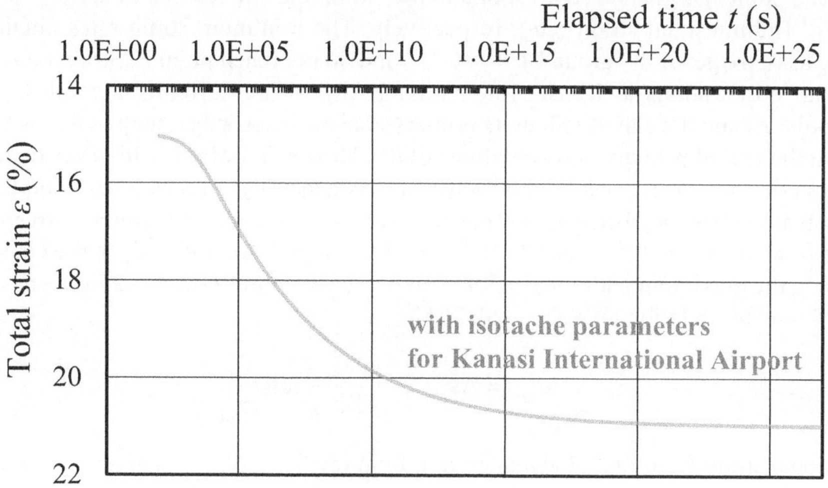

Figure 11.3 Calculated result of long-term consolidation settlement as a relationship between total strain and elapsed creep time.

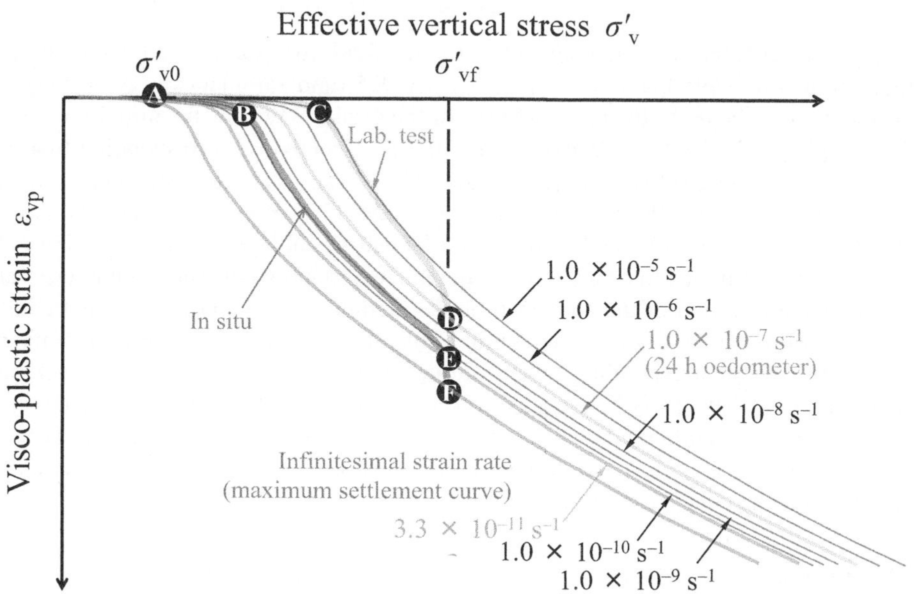

Figure 11.4 Schematic illustration of compression paths for incremental loading consolidation in the laboratory and in situ behaviour. (Watabe et al. 2012.)

A schematic illustration of the compression curve, which would be obtained from a long-term consolidation test in the laboratory ($A \rightarrow C \rightarrow D \rightarrow E$) and in situ ($A \rightarrow B \rightarrow E \rightarrow F$) for an incremental loading from the effective overburden stress, σ'_{v0}, to a vertical pressure, σ'_{vf}, in the post-yield domain is shown in Figure 11.4 (after Watabe et al. 2012). The compression curves corresponding to several strain rates, i.e. isotache compression curves, are superimposed. The curve for the infinitesimal strain rate represents the maximum potential compression of the clay. The points *D*, *E*, and *F* correspond to the strain rates of $1.0\times10^{-7}\mathrm{s}^{-1}$, $1.0\times10^{-11}\mathrm{s}^{-1}$, and $1.0\times10^{-\infty}\mathrm{s}^{-1}$ (infinitesimal strain rate), respectively. The minimum strain rates obtained from laboratory tests are generally around $1.0\times10^{-9}\mathrm{s}^{-1}$ and do not reach point *E*. Strain rates obtained from in situ observations at Kansai International Airport are currently around $3.3\times10^{-11}\mathrm{s}^{-1}$ (close to point *E*) and the clay is still in its primary consolidation stage; they will thus be smaller than this at the end of primary consolidation (EOP). However, point *F* will never be reached.

In engineering practice, consolidation settlement is generally estimated based on the $e-\log\sigma'_v$ curve obtained from a 24-hours incremental loading oedometer test, which corresponds to a strain rate of about $1.0\times10^{-7}\mathrm{s}^{-1}$ (point *D*). Using the compression index C_c at the consolidation pressure σ'_{vf}, the maximum additional settlement $\Delta\varepsilon_{ult}$ (from point *D* to point *F*) can be estimated using the geometric relation (Watabe et al. 2012).

$$\Delta\varepsilon_{\mathrm{ult}}=\Delta\varepsilon_{D\rightarrow F}=\frac{C_c}{1+e_0}\log\frac{\sigma'_{\mathrm{p0}}}{\sigma'_{\mathrm{pL}}}\tag{11.17}$$

The additional strain from *D* to a given strain rate in the field, $\Delta\varepsilon_{Field}$, is given by the following equation.

$$\Delta\varepsilon_{\mathrm{Field}}=\Delta\varepsilon_{D\rightarrow E}=\frac{C_c}{1+e_0}\log\left[\frac{\sigma'_{\mathrm{p0}}}{\sigma'_{\mathrm{pL}}}\left\{\frac{1}{1+\exp\left(c_1+c_2\ln\dot{\varepsilon}_{\mathrm{Field}}\right)}\right\}\right]\tag{11.18}$$

Here, $\dot{\varepsilon}_{\mathrm{Field}}$ represents the strain rate considered in the field, for example, at the end of primary consolidation in situ. Note that $\sigma'_{\mathrm{pL}}/\sigma'_{\mathrm{p0}}=0.7$, $c_1=0.935$ (and automatically $c_2=0.111$) can be used, as mentioned above. In the case of Ma12 of the Osaka Bay clay, the initial void ratio, e_0, and C_c are typically 2.2 and 1.0 (Watabe et al. 2012), respectively, corresponding to a natural water content of around 100%. Using equation (11.17) $\Delta\varepsilon_{D\rightarrow F}$ can be estimated to be 0.048 (4.8%). Furthermore, since the in situ strain rate for Ma12 at Kansai International Airport is typically $3.3\times10^{-11}\mathrm{s}^{-1}$, as evaluated from the settlement of sublayers, using equation (11.18), then $\Delta\varepsilon_{D\rightarrow E}$ can be estimated to be 0.025 (2.5%), which is roughly consistent with observations.

For practical application of the abovementioned method, useful charts to predict additional strain (creep strain) can be drawn as shown in Figure 11.5. Figure 11.5a shows the maximum additional strain $\Delta\varepsilon_{ult}$ corresponding to an infinitesimal strain rate, i.e. $\dot{\varepsilon}_{vp}$ approaching 0.0, from a strain rate of $1.0\times10^{-7}\mathrm{s}^{-1}$ calculated as a function of $C_c/(1+e_0)$ using equation (11.17). Figure 11.5b shows the additional strains in the field $\Delta\varepsilon_{\mathrm{Field}}$ calculated as a function of $C_c/(1+e_0)$ and in situ strain rate $\dot{\varepsilon}_{\mathrm{Field}}$ using equation (11.18). These charts are simple and practical for applying the isotache concept in engineering practice.

11.2.3 Prediction of differential settlement using construction management records

Dredged soil dumped into a reclamation site mainly consists of clay and silt, typically in an ultra-soft state with very high water content. To use the dredged soil deposits as a foundation ground, prefabricated vertical drains (PVDs) are installed to improve the ground and accelerate consolidation, followed by preloading via earth fill or vacuum consolidation. Projects such as the offshore expansion project at Tokyo International (Airport Haneda Airport, Nakada et al. 1997) and the construction of Kita-Kyushu Airport (Terashi and Katagiri 2005) have applied these techniques. Recently, vacuum consolidation technology has also been applied to some dredged soil deposits.

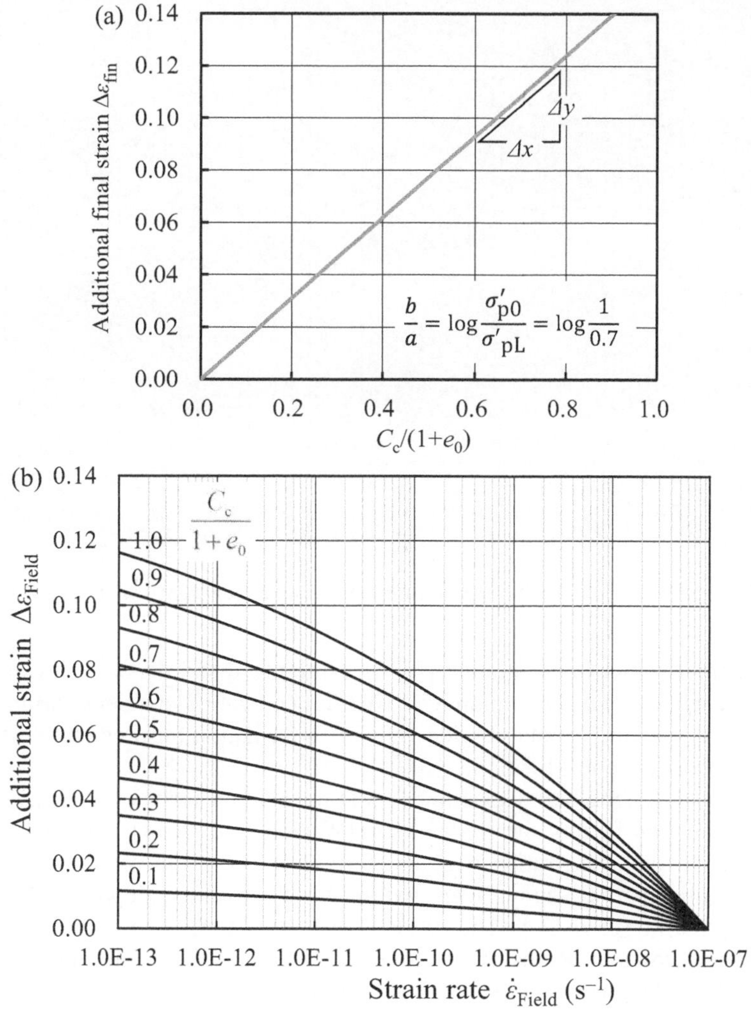

Figure 11.5 (a) Maximum additional strain corresponding to an infinitesimal strain rate as a function of $C_c/(1+e_0)$. (Watabe and Leroueil 2015.) (b) Additional strains in the field calculated as a function of $C_c/(1+e_0)$ and in situ strain rate. (Watabe and Leroueil 2015.)

Dredged soil dumped into a reclamation facility is generally heterogeneous. When reclamation is constructed using hydraulic transportation through pipelines, coarser particles tend to settle around the outlets, while fine particles settle further away, resulting in significant grain size segregation. Applying a preload to the dredged soil deposits may result in unexpected differential settlement as shown in Figure 11.6.

As part of construction management, hydraulic pressure for operating the mandrel in PVD installation is recorded using a data acquisition system. However, the recorded data is generally not used unless there is an operational problem. Given the dense arrangement of PVD installations and the continuity of the data in the depth direction, this dataset could potentially be used to identify partial sandy layers caused by the grain size segregation. By transforming

Figure 11.6 An example of differential settlement due to soil segregation at Kita-Kyushu Airport.

the data into a three-dimensional stratigraphy of the dredged soil deposits, it can be used in a similar way to conduct cone penetration tests (CPTU) conducted in a dense arrangement. This three-dimensional stratigraphy can provide valuable information for predicting differential settlement derived from the consolidation.

The case study described below examines PVD installation for vacuum consolidation conducted in a reclamation facility for dredged soil (Watabe et al. 2014). The hydraulic pressure data set for the mandrel penetration was transformed into a depth profile of the penetration resistance to identify partial sandy layers within the dredged soil deposits. The reclamation facility was located in Tokyo Bay. The original seabed was at a depth of −12 m, and dredging was conducted down to depths of −17 to −23 m. Dredged soils from other sites were then brought in to fill the facility, reaching an elevation of around ±0 m. Below the depth of −17 to −23 m, a very high plastic Holocene clay layer, approximately 20-m thick was deposited as the original seabed.

In the vacuum consolidation process at this site, the surface of the dredged soil was used as a seal layer instead of an impermeable sheet (Fujii et al. 2002, Yoneya et al. 2003, Chai et al. 2008). This ground improvement technology is illustrated in Figure 11.7. The PVDs were not embedded into the sandy bearing layer (below −36 m) to apply the vacuum to their tips. Similar to the tip resistance, q_t, of the CPTU test, the penetration resistance of PVD, p_m, is defined as the penetration force of the mandrel, P, divided by the cross-sectional area of the mandrel tip, A, of 0.02365 m^2.

A cone penetration test (CPTU) was conducted at the centre of working blocks A2 and A3 prior to the PVD installation. The depth profiles of the cone tip resistance of CPTU, q_t, are shown in Figure 11.8, alongside the depth profile of the penetration resistance of the PVD mandrel, p_m, obtained at the nearest locations of CPTU. The depth profile of the mandrel penetration resistance shows a similar trend to that of the cone tip resistance. For example, a high p_m value corresponds to a high q_t value, indicating the presence of a sandy layer. These results indicate

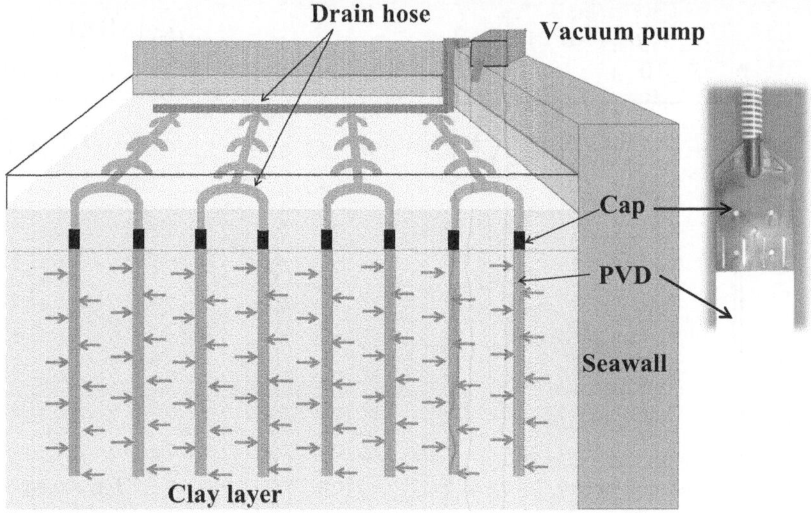

Figure 11.7 Illustration of ground improvement by PVD with vacuum consolidation. (Watabe et al. 2014.)

that the depth profile of the penetration resistance of the mandrel, p_m, can be used to identify soil types.

Partial sandy layers identified from the depth profile of penetration resistance are mapped in Figure 11.9. In this figure, the left- and right-hand sides correspond to the working blocks A2 and A3, respectively. Given the significant heterogeneity of the dredged soil deposit, a high-resolution, three-dimensional distribution of the partial sandy layers aligned with the pitch of the drain installation was considered valuable information for calculating differential settlement.

The planar distribution of settlement by vacuum consolidation (including self-weight consolidation) obtained by comparing the surface profiles levelled by bathymetry measured before and after the vacuum consolidation is shown in Figure 11.10. This compares with Figure 11.9. It is seen that the regions with the thicker sandy layer correspond to the smaller consolidation settlement.

The consolidation settlement across the entire A2 and A3 area was calculated, and the planar distribution of surface settlement is shown in Figure 11.11. The calculation details are described in Watabe et al. (2014). The calculations were carried out at intervals of 3.6 m (as the PVD installation arrangement was 1.8 m in a square pattern, and data were alternately used), whereas the bathymetry observation data were collected at every 10 m (Figure 11.10). Since the resolution of planar distributions differs between Figures 11.10 and 11.11a, these two figures present different impressions and cannot be directly compared. To address this, the calculated planar distribution was averaged at every 10.8 m (= 3 × 3.6-m square) as shown in Figure 11.11b and compared to Figure 11.10.

A very similar planar distribution is observed in Figures 11.10 and 11.11b. This indicates that the three-dimensional distribution of the partial sandy layers can be identified using the penetration resistance of the mandrel during PVD installation, and this three-dimensional distribution is highly useful for predicting the planar distribution of consolidation settlement, i.e. differential settlement.

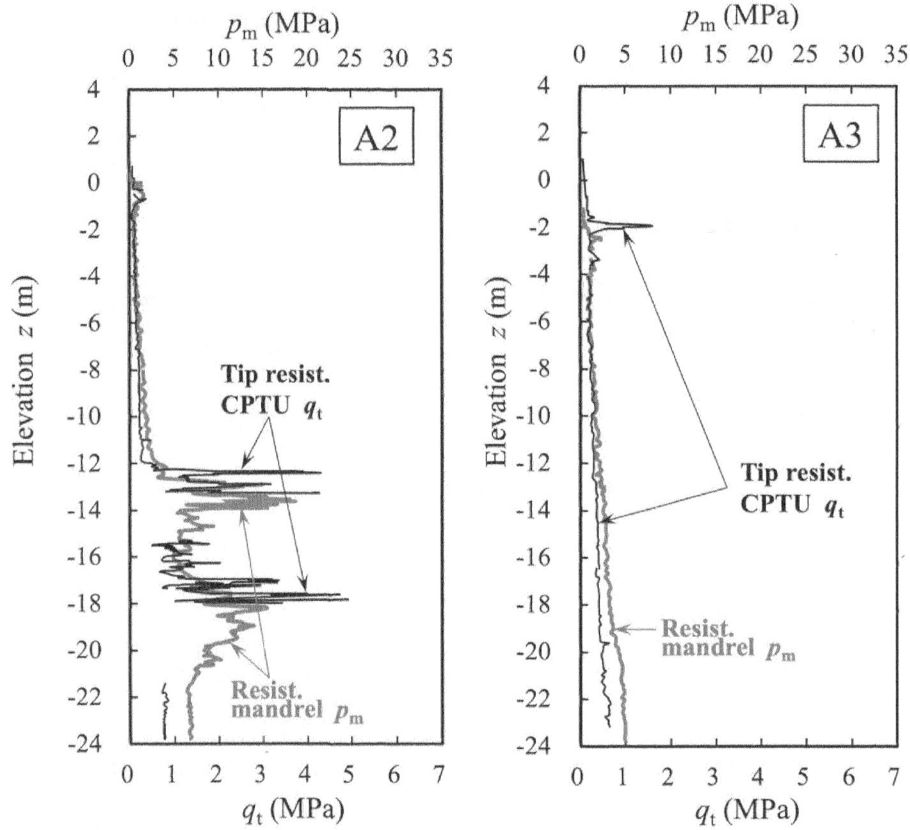

Figure 11.8 Comparison between penetration resistance of mandrel in PVD installation (p_m) and cone tip resistance of CPTU (q_t). (Watabe et al. 2014.)

11.3 Ground improvement technologies

Ground improvement is necessary for transforming soft soil deposits into a suitable foundation ground for construction. The major objectives of ground improvement are classified into four categories, as outlined below. Items (i), (ii), and (iii) are primarily applicable to clayey soil deposits, while item (iv) applies mainly to sandy soil deposits.

i Increasing shear strength: This enhances the stability of a structure
ii Reducing compressibility: This helps reduce residual settlement of a structure
iii Lightening and strengthening: This improves both the stability and reduces residual settlement of a structure
iv Increasing liquefaction resistance: This contributes to anti-seismic design of facilities

To achieve the objective 'Increasing shear strength', there are three main principles (Sections 11.3.1–11.3.3), to achieve the objective 'Reducing compressibility', there are two main principles (Sections 11.3.4 and 11.3.5), to achieve 'Lightening and strengthening', there is one main principle (Section 11.3.6), and to achieve 'Increasing liquefaction resistance' in sandy soil deposits, there are three main principles (Sections 11.3.7–11.3.9):

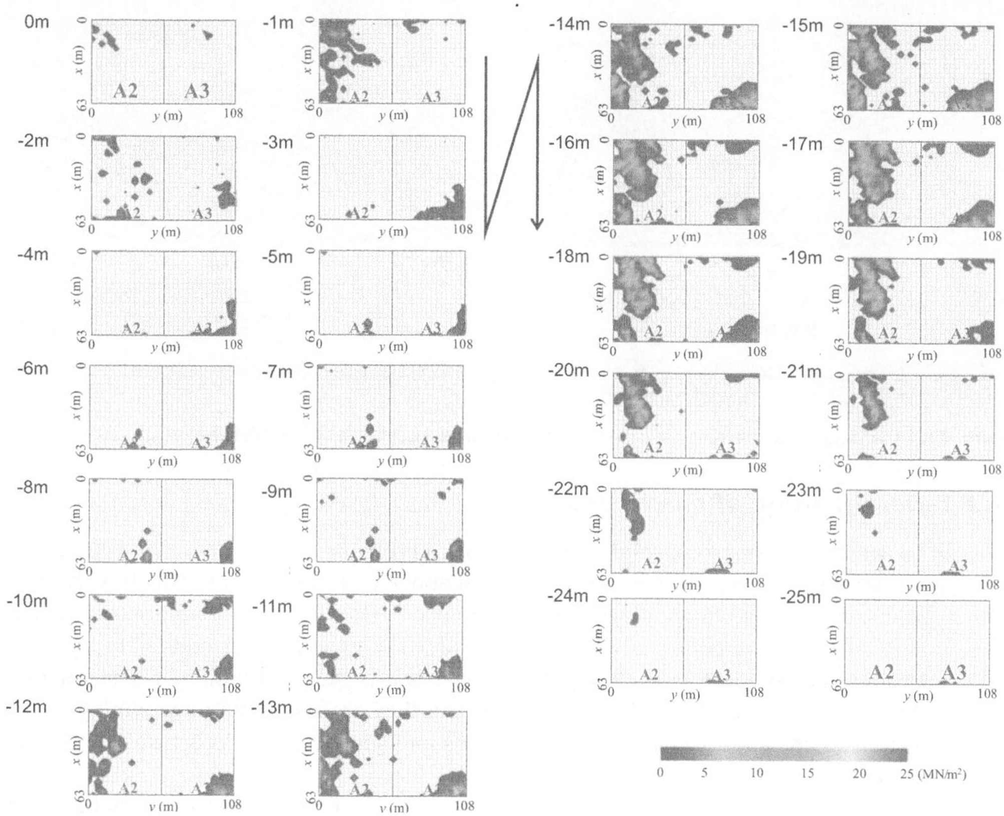

Figure 11.9 Distribution of identified partial sandy layers at every 1-m depth. (Watabe et al. 2014.)

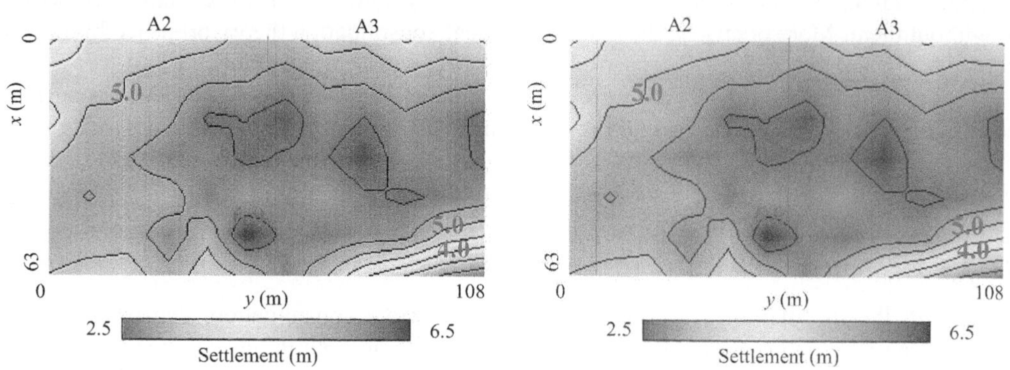

Figure 11.10 Observed planar distribution of settlement by vacuum consolidation. (Watabe et al. 2014.)

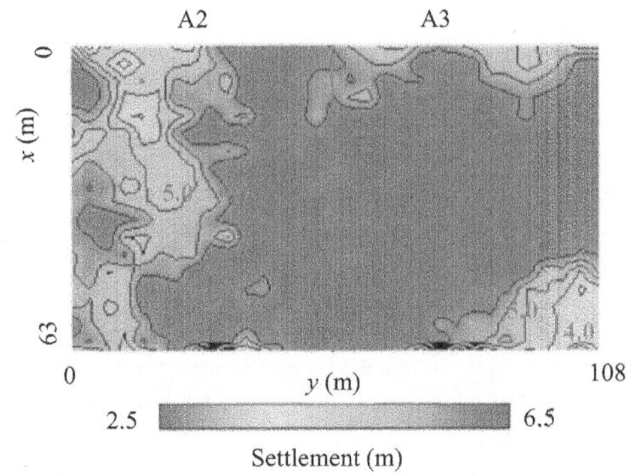

(a) Calculated at intervals of 3.6 m

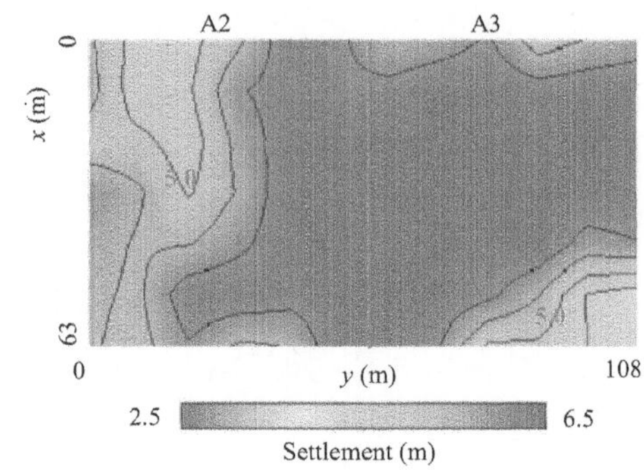

(b) Averaged at every 10.8 m

Figure 11.11 Planar distribution of calculated surface settlements. (Watabe et al. 2014.)

11.3.1 Densification of clayey deposits

This can be achieved through consolidation; however, a problem is that it generally takes a very long time, sometimes several decades, as mentioned above (see equation (11.8)). To solve this, vertical drains are installed in the clayey deposits to accelerate the consolidation process. Two types of vertical drains are commonly used worldwide. The most popular method today is the PVD as an industrial product, while the conventional one is the sand drain (SD) as a natural material. Previously, SD had been often used; however, they are less commonly employed now due to restrictions on sand extraction from riverbeds and sand mining. As a result, PVDs are now preferred.

Vertical drains (whether SD or PVD) are installed in clayey soil deposits to shorten the maximum drainage distance for consolidation (see Figure 11.7). For example, in the case of reclamation above a 10-m thick clay deposit (see Figure 11.1), the maximum drainage distance is 5 m without vertical drains. When vertical drains are installed, the maximum drainage distance is reduced to approximately 0.5 m. As a result, the required consolidation time, for example, up to a certain degree, 90% consolidation, becomes 100 times shorter with vertical drains compared to without them. More precisely, the axi-symmetrical consolidation theory proposed by Barron (1948) is applied to calculate the consolidation process with vertical drains.

A scene of the installation of SD in offshore construction work and that of PVD in onshore construction work are shown in Figures 11.12 and 11.13, respectively. In the case of ground improvement for muddy soil deposits, such as dredged clay dumped into a waste-soil reclamation facility, PVDs are used to ensure its continuity.

11.3.2 Replacement with good material

Even with the installation of vertical drains to accelerate consolidation, densifying clayey deposits can take a long time, often several years, including both the installation of the vertical drains and the consolidation process. Additionally, the undrained shear strength, c_u, which is determined by the effective overburden stress, remains an issue before the construction of a facility. To address this problem, replacing the clayey material with sand is one of the solutions, as the drained shear strength can be used to evaluate the stability of the soil. The shear strength increases as the construction progresses. The displacement method was initially used, but it was difficult to control the cross-section. Eventually, dredging the clay deposit and filling it back

Figure 11.12 A scene of the installation of SD in offshore construction work.

Figure 11.13 A scene of the installation of PVD in onshore construction work (inset: cross-section of PVD).

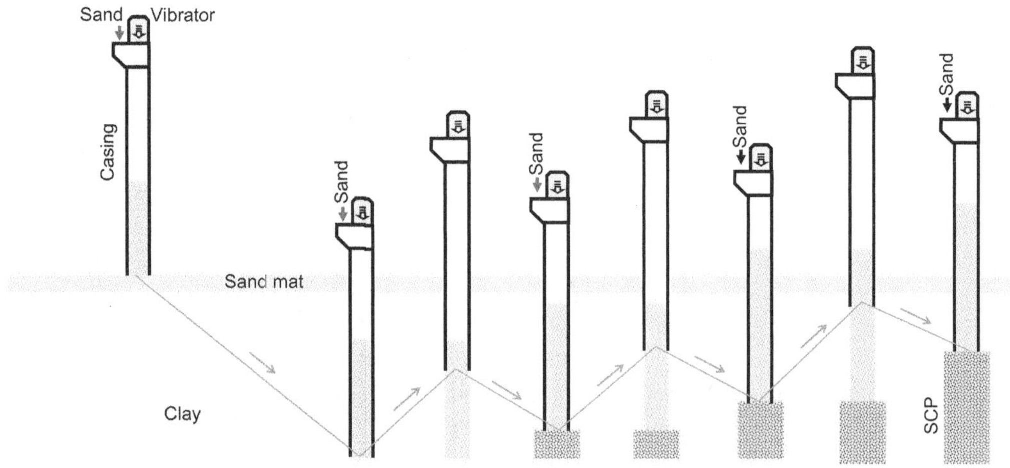

Figure 11.14 Illustration of the procedure of SCP installation.

with sand became a more common practice. However, dredging caused turbidity issues in the marine environment. Today, the sand compaction pile (SCP) method is the most widely used technology for replacing sand. The SCP can be installed without causing turbidity, and it allows for better control over the cross-sectional shape of the structure.

The procedure of SCP installation is illustrated in Figure 11.14. The casing pipe, with a diameter (0.4–1.2 m) and filled with sand at the tip, is vibrated into the ground to the target depth. The casing pipe is then pulled up with vibration, allowing the sand to be released. Following this, the casing pipe, still filled with sand, is vibrated back down to compact the sand into a larger diameter, ranging from (0.7–2.0 m). This process is repeated until the ground surface is reached. A scene of the installation of SCP in offshore construction work and that in onshore construction work are shown in Figures 11.15 and 11.16, respectively.

11.3.3 Solidification

In situ solidification by using cement (or lime) is an effective technique for achieving high shear strength. The solidification is relatively quick, typically taking a few weeks. Shear strength can be controlled by adjusting the mix proportions. When solidification is applied to deeper layers, it is referred to as a deep mixing method (DMM) or cement deep mixing method (CDM). For shallower layers, it is known as the shallow mixing method, which is mainly used to ensure trafficability for ground improvement machines. In the case of marine sediments, blast furnace cement is preferred over ordinary Portland cement to prevent elution of hexavalent chromium.

The procedure of DMM column installation is illustrated in Figure 11.17. Mixing blades penetrate the ground to the target depth with rotation, and then they withdraw upward. During this procedure, cement slurry (or sometimes cement power) is injected either during penetration or during withdrawal. An installation of a DMM column in offshore construction work and an example of a mixing blade system are shown in Figures 11.18 and 11.19, respectively. Typically, machines used for DMM have at least two shafts rotating in opposite directions to counterbalance the rotational forces. DMM columns are usually arranged in an overlapping pattern, forming either a solid block or an aligned wall. Due to the high strength of these structures, DMM improvements are designed as stable reinforcement within the ground.

Figure 11.15 A scene of the installation of SCP in offshore construction work.

Figure 11.16 A scene of the installation of SCP in onshore construction work.

In addition to the previously mentioned in situ solidification, pre-mixed solidification is also a useful technique, particularly when using cement-treated soil as backfill material. The technology is named cement-treated soil. It involves mixing dredged clay with cement before placement on site. Due to the high water content of the dredged clay, cement-treated soil is considered a lightweight soil. If further lightness is required, air-foam can be added to the

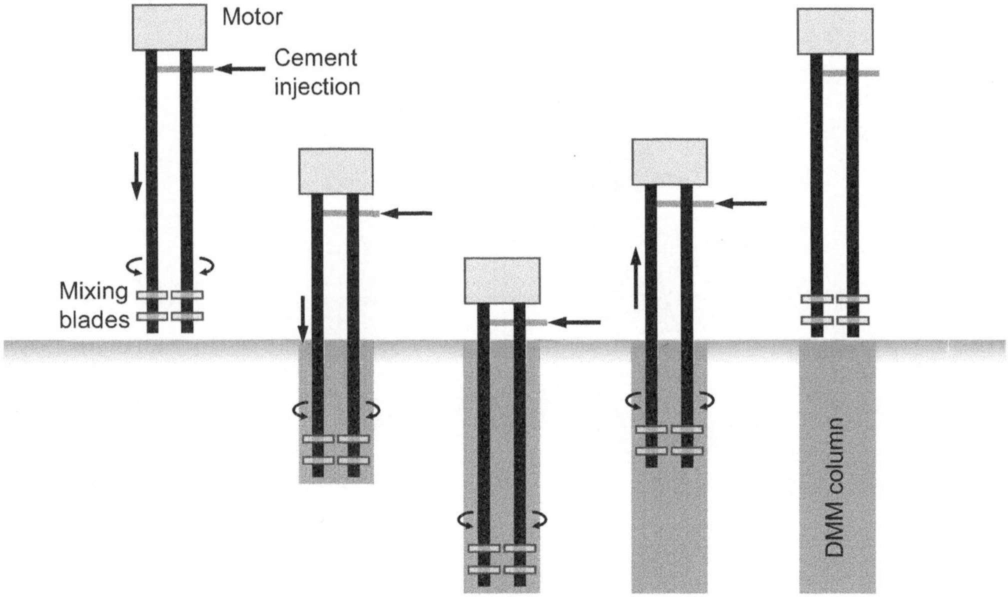

Figure 11.17 Illustration of the procedure of a DMM column installation.

Figure 11.18 A scene of the installation of a DMM column in offshore construction work.

cement-treated soil mixture, creating air-foam-treated lightweight soil. These will be described in more detail later.

Figure 11.19 An example of a mixing blade system.

11.3.4 Densification of clayey deposits

Densification is the most effective ground improvement technology for minimising residual settlement. For clayey soil, densification is primarily achieved by consolidation. Therefore, many of the ground improvement technologies used to reduce compressibility are common with or are the same as those used to increase shear strength, such as SD and PVD, which accelerate consolidation. However, to effectively reduce compressibility, preloading is required to create an over-consolidated state. This is typically achieved using either earth filling or vacuum preloading.

For natural clay deposits, both SD and PVD can be used as vertical drains; however, for dredged clay dumped at a waste-soil reclamation facility, PVD is preferred, as it ensures continuity even in slurry state clay deposits.

11.3.5 Solidification

Lightweight soils, such as cement-treated soil and air-foam-treated soil, are useful to reduce residual settlement. When used as backfill, they help minimise the incremental consolidation pressure caused by earth filling or land reclamation.

The key principle for achieving both 'lightening and strengthening' is solidification through cement hydration. Cement-treated soil plays a crucial role in this process.

11.3.6 Cement-treated soils

The effective utilisation of dredged clay is a growing social demand, as it has become increasingly difficult to find a suitable site for disposal facilities for dumping dredged soils. The most common technology is to mix dredged clay with cement and use it as backfill or embankment material. The shear strength can be controlled by adjusting the mix proportion. Due to the high water content of dredged clay, cement-treated soil is lighter compared to normal soils. The cement-treated soil is not always classified as a lightweight soil, in many cases; however, it should be considered as one.

For large-scale reclamation, using a conventional mixing plant is inefficient. A more effective method is the pneumatic flow mixing method, as illustrated in Figure 11.20. In this method, mud slurry is transported through the pipeline, but mud slurry is divided by pressurised air. Therefore, the mud slurry is always making plug flow. During transportation through the pipeline, cement fluid is also injected into the slurry. Just before placing the mixture at the construction site, pressurised air must be removed. This method has been successfully used in major projects, including the northern section of Chubu Centrair International Airport (Chubu Airport), where cement-treated soil served as the main reclamation material (Kitazume and Satoh 2003), and the D-runway of Tokyo International Airport (Haneda Airport) where it was used as backfill for the seawall (Watabe and Noguchi 2011).

If further weight reduction is needed, a lightweight soil mixture consisting of dredged clay, cement, and air-foam, known as air-foam-treated lightweight soil, is highly effective. For coastal construction, the bulk density must be slightly higher than that of water. Because quality control for the air-foam-treated lightweight soil, particularly in its bulk density, is very sensitive, a plant for mixing is generally used for high-level quality control. Therefore, the pneumatic flow mixing method is not used for the air-foam-treated lightweight soil. Notable large-scale constructions that have used **air-foam-treated lightweight soil** include the restoration of the caisson-type quay wall at Kobe Port Island, which was damaged by the Hanshin-Awaji earthquake (Figure 11.21, Watabe et al. 2009), and the interface structure connecting the piled-pier section and reclamation section of the D-runway at Tokyo International Airport (Haneda Airport, Figure 11.22, Watabe and Noguchi 2011).

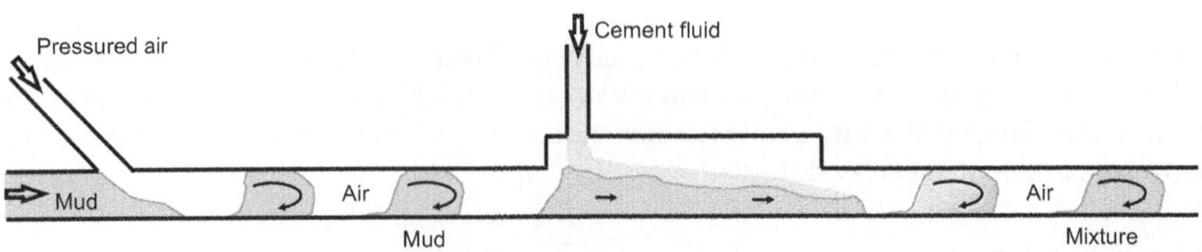

Figure 11.20 Illustration of the pneumatic flow mixing method.

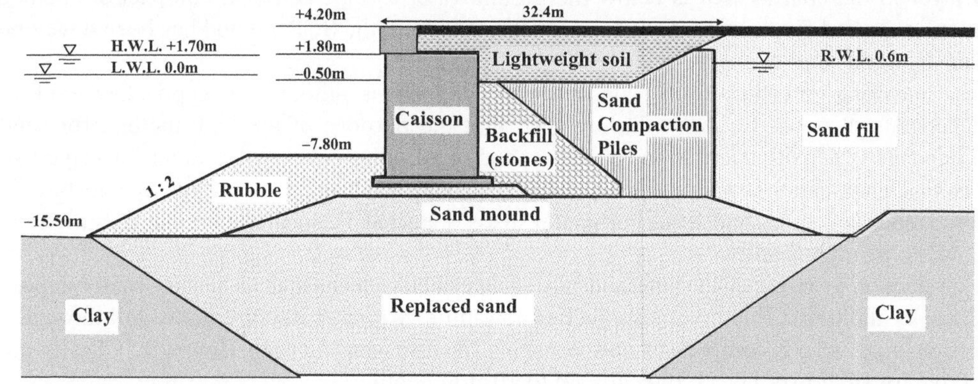

Figure 11.21 A cross-section of the restoration work of the quay at Kobe Port Island. (Watabe et al. 2009.)

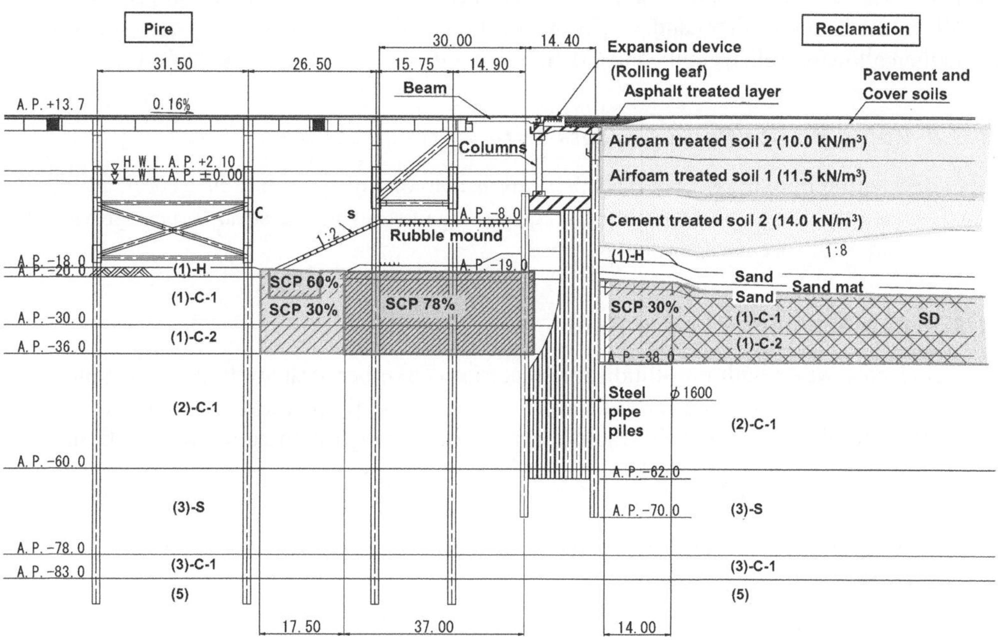

Figure 11.22 Cross-section of the joint structure between the reclamation and pier sections. (Watabe and Noguchi 2011.)

11.3.7 Densification for sand

Densification is an effective method for preventing liquefaction, as dense soil exhibits positive dilatancy in shear behaviour. Compaction and consolidation are useful geotechnologies to densify sandy soil deposits and clayey soil deposits, respectively. However, compaction is only available for unsaturated sandy soil above the groundwater level. Since the liquefiable sandy soil layer in the coastal area is below the groundwater level, traditional compaction methods cannot be applied for densification. To solve this problem, the SCP method has been developed to densify sandy soil deposits.

The installing procedure for SCP in sandy soil deposits is similar to its application in clayey soil deposits. It needs to be noted, however, that the purpose of the SCP method for sandy soil deposits is densification of surrounding sandy soil rather than replacement for clayey soil deposits. Densification in sandy soil is achieved through the displacement-type SCP installation method, and then an increment of horizontal stress due to SCP installation remains in the sandy soil due to its high rigidity.

Compaction grouting (CPG) method is also an effective technique to densify sandy deposits (Adachi et al. 2010). This involves injecting low-fluidity mortar into the sandy soil through an injection pipe, which compresses and densifies the surrounding soil. However, CPG is only effective for shallower layers, typically up to 10 m in depth.

The vibro-floatation (VF) method is also a well-known technique for densifying sandy soil deposits. However, its mechanism differs significantly from that of SCP. The SCP method compacts sand columns by displacing the original sandy soil through vibration, increasing densification and horizontal earth pressure in the in situ sandy soil. CPG is also a similar principle to this. The VF method, on the other hand, works by inserting a vibrator into the sandy deposit, liquefying it, and then allowing self-weight consolidation to compact the soil after the vibrator is removed.

11.3.8 *Rapid dissipation of excess pore water pressure*

Rapid dissipation of excess pore water pressure is also one of the solutions to prevent liquefaction. Because very high permeability is required against strong seismic forces, gravel drains (GD) instead of SD are used for this purpose.

11.3.9 *Replacement of pore fluid*

Replacing pore water with non-fluidisable material is another strategy to prevent liquefaction. Silica glass is commonly used for this purpose, as it can easily permeate soil pores and prevent fluidisation. Replacement of pore water is carried out through permeation or injection, a technique known as chemical grouting.

11.4 Integration of ground improvement technologies—a case study

In the previous sections, various ground improvement technologies have been described, along with their principles and applications. However, in real construction design, it is necessary to consider that the right ground improvement method is applied in the right location. A typical example of this can be seen in the cross-sections of the D-runway at the Tokyo International Airport (Haneda Airport). This hybrid structure consists of both a reclamation section and a piled-pier section (Watabe and Noguchi 2011). The construction site was partly located within the mouth of the Tama River, where reclamation was prohibited due to environmental regulations. The ground improvement technologies applied to the reclamation section can be classified into three principles:

a Densifying the clay deposits by consolidation with vertical drains (SD: sand drain)
b Replacing the clay deposits with sand (SCP: sand compaction pile)
c Strengthening the clay deposits through solidification using the DMM method (CDM).

Typical cross-sections are illustrated in Figures 11.23 and 11.24. Figure 11.23 shows a general cross-section of the mild slope rubble seawall, the soft clay deposits inside the reclamation section are improved by SD to accelerate consolidation. Under the rubble seawall is improved by SCP with a replacement area ratio of 30%. Due to the low replacement area ratio (30%), the SCP columns also served as vertical drains. Figure 11.24 shows a seawall with an abutment for the approach light bridge. The soft clay deposits inside the reclamation section were improved using SD. Under the rubble seawall with the abutment, DMM was applied for greater stability. Since DMM is an expensive ground improvement method, SCP with a low replacement area ratio was installed on both sides of the DMM region to optimise cost and efficiency. In both cross-sections, cement-treated soil was used as backfill material through the pneumatic flow mixing method, ensuring shear strength while maintaining lightness. Additionally, the use of cement-treated soil facilitated the recycling of dredged clay, reducing the demand for sand as a reclamation material.

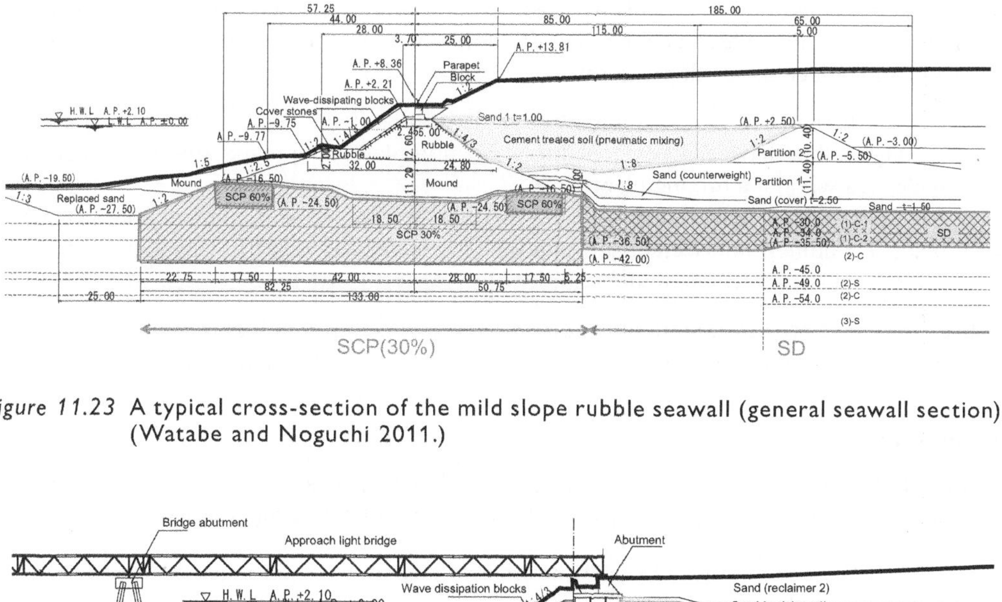

Figure 11.23 A typical cross-section of the mild slope rubble seawall (general seawall section). (Watabe and Noguchi 2011.)

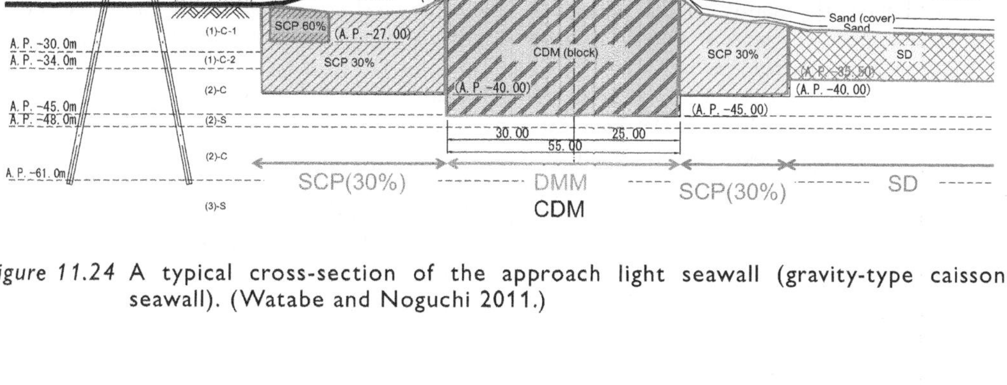

Figure 11.24 A typical cross-section of the approach light seawall (gravity-type caisson seawall). (Watabe and Noguchi 2011.)

11.5 Conclusion

The ground improvement technologies introduced in this chapter can be summarised as below:

1. As conventional ground improvement methods for land reclamation, three major technologies are based on three different principles:
   - i Combination of vertical drains (SD/PVD) and preloading to accelerate consolidation settlement. Prediction of differential settlement had been impossible in the previous days; however, in the case of PVD, it is predictable today by using construction management data (penetration resistance of PVD).
   - ii SCP to replace soft soil with compacted stiff sand columns. Note here that the main objective of SCP installation is replacement for clayey ground and densification for sandy ground, respectively.
   - iii DMM to solidify soft soil by in situ mixing with cement/lime. The target shear strength is much higher than the following premix soil.
2. Cement-treated soil (premix soil) is an advanced technology to conduct rapid construction and provide a value-added geomaterial. The following two technologies are widely used as backfills to reduce the earth pressure:
   - i Cement-treated soil which is a mixture of dredged soil and cement. The pneumatic flow mixing method has been developed to efficiently place the premix soil into the site in a wide area without a mixing plant.
   - ii Air-foam-treated lightweight soil is a mixture of dredged soil, cement and airfoam to place a solidified lightweight soil into the site, mainly as backfill. Lightness is an additional value to contribute to the stability of the seawall structure.
3. In a large-scale project, integration of ground improvement technologies is important in consideration of that the right ground improvement method is applied in the right location.

References

Adachi, M., Shinsaka, T., Kurosaka, M. and Zen, K. 2010. Case history of the compaction grouting method as countermeasure against liquefaction. Geotechnical Engineering Magazine, 58(11), 26–29. (in Japanese)

Barron, R.A. 1948. Consolidation of fine-grained soils by drain wells. Transactions of American Society of Civil Engineers, 113, 718–754.

Chai, J.C., Miura, N. and Bergado, D.T. 2008. Preloading clayey deposit by vacuum pressure with cap-drain: Analyses versus performance. Geotextiles and Geomembranes, 26(3), 220–230.

Fujii, A., Tanaka, H., Tsuruya, H. and Shinsha, H. 2002. Field test on vacuum consolidation method by expecting upper clay layer as sealing-up material. Proceedings of the Japanese National Symposium on Recent Research and Practice on Clayey Grounds – From Observation of Microstructure to Ground Improvement for Ultra-Soft Clays, Japanese Geotechnical Society, Tokyo, 269–274. (in Japanese)

Kitazume, M. and Satoh, T. 2003. Development of pneumatic flow mixing method and its application to Central Japan International Airport construction. Ground Improvement, 7(3), 139–148.

Leroueil, S., Kabbaj, M., Tavenas, F. and Bouchard, R. 1985. Stress-strain-strain rate relation for the compressibility of sensitive natural clays. Géotechnique, 35(2), 159–180.

Nakada, H., Akimoto, K., Kanazawa, H., Tsuji, Y. and Inada, M. 1997. Haneda Airport offshore expansion project. Proceedings of the ICE – Civil Engineering, 120(3), 90–100.

Šuklje, L. 1957. The analysis of the consolidation process by the isotache method. Proceedings of the 4th International Conference on Soil Mechanics and Foundation Engineering, London, Vol. 1, 200–206.

Terashi, M. and Katagiri, M. 2005. Key issues in the application of vertical drains to a sea reclamation by extremely soft clay slurry. Ground Improvement—Case Histories, Elsevier *Geo-Engineering Book Series*, Edited by Indraratna and Chu, Elsevier, Vol. 3, 119–143.

Watabe, Y. and Leroueil, S. 2015. Modelling and implementation of isotache concept for long-term consolidation behavior. International Journal of Geomechanics, ASCE, 15(5), A4014006. https://doi.org/10.1061/(ASCE)GM.1943-5622.0000270

Watabe, Y. and Noguchi, T. 2011. Site-investigation and geotechnical design of D-runway construction in Tokyo Haneda Airport. Soils and Foundations, 51(6), 1003–1018. https://doi.org/10.3208/sandf.51.1003

Watabe, Y., Shinsha, H., Yoneya, H. and Ko, C. 2014. Description of partial sandy layers of dredged clay deposit using penetration resistance in installation of prefabricated vertical drains. Soils and Foundations, 54(5), 1006–1017. https://doi.org/10.1016/j.sandf.2014.09.006

Watabe, Y., Tanaka, M., Saegusa, H., Shinsha, H. and Tsuchida, T. 2009. Long-term properties of air-foam-treated lightweight soil made from dredged clay. *Journal of ASTM International*. https://doi.org/10.1520/JAI102134

Watabe, Y., Udaka, K. and Morikawa, Y. 2008. Strain rate effect on long-term consolidation of Osaka bay clay. Soils and Foundations, 48(4), 495–509. https://doi.org/10.3208/sandf.48.495

Watabe, Y., Udaka, K., Nakatani, Y. and Leroueil, S. 2012. Long-term consolidation behavior interpreted with isotache concept for worldwide clays. Soils and Foundations, 52(3), 449–464. https://doi.org/10.1016/j.sandf.2012.05.005

Yoneya, H., Shiina, T. and Shinsha, H. 2003. Improvement of soft ground applying vacuum consolidation method by expecting upper clay layer as sealing-up material. Soft Ground Engineering in Coastal Areas (Proceedings of the Nakase Memorial Symposium, Yokosuka, Japan, 2002), 275–293.

Chapter 12

Underpinning and strengthening of an existing port infrastructure to allow the installation of a high-capacity vessel synchrolift

Paulo Matias Ramos

12.1 Introduction

The naval base described in this project includes a naval repair sector occupying approximately 16 ha. To support the maintenance and repair of the operational vessels, the base was equipped with a lifting platform capable of handling ships up to 3,000 tons, facilitating their transitions between the sea and the port platform.

Over time, deformations in the quay wall structures and corrosion in the steel and electromechanical elements of the lifting platform prompted the responsible authorities to initiate an international tender for rehabilitation and reinforcement solution.

The main objective was to facilitate the installation of a new 9,000-ton capacity synchrolift, aligning with the size and capacity demands of the existing operational vessels. The key requirements outlined in the tender specifications included the following:

- The design of onshore and offshore reaction structures, to support a vessel-lifting platform measuring 25 m × 145 m and a 9,000-ton capacity.
- Reinforcement of the current quay walls, enabling the dredging procedures ranging from 3 to 8 m in depth within the new berthing quay and the new lifting platform areas.
- Development of a 145 m length transference zone for ship transfer operations between the lifting platform and five designated parking areas for vessel construction, repair, and maintenance, with lengths ranging from 100 to 150 m.
- Design of a 623 m circulation path for two overhead cranes with a lifting capacity of 40 tons, at a 50 m distance.
- Construction of support and control buildings, as well as extensive network grid systems and operational equipment.

The thorough analysis of these requirements, considering the available areas and the existing constraints, resulted in the occupational layout shown in Figure 12.1.

12.2 Soil characterization and evaluation of existing port infrastructures

The assessment of the most suitable geotechnical/structural solutions to overcome the tender requirements began with the evaluation of the existing infrastructures and of the geomechanical characteristics of the soil. Figure 12.2 shows the stability design analysis of the existing quay

DOI: 10.1201/9781003530220-15

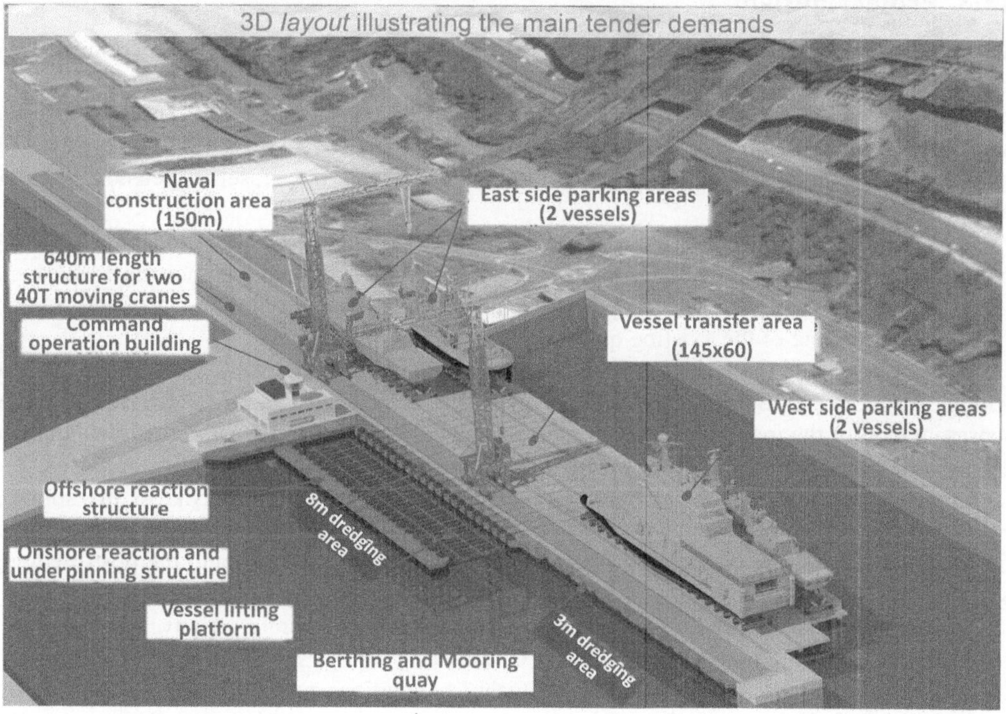

Figure 12.1 3D image of the conceived intervention layout.

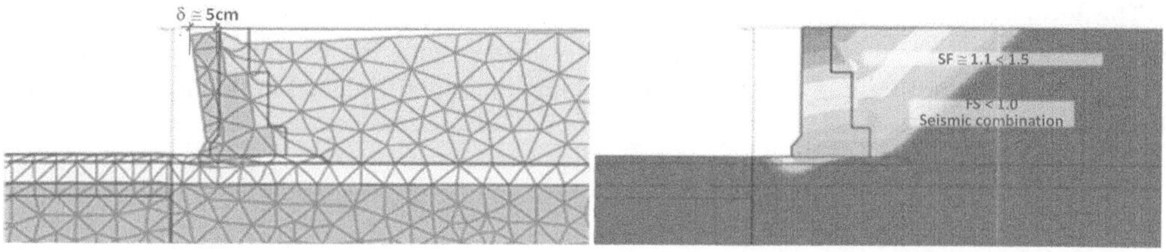

Figure 12.2 Numerical simulation of the existing quay structures (SF ≈ 1.1 => very low).

walls, whose base lies at the seabed level (−10.00 ZH). This assessment was based on a comprehensive geotechnical survey, including 25 boreholes with various characterization procedures, complemented by laboratory tests.

The data collected was summarized in geotechnical reports, which formed the basis for the interpretation and assignment of geomechanical parameters used in subsequent calculations. The survey results identified the following soil layers.

- C2 – Heterogeneous port backfills: Ø'=25°; $c'=5$ kN/m^2; $\gamma=19$ kN/m^3;
- C3A – Clay with sand intercalations (15<NSPT<30): Ø'=25°; c'=30 kN/m^2; $\gamma=20$ kN/m^3;
- C3B – Silty clays (30<NSPT<60): Ø'=25°; $c'=50$ kN/m^2; $\gamma=20$ kN/m^3;
- C4 – Compact marls (NSPT>60): Ø'=30°; $c'=50$ kN/m^2; $\gamma=20$ kN/m^3.

12.3 Project design

The assessment of the existing quay, revealing its imminent rupture condition, confirmed the complexity of the tender demands. In addition to rehabilitating, retrofitting and strengthening a highly sensitive infrastructure, with poor safety conditions, to comply with current regulation standards, this project involved the reinforcement of the port's backfill and of the pre-existing quay to cope with the substantial load increase resulting from the vessel suspension, transfer, and parking operations.

Furthermore, the project vessel draft specifications, the height (thickness) of the new lifting platform and the characteristics of the new transfer system (including supporting trestles and bogies), dictated the necessity to deepen the seabed by 8 m, from −10.00 to −18.00 ZH, requiring the underpinning of the existing quay walls.

Even though the singularities of the synchrolift operations impose a wide range of load actions whose description is beyond the scope of this chapter, Figure 12.3 provides a visual representation of the maximum loads managed by all 38 suspension winches (19 onshore and 19 offshore) during the new vessel-lifting operations.

Alongside the load increment due to vessel-lifting and transfer procedures, the solution needed to address the substantial ≈21.2 m earth pressure differential arising from the differential between the level of the elevation of the new port platform (+3.20 ZH) and the post-dredging seabed level (−18.00 ZH).

Furthermore, it is essential to emphasize the significance of seismic activity, a decisive factor in the design for Ultimate Limit States. Considering all the specified criteria and following extensive research, evaluating numerous technical solutions and construction scenarios, the analysis led to the implementation of an integrated structural approach, where the new transfer and parking areas function as integral load-bearing components, along with the existing quay walls and their innovative underpinning solution.

The designed underpinning structure consists of a retaining wall curtain formed by Ø1,200 mm cast-in-place bored piles with a 1.35 m spacing, acting with the complementary stabilizing resistance of the existing quay walls and by a plain concrete fill placed in the 6 m gap between these two structures.

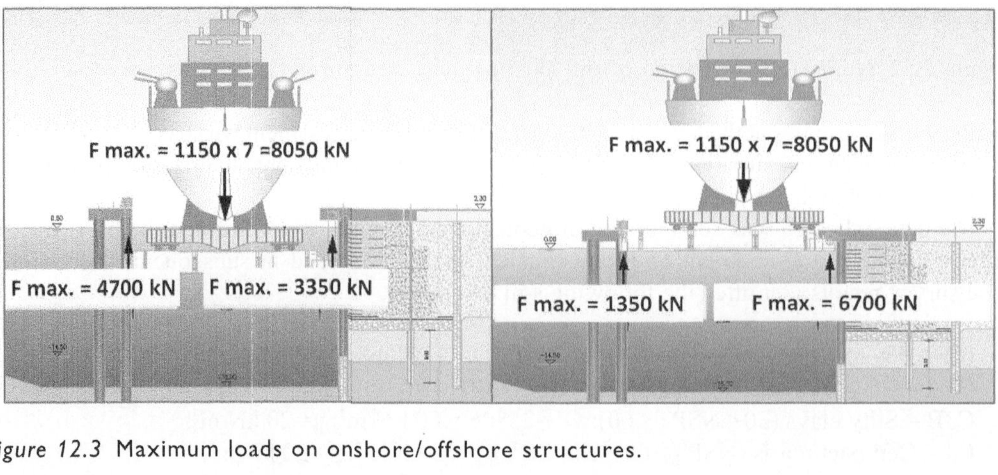

Figure 12.3 Maximum loads on onshore/offshore structures.

The massive loadings resulting from ship transfer and parking led to the use of a transfer system carried out by bogies, circulating on rail tracks positioned over a grid of reinforced concrete beams (2.0 m × 1.8 m). Along with the reinforced concrete platform slabs, this grid of high-capacity beams forms a horizontal diaphragm element that is supported by Ø1,500 mm bored piles, reaching depths compatible with suitable penetration into the marl substrate, occurring from a depth of 16 m. This diaphragm allows for the joint mobilization of the passive soil earth pressure along the pile foundation shafts that, functioning together with the underpinning structure, proved to be capable of counteracting the ≅ 21.2 m active earth pressures resulting from dredging procedures.

The joint interaction between existing and strengthening structures imposes a suitable interconnection between all interface areas, capable of ensuring their monolithic behaviour. This interconnection proved to be essential for the overall solution stability, where the bond between structural elements was achieved by suitable dowel steel bars, attached to the existing quay walls through epoxy resins and to the new reinforced concrete structural elements by means of embedded anchoring lengths of the reinforcing bars.

Over the existing quay, the impossibility of large diameter deep foundation drilling across the existing concrete structures led to the adoption of micropiles, suitably spaced according to their resistance and with the applied load values.

These micropiles (N80 − Ø177.8 × 15 mm) were grouted into the marl substrate through a repetitive and selective injection system.

The described solution is generically shown in Figure 12.4, by the schematic illustration of the lifting platform suspension operations over the offshore and onshore reaction structures, allowing the vessel ascent that precedes its transfer to the parking areas.

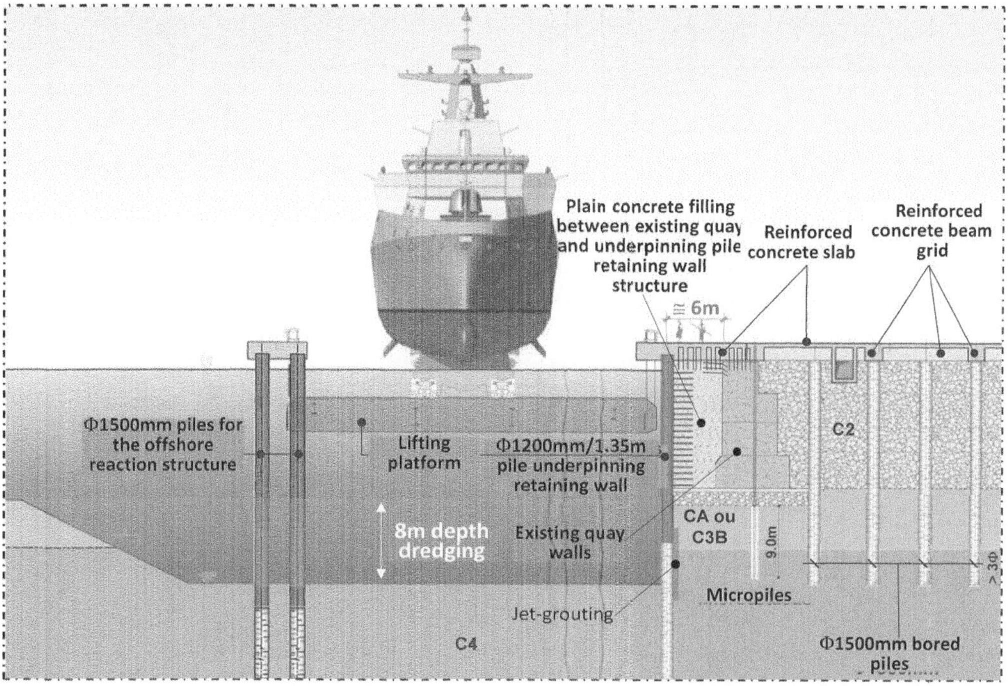

Figure 12.4 Typical section cut of the described solution.

The design of the adopted solution was performed with 2D and 3D nonlinear numerical simulations, using different software and conducting sensitivity analyses by judicious comparison of the obtained results. The 2D models were developed using the software PLAXIS, where the soil is simulated through finite elements governed by Mohr-Coulomb failure criteria, and the remaining elements (micropiles, quay walls, backfill, piles, beams, and slabs) through linear/elastic behaviour elements. In these models, it was possible to simulate the construction phasing and determine the envelope of the combined design stresses, forces, deformations, and safety factors for each intervention stage (Figure 12.5).

In addition to the 2D numerical simulations, the overall design was complemented by several 3D nonlinear analysis, performed in the software SAP2000, like the model shown in Figure 12.6, developed for the lifting/transfer areas. These models numerically simulated the soil/structure interaction through nonlinear spring elements, with linear stiffness corresponding to the soil's Winkler modulus and plastic behaviour corresponding to the value of the passive earth pressure in depth. The steel, plain concrete, and reinforced concrete elements were simulated by linear/elastic finite elements.

With the developed numerical calculation models, properly adapted to the characteristics and particularities of the various structures and intervention areas, it was possible to determine the maximum values of forces, stresses, and displacements for all Ultimate Limit State and Service Limit State combinations, by numerical simulation of all structural elements and interconnections, considering all construction phases.

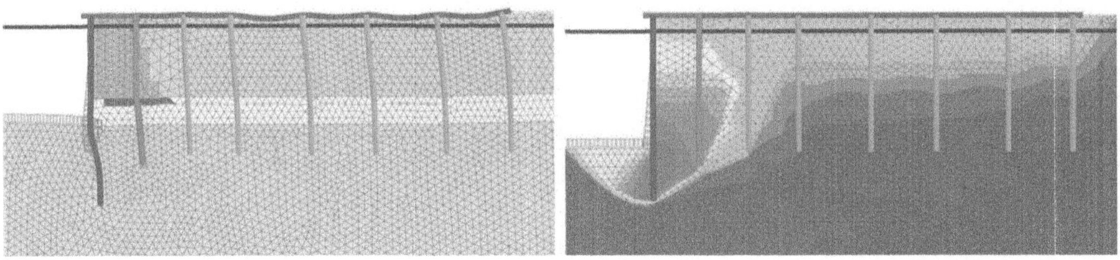

Figure 12.5 Calculation deformations (δ max horiz. = 1.7 cm) and failure surfaces of the adopted solution (SF ≈ 2.6).

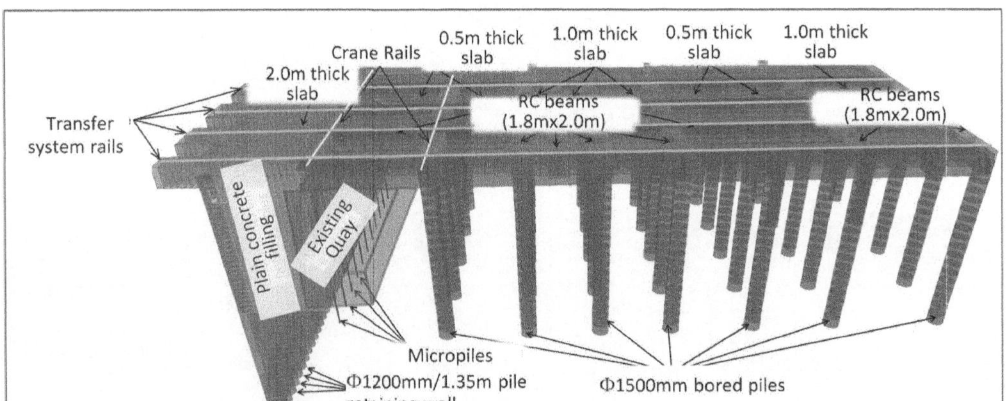

Figure 12.6 3D model for the transfer zone structural body.

In the safety analysis for seismic combinations, the capacity of the foundation piles was performed, foreseeing the possibility of the occurrence of liquefaction phenomena in depth, along the ≈10 m thickness port embankment layer. This landfill layer was built with a highly heterogeneous sandy/clayey matrix, which, despite not fitting into soils characterized by poorly graded/monogranular sands, with high liquefaction potential, has the water level at its top surface. Along with high seismic accelerations, this particularity potentially increases pore pressure to values that might trigger the liquefaction phenomena, which has a massive impact on foundation pile design (Figure 12.7).

In addition to the global numerical simulations, the calculations also involved numerous additional lower scale models, performed to address more specific situations. Figure 12.8 presents two images of one of these supplementary models.

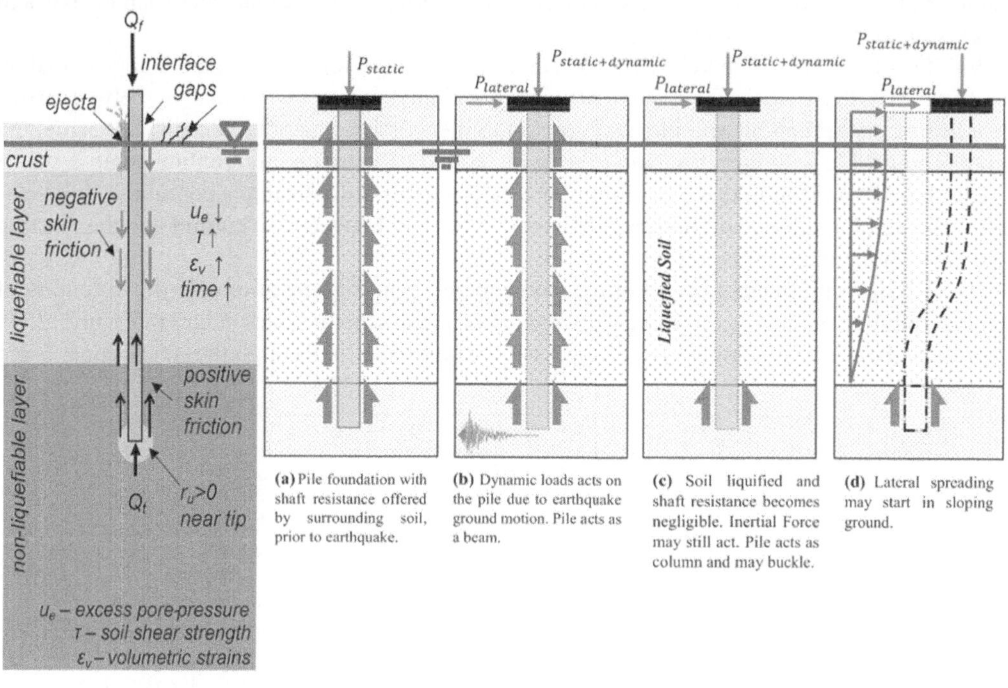

Figure 12.7 The liquefaction phenomena effect on foundation piles.

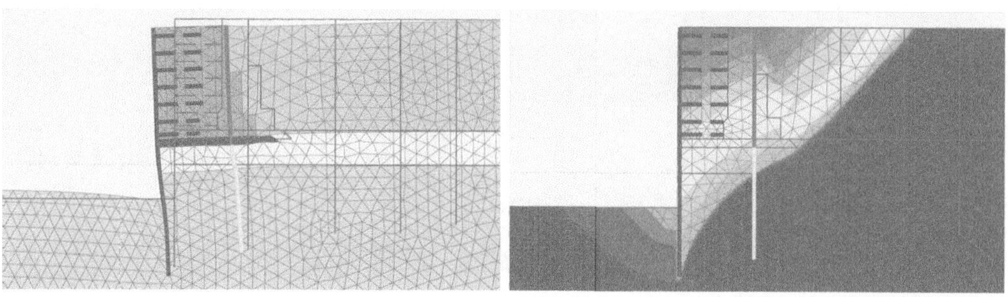

Figure 12.8 Evaluation of the underpinning solution stability due to short-term dredging effects.

In this case, the calculations were performed to simulate the solution's stability immediately after the dredging procedures. In this construction phase, it was mandatory to have the micropiles for existing quay foundation strengthening, the pile retaining wall curtain and the plain concrete stabilization filling, fully concluded. However, the connection to the platform slab, which establishes the stabilization diaphragm of the solution in its final stage, was yet to be completed, which imposed additional verifications.

12.4 Construction methods and phasing

The underpinning of the existing quay walls, whose main operations are represented in Figure 12.9, began with the execution of micropiles foundation strengthening elements, followed by sequential vibro-driving of a curtain of permanent steel casing tubes, with an internal diameter of Ø1,200 mm and a thickness of 12 mm, ensuring a 6 m distance between its axis and the existing quay wall elements.

With lengths of approximately 16 m (5 m driven into the ground => casing toe approximately at −15.00 ZH), these steel casing elements were positioned with a 1.35 m spacing between axes, establishing a steel combi-wall element parallel to the existing quay that, despite not achieving depths compatible with the subsequent dredging, covered the entire intervention area.

The "hook" connection that assembles these casing tubes, establishing a continuous joint element, allowed for subsequent underwater concrete placing between the curtain and the existing quay, preventing material leakage into the sea.

The drilling, reinforcement positioning, and concrete pouring operations for the execution of the underpinning bored pile retaining wall, presented in the schemes and images of Figure 12.10, were carried out with machinery operating from the port backfill, through the steel casing, using

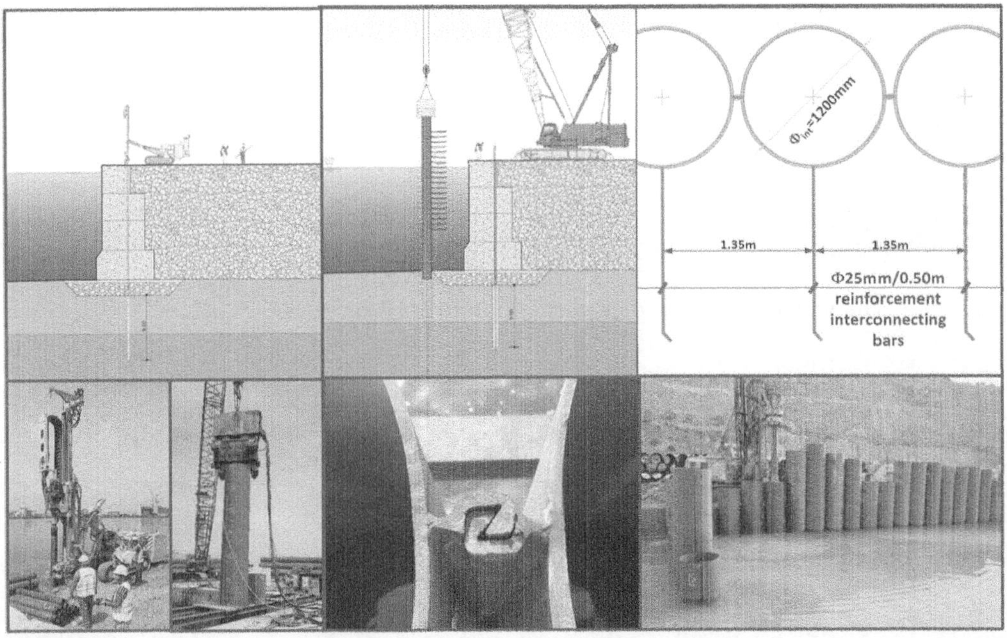

Figure 12.9 Schemes and images of micropile execution and steel casing driving.

Figure 12.10 Schemes and images of the construction operations of the bored piles of the underpinning retaining wall.

drilling equipment attached to powerful cranes or a suspended fly-drill equipment, reaching depths compatible with the −26.00 ZH level.

Following the defined phasing sequence, the construction works proceeded with the filling of the gap between the underpinning pile retaining wall and the existing quay walls, through submerged pouring of plain concrete, followed by the execution of jet-grouting columns that filled the 15 cm gap between adjacent piles, located below the base of the casing tubes (and their connecting elements), exposed after dredging procedures of the seabed (from −10.00 to −18.00 ZH).

These operations were executed along with the installation of interconnection dowel bars, which were sealed with epoxy resins or incorporated into the new structural elements, designed to ensure a monolithic behaviour between the piles, the existing quay, the plain concrete filling, and the reinforced concrete structural slab/beam grid (Figure 12.11).

Following the previously described design established phasing procedures, the conclusion of the primary underpinning reinforcement retaining wall, including the plain concrete filling, allowed to begin the dredging operations (Figure 12.12), to deepen the seabed in 8 m, achieving the required clearance for the project vessel draft, including the height of the synchrolift system's platform structure and the vessel supporting berth beams (trestles).

These dredging operations involved the removal of marl material from the C3A, C3B, and C4 geotechnical layers, occurring under approximately 1.0 m of sedimentary material.

The construction of the 411 foundation piles (72 Ø1,200 and 339 Ø1,500 mm) for the reinforced concrete beam foundation for the crane runway and the ship transfer/parking platforms

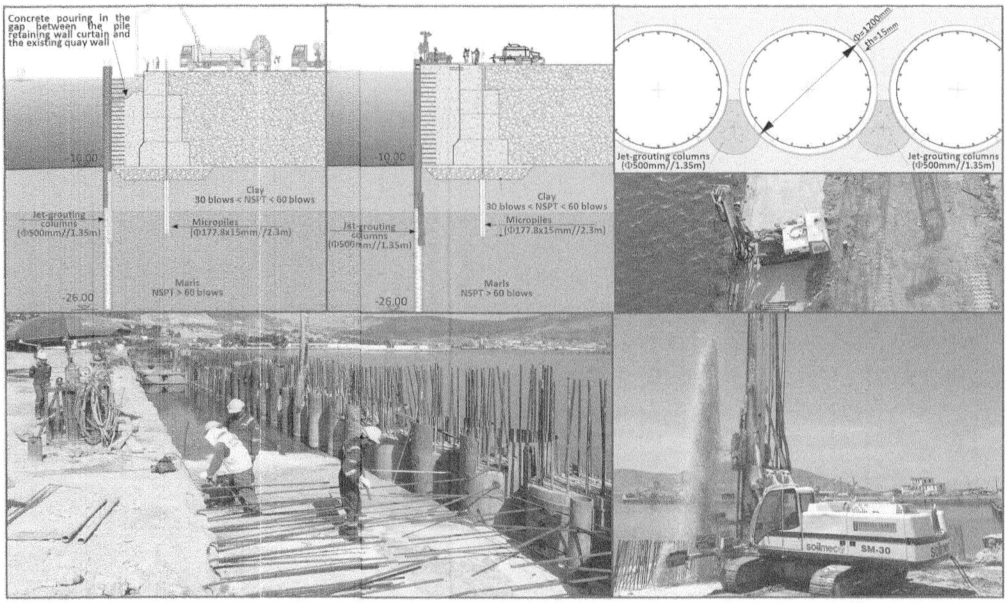

Figure 12.11 Concrete pouring in the gap between the pile retaining wall curtain and the existing quay wall and execution of the jet-grouting columns.

Figure 12.12 Dredging operations reaching maximum depths of 8 m, up to the level −18.00 ZH.

was executed simultaneously with the previously described, quay underpinning and foundation strengthening operations.

As illustrated in the schematic drawings and images of Figure 12.13, these structural elements are located on the port backfill areas. Ensuring a minimal penetration of 3Ø into the marl formations, the piles have variable lengths between 21 and 25 m and were designed to support vertical loads and to provide a supplementary horizontal stability solution through the mobilization of passive earth pressure along their shafts.

The conclusion of all dredging operations allowed the beginning of the construction of the offshore reaction structure, located approximately 28 m from the alignment of the quay's underpinning pile curtain.

Figure 12.13 Execution of foundation piles for the ship transfer and parking platform.

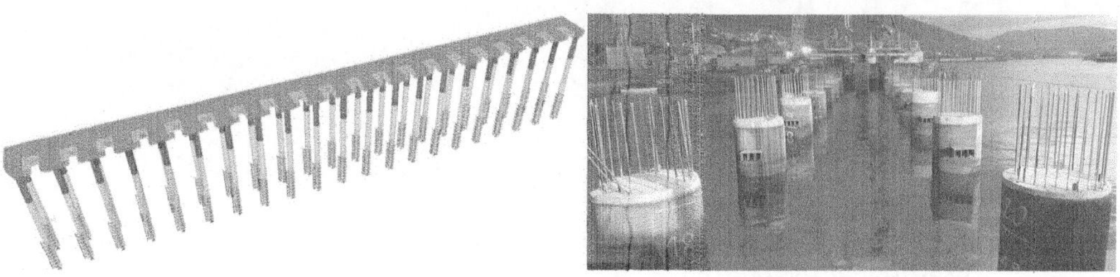

Figure 12.14 Design model and pile execution of the offshore structure, using floating platforms.

This structure consists of a reinforced concrete 1.6 m thickness slab, placed over 42 bored Ø1,500 mm piles, executed by drilling through steel casing elements, previously introduced into the ground (Figure 12.14).

The vibro-driving of the steel casing elements, the drilling, reinforcement placement, and concrete pouring for the execution of the offshore piles followed similar procedures to those described earlier, with the particularity of being carried out by equipment operating on floating platforms, using specific structures to ensure the accurate location and verticality of casing driving/positioning operations (Figure 12.15).

The drilling, reinforcement placing, and tremie concreting operations were performed using temporary bracing structures, installed to assure the accurate pile positioning and to minimise the negative impact of deformation and vibrating effects resulting from tide-wind-generated loads on fresh concrete, before the bored pile cast-in-situ concrete hardening (Figure 12.16).

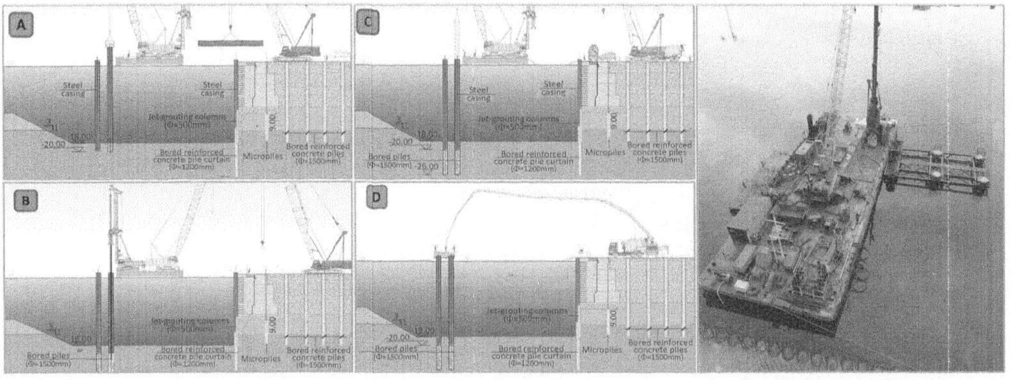

Figure 12.15 Generic offshore pile execution phasing.

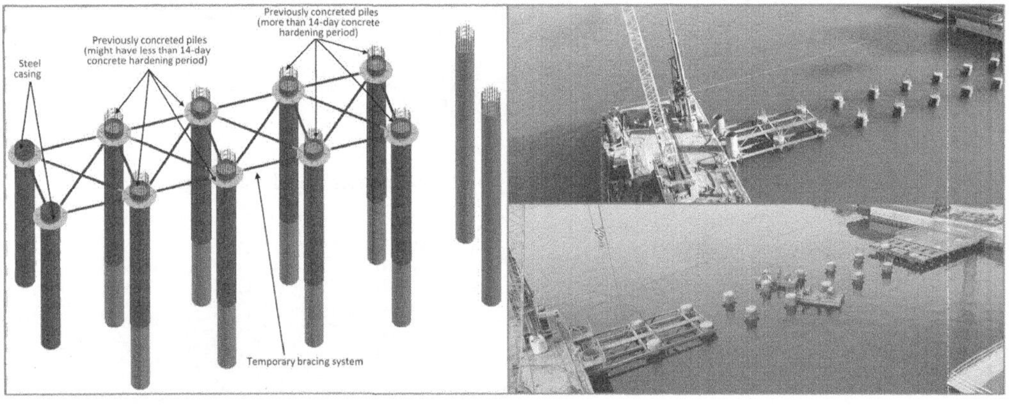

Figure 12.16 Temporary steel casing bracing system to allow the subsequent bored cast-in-situ execution process.

12.5 Testing and monitoring procedures

The validation of the pile load-bearing capacity, assessing the geomechanical parameters and the methodologies adopted in their design, was made by static load tests. These tests were carried out on 4 Ø1,500 mm piles, chosen according to the site geotechnical particularities of the different geomechanical calculation profiles, used in the vast area of this intervention.

The 4 Ø1,500 mm foundation static pile load tests were performed with a load value of 9,210 kN, corresponding to the maximum absolute vertical pile reaction value, resulting from the envelope of all Service Load Combinations (SLSs), and involved the implementation of a robust reaction structure, composed of reinforced concrete pile caps, steel beams/profiles, and pre-stress bars, intended to transmit the test load, imposed by hydraulic jacks of suitable capacity, to two additional reaction piles, designed according with que equilibrium tensile efforts (two additional piles per test).

The design of the reaction structure and the test plan, illustrated in Figures 12.17 and 12.18, respected the stipulations of EC7 and the guidelines set out in « NF P 94-15-1- Essai Statique de Pieu Isolé sous un Effort Axial ».

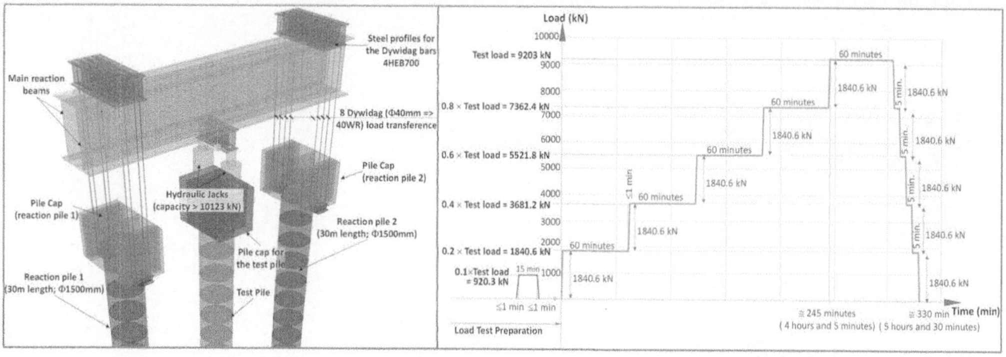

Figure 12.17 Reaction structure and loading levels diagram for each static control tests carried out on the foundation piles of the transfer and parking platforms.

Figure 12.18 Images of the reaction structure used in the static load tests for the foundation piles.

Since they did not take the piles to failure, which would imply the adoption of substantially higher loads (approximately 35,000 kN), these static load tests did not allow the accurate parametrization of the values used for equivalent end bearing (qc) and shaft (qs) pile resistance.

However, the small deformations obtained, less than 2 mm, unequivocally attested to the resistant capacity and good performance of the piles for their maximum service demands, validating their use as foundation elements for a transfer and parking platform for the vessels operating on a 9,000-ton capacity synchrolift:

$$N \text{ test piles } \text{Ø}1500 = N_{max}.\text{ service} = 9210 \text{ kN} \Rightarrow \text{Dmax. test} = 2 \text{ mm}$$

The foundation micropiles' resistance was also assessed by the execution of two static control tests. These tests were carried out using a vertical tensile and compression load corresponding to the value of their maximum axial stress resulting from the envelope of all Service Load Combinations (SLSs), with a value of 1,410 kN.

These tests were performed using a specifically built reaction structure, composed of reinforced concrete pile caps, steel profiles, and pre-stress bars, designed to transmit the test loads (traction or compression), imposed by a hydraulic jack of suitable capacity. In both tests (traction and compression), the reaction elements were formed by four micropiles, two belonging to the definitive structure and two supplementary.

Figures 12.19 and 12.20 show the load increase/decrease phasing graphic and an image of the reaction structures used, which followed the guidelines stipulated in EC7 (7.5.2.1.), adapting

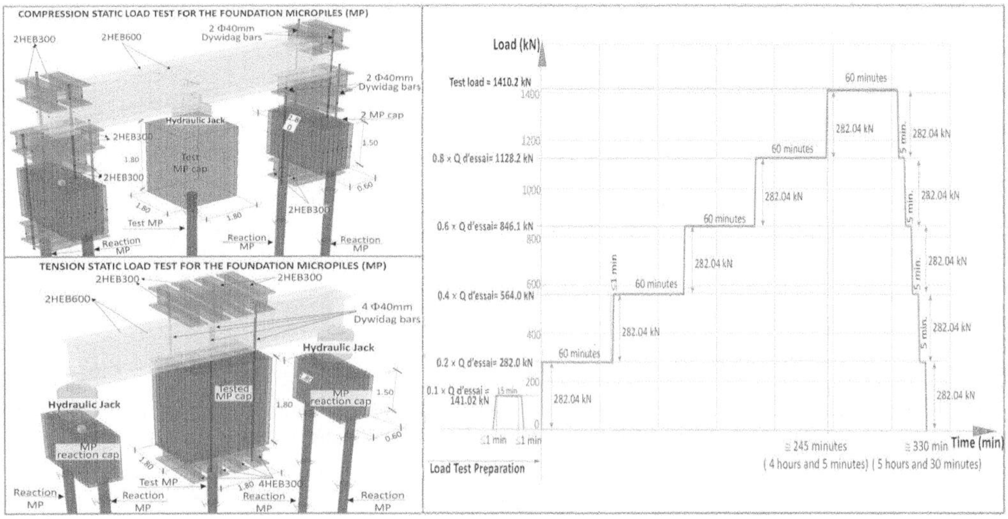

Figure 12.19 Reaction structures and loading level diagram of the static control tests carried out on the foundation micropiles of the crane transfer and circulation platforms.

Figure 12.20 Reaction structure used in the static compression micropile load tests.

them to the provisions of the standard « NF - P 94-150-1- Essai statique de pieu isolé sous un effort axial – en compression » and the standard « NF- P 94-150-2- Essai statique de pieu isolé sous un effort axial – en arrachement ».

Since they did not impose loads until micropile failure, these tests did not allow for measuring the value of unit lateral soil friction (qsi) used in the design. However, the low value of deformations obtained, with a maximum value of about 3 mm, unequivocally assesses the resistant capacity and good performance of the micropiles for maximum service loadings, validating the service demands for transfer and parking operations of 9,000 ton vessels and operational requirements for the two mobile 40 ton at a 50 m radius cranes:

$$N \text{ micropile test} = N_{max} \cdot \text{service} = 1410 \text{ kN} \cdot \Rightarrow \text{Dmax test} = 3 \text{ mm}$$

In addition to the static control tests, a complementary monitoring plan was implemented, which included the installation of inclinometers and topographic targets. The monitoring readings allowed deformation control throughout the 8 m seabed deepening, obtained by dredging operations, revealing that, after dredging conclusion (reaching the −18.00 ZH level), the retaining wall structure had maximum horizontal displacements, at its top level, of approximately 12 mm.

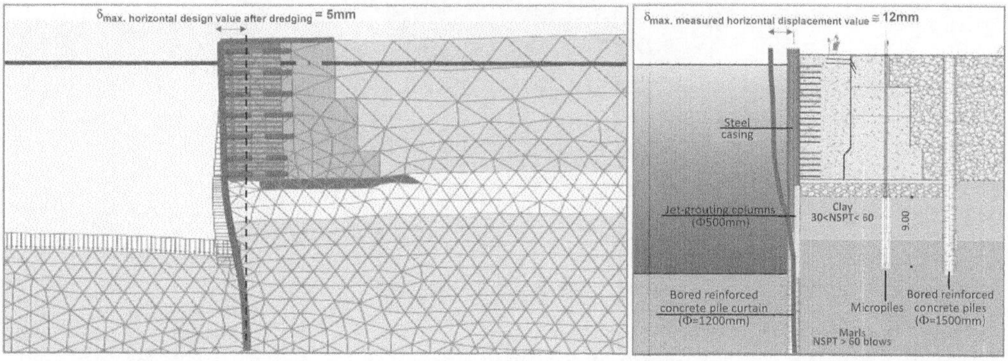

Figure 12.21 Maximum design and measured horizontal values for the retaining wall quay structure, achieved after conclusion of the 8 m seabed deepening obtained by the dredging operations.

This deformation is substantially lower than the deformations obtained in design calculations and is well below the established alert (1.2 × δmax. design horizontal displacement = 30 mm) and alarm (1.5 × δmax. design horizontal displacement = 38 mm) values (Figure 12.21):

Dmax. hor. after dredging = 12 mm < Dmax. hor. design displacement = 25 mm < D alert value = 30 mm < D alarm value = 38 mm

12.6 Conclusions

The imposition of new exploitation requirements on old maritime structures generally implies the execution of complex and expensive structural interventions. In this context, the execution of reinforcement and underpinning interventions requires the adoption of multidisciplinary solutions and suitable construction methodologies, capable of overcoming existing constraints and achieving the structural behaviour required by the new operating conditions.

Whenever the geomechanical characteristics of the soils do not allow element driving, such as sheet piles or any other kind of driven piles, until the required depths for underpinning structures, the option for bored cast-in-place piles may be advantageous. This methodology, which uses drilling with soil extraction, overcomes many of the constraints imposed by driving limitations, enabling the achievement of the design depths through the use of suitable drilling tools.

The extreme constraints of this intervention required the design of an integrated solution where the overall equilibrium is achieved throughout the joint functioning of both existing and reinforcing structural elements. It is imperative to highlight that this global integrated system required a thorough analysis of all material rheological properties, evaluating the structural behaviour of both existing and strengthening elements, considering the geomechanical characteristics of the soil and of the structural interaction between all components. This option, which proved to be crucial in the success achieved in this complex project, is shown in Figures 12.22 and 12.23.

The combined structural behaviour between the underpinning pile curtain, the existing quay walls, the plain concrete filling, jet-grouting columns, micropiles, and foundation piles, all interconnected by a top diaphragm consisting of reinforced concrete beams and slabs, made it possible to obtain a resistance capacity compatible with a significant load increase and a significant dredging, enabling the project vessel-lifting and transfer operations, such as shown in Figure 12.23, meeting the current regulation requirements and achieving significant safety factors.

Figure 12.22 Project site after conclusion of the control building and installation of the synchrolift.

Figure 12.23 Approach, lifting, and transfer of an 8,500-ton vessel.

Bibliography

CEN. 2010. *Eurocode 7: Geotechnical design—Part 1: General rules, EN 1997-1:2010*. European Committee for Standardization, Brussels.

Matias Ramos, P. and Esteves, L. 2024. Underpinning and strengthening of an existing port infrastructure to allow 8 m depth dredging for the installation of a high-capacity vessel synchrolift. In: Guerra, N., et al. (eds.), *Geotechnical Engineering Challenges to Meet Current and Emerging Needs of Society. Proceedings of the XVIII European Conference on Soil Mechanics and Geotechnical Engineering (*Lisbon, *26–30 August 2024). XVIII ECSMGE 2024.*

NF P 94-15-1. 1999a. *Essai Statique de Pieu Isolé sous un Effort Axial – Partie 1: En compression.* AFNOR.

NF P 94-15-1. 1999b. *Essai Statique de Pieu Isolé sous un Effort Axial – Partie 2: En traction.* AFNOR.

Ziotopoulou, K., Sinha, S.K. and Kutter, B.L. 2022. Liquefaction-induced downdrag on piles: Insights from a centrifuge and numerical modeling program. In: Wang, L., Zhang, J.M. and Wang, R. (eds.), *Proceedings of the 4th International Conference on Performance Based Design in Earthquake Geotechnical Engineering (Beijing 2022). PBD-IV 2022. Geotechnical, Geological and Earthquake Engineering*, vol. 52. Springer, Cham. https://doi.org/10.1007/978-3-031-11898-2_40

Chapter 13

The port of Leixões (Portugal)

Over 120 years of knowledge

Hugo Guedes Lopes and Joel Silva

13.1 Introduction

The northwest Iberian coastline is shaped by numerous rivers, inlets, and bays. The shoreline has a dynamic, heterogeneous, and mixed coastal and maritime environment (e.g., Mota-Oliveira et al. 1982; Dias et al. 2000, Veloso-Gomes et al. 2004, Pérez-Alberti et al. 2013, Pires et al. 2016 and references therein). From a rocky shores or platforms to a variety of environments, the region includes sandy beaches and coastal protection works (e.g., Mota-Oliveira et al. 2000, Veloso-Gomes and Taveira-Pinto 2003, Pires et al. 2014, 2016). Furthermore, coastal erosion results from the dynamic nature of its coastal zones, anthropogenic interactions, and influences, such as coastal interventions and littoral occupation, as well as the effects of climate change (Coelho et al. 2009). For this reason, keen and integrative studies for harbour and maritime geotechnics are essential (CIRIA et al. 2007).

The Port of Leixões is located in mainland Portugal, in the northwest of the Iberian Peninsula, approximately 4 km north of the mouth of the River Douro (near the Porto and Gaia cities), facing the Atlantic Ocean. The port is surrounded by the municipalities of Leça da Palmeira, to the North, and Matosinhos to the South, and close to the Porto, the second-largest city in Portugal.

The Port of Leixões was constructed in the 19th century, as an alternative to the Douro port in Porto (Alves and Brògueira-Dias 2001, Sousa and Alves 2002). Regardless of the importance of Douro port in the XV and XVI centuries, and later in the 18th century with the *Port Wine* trade, it had several restrictions, such as stream floods in the winter, a river mouth with strong currents, a rocky bottom (with a maximum depth of 5 m), and a dynamic sand spit (Figure 13.1). As ships grow larger, the entrance to the Douro river mouth became even more dangerous and risky, and thus an alternative had to be found (Alves and Brògueira-Dias 2001, Brògueira-Dias and Alves 2010).

The first studies of the new Port of Leixões dated back to the 15th century, but it was in the 18th century that substantial efforts were made to deliver a viable alternative to Douro port (Figure 13.2). Due to the magnitude of the project, it was considered one of the major and emblematic engineering works of the 19th century in Portugal (Brògueira-Dias and Alves 2010), (Figure 13.3).

The entrance to the river Douro and its sand spit have retreated by more than 600 m over the past 150 years, leading to more than 30 years of studies to define the solution, which was implemented at the beginning of the 21st century. This solution aimed to stabilise the mouth of the river and improve the safety conditions on both riverbanks (Veloso-Gomes et al. 2004, Gomes et al. 2005). However, the shallow depth at the river entrance (−5.0 m) remains a limitation for larger vessels, confirming the good decision made in the 19th century.

DOI: 10.1201/9781003530220-16

Figure 13.1 The port of Douro and river floods. *From Lopes and Dias (2021), published under CC BY 4.0.*

Since its construction, the port has undergone numerous improvements, particularly expansion towards the sea and the excavation of new docks inland before 1936. In 1936, the Port Authority (now APDL) decided to construct a large, submerged breakwater with approximately 1000 m in length, outside the old port protection structure to protect the entrance against storms. The construction of the submerged breakwater significantly improved operational conditions within the port. However, to protect the oil terminal, from the direct impact of waves that overtopping the submerged breakwater, it was raised in 1969. The solution, as built, consisted of a 40-tonne tetrapod slope, built over the 90-tonne blocks of the submerged breakwater, along with a concrete superstructure to protect the oil terminal access platform against overtopping. The initial submerged breakwater had two main objectives: to create the proper conditions for port operations and to enhance the safety of ships within the port basin. As a complement, access conditions to the port were to be improved, aiming to increase the number of days each year during which the ships could safely enter. However, the stability of the tetrapods, which had been satisfactory in the model tests, proved to be less effective in the prototype. Apart from the fractures and/or the sliding of tetrapods along nearly the entire length of the breakwater, two particularly vulnerable areas were identified: the head of the breakwater and the oil terminal section. The damage was so significant that by 1972, two years after the construction was completed, a recharge of 1000 tetrapods was required (compared to the 2868 tetrapods originally used). In response to the

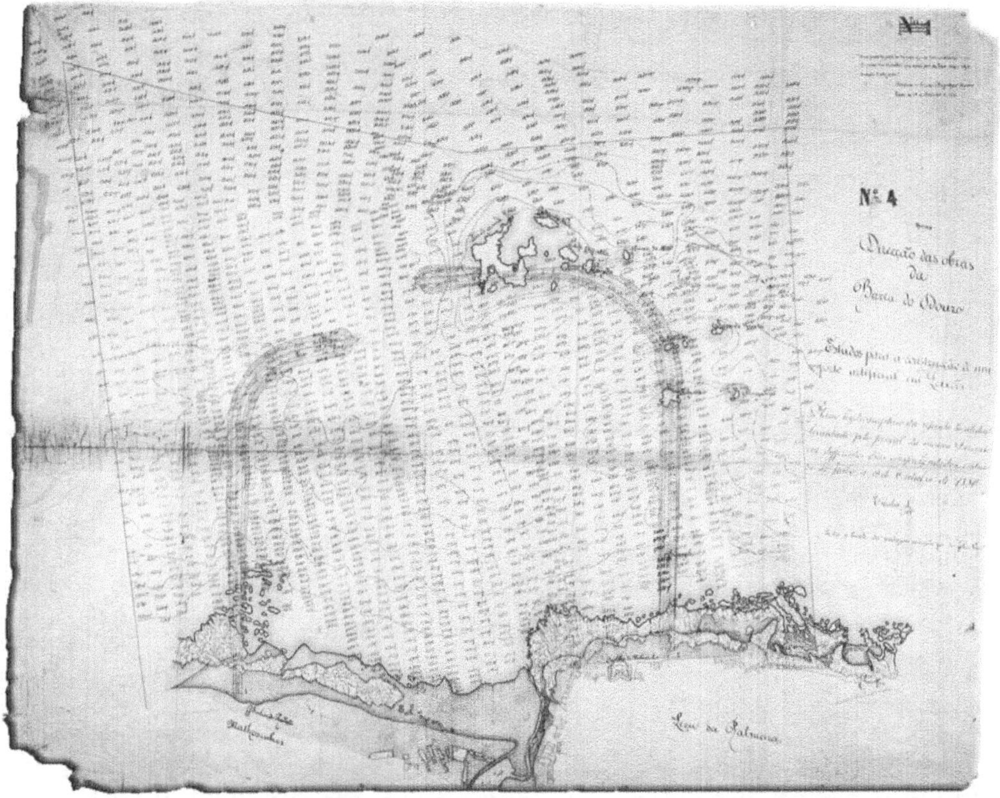

Figure 13.2 Layout of the port of Leixões breakwater. *From Lopes and Dias (2021), published under CC BY 4.0.*

damage caused by the storms of 1973 and 1974, a total of 1010 additional tetrapods were placed. The repair works also included reinforcing the southern limit of the breakwater. This reinforcement involved replacing the tetrapods below the +1.0 m Chart Datum level with 40-tonne cubic blocks. Above this level, a concrete-steeped structure built was constructed "in situ." The head of the breakwater appeared to be very vulnerable, with the damages exceeding expectations. As a result, a detached breakwater was built in 1982 (details in Lopes 2013).

Finally, the current mission of APDL is to manage the ports of Douro, Leixões and Viana do Castelo, as well as the inland waterway of the Douro River. This includes the economic exploitation, conservation and development of these areas, as well as the powers assigned to the port authority.

13.2 Harbour geotechnics: key studies on the Port of Leixões construction, problems, and rehabilitation

Over the past decades, the Port of Leixões has undergone several improvements, primarily involving the expansion towards inland with the excavation of new docks, the extension and rising of the original breakwater, and the construction of a new submerged breakwater to protect the head of the breakwater (details in Lopes 2013). Additionally, several studies have been published in recent years concerning the Leixões port area and related geotechnical issues (e.g., APDL

Figure 13.3 The port of Leixões at the beginning of the 20th century. *From Lopes and Dias (2021), published under CC BY 4.0.*

1987, Brògueira-Dias 1992, Brògueira-Dias and Coutinho 1998, Veloso-Gomes et al. 2009, Lopes et al. 2013, Rosa-Santos and Taveira-Pinto 2013, Pinto et al. 2013 and references therein).

13.2.1 Dock n°1

Dock n°1 was constructed between 1932 and 1940 (Figure 13.4). This dock was entirely excavated onshore, upstream the River Leça, which had to be diverted to allow for the construction. The deviation was guaranteed through two branches, one to the north with a twin aqueduct and another to the south with an artificial channel.

Prior to construction, a geotechnical survey was conducted using Degousé rotation and manual percussion, with an array measuring 80 × 180 m. Generally, the bedrock was detected at a depth of −10.0 m ZHL (Leixões chart datum). However, there were two regions, one in the northeast dock, approximately 100 m in length, and another in the southwest dock, measuring 80 m, where the bedrock was only encountered at −20.0 m ZHL.

Two different technical solutions were defined for the quay wall: Solution 1 – a continuous gravity wall, implemented whenever the bedrock was above −11.0 m ZHL; and Solution 2 – a pier structure with arches founded on the bedrock, using the Havage method (Figure 13.5).

However, by the 1970s, solution 2 began to show signs of fragilities. The arches started to fall into disrepair, although the exact causes were not fully understood. Bad/poor mortar behaviour and ground movement beneath the pillars were pointed out as potential contributing factors.

Figure 13.4 Construction of dock number 1. *From Lopes and Dias (2021), published under CC BY 4.0.*

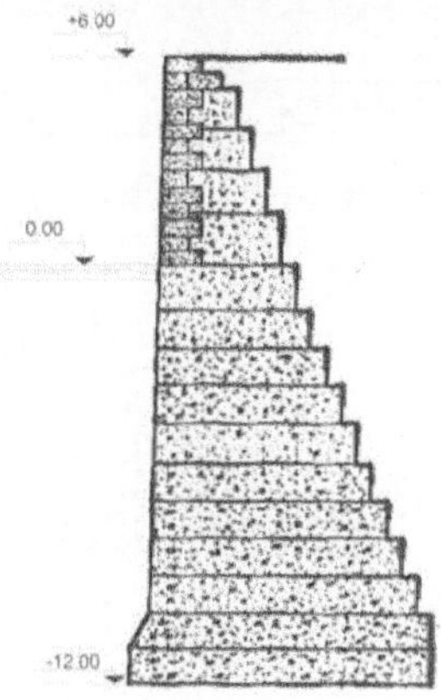

Solution 1

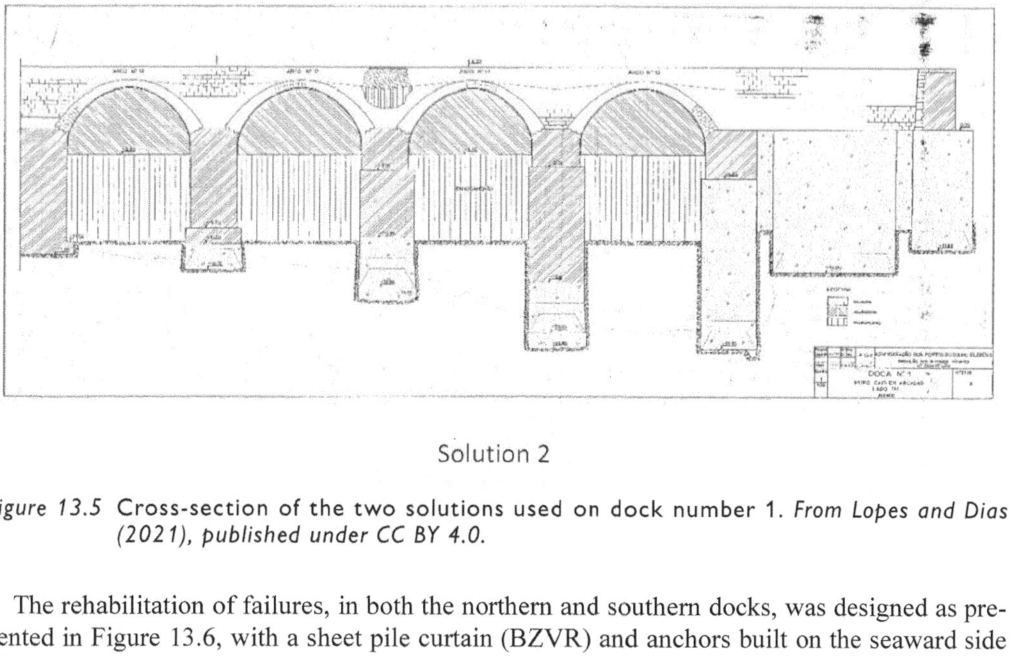

Solution 2

Figure 13.5 Cross-section of the two solutions used on dock number 1. *From Lopes and Dias (2021), published under CC BY 4.0.*

The rehabilitation of failures, in both the northern and southern docks, was designed as presented in Figure 13.6, with a sheet pile curtain (BZVR) and anchors built on the seaward side of the old quay.

In the 1990s, the continuous gravity walls also began to present some problems in the southeast region of Dock 1, in an extension of 140 m. Caverns were detected in the quay wall, which were overcome with the solution shown in Figure 13.7, involving a sheet pile curtain and oblique ground anchors. On the northern side of the dock, the rehabilitation included a 15 m extension of the embankment to accommodate the installation of Leixões' first RO-RO terminal.

As part of this rehabilitation, an impressed current cathodic protection system was installed to reduce corrosion on the steel sheet piles and the associated concrete capping beam.

Despite these extensive efforts, including the protection of the ground anchors, new problems have begun to emerge in these structures, eventually related with the ageing of the structures, with the increase of the loads on the quays and embankments, as well as eventual ground movements.

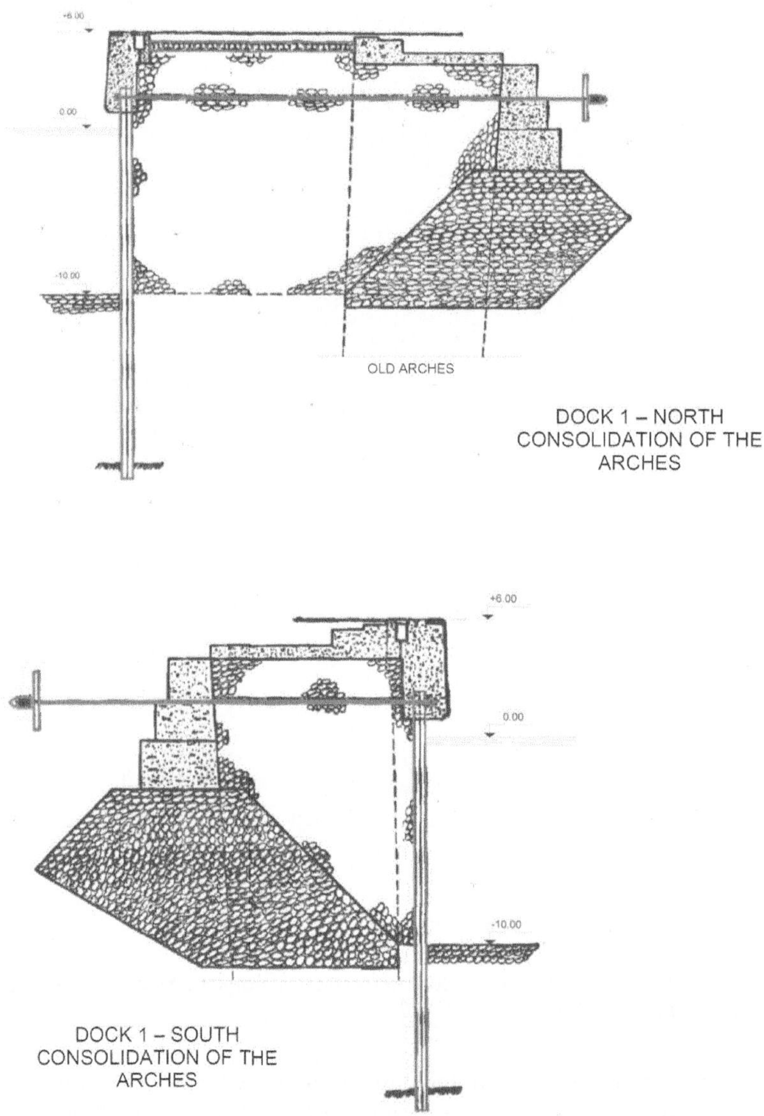

Figure 13.6 Rehabilitation project of dock number 1- north and south. *From Lopes and Dias (2021), published under CC BY 4.0.*

13.2.2 Dock n°2

With the expansion of port operations, the need for new quays and embankments arose, leading to the design and construction of Dock n°2. Following the same approach used for Dock n°1, Dock n°2 was excavated inland between 1958 and 1966. The construction of the dock, resulted in the interruption of the connection between the northern and southern parts of the city of Matosinhos, necessitating the construction of a drawbridge in a region where the bedrock was located at a depth of −25.0 m ZHL. This location was essential to maintain, as it connected two important social areas of the city.

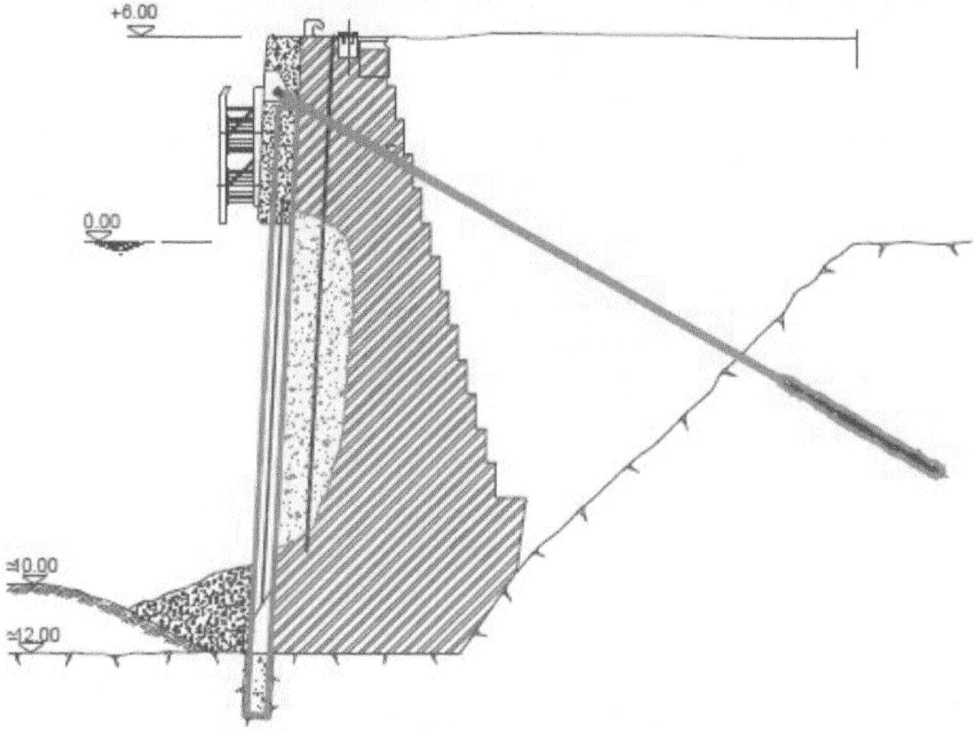

Figure 13.7 Rehabilitation of the gravity wall of dock number 1- south. *From Lopes and Dias (2021), published under CC BY 4.0.*

The geotechnical survey revealed two distinct scenarios. On the southern side, the geotechnical conditions were favourable, with good-quality materials and the bedrock at a depth less than −11.0 m ZHL. On the northern side was characterised by thick alluvial layers with poor geotechnical properties, and the bedrock was located deeper than −16.0 m ZHL, due to its position along the natural meandering path of the River Leça in this region.

Based on the geotechnical survey undertaken, three different solutions were defined:

a Solution 1: A gravity wall that was implemented whenever the bedrock was above −11.0 m ZHL (South dock, 700 m in length, with no problems registered to date);
b Solution 2: A pile quay wall (without a+ retention wall) – northwest Dock (300 m),
c Solution 3: A sheet pile quay wall with a retention wall (sheet pile curtain) – Northeast Dock (370 m) (Figure 13.8).

The pile solutions were defined where the bedrock was located below −16.0 m ZHL.

However, in the 1970s, significant displacements were registered on the crest of the northwest quay wall beam (300 m), including cracks, displacements of up to 50 cm, settlements of up to 25 cm on the embankment, and failure of transverse beams. This led to the inoperability of the quay wall for more than 20 years. Later analyses assumed that the failure was a consequence of a steep slope embankment beneath the quay (1:2.5), associated with an inadequate geotechnical survey that only included 8 boreholes. This survey failed to anticipate for the

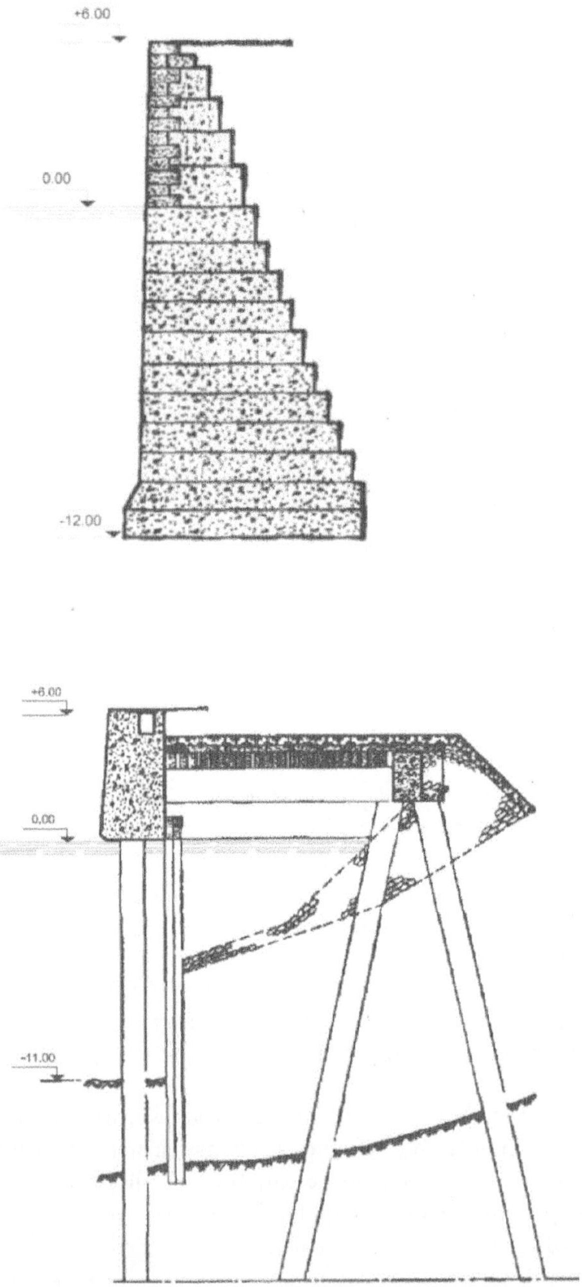

Figure 13.8 Cross-sections of dock number 2 quays. *From Lopes and Dias (2021), published under CC BY 4.0.*

thick sludge layers, which were subsequently identified and contributed to the failure of the piled structure.

In the late 1990s, the rehabilitation of Dock n°2 included the northwest and northeast quays. The reinforcement involved the addition of a set of four new piles transversely and a slab on top, which was embedded in the existing capping beam (Figure 13.9). No problems have been reported since this rehabilitation.

13.2.3 Dock n°4

The Dock n°4 was constructed between 1974 and 1983, utilised two distinct structural solutions (Figure 13.10), based on comprehensive geophysical and geotechnical surveys, which included sampling and laboratory tests:

a Solution 1: A gravity quay wall, where the bedrock was approximately −12.0 m ZHL (south and west quay – total of 600 m);
b Solution 2: A sheet pile structure quay wall, implemented where the bedrock was below −12.0 m ZHL (north, east and southeast quays – total of 700 m).

The survey identified muddy layers exceeding 25 m in thickness, as well as weathered granites (W_{4-5}). During construction, piles of over 50 m were driven, revealing significant heterogeneity in the ground conditions, with sudden variations in the pile-driving depth.

Despite the extensive in situ studies and tests conducted, the careful monitoring of the execution process revealed unexpected ground behaviour during the dredging process and the pre-casting of blocks. This led to the need to rehabilitate part of the south dock (Figure 13.11) and to the execution of an additional 100 m of reinforced pile structure quay on the southeast

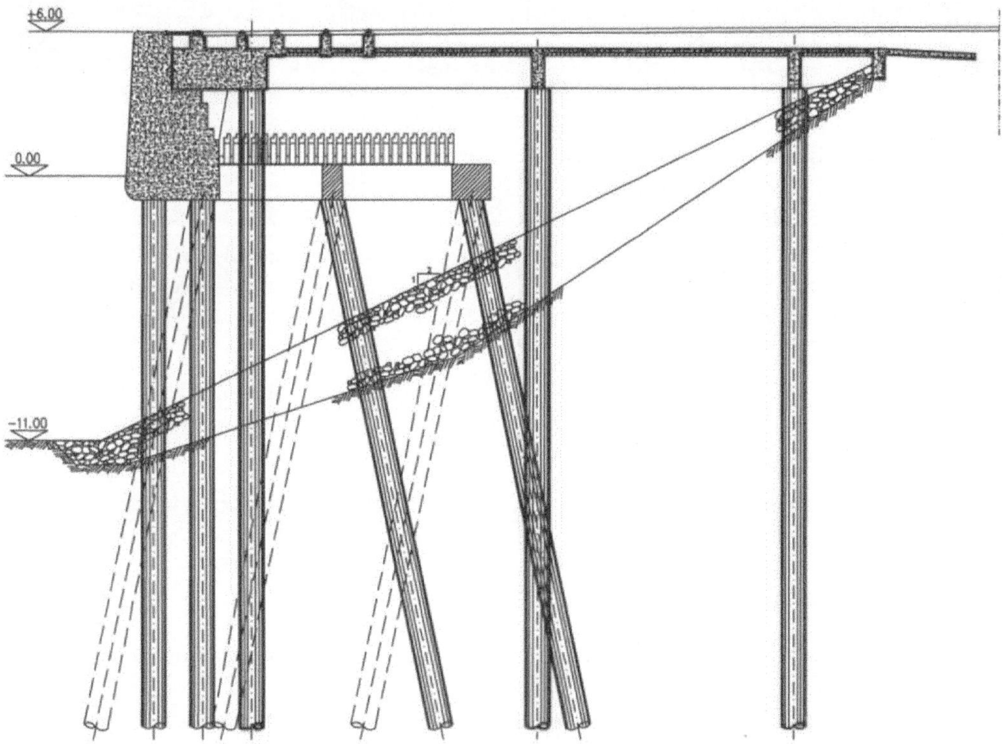

Figure 13.9 Rehabilitation solution applied on the dock 2 south. *From Lopes and Dias (2021), published under CC BY 4.0.*

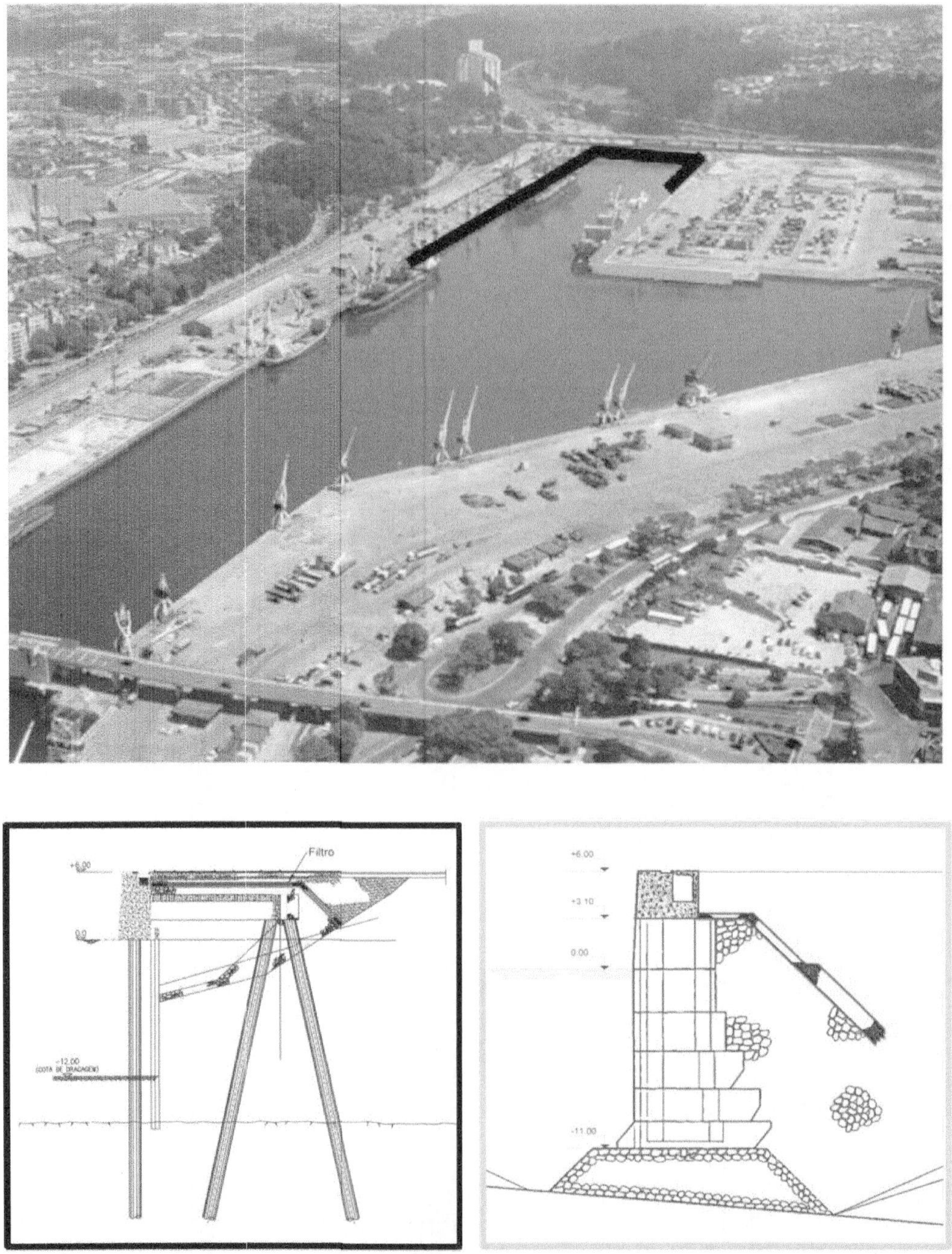

Figure 13.10 Aerial view and cross-sections of the dock number 4 quays. *From Lopes and Dias (2021), published under CC BY 4.0.*

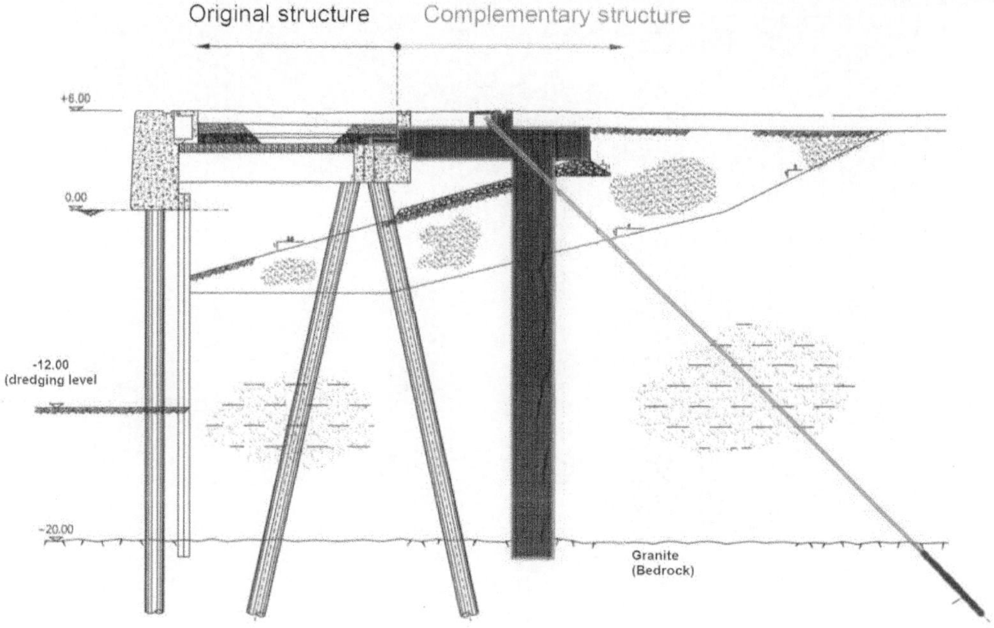

Figure 13.11 Rehabilitation of first 100 m of the south-east quay of dock number 4. *From Lopes and Dias (2021), published under CC BY 4.0.*

side of the dock. Furthermore, partial rehabilitation of the foundations of the highway bridge on the east side of the dock was required.

13.3 South container terminal

The geology on this area of the port is characterised by the presence of thick layers of alluvial sludge at shallow depths, which occupy the fossil valley of the Leça River. The results of the geotechnical survey demonstrated that these soils were likely to present significant compressibility.

The geotechnical survey recommended mitigating potential stability problems with the railway foundation by designing a pile solution that would extend down to the bedrock.

Taking into consideration: (i) The bad experience on the site where the blocks were pre-casted; (ii) That requirement for a 5 tonne/m^2 embankment for the future terminal; (iii) he need for adequate stability for cranes under rails. A detailed pre-consolidation programme was developed, to anticipate the expected +1.0 m settlement of the embankment, which could be asymmetric and potentially damage the piles, limiting the future operation of the terminal.

The development of the pre-consolidation programme included the following main objectives:

1 Alluvial non-consolidated thick formations could lead to important settlements, eventually asymmetric, that would be incompatible with the normal operation of the terminal;
2 Pre-consolidation of the embankment ($\bar{U}$ = 90%) could be achieved using vertical geo-drains and temporary embankments (Figures 13.12 and 13.13);

Figure 13.12 Excavators, guiding towers, and geo-drains installed. *From Lopes and Dias (2021), published under CC BY 4.0.*

3 Design of the geo-drains grid defined to guarantee maximum pre-settlement over the shortest period possible;
4 A monitoring programme designed to control the settlements process, using different equipment, such as inclinometers, piezometers, and levelling plates.

Figure 13.13 Artificial embankments used. *From Lopes and Dias (2021), published under CC BY 4.0.*

13.3.2 Pre-consolidation project

Based on Terzaghi Abacus, which relates the consolidation degree Ū (assumed as dimensionless consolidation) with the time factor ***Tv*** and ***T'***, with the diameter's relation ***n*** between diameters ***D*** of the influence area of the drain and ***d*** of the drain itself. The project assumed the following equations:

A Total settlement

$$\Delta H = mv.\Delta p.H \tag{13.1}$$

B Time, considering natural settlement

$$Tv = t.\frac{Cv}{\left(\frac{H}{2}\right)^{-2}} \tag{13.2}$$

C Time, using consolidation system

$$T' = 4.Cr.\left(\frac{t}{d}\right)^{-2} \tag{13.3}$$

where

ΔH = total settlement;
Δp = Effective stresses variation;
H = thickness of the compressible layer;
t = time;
d = theoretical diameter of the drain used;
Tv = Time factor considering natural settlement;
T' = Time factor considering the consolidation system;
mv = volumetric compressibility coefficient;
Cv = Vertical consolidation coefficient;
Cr = Radial consolidation coefficient.

The geotechnical survey included 28 CPT tests, 24 Standard Penetration Test (SPT and Vane Tests), and coring for laboratory analysis (Atterberg limits, granulometry, specific weight, humidity, non-limited compression, tri-axle compression, and oedometer test with vertical and radial drainage). The results revealed considerable variability in the geotechnical characteristics, but enabled the definition of the most probable values for the following parameters:

East side of the embankment

$mv = 3.0 \times 10^{-2}\,\text{cm}^2/\text{kg}$
$Cv = 4.0 \times 10^{-4}\,\text{cm}^2/\text{s}$
$Cr = 16.0 \times 10^{-4}\,\text{cm}^2/\text{s}$

West side of the embankment

$mv = 5.0 \times 10^{-2}\,\text{cm}^2/\text{kg}$
$Cv = 2.5 \times 10^{-4}\,\text{cm}^2/\text{s}$
$Cr = 25.0 \times 10^{-4}\,\text{cm}^2/\text{s}$

The analysis of the results estimated that total settlements could range from 4 to 90 cm, and that pre-consolidation could be achieved in 3 months using a grid of geo-drains (Mebradrain 7404) with diameters ranging from 1.1 to 2.0 m, depending on the area considered. To guarantee the desired settlement, artificial sand embankments with a height of 3.0 m were used to simulate the 5 tonne/m^2 load expected during the terminal operation. The remaining heights of these embankments were moved for a different location whenever 90% of the settlement was reached.

13.3.3 Execution of the pre-consolidation project

The geo-drains were installed using an excavator, which was properly adapted with a guiding tower over 30 m, enabling the installation of the geo-drains with the non-woven polypropylene mattress (Figure 13.14).

In the area of the future terminal, piezometers, levelling plates, and inclinometers were installed and properly monitored throughout the entire pre-consolidation process. This also included several geotechnical campaigns conducted before, during, and after the intervention.

The levelling plates provide good and expeditious results, while the piezometers also delivered good results, though they were costly and delayed in comparison to the levelling plates.

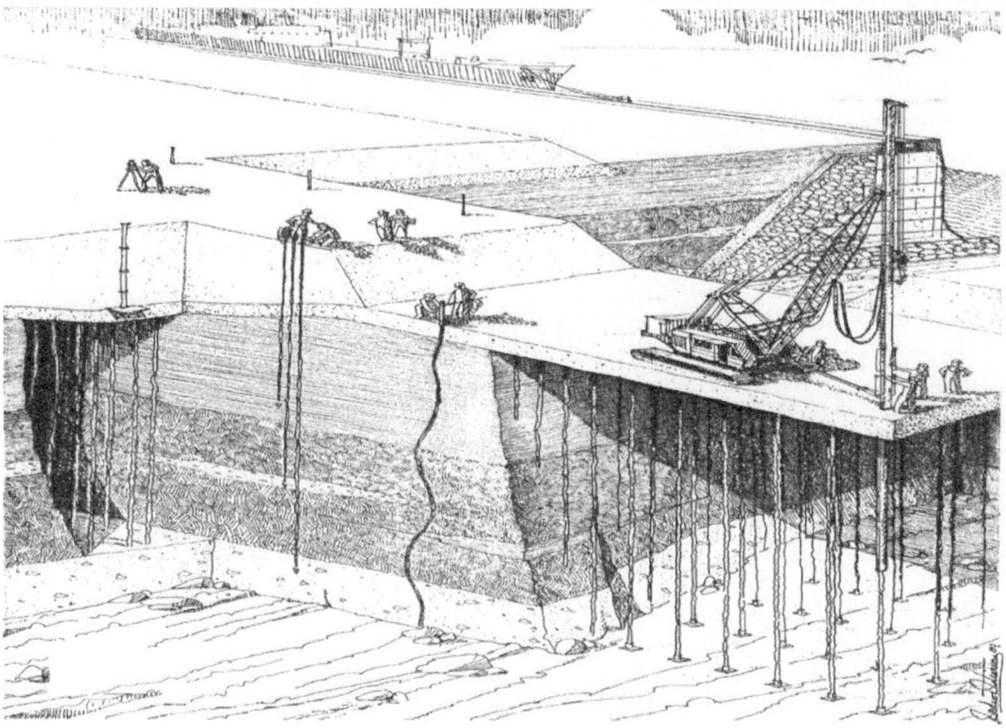

Figure 13.14 Pre-consolidation works and monitoring systems. *From Lopes and Dias (2021), published under CC BY 4.0.*

Despite these challenges, the piezometers were essential for validating the results of the levelling plates. Unfortunately, the inclinometers did not yield feasible results and were also expensive and delayed.

13.3.4 Beams of the crane rails

As previously mentioned, due to geotechnical vulnerabilities, the beams supporting the railways were founded on piles with a diameter of 800 mm, using a Benoto EDF 1180 equipment, except for the seaside rails, which were directly supported by the quay's capping beam. Those piles were executed after the pre-consolidation was completed, prevent any horizontal soil movements from damaging the piles.

Over the past 30 years of service at the container terminal, some displacements have been registered on the gravity quay wall. This has induced changes in the rail gauge, which has led to the need for stabilisation and correction work.

13.3.5 Stabilisation of the quay wall in blocks of dock 4 (south)

In 2020, APDL began an intervention on the quay wall blocks of Dock No. 4 (South).

As mentioned earlier, the quay wall in question, composed of overlapping simple concrete blocks (Figure 13.15) showed signs of movement at a certain stage in its service life, indicating a possible phenomenon of global instability. Subsequently, and with the aim of mitigating the

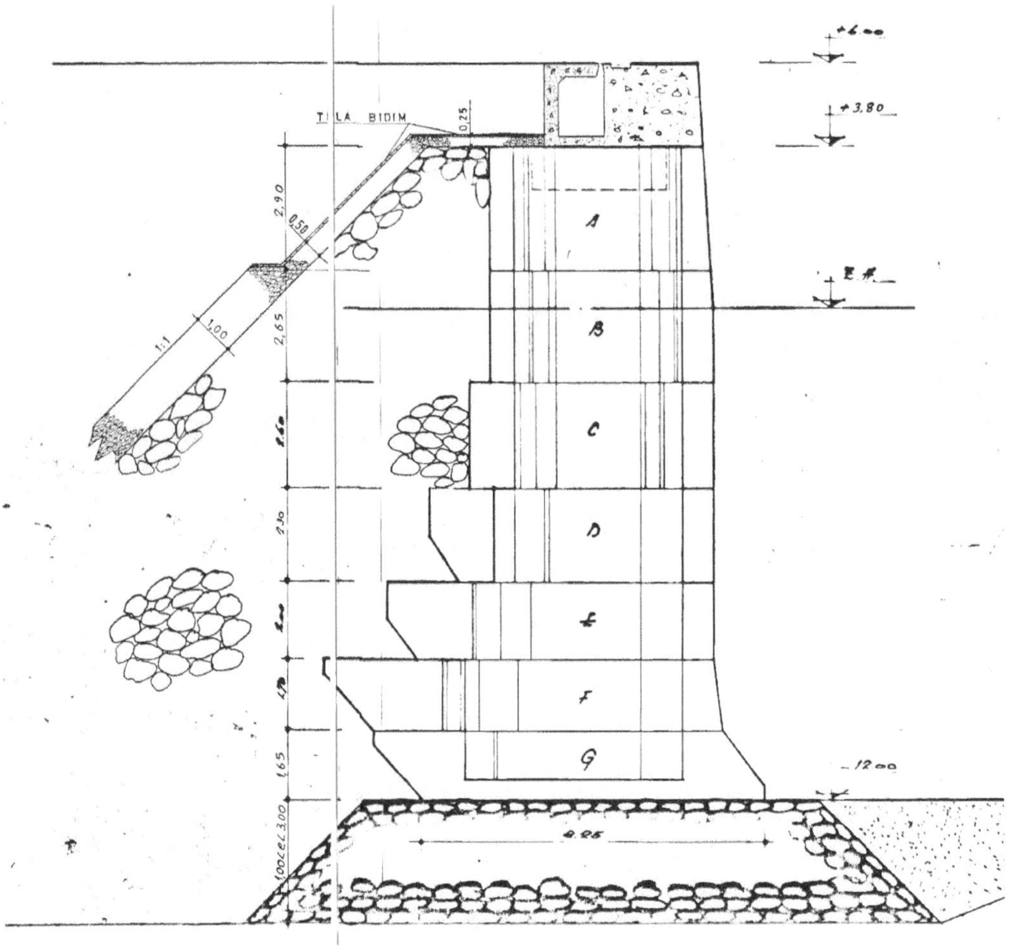

Figure 13.15 Cross-sections of quay wall in blocks of dock 4 (south).

aforementioned movements, a stabilisation solution was considered that would minimise disturbance to the terrain, be easy to implement, and reduce the intervention time, and consequently, minimising the impact on the pier's normal operations.

In the intervened area, the quay wall is founded at −12 m, on a rockfill prism of variable height, depending on the local geology and the depth at which the granite massif is located. The pier structure consists of concrete block columns, prefabricated and shaped like an "I" in plan, 3.4 m wide, with lengths between 5 and 9.25 m and heights between 1.65 and 2.9 m. The gaps between blocks are filled by rockfill, and the structure is topped with a reinforced concrete superstructure 3.75 m wide and 2.2 m high, where the gutter for installing the crane rail that operates on the pier is located. The pier floor is at +6 m. In 2005, following the observation of certain pathologies, namely the sliding of juxtaposed blocks, an intervention was carried out to join the blocks of each column, forming a monolithic structure. This was achieved by drilling vertical holes and installing and sealing metal profiles.

Recent inspection concluded that significant displacements were being recorded in the transition zone between the two foundation solutions (Figure 13.16). Specifically, measurements taken from cores installed at the top of the quay wall showed a cumulative movement of approximately 29 mm by 2018, with a noticeable increase in movement since 2016. The joint analysis of this information, along with a visual inspection of the blocks, making up the quay wall, led to the conclusion that the movements were likely caused by a global movement of the structure, rather than relative displacement between the blocks. An underwater inspection validated the existence of a ditch in front of the pier, along with the loss of contact area at the base of the block pier wall. This resulted in a reduction of confinement in the foundation material and an increase in stresses on the soil, potentially leading to settlements at the foundation level. This, in turn, could cause the structure to rotate inward towards the dock due to the eccentricity of the loading.

Figure 13.16 Registration of observed differential deformations.

To enhance the overall stability of the structure, it was decided to execute a series of pre-stressed anchors using the self-drilling system. These active elements, spaced 4.0 m apart, were therefore installed at a depth and inclination that avoided intersection with the rear filter and allowed their sealing in the rock mass. Its connection to the quay wall structure was carried out via a reinforced concrete header beam at floor level, which was structurally connected to the underlying blocks with steel bolts sealed with non-retractable mortar. Due to the positioning of the active anchors, the vertical component of these anchors also contributed to the stability of the quay wall by promoting a stabilising moment relative to its base. The free length was adjusted for each anchorage according to local geological conditions, namely the depth of the rock mass. The maximum free length of the anchors was 50 m, with a minimum sealing length of 6.0 m in the rock mass (Figure 13.17).

The anchors were made of 6 0.62″ cords (fptk = 1860 MPa) inside a self-drilling hollow bar ANP H2400-108 mm (fy > 680 N/mm^2), with a 200 mm diameter lost bit and tungsten ballistic buttons. pre-stressed for a payload value of 700 kN. Regarding the cement grout injection process, with a minimum dynamic pressure of 5 bar, a grout with a W/C ratio of 0.7 was used during drilling and with a W/C ratio of 0.4 for the definitive sealing of the anchorage, with pressures of up to 10 bar being used in the materialisation of the bulb. The installation of the combined system also included the use of bar centralisers and wedge plates with a metal horn, sealed against. the self-drilling bar in the connection area to the header beam (Figure 13.18).

To assess the behaviour of the pier wall during the execution of the reinforcement work and, subsequently, throughout its useful life, electrical load cells were installed in some of the anchorages to measure variations in installed force. Additionally, topographic markers were placed next to the crown of the quay wall to monitor the altimetric and planimetric movements.

Given the characteristics of the work, it was recommended that the devices installed on the quay wall and the header beam be monitored fortnightly during the execution of the work. After completion, readings would be taken quarterly for the following three years, with adjustments in frequency depending on the observed results.

The anchor holes were drilled using sonic and vibro-drilling techniques, which allowed high yields to be obtained during execution (Figure 13.19).

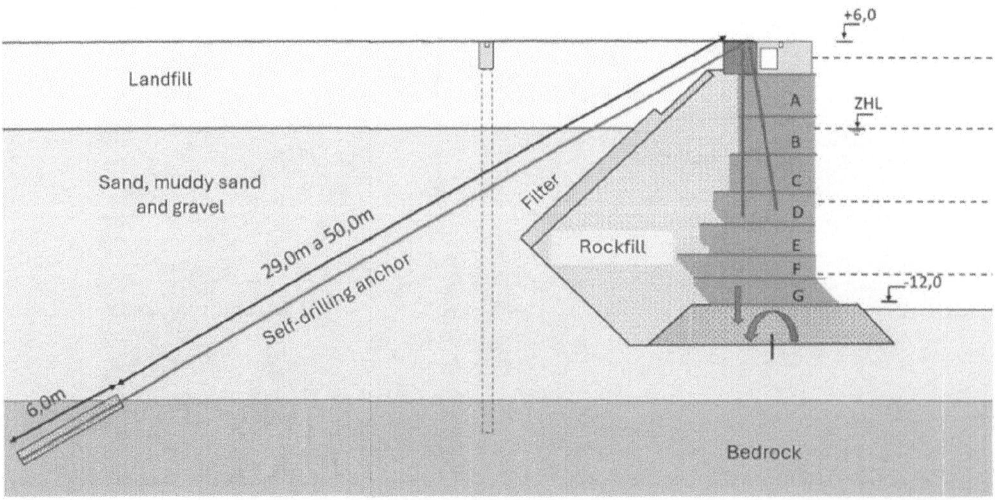

Figure 13.17 Stabilisation solution – Type cross-section.

Figure 13.18 Details of the self-drilling anchor.

The innovative anchoring technology used in this project allowed for the execution of the required elements from the embankment, minimising ground disturbance and ensuring compatibility with ship operations at the pier. The initial phase included preliminary tests on the anchorages, which validated the design assumptions related to the sealing bulb of the aforementioned anchorages.

13.4 Final remarks

The port of Leixões has completed more than 12 years of existence. This artificial port extends 2 km into the ocean and more than 3 km inland (Figure 13.20). This work focused primarily on the inland stretch where most geotechnical problems arise. Over the years, significant effort has been dedicated to monitoring and surveying the port structures; before, during and after interventions, but there were always issues that were not anticipated. For this reason, the port of Leixões confirms the importance of ongoing research and the development of innovative processes to enhance knowledge in the evolving field of harbour and marine geotechnics. However, it is crucial to preserve the historical understanding of the design, planning, and construction of the port. The construction of the quays and embankments over the fossil valley (in geological sense) of the river Leça has strongly contributed to the geotechnical problems that were registered over the years on the port. In fact, important harbour geotechnical lessons were attained at the port of Leixões. In addition, the port of Leixões continues to advance technological and intellectually, embracing a key philosophy: "We cannot fight Nature, we should live with It", or in the impressive words of McHarg (1992) we shall "Design with Nature."

Figure 13.19 Execution of the self-drilling anchorage, using the vibro-drilling technique.

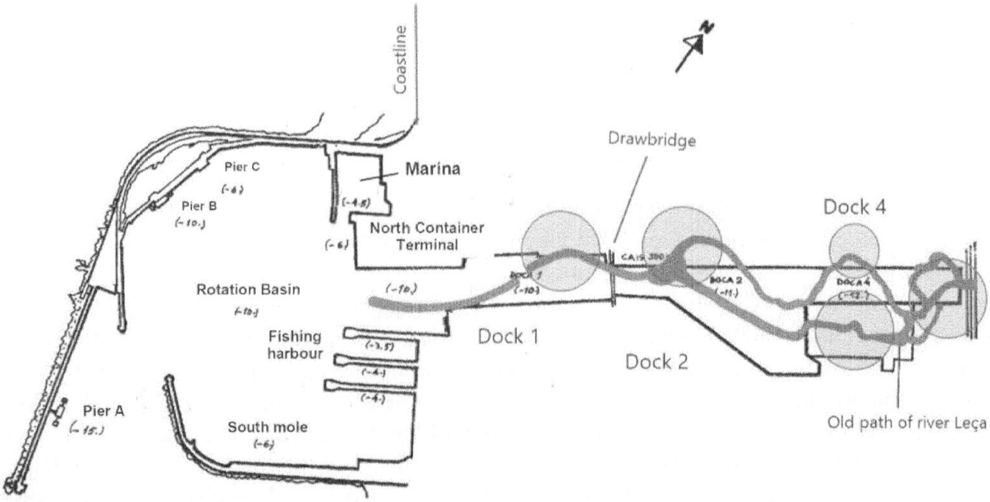

Figure 13.20 The port of Leixões and the old fossil valley of River Leça. *From Lopes and Dias (2021), published under CC BY 4.0.*

References

Alves J.F. and Brògueira-Dias E. 2001. O fio de água: o porto e as obras portuárias (Douro-Leixões). *História, Revista da Faculdade de Letras, Universidade do Porto*, 2(III Série):93–106. (in Portuguese)

APDL. 1987. Terminal de Contentores TC4S – 1ª Fase – Nota Descritiva dos Trabalhos em Curso. APDL, Leça da Palmeira (Unpublished report). (in Portuguese)

Brògueira-Dias E. 1992. Port of Leixões: Significance of geological conditions in dock construction. In: *A Geotecnia no Mundo Técnico: Problemas Concretos Surgidos em Obra*, APDL, Leça da Palmeira. (in Portuguese)

Brògueira-Dias E. and Alves J.F. 2010. Ports, policies and interventions in ports in Portugal – 20th century. *Cahiers de la Méditerranée*, 80:41–64.

Brògueira-Dias E. and Coutinho A. 1998. Dragagens no Porto de Leixões e Barra do Douro (Década de 90) e alimentação artificial da faixa costeira adjacente. In: *Seminário sobre Dragagens, dragados e ambientes costeiros*, Eurocoast , Universidade do Porto, pp. 123–133. (in Portuguese)

CIRIA, CUR and CETMEF. 2007. *The rock manual: The use of rock in hydraulic engineering*. 2nd Edition. C683. CIRIA – Centre for Civil Engineering Research, London, UK, 1200 p.

Coelho C., Silva R., Veloso-Gomes F. and Taveira-Pinto F. 2009. Potential effects of climate change on northwest portuguese costal zones. *ICES Journal of Marine Science*, 66(7):1497–1507.

Dias J.M.A., Boski T., Rodrigues A. and Magalhães F. 2000. Coast line evolution in Portugal since the Last Glacial Maximum until present: A synthesis. *Marine Geology*, 170(1):177–186.

Gomes A., Ramos-Pereira A., Araújo M.A., Sousa A. and Veloso-Gomes F. 2005. Douro's estuary dynamics. In: A. Ramos-Pereira, J. Trindade and M. Neves (Eds.), *Portugal: Coastal Dynamics – field trip guide A1, Sixth International Conference of the IAG, Zaragoza*, pp. 11–15. https://repositorio.ulisboa.pt/entities/publication/34a60a38-ffd7-4e1b-a75f-f46394241c84

Lopes H.G. 2013. Experimental analysis of the interaction between waves and a rubble-mound breakwater protected by a detached breakwater, using video imagery techniques. The port of Leixões case. University of Porto (PhD thesis). https://repositorio-aberto.up.pt/handle/10216/73972

Lopes H.G., Taveira-Pinto F., Veloso-Gomes F., China R., Cabral J., Gadelho J., Neves M.G. and Reis M.T. 2013. Application of image processing tools on physical model tests. In: *Proceedings of the 6th SCACR – International Short Course/Conference on Applied Coastal Research*, LNEC, Lisbon, 11 pp. (CD-Rom)

Lopes H.G. and Dias E.B. 2021. The port of Leixões (Portugal): over 120 years of knowledge. *Geotechnical Research,* 8(2):42–53, https://doi.org/10.1680/jgere.18.00015

McHarg I.L. 1992. Design with Nature. 25th anniversary edition, Wiley Series in Sustainable Design, John Wiley & Sons: New York.

Mota-Oliveira I., Valle A.F. and Miranda F. 1982. Littoral problems in the Portuguese west coast. *Proceedings Coastal Engineering*, 3:1951–1969.

Mota-Oliveira I.B., Trigo Teixeira A. and Sanches do Valle A. 2001. West coast of Portugal (Espinho): A comparison between project predictions and reality. In: B. Edge (Ed.), *Proceedings of the Coastal Engineering'2000*, Sydney, Australia, ASCE, pp. 3552–3565.

Pérez-Alberti A., Pires A., Freitas L. and Chaminé H.I. 2013. Shoreline change mapping along the coast of Galicia (NW Spain). *Maritime Engineering – Proceedings of the Institution of Civil Engineers*, 166(3):125–144.

Pinto A., Pita X., Neves M. and Vaz J. 2013. Unusual geotechnical solutions at the Leixões Cruise Terminal. In: *Proceedings of the 18th International Conference on Soil Mechanics and Geotechnical Engineering*, Paris, pp. 2075–2078.

Pires A., Chaminé H.I., Piqueiro F., Pérez-Alberti A. and Rocha F. 2016. Combining coastal geoscience mapping and photogrammetric surveying in maritime environments (Northwestern Iberian Peninsula): Focus on methodology. *Environmental Earth Sciences*, 75:196.

Pires A., Chaminé H.I., Piqueiro F. and Rocha F. 2014. Coastal geo-engineering techniques for the assessment of rock armour structures. *Marine Georesources & Geotechnology*, 32(2):155–178.

Rosa-Santos P.J. and Taveira-Pinto F. 2013. Experimental study of solutions to reduce downtime problems in ocean facing ports: The Port of Leixões, Portugal, case study. *Journal of Applied Water Engineering and Research*, 1(1):80–90.

Sousa F. and Alves J. 2002. *Leixões: uma história portuária*. Edição APDL – Administração dos Portos do Douro e Leixões SA., Leça da Palmeira. (in. Portuguese)

Veloso-Gomes F. and Taveira-Pinto F. 2003. Portuguese coastal zones and the new coastal management plans. *Journal of Coastal Conservation*, 9(1):25–34.

Veloso-Gomes F., Taveira-Pinto F., Lopes H.G. and Brògueira-Dias E. 2009. Overtopping on breakwaters of complex cross sections. In: L. Franco, G. Tomasicchio and A. Lamberti (Eds.), *Proceedings of the 5th Coastal Structures International Conference*, Venice, Italy, World Scientific, Singapore, pp. 1596–1607.

Veloso-Gomes F., Taveira-Pinto F., Neves L., Barbosa J.P. and Coelho C. 2004. Erosion risk levels in the NW Portuguese coast: The Douro mouth – Cape Mondego stretch. *Journal of Coastal Conservation*, 10:43–52.

Chapter 14

The use of the observational approach in the risk management of an excavation slope on the A1 Motorway near Condeixa, during the operation phase

Joana Carreto, Isabel Gonzalez, Joana Ferreira Martins, Miguel Quintela Cruz, Eduardo Mira Fernandes, and Laura Caldeira

14.1 Introduction

The case study presented in this chapter refers to a slope on the northern motorway, A1, situated between km 188+600 and km 189+100 of the Condeixa – Coimbra Sul sub-section (Figure 14.1). At this location, the A1 alignment follows a south-north orientation and runs approximately parallel to the Frades River. Figures 14.2 and 14.3 show images of the current situation.

The motorway section extends along a mid-slope, as illustrated in the topographic plan in Figure 14.4. The motorway design, initiated in 1978, originally included an excavation up to 10 m deep at the axis, with slopes inclined at 1:2 (V:H) and an intermediate berm. Subsequently, to ensure that the EN 110-2 road, located on the eastern slope, would not be affected, the berms were removed, and the excavation slopes were steepened to a 1:1.5 (V:H) inclination and, in some cases, to 1:1 (V:H).

In October 1980, the first major landslide occurred near km 189+000, disrupting the EN 110-2 road and affecting approximately 150 m. In response, drainage blankets and rock buttresses were constructed to improve ground stability and reduce pore water pressures, as depicted in Figure 14.5. A drainage trench was created at the slope toe to serve the same purpose.

On December 8th of the same year, heavy rainfall triggered further slope instability, damaging the previously destabilised area and a building on the eastern side. As a result, the EN 110-2 road was closed.

After this landslide, a crack, known as the "Pine Forest crack," formed to the east of the buildings. The crack was approximately 300 m long, running roughly parallel to the motorway, 150 m from its alignment, and several metres in depth. Figure 14.6 shows the location of this crack.

Immediate action was taken to prevent the progression of the slope failure, which posed as a threat to the surrounding buildings. These involved removing the materials placed at the landslide toe and replacing them with rockfill and filter layers. The primary objective was to improve drainage conditions on the slope by excavating permeable layers and applying higher-strength materials. The original geometry of the excavation slope was restored, and the embankment beneath the motorway platform was rebuilt. The EN 110-2 road was reinstated at the top of the slope, with a 9-m-wide lane.

Stabilising the west-facing slope between km 188+780 and km 189+025 required approximately 32,500 m^3 of rockfill. Approximately 12,500 m^3 of rockfill were placed north of km 189+025 and up to km 189+150.

The difficult local conditions, including the deposition of enormous amounts of fine soils near the toe of the landslide, and the need to implement stabilisation measures, prevented the

DOI: 10.1201/9781003530220-17

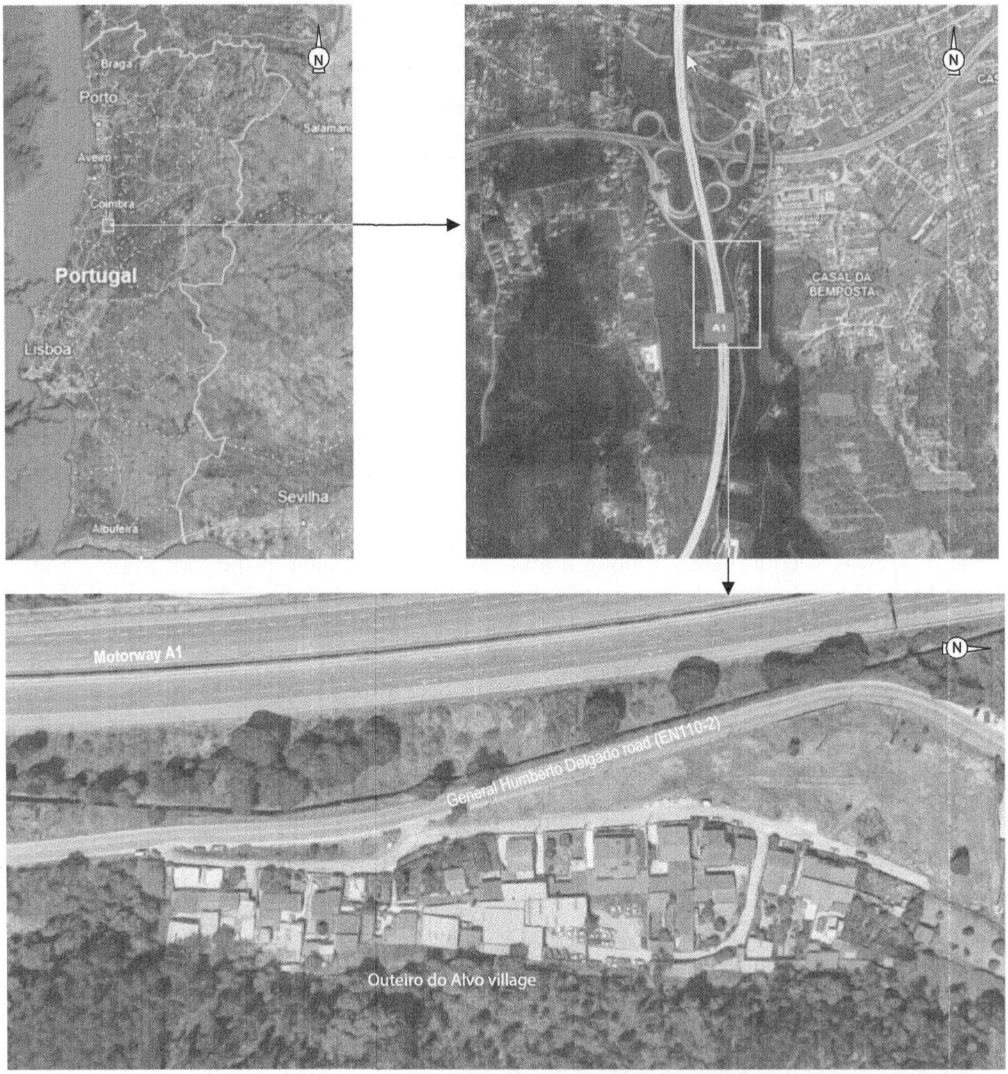

Figure 14.1 Location of the A1 excavation slope at km 189 of the Condeixa Sul – Coimbra sub-section (aerial photograph taken from Google Earth on 02/09/24).

excavation from reaching the required depth to achieve proper drainage. As a result, additional landslides were recorded a few months later, in mid-1981.

Due to poor performance of the stabilisation works, the geometry of the motorway layout in this section had to be modified. The south-north motorway lane was raised approximately 1.5 m higher than the opposing lane, which served as a key stabilising element.

Recognising the impact of high pore water pressures in landslides, a subsurface drainage system was designed in July 1981, as shown in Figure 14.7. This system included three wells with projected depths of 9 to 10 m, six sub-horizontal drain holes with a maximum length of 30 m arranged in a fan configuration, and a drainage gallery with a 35-m extension. Once sufficient drainage was obtained with sub-horizontal drain holes, the high-water pressures from

Figure 14.2 West-to-east view of the slope.

Figure 14.3 Southward view of the motorway and its slope from its crest.

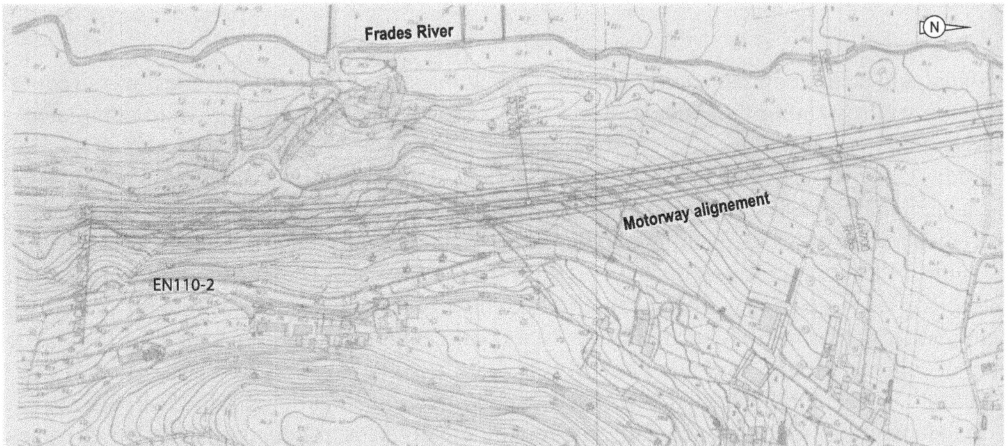

Figure 14.4 Topographic plan from the Detailed Geological and Geotechnical Study, Execution Design Phase, April 1978. (Adapted from Geocontrole 1981.)

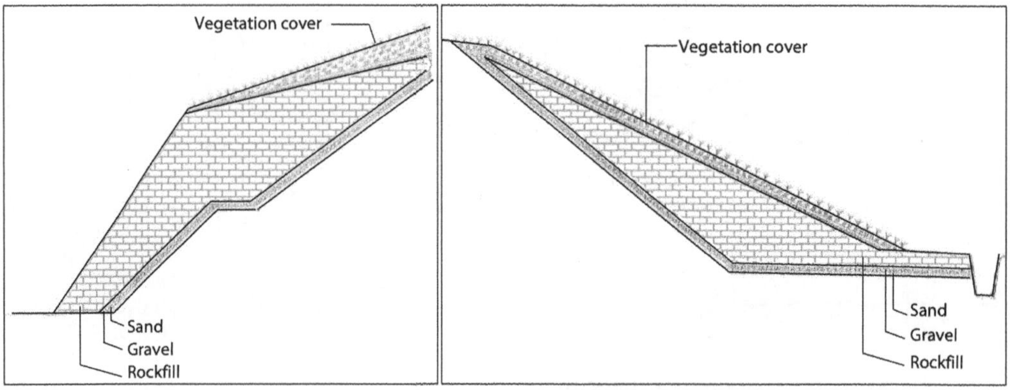

Figure 14.5 Typical cross-sections of rock buttresses, constructed after the October 1980 landslide (Adapted from Geocontrole 1981.)

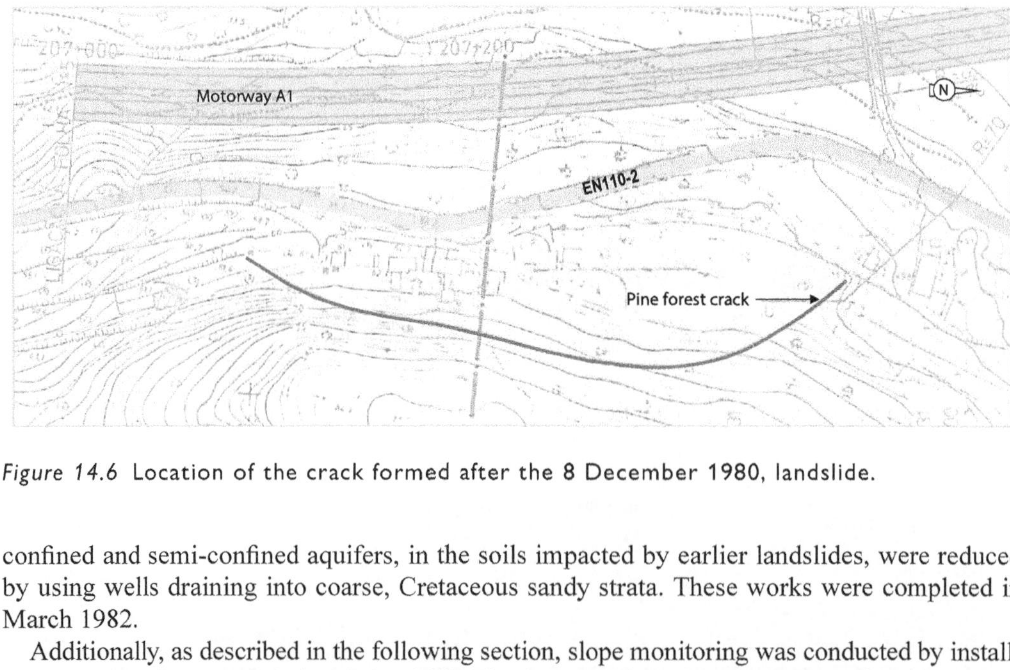

Figure 14.6 Location of the crack formed after the 8 December 1980, landslide.

confined and semi-confined aquifers, in the soils impacted by earlier landslides, were reduced by using wells draining into coarse, Cretaceous sandy strata. These works were completed in March 1982.

Additionally, as described in the following section, slope monitoring was conducted by installing several inclinometers in boreholes drilled for the geotechnical investigation of the site. These instruments were distributed throughout the slope, including the village of Outeiro do Alvo.

Figure 14.8 shows cross-sections A and B (with their locations provided in Figure 14.7) illustrating the slope geometry after completion of the works. The volume of rockfill placed was significantly higher in cross-section B due to its poorer stability conditions there (no landslides were documented in the northern area).

In 1984, seven sub-horizontal boreholes, each with an average length of 35 m, were drilled in the slope's northern area, reaching hydrogeologically very productive sandy strata. These boreholes appeared to stop draining after a certain period; thus, an attempt to unclog them in 1996 was unsuccessful.

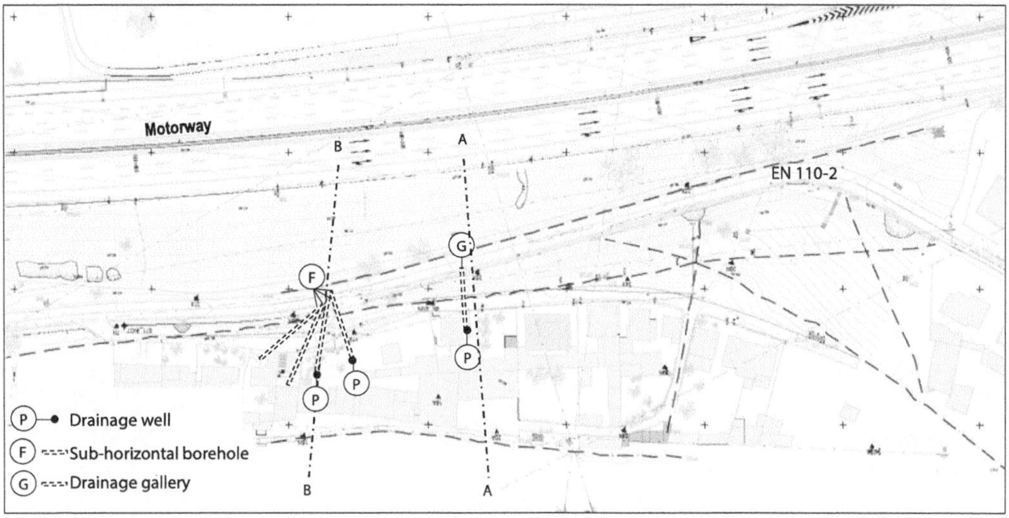

Figure 14.7 Plan to illustrate the location of the proposed subsurface drainage system. (Adapted from Geocontrole 1981.)

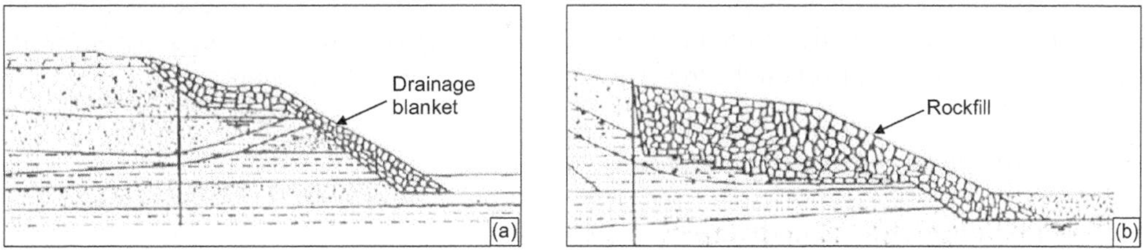

Figure 14.8 Slope geometry after stabilisation work: (a) Cross-section A; (b) Cross-section B.

Following the previous year's heavy rainfall, four drainage pipes were installed early in 1997, intersecting the rockfill up to a depth of 18 m, as well as the silty-clay soils below. The drainage holes ranged from 45 to 75 m in length. The exact locations are unknown.

To further improve slope drainage in 2000, four drainage wells, P1 through P4, were strategically positioned based on groundwater level indicators. The wells were lined with rigid PVC pipes, wrapped in graded gravel, and had an outer diameter of 140 mm.

On 23 October 2001, a solution was presented for the construction of two rows of groundwater pumping wells (11 wells in one row and 4 wells in the other), as depicted in Figure 14.9. The pumping wells would reach the Cretaceous formations. The implementation of this new system was delayed and only completed in 2006.

One drainage row with 10 wells was installed to the east of the buildings, while the other with four wells was positioned west of the first row but east of the EN 110-2 road. The wells were equipped with NOWAL pumps and had a depth of 25 m in the west row and 30 m in the east row. When sensors, located 2.5 m above the pumps, encounter water, pumping starts automatically, triggering the discharge mechanism for the incoming flow.

In 2008, the motorway was widened from 2+2 to 3+3 lane configuration, which included excavations of up to 4 m in height in its central divider to minimise interference with the

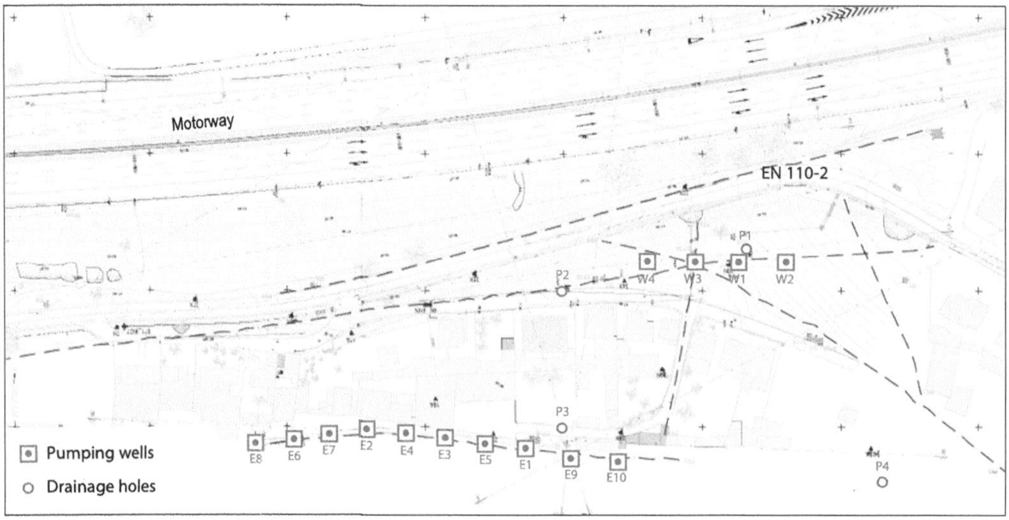

Figure 14.9 Location of the pumping systems and drainage wells.

excavation slope. This expansion came at the expense of enlarging a single lane: Porto/Lisbon (on the western side of the motorway, opposite the destabilised slope). The excavations in the middle divider were intended to build a retaining wall between the two lanes. These works appear to have had an impact on the lower elevations of the monitored slopes.

14.2 Geological-geotechnical framework

14.2.1 Lithological description

The A1 alignment required the excavation of sandstones and shales with a uniform stratigraphy and a small, but westward dip. Based on the geotechnical investigation, formations were divided into three main categories:

1. Recent materials, most likely from the Quaternary (Qsa). This group consists of layers of brown to reddish sandy clays with abundant gravel and rolled cobbles, transitioning into layers with numerous irregularly shaped cobbles embedded in a primarily sandy matrix. These materials appear to be recent fill or the result of mass movement from an upstream slope.
2. Superficial sandy levels with varying thicknesses, possibly related to alluvial terraces. This category includes medium to coarse sands, with finer layers in a laminar structure, as well as some decompressed superficial clay levels. The grain size of these deposits varies, indicating that differing energy levels existed in ancient watercourses.
3. Taveiro Formation (Upper Cretaceous, C2Ta), which is predominantly composed of red to brown clays. These clays can be siltier or sandier. Fine sands and silty sands are also present, albeit in lesser quantities, and form small lenticular deposits.

14.2.2 Electrical resistivity profiles

Electrical resistivity profiles were obtained across the area, and the results were consistent with the presence of both more permeable materials (such as limestone rockfill) and less permeable materials (clays). The profile obtained along the access road to the Outeiro do Alvo buildings (see Figure 14.1) provided an approximate indication of the physical boundary between the "in situ" materials (clays, etc.) and the limestone rockfill used during the 1980 landslide. Furthermore, on the former EN 110-2, where a substantial depression of the platform was observed (near inclinometer 32X), the electrical resistivity profile revealed a significant zone extending in depth of materials with low permeability and high saturation potential.

14.2.3 Laboratory characterisation

The laboratory test results from samples collected from boreholes in the area reveal soils with a significant fine component (between 45% and 99%), as well as some sandy materials containing 6%–12% of fines. Fine soils have high-plasticity indexes (more than 15). Natural water content ranges from 12% to 22%, which is consistent with the materials and local geological background. The methylene blue test results ranged from 0.2 to 9.1 g/100 g of soil, with values greater than 3 g/100 g of soil related to samples collected from boreholes drilled on the A1 platform.

X-ray diffraction tests indicated that the presence of expansive clay minerals (smectites) is more prevalent in the materials below the A1 platform and in the northern part of the excavation slope of EN110-2.

The consolidated undrained triaxial test results reveal the clayey and varied nature of the samples, with shear stress intercept values ranging from 5 to 100 kPa and an effective angle of internal friction at peak state between 15° and 35°. The critical state strength parameters were not determined.

14.2.4 Geological-geotechnical and hydrogeological context

The geotechnical site investigations, which included both field and laboratory testing, revealed the following:

- Outeiro do Alvo (view location in Figure 14.1) is located on Quaternary formations with rolled quartz pebbles and cobbles of different sizes, embedded in a clay-dominated soil matrix. The Taveiro Formation lies beneath the Quaternary deposits, and its soils are predominantly clayey and silty.
- Most of the soils in the area are lean clays, although there are also high-plasticity clays, silty sands, and clayey sands, albeit in smaller quantities. Notably, under the A1 platform, particularly in the South/North direction, high-plasticity clays with montmorillonite have been identified, indicating expansive potential.
- Drainage blankets and rockfill material were placed in the area stretching from the access road to Outeiro do Alvo to the foundation of the A1 platform in a south/north orientation.
- A significant lateral variation in geological materials was detected, along with some sudden depth differences between clayey and sandy soils (occasionally coarse).
- Water contents obtained from the examined materials suggest a continuous recharging through seepage and infiltration from high-permeability Quaternary deposits, with the underlying Cretaceous formations absorbing these waters.

Figure 14.10 shows a geological cross-section that illustrates the nature of the ground following the 1980 destabilisation. This figure depicts the complex arrangement of geological materials with disturbed deposits (dv) reaching a depth of 20–25 m. These deposits are composed

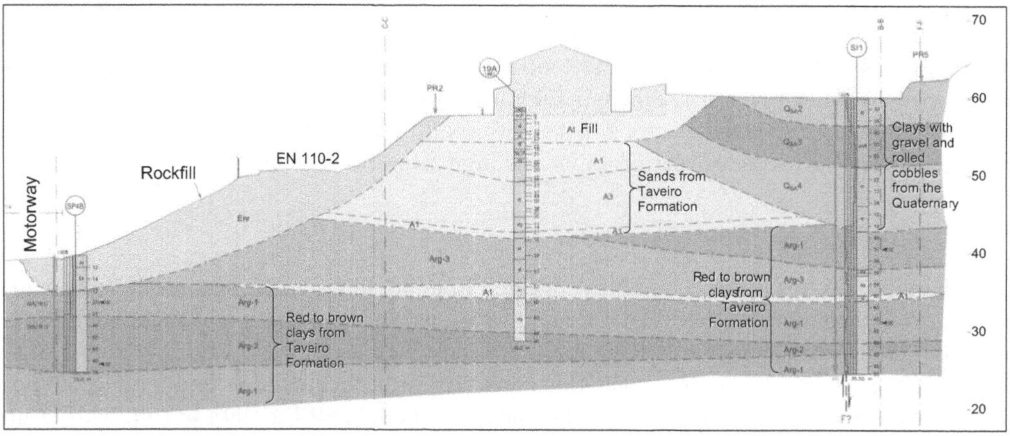

Figure 14.10 Geological Cross-section. (Adapted from Elsamex 2020.)

primarily of clayey silts and silty sands, with some gravel. The weakest soils (clayey silts) are found towards the bottom of the deposits. Beneath these deposits, and upstream of the slope, there are several "in situ" layers from the Quaternary (Q) and Cretaceous periods.

From a hydrogeological perspective, productive terrains are represented by terrace deposits (Q), disturbed deposits (dv), and the coarse Cretaceous facies. The hydrogeological potential of the terrace deposits (Q) is significant. The terrains designated as (dv) are conducive to the formation of highly productive aquifers due to their generally high permeability. They are strongly replenished by both the terrace deposits (Q) and the Cretaceous formations. The base of the deposit is sealed by the fine-grained Cretaceous formations.

According to the literature, the geometry of this deposit concurs with preferential flow to the north, supplying the old spring located near the EN 110-2. Conversely, the somewhat random distribution of fine-grained strata within the deposit creates unique seepage or groundwater retention conditions.

14.3 Slope monitoring

Following the described landslides, field instrumentation was installed on the motorway slope and berms, along the EN110-2 road, and in the residential area of Outeiro do Alvo, to monitor its behaviour. Over the past 42 years, 54 boreholes have been drilled, and there are currently 28 operational inclinometers and 7 piezometers. Over time, the monitoring system underwent several modifications to both supplement the acquired data and to replace instruments that became inoperative for various reasons.

Monitoring campaigns have been conducted throughout the last four decades by Brisa's Geotechnical Department and the National Laboratory for Civil Engineering, which was contracted by Brisa in 1988 and continues to provide ongoing assistance. The frequency of monitoring campaigns has remained about three times per year, with a significant archive of data accumulated over the years.

The following sections describe the monitoring system and its results at various stages of the motorway construction and operation.

14.3.1 Slope monitoring from October 1980 to 1983

Following the first major landslide in October 1980, the building cracks were instrumented and measured periodically. Crack gauges revealed an ongoing increase in movement rates, reaching 0.58 mm/day by early November 1980, coinciding with an increase in rainfall. Topographic markers were placed on the slope in November 1981, while stabilisation and drainage works were still in progress, as shown in Figure 14.11.

Figure 14.12 presents the evolution of vertical displacement rates for several topographic markers, showing that, after an apparent stabilisation trend in December 1981, displacement rates increased once again in January 1982. Displacement rates then decreased to roughly 9.5 mm/year between the end of January and the end of March 1982, when the first subsurface drainage system was built.

One month after the initial markers were installed, in December 1981, the system was upgraded with the addition of inclinometers, as shown in Figure 14.11. Cross-sections A and B in Figure 14.13, with their locations indicated in Figure 14.11, show the horizontal displacement profiles of inclinometers 16, 17, 19, 21, and 24 during November 1983.

With the exception of inclinometer 19, all the others revealed a zone of localised shearing strain at depths ranging from 6.3 to 20 m, revealing an identifiable slip surface. The horizontal displacement profile of inclinometer 19 suggests stress relief, in accordance with its proximity to the slope crest. In this inclinometer, the localised slip surface was not found. However, it may be beneath it.

A comprehensive analysis of these results leads to the conclusion that a significant proportion of the deformations occurred between December 1981 and April 1982, coinciding with a period of heavy rainfall as well as with the completion of the subsurface drainage system.

Until November 1982, the greatest displacements in depth were between 17 and 36 mm. From then until November 1983, the date of the last measurement shown in Figure 14.13,

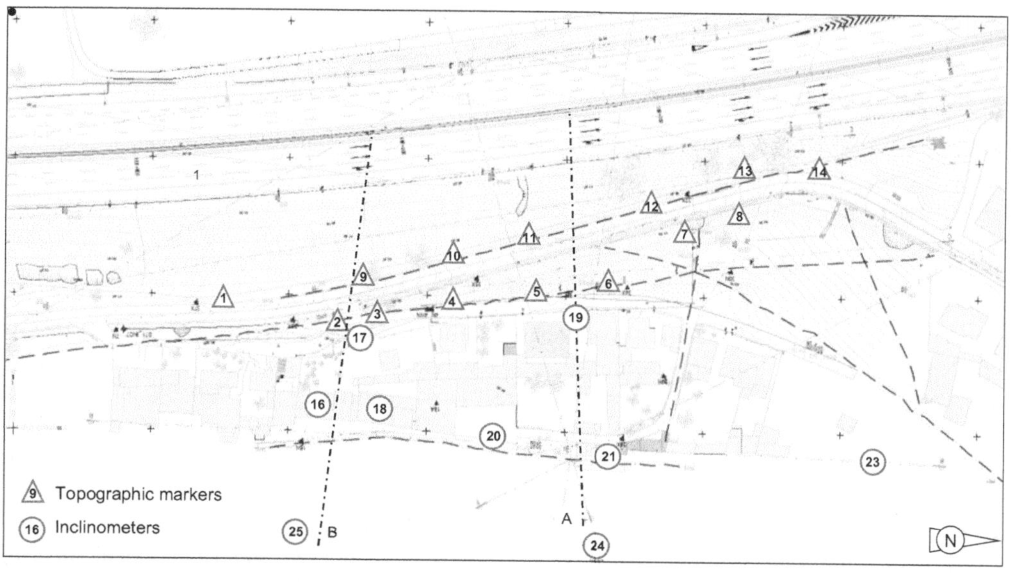

Figure 14.11 Approximate location of the topographic markers installed in November 1981 and of the inclinometers installed in December 1981. (Adapted from Elsamex 2020.)

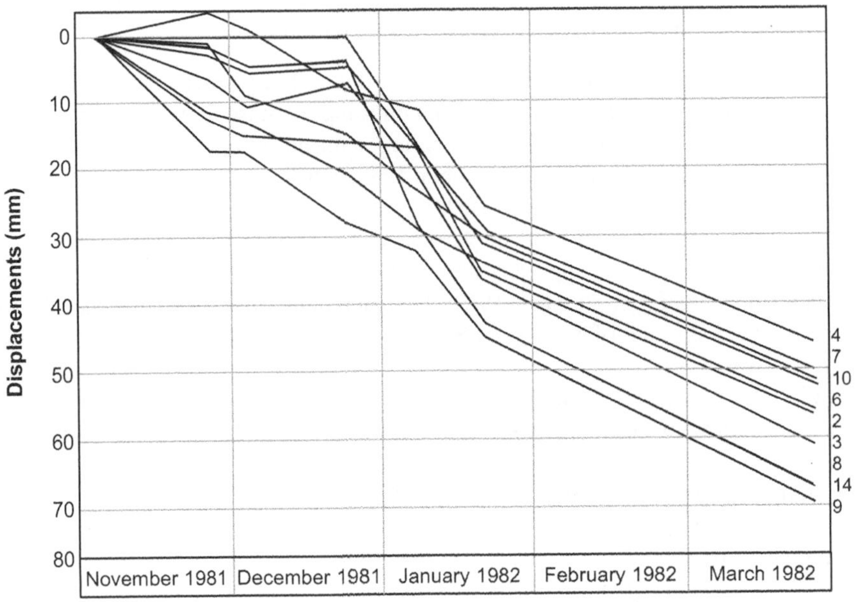

Figure 14.12 Vertical displacements of the topographic markers installed in November 1981.

the maximum displacement did not exceed 39 mm, suggesting a significant displacement rate reduction. During this period, the highest rate was recorded in inclinometer 20 at 3.8 mm/year.

Monitoring measurements enabled to identify the depth and area affected by slope instability and to conclude that, despite the reduction in the displacement rate, the situation had not yet stabilised.

14.3.2 Monitoring between 1983 *and* 1987

Between 1983 and 1987, inclinometer measurements were unavailable. However, it is known that during this period, seven drain holes and three drain wells were constructed, purportedly to stabilise the slope.

14.3.3 Monitoring between 1987 *and* 1995

From 1987 to 1995, no stabilization works or any other works were carried out on this slope.

In November 1987, a reference measurement was carried out on the instruments 7, 17A, 18A, 19A, 20A, 21A, 23A, and 26A (see Figure 14.14). Analysing the available charts, it was determined that within this group of instruments, inclinometers 7, 19A, and 26A moved off their base. The maximum accumulated displacements measured in the remaining inclinometers (see Figure 14.15) between November 1987 and March 1995 ranged from 7 mm in inclinometer 21A to 30 mm in inclinometer 18A. Rates varied from 0.9 mm/year to 4.1 mm/year over this period, with some values higher than those recorded between 1982 and 1983. Local shearing was detected in all these inclinometers, with depths ranging from 9.5 to 21.9 m, consistent with previous records.

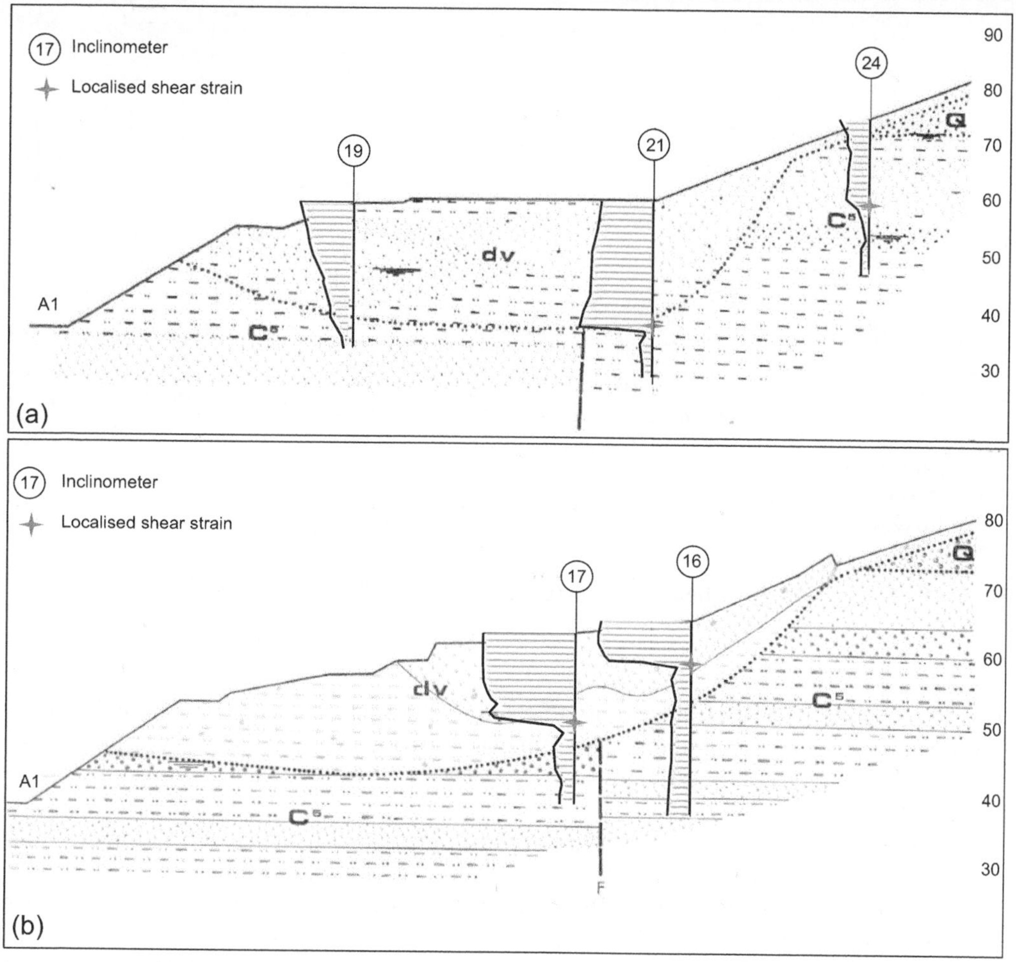

Figure 14.13 Horizontal displacement in November 1983, measured by inclinometers in cross-sections: (a) A–A′; (b) B–B′.

14.3.4 Monitoring between 1995 *and 2006*

Figure 14.15 shows the horizontal displacements observed between 1995 and 1998 in inclinometres installed in 1987. During this time, displacements in inclinometers 17A, 18A, 20A, and 23A increased at a faster rate than the preceding period. The most significant increase was observed in instrument 17A, which increased from 2.4 to 6.25 mm/year, followed by 20A, which increased from 3.1 to 5.7 mm per year.

At the beginning of 1996, new inclinometers were installed to replace those that had become inoperative and supplement the monitoring system, namely instruments 4N, 7N, 16N, 24N, 25N, 28N, 29N, and 30N (see Figure 14.14). In 1997, four drain holes were built in response to the heavy precipitation in 1996. Three years later, in 2000, four drainage wells were built. In the same year, the monitoring system was expanded with eight inclinometers labelled A, B, C,

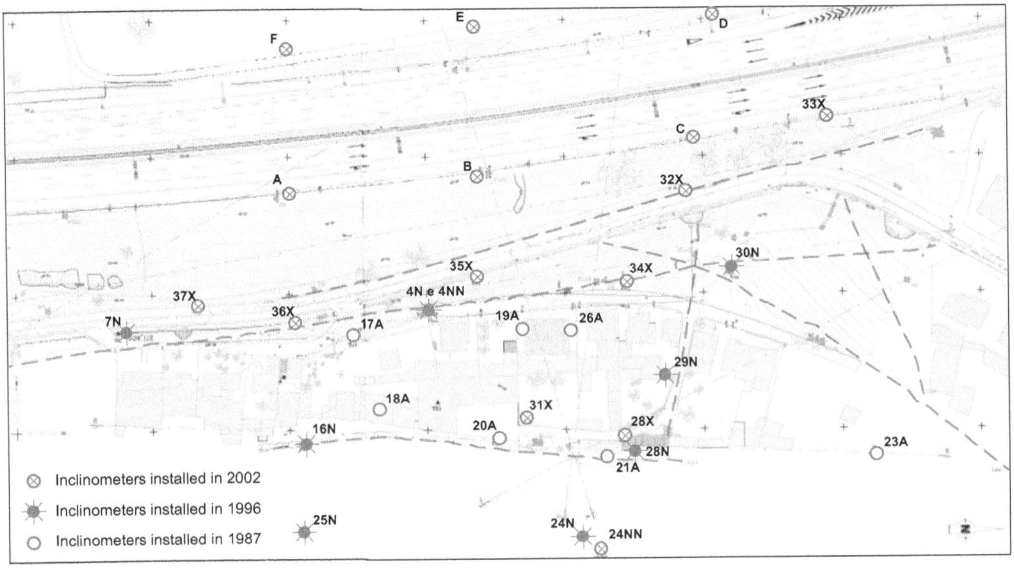

Figure 14.14 Location of inclinometers installed between 1987 and 2002.

D, E, F, 4NN, and 24NN. At the end of 2002 and the beginning of 2003, due to unfavourable movements in most of the monitoring instruments, new inclinometers were installed with the following designations: 28X, 31X, 32X, 33X, 34X, 35X, 36X, 37X, and CX, with their locations also shown in Figure 14.14.

Figures 14.16 and 14.17 show the magnitude and direction of the lateral displacement vector at various depths for some of the most relevant inclinometers.

Among inclinometers installed in 1996, a shear surface was identified only in instrument 28N. The profile of inclinometer 7N suggested the presence of a void, while the bases of inclinometers 4N, 25N, and 29N were not on rigid ground, complicating interpretation. Similarly, for the inclinometers installed in 1997, the available data confirmed that instruments A, D, E, F, and 24NN did not have a fixed base.

From 1998 to October 2006, the analysis of Figures 14.16 and 14.17 shows that the higher rate of horizontal displacements occurred during the rainy seasons of 2000–2001 and 2002–2003. In the 2000–2001 season (up to March), the quantity of rainfall was around 178% of the average annual value (data from Coimbra-Cernache and "Climatological Normals of Coimbra, 1961–1990," from the Institute of Meteorology). For the 2002–2003 season (up until March), rainfall was around 160% of the average annual amount.

During these periods, horizontal displacements were measured in inclinometers where significant localised shear strain had already been observed, with each new campaign following a similar pattern to the preceding ones. While the displacement increments were considerable in both situations, they were significantly more pronounced in 2000–2001 (up until March 2001) than in 2002–2003. The magnitude of displacement increments measured in inclinometers with deep localised shear strain was approximately half that of the first case. In the first period, displacement increments ranged from 8 to 23 mm in inclinometers 28N, 23A, C, 20A, 4NN, 18A, and 17A, whereas in the second period, increments ranged from 4 to 16 mm in the same

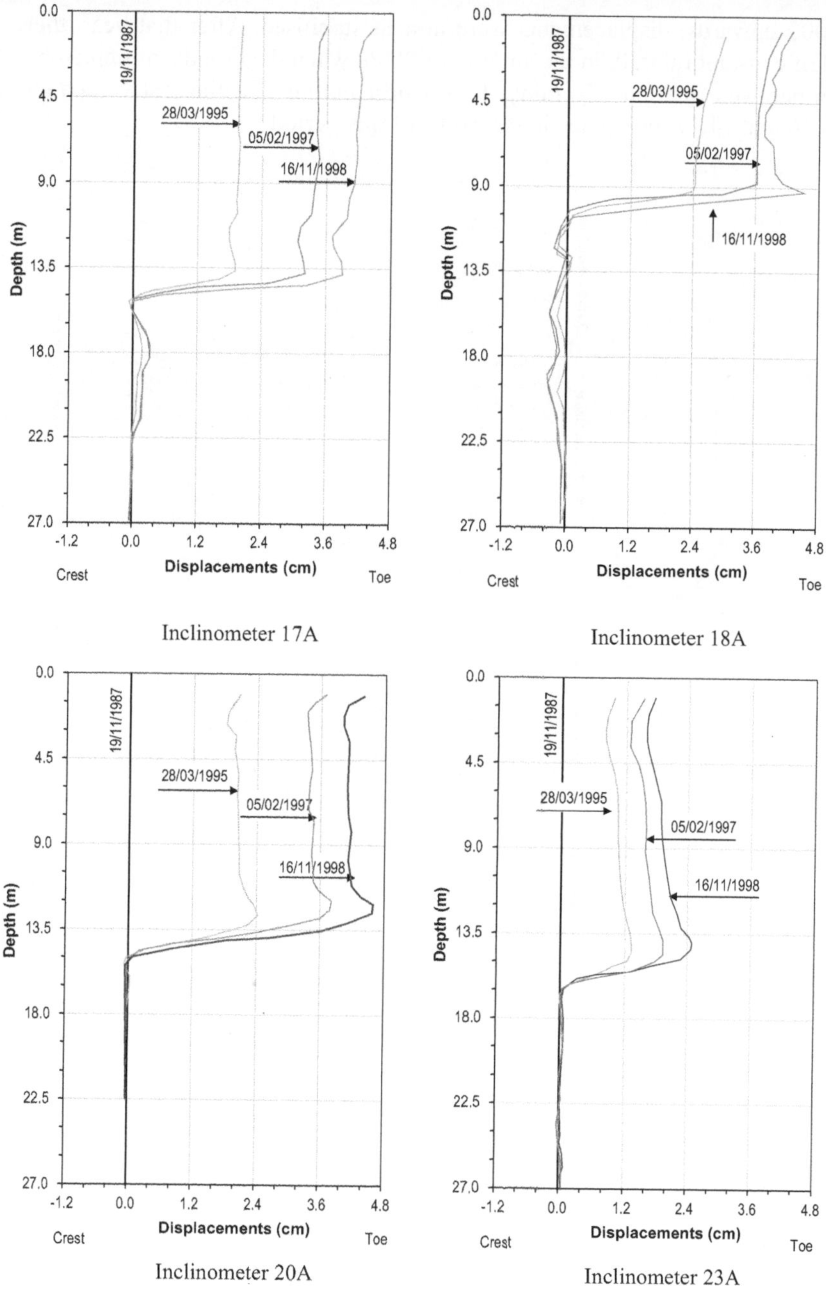

Figure 14.15 Inclinometer profiles at different times up to 1998, with reference measurements taken in November 1987.

instruments, with the exception of inclinometer 28X, which replaced the out-of-service 28N. Values in the second period were between 43% and 50% of those recorded in the first period, with the exception of inclinometer 18A, which reached 70%.

From 2003 onwards, displacements were almost stabilised. After that year, there was just one period of constant rainfall, in the autumn of 2006, when the months of September, October, and November were particularly rainy. It is worth noting that the last campaign shown in Figures 14.16 and 14.17 took place in the midst of this period.

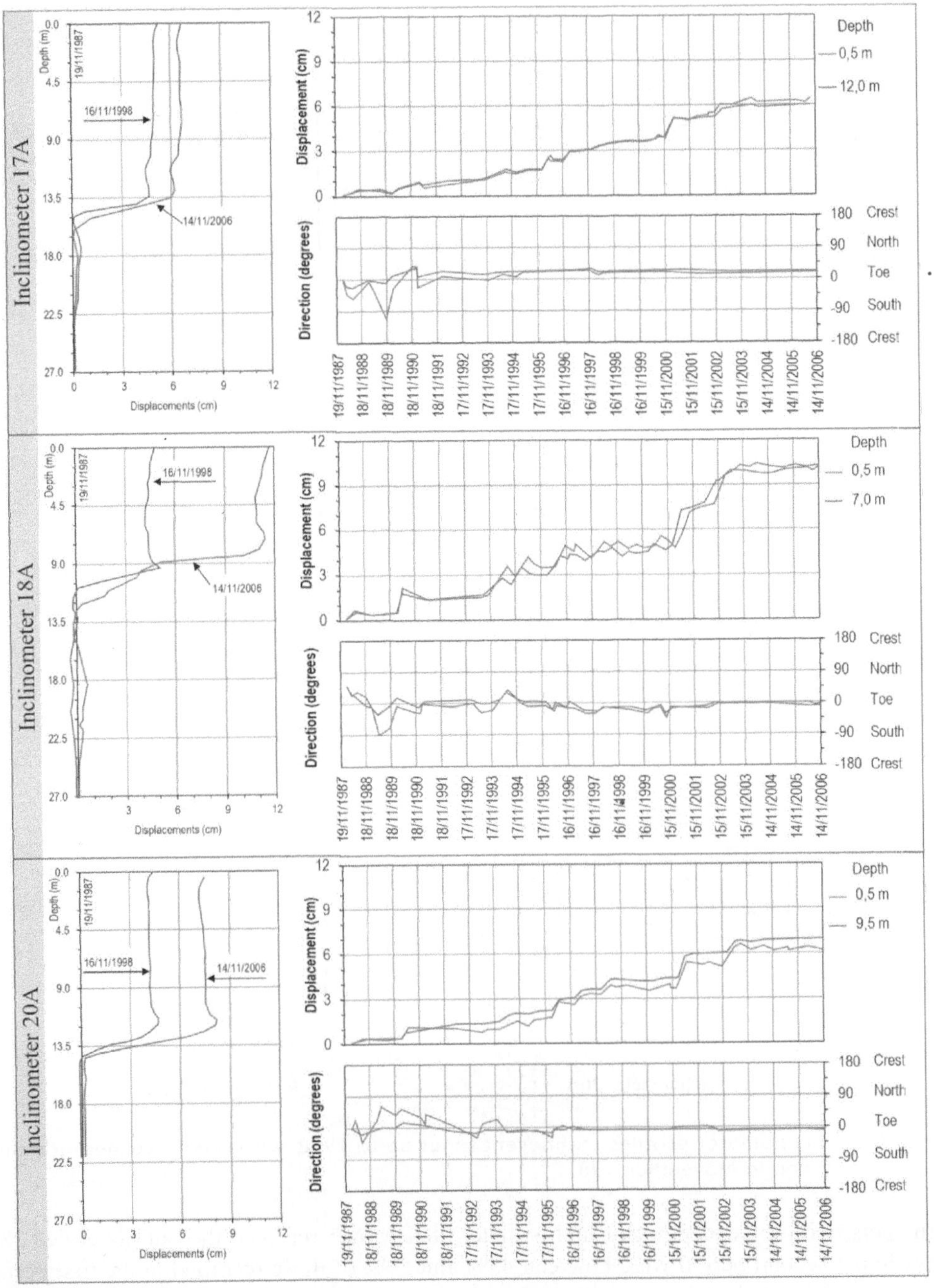

Figure 14.16 Magnitude and direction of the lateral displacement vector at two different depths from 1987 to 2006: inclinometers 17A, 18A, and 20A.

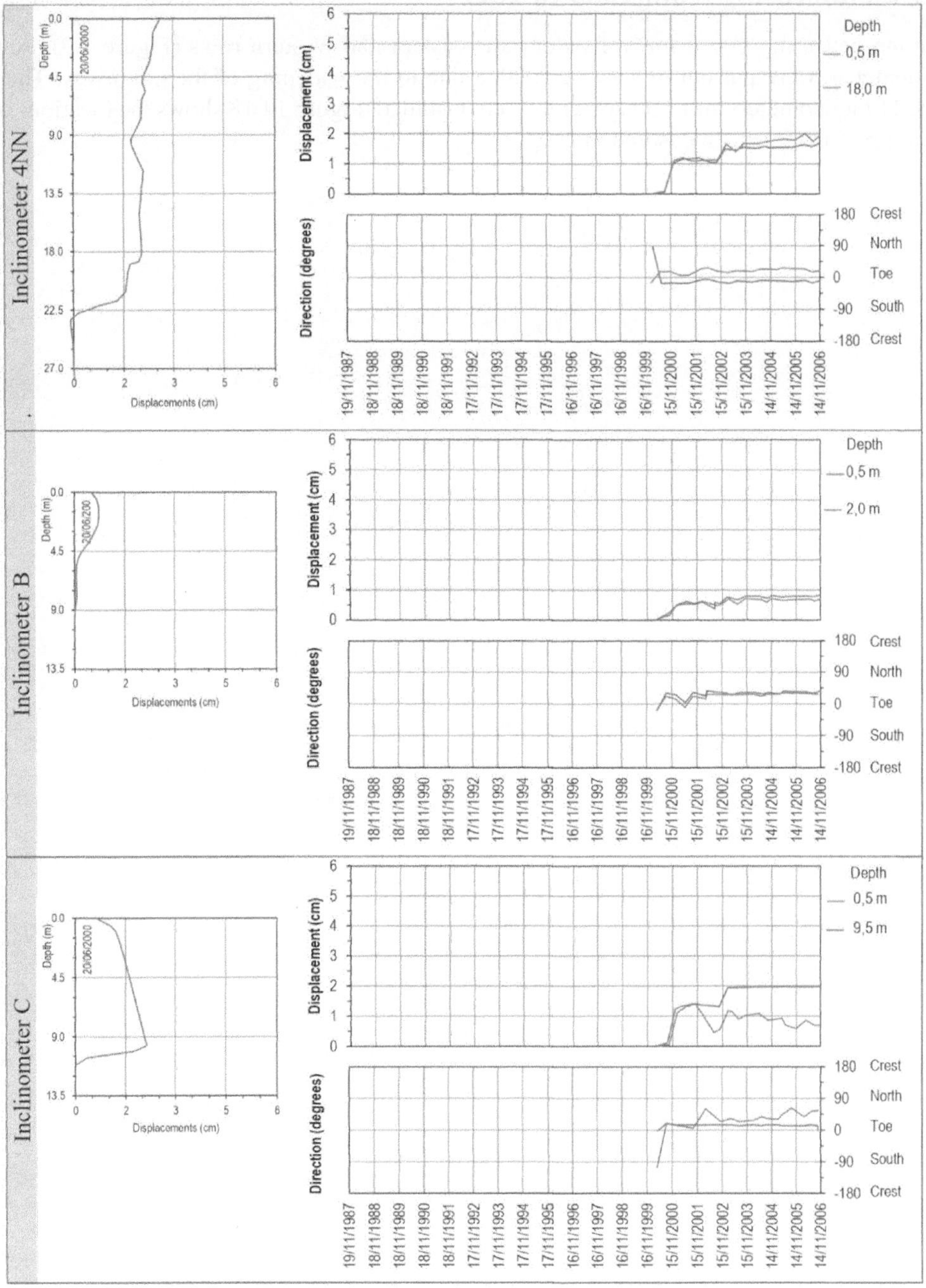

Figure 14.17 Magnitude and direction of the lateral displacement vector at two different depths from 2000 up to 2006: inclinometers 4NN, B, and C.

14.3.5 Monitoring from 2006 to 2022

Pumping began in 2006 from wells along the eastern and western rows (Figure 1.9). Several inclinometers were put out of service in 2008 due to the widening of the motorway. Then, in 2018, 11 inclinometers and 7 piezometers were installed. Figure 14.18 shows the locations of all monitoring instruments measured in 2022.

Figure 14.18 Location of monitoring instruments in 2022.

Figures 14.19 and 14.20 show the magnitude and direction of lateral displacement at various depths relative to the reference geometry, obtained in 2006, for the most representative inclinometers.

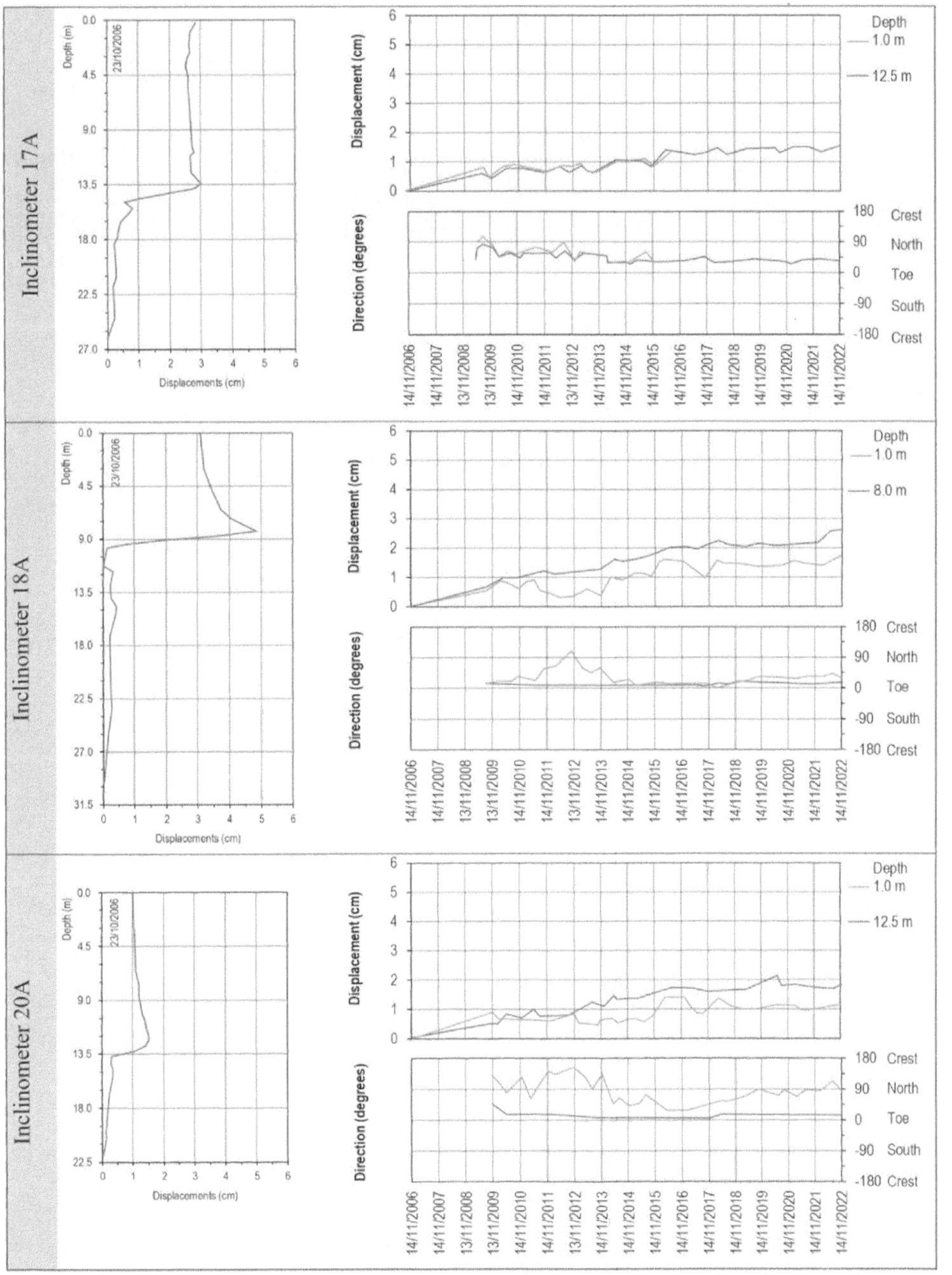

Figure 14.19 Amount and direction of the lateral displacement vector at two different depths up to 2022: 17A, 18A, and 20A.

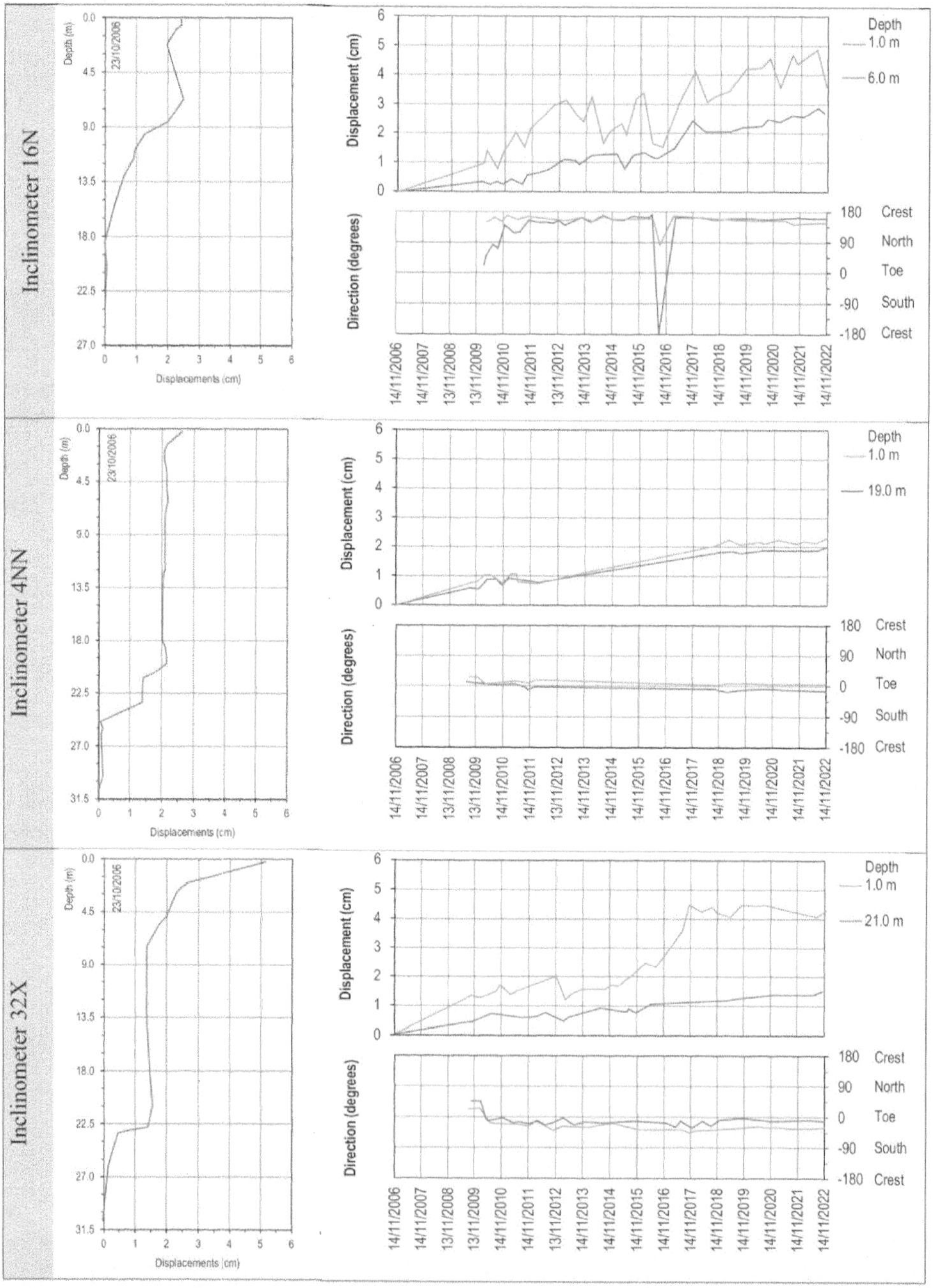

Figure 14.20 Amount and direction of the lateral displacement vector at two different depths up to 2022: 16N, 4NN, and 32X.

Figure 14.21 depicts the direction of displacement vectors above the localised shear plane identified by inclinometer measurements in 2022. Only instruments in which this surface was clearly detected are shown. It should be noted that displacements observed in 2022, using

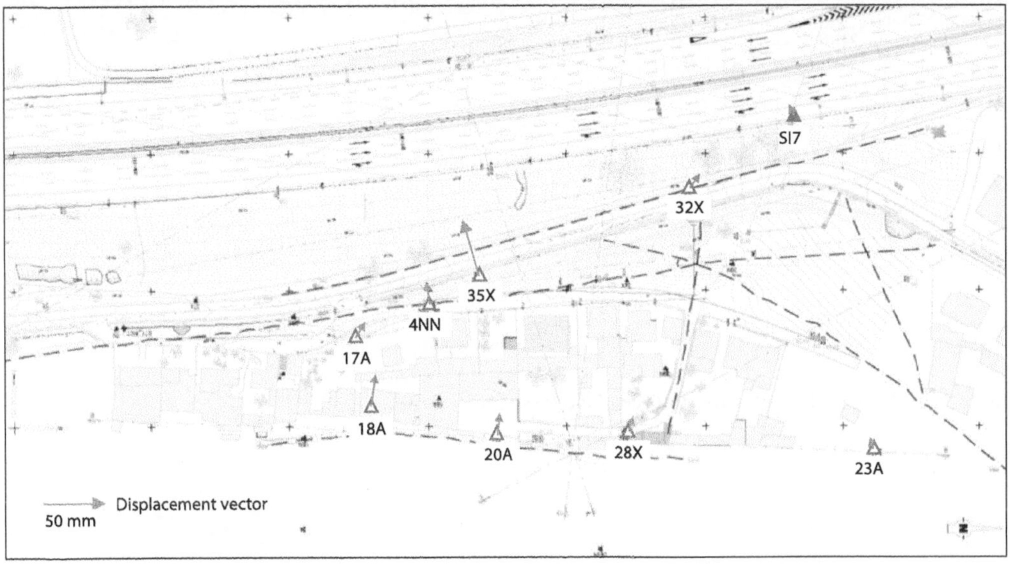

Figure 14.21 Lateral displacement vectors of the slip surface in 2022.

inclinometers installed in 2018, were small, making it difficult to objectively identify any localised shear strain.

Except for the inclinometer 32X, displacement rates remained either steady or decreased between 2006 and 2022. From 2006 to 2011, values ranged from 0.9 mm/year (28X) to 5.6 mm/year (35X), then decreased to 0.8 mm/year (17A) to 2.3 mm/year (35X). Between 2011 and 2016, they further reduced to 0.3 mm/year (28X) to 1.9 mm/year (35X) from 2016 to 2021. In contrast, displacements at the top of inclinometer 32X began to increase in 2015 and stabilised by the end of 2017.

Rainfall exceeded the average recorded in Portugal only in 2014, with no obvious impact on the evolution of slope movements. The reduction in rainfall could explain the decreasing rate, which coincided with the pumping system's operation.

Figure 14.22 shows the location of the slip surface in cross-section B–B′, whereas Figure 14.23 shows the maximum and minimum water levels measured using inclinometers and piezometers along the same profile. Figures 14.24 and 14.25 show the same elements for cross-section C–C .

Like the preceding period, this also shows a zone of localised shear strain at significant depth east of the motorway. This localised shear strain appears to be related to a slip surface that confines a moving mass of soil above, which is moving towards the motorway (the slope toe). A shear surface wasn't found in some inclinometers positioned at intermediate distances from the slope toe. Geographically, this moving mass begins to the east of the LE–LE′ cross-section, stops west of the motorway's eastern berm, and appears to be confined to the south by the D–D′ cross-section. The slip surface was observed in two inclinometers, 23A and SI7, but not in the SI4, which has a short observation period.

By comparing the location of the slip surface to the results of the geological-geotechnical survey, it is possible to conclude that the soils involved are high-plasticity clays, sometimes with expansive minerals and sometimes with slickensides, or more sandy materials with low recovery rates or SPT test results lower than those obtained in the underlying layers.

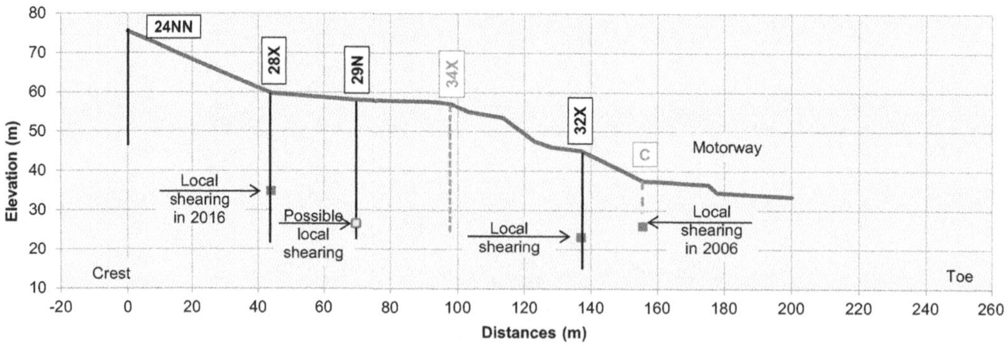

Figure 14.22 Location of slip surface. Cross-section B–B′.

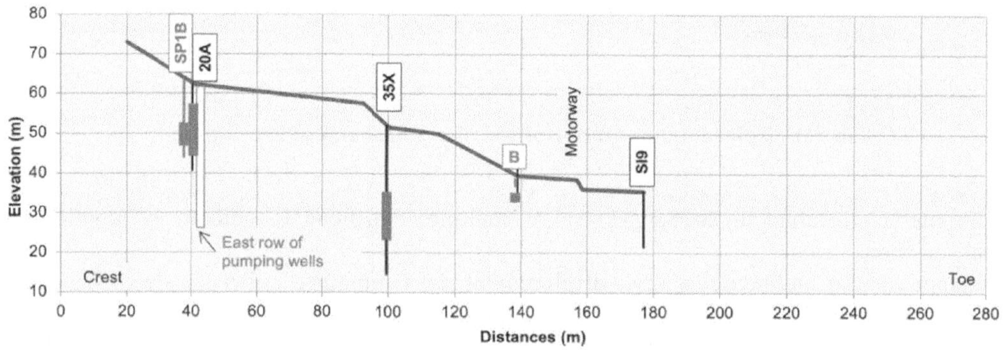

Figure 14.23 Water levels and pore water pressures measured in inclinometers and piezometers (minimal and maximal levels): cross-section B–B′.

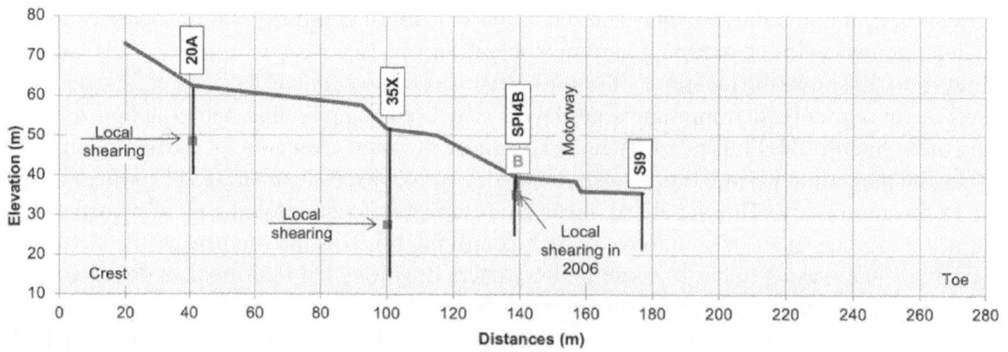

Figure 14.24 Location of slip plane. Cross-section C–C′.

The water table in cross-sections B–B′ and C–C′ is above the local shearing surface. Pore pressures are high in these profiles, contributing considerably to the occurrence of movements, which explains why the rate of movement fluctuates over time, especially in given seasonal variation.

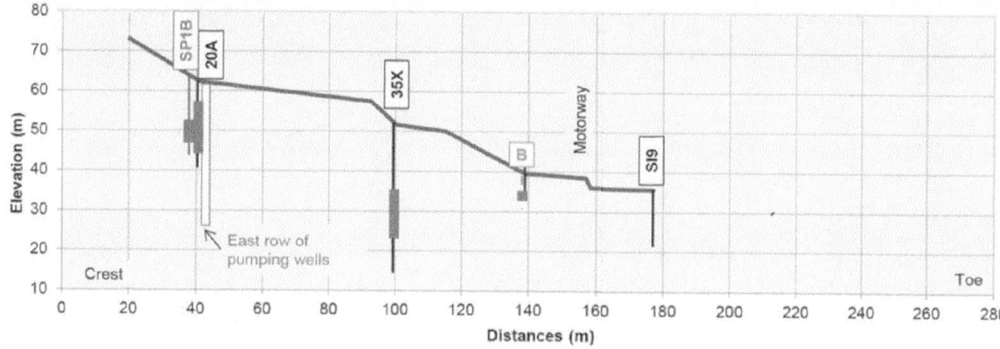

Figure 14.25 Water levels and pore water pressure measured in inclinometers and piezometers (minimal and maximal levels): Cross-section C–C′.

Piezometers placed at shallower depths (SP1B, SP2B, and SP3B) recorded higher water levels than those with intake zones at greater depths. Similarly, the evolution of water levels is different in these two sets of piezometers. Piezometers placed at shallower depths are more sensitive to rainfall variations, whereas those installed at lower elevations show fewer fluctuations in water level. These results indicate the presence of two different, noncommunicating groundwater levels.

Water levels in pumping wells, both east and west, were found to be like those of piezometers SP1A, SP1B, and SP3A, which have the intake zone at the same level as the well's base. Based on this, it is likely that the wells drain deeper water levels but not shallower ones, or that their impact on these water levels is restricted.

The water level in the southern zone is lower than in the northern zone, which may explain the lack of evidence of a deep slip surface in this location.

14.4 Conclusions

Monitoring data gathered since the early 1980s along the slope at km 189.000 of the North Motorway show the existence of huge mass soil movements restricted by a surface with localised shearing at depths that, in some instances, reach 20 m. The same data revealed the presence of a high-water level above the surface, which is most likely one of the causes of the continued creep movements.

The use of the observational method resulted in the construction of many subsurface drainage systems that helped restrict the rate of displacement, avoiding the need for extensive and costly stabilisation measures. After four decades, a progressive reduction in movements has been recorded, indicating that this is a successful and long-lasting example of using the observational approach to risk manage slopes in the post-failure phase.

References

Elsamex 2020. Estabilização da escavação ao km 189+000 da A1 – Autoestrada do Norte (sentido Lisboa/Porto), no sublanço Condeixa-Coimbra Sul. Relatório das condições existentes com análise de estabilidade. Estudo Geológico-Geotécnico.

Geocontrole 1981. Auto-Estrada Condeixa-Coimbra. Escavação entre os km 207+000 e 207+500. Memorandum de 1980 e 1981.

Index

Note: **Bold** page numbers refer to tables; *italic* page numbers refer to figures.